T0348585

VOLUME SIXTY THREE

ADVANCES IN
MICROBIAL PHYSIOLOGY
Microbial Globins - Status and
Opportunities

VOLUME SIXTY THREE

ADVANCES IN
MICROBIAL PHYSIOLOGY
Microbial Globins - Status and
Opportunities

Edited by

ROBERT K. POOLE
West Riding Professor of Microbiology
Department of Molecular Biology and Biotechnology
The University of Sheffield
Firth Court, Western Bank
Sheffield, UK

AMSTERDAM • BOSTON • HEIDELBERG • LONDON
NEW YORK • OXFORD • PARIS • SAN DIEGO
SAN FRANCISCO • SINGAPORE • SYDNEY • TOKYO
Academic Press is an imprint of Elsevier

Academic Press is an imprint of Elsevier
32 Jamestown Road, London NW1 7BY, UK
The Boulevard, Langford Lane, Kidlington, Oxford, OX5 1GB, UK
Radarweg 29, PO Box 211, 1000 AE Amsterdam, The Netherlands
225 Wyman Street, Waltham, MA 02451, USA
525 B Street, Suite 1800, San Diego, CA 92101-4495, USA

First edition 2013

ISBN: 978-0-12-407693-8
ISSN: 0065-2911

For information on all Academic Press publications
visit our website at store.elsevier.com

Printed and bound by CPI Group (UK) Ltd, Croydon, CR0 4YY
13 14 15 16   11 10 9 8 7 6 5 4 3 2 1

Working together
to grow libraries in
developing countries

www.elsevier.com • www.bookaid.org

CONTENTS

Contributors   ix
Preface   xi

1. **Haem-Based Sensors: A Still Growing Old Superfamily**   1
   Francesca Germani, Luc Moens, and Sylvia Dewilde

   1. Introduction and Background: Why *Haem-Based* Sensors?   2
   2. Functions of the Haem-Based Sensors   3
   3. The Evolution of the Globins   28
   4. Conclusions and Future Perspectives   29
   References   30

2. **The Diversity of 2/2 (Truncated) Globins**   49
   Alessandra Pesce, Martino Bolognesi, and Marco Nardini

   1. Introduction   50
   2. Fold and Fold Variation in 2/2Hb Groups I, II, III   52
   3. The Haem Environment   54
   4. Tunnels and Cavities Through 2/2Hb Protein Matrix   60
   5. Proposed Functions for the 2/2Hb Family   66
   6. Conclusions   71
   References   72

3. **Protoglobin: Structure and Ligand-Binding Properties**   79
   Alessandra Pesce, Martino Bolognesi, and Marco Nardini

   1. Introduction   80
   2. Overall Structure   82
   3. The Two-Tunnel System   85
   4. The Haem Environment   87
   5. Biochemical and Functional Characterisation   90
   6. Conclusions   93
   References   94

4. **The Globins of *Campylobacter jejuni***   97
   Mariana Tinajero-Trejo and Mark Shepherd

   1. Overview   99
   2. NO and RNS in Biology   100

3. Oxygen and Reactive Oxygen Species in Biology   107
4. Respiratory Metabolism of *Campylobacter*   108
5. *Campylobacter* Single-Domain Globin, Cgb: Functional and Structural
   Characterisation   111
6. *Campylobacter* Truncated Globin, Ctb: Functional and Structural
   Characterisation   120
7. Control of Globin Expression   123
8. Integrated Response Involving Cgb and Ctb   131
9. Conclusions   132
Acknowledgements   133
References   133

5. **Haemoglobins of Mycobacteria: Structural Features
   and Biological Functions**   **147**
   Kelly S. Davidge and Kanak L. Dikshit

   1. Introduction: Discovery of Haemoglobins in Mycobacteria   148
   2. Co-occurrence of Multiple Hbs in Mycobacteria   149
   3. trHbs of Mycobacteria: Small Hbs with Novel Characteristics   152
   4. Conventional and Novel FlavoHbs of Mycobacteria   173
   5. Biological Functions of trHbs and FlavoHbs   182
   6. Concluding Remarks   189
   References   189

6. **The Globins of Cyanobacteria and Algae**   **195**
   Eric A. Johnson and Juliette T.J. Lecomte

   1. Overview   196
   2. Historical Perspective   199
   3. Phylogeny   204
   4. Physiological Characterization   217
   5. *In Vitro* Characterization   226
   6. Needs and Opportunities   254
   References   262

7. **The Dos Family of Globin-Related Sensors Using PAS
   Domains to Accommodate Haem Acting as the Active
   Site for Sensing External Signals**   **273**
   Shigetoshi Aono

   1. Introduction   275
   2. General Properties of PAS Domains   276

3. Single-Component Systems for Transcriptional Regulation   276
4. Two-Component Systems for Transcriptional Regulation   281
5. Haem-Containing PAS Family for the Regulation of Diguanylate
   Cyclases and Phosphodiesterases   298
6. Haem-Containing PAS Family for Chemotaxis Regulation   307
7. Conclusion   316
References   317

**8. The Globins of Cold-Adapted *Pseudoalteromonas haloplanktis*
   TAC125: From the Structure to the Physiological Functions   329**

Daniela Giordano, Daniela Coppola, Roberta Russo, Mariana Tinajero-Trejo,
Guido di Prisco, Federico Lauro, Paolo Ascenzi, and Cinzia Verde

1. The Polar Environments   331
2. Phylogeny and Biogeography of Cold-Adapted Marine Microorganisms   334
3. The Role of Temperature in Evolutionary Adaptations   337
4. Bacterial Globins   340
5. The Antarctic Marine Bacterium *Pseudoalteromonas haloplanktis*
   TAC125: A Case Study   348
6. *P. haloplanktis* TAC125 Globins   358
7. Conclusion and Perspectives   372
Acknowledgements   374
References   374

9. Microbial Eukaryote Globins   391

Serge N. Vinogradov, Xavier Bailly, David R. Smith, Mariana Tinajero-Trejo,
Robert K. Poole, and David Hoogewijs

1. Overview   393
2. Historical Perspective   394
3. Globin Nomenclature   395
4. Microbial Eukaryote Globins   397
5. Phylogenetic Relationships   406
6. Globin Function   420
7. Conclusion   434
Acknowledgements   434
References   436

Author Index   *447*
Subject Index   *483*

CONTRIBUTORS

Shigetoshi Aono
Okazaki Institute for Integrative Bioscience & Institute for Molecular Science, National Institutes of Natural Sciences, Okazaki, Japan

Paolo Ascenzi
Institute of Protein Biochemistry, CNR, Naples, and Interdepartmental Laboratory for Electron Microscopy, University Roma 3, Rome, Italy

Xavier Bailly
Marine Plants and Biomolecules, Station Biologique de Roscoff, Roscoff, France

Martino Bolognesi
Department of Biosciences, University of Milano, and CIMAINA and CNR Institute of Biophysics, Milano, Italy

Daniela Coppola
Institute of Protein Biochemistry, CNR, Naples, Italy

Kelly S. Davidge
School of Life Sciences, Centre for Biomolecular Sciences, University of Nottingham, University Park, Nottingham, United Kingdom

Sylvia Dewilde
Department of Biomedical Sciences, University of Antwerp, Universiteitsplein 1, B-2610 Antwerp, Belgium

Guido di Prisco
Institute of Protein Biochemistry, CNR, Naples, Italy

Kanak L. Dikshit
CSIR-Institute of Microbial Technology, Chandigarh, India

Francesca Germani
Department of Biomedical Sciences, University of Antwerp, Universiteitsplein 1, B-2610 Antwerp, Belgium

Daniela Giordano
Institute of Protein Biochemistry, CNR, Naples, Italy

David Hoogewijs
Institute of Physiology and Zürich Center for Integrative Human Physiology, University of Zürich, Zürich, Switzerland

Eric A. Johnson
T.C. Jenkins Department of Biophysics, Johns Hopkins University, Baltimore, Maryland, USA

Federico Lauro
School of Biotechnology & Biomolecular Sciences, The University of New South Wales, Sydney, New South Wales, Australia

Juliette T.J. Lecomte
T.C. Jenkins Department of Biophysics, Johns Hopkins University, Baltimore, Maryland, USA

Luc Moens
Department of Biomedical Sciences, University of Antwerp, Universiteitsplein 1, B-2610 Antwerp, Belgium

Marco Nardini
Department of Biosciences, University of Milano, Milano, Italy

Alessandra Pesce
Department of Physics, University of Genova, Genova, Italy

Robert K. Poole
Institute of Biology and Biotechnology, Department of Molecular Biology and Biotechnology, The University of Sheffield, Sheffield, United Kingdom

Roberta Russo
Institute of Protein Biochemistry, CNR, Naples, Italy

Mark Shepherd
School of Biosciences, University of Kent, Canterbury, United Kingdom

David R. Smith
Department of Biology, Western University, London, Ontario, Canada

Mariana Tinajero-Trejo
Department of Molecular Biology & Biotechnology, The University of Sheffield, Sheffield, United Kingdom

Cinzia Verde
Institute of Protein Biochemistry, CNR, Naples, and Department of Biology, University Roma 3, Rome, Italy

Serge N. Vinogradov
Department of Biochemistry and Molecular Biology, Wayne State University School of Medicine, Detroit, Michigan, USA

PREFACE

Haem proteins play vital and diverse roles in virtually all life forms. The history of microbiology shows us that respiration, aerobic and anaerobic, was a major focus in the heady early days of bacterial metabolism. In the Third Edition of 'Bacterial Metabolism' (1949)—Marjory Stephenson's classic text first published in 1930—'Respiration' is the first chapter after the Introduction and contains much detail on the components of the 'oxidising system'. Haem proteins and cytochromes, the activation of oxygen, and the diversity of cytochromes were all covered. However, in microorganisms, the research area of O_2-binding proteins has now become dominated by studies of globins and globin-related proteins—proteins that get no mention in 'Bacterial Metabolism' (although leghaemoglobin did warrant a brief account).

Now, every 2 years, the amazing progress in this field has been evident by presentations at the Oxygen-Binding Proteins (O2BiP) held across Europe at venues that include Naples, Aarhus, Antwerp and, most recently (2012), Parma. It was this last meeting that led to the idea of a volume to mark some of these achievements. The focus there and here is bacterial and archaeal proteins, where the most dramatic progress has been made.

The first accounts of globins in microbes were those of Keilin in the 1930s, who demonstrated distinctive redox-dependent absorbance bands by use of his hand spectroscope. With no confirmation of the chemical nature of the pigments he observed, he correctly assigned strong bands in yeast, moulds and protozoa to haemoglobin (Keilin, 1953; Keilin & Ryley, 1953; Keilin & Tissieres, 1953). Then, in the 1970s, Oshino, Chance and others investigated in biochemical detail the yeast pigment and showed it to comprise not only a haem domain but also a flavin chromophore (Oshino et al., 1973; Oshino, Oshino, Chance, & Hagihara, 1973); this was the first description of a flavohaemoglobin. Recognition of haemoglobins in bacteria was, however, much later. Webster published many important papers in the 1970s describing a 'soluble cytochrome o' in the microaerobic bacterium *Vitreoscilla*. The pigment was observed in intact cells as a stable oxygenated species. Webster and others later published the sequence of this protein (Wakabayashi, Matsubara, & Webster, 1986), showed it was a globin and thus provided the first definitive proof that haemoglobins occur also, and unexpectedly, in bacteria (Perutz, 1986). The first nucleotide sequence of a bacterial globin was that of Hmp, the bacterial homologue of Keilin's and Chance's yeast protein. The sequence (Vasudevan et al., 1991) clearly revealed the chimeric nature of the protein with an N-terminal monohaem domain and a C-terminal flavin domain. More detailed accounts of the development of our now extensive knowledge of bacterial flavohaemoglobins (Forrester & Foster, 2012) can be found elsewhere. Concomitantly, a shorter than usual 'truncated' globin, GlbN, was observed in the cyanobacterium *Nostoc commune* and related cyanobacteria (Potts, Angeloni, Ebel, & Bassam, 1992) and a haemoprotein with kinase activity, and therefore, sensor function was found in *Rhizobium meliloti* (Gilles-Gonzalez, Ditta, & Helinski, 1991). Additional genomic information revealed the presence of single-domain globins homologous to the globin domain of flavohaemoglobins (Wu, Wainwright, & Poole, 2003) a family of globin-coupled sensors involved in chemotaxis in Archaea and bacteria (Hou et al., 2000) and related single-domain protoglobins (Hou et al., 2001).

These important discoveries paved the way for the advances described in this volume that comprises a selection of illustrative topics and system. The editor asks to be forgiven for omitting other important examples. Here are contributions covering single-domain globins lacking flavin, truncated globins in diverse bacteria including pathogens, as well as cyanobacteria and algae, truncated globins, haem-based sensors and protoglobin. The volume will also highlight the remarkable number of globin (gene) sequences now known but also attempts to discern, predict or guess what function(s) they may fulfil. This task is fraught with uncertainty and difficulty. The number of globin sequences is vast, yet the volume of experimental data on which to base any meaningful prediction of the significance of globin distribution between physiological groups or genera of bacteria is minimal. In my opinion, the 'gold standard' is the exploration of globin function by mutagenesis and phenotypic characterization. Sadly, the frequency of published papers is in overwhelming favour of those describing sequences and structures and studies of function that can be observed when the globin is expressed in another host. Hopefully, this volume will prompt telling experiments that can tell us more about function, that is, physiology.

REFERENCES

Forrester, M. T., & Foster, M. W. (2012). Protection from nitrosative stress: A central role for microbial flavohemoglobin. *Free Radical Biology & Medicine, 52*, 1620–1633.
Gilles-Gonzalez, M. A., Ditta, G. S., & Helinski, D. R. (1991). A haemoprotein with kinase activity encoded by the oxygen sensor of *Rhizobium meliloti*. *Nature, 350*, 170–172.

Hou, S. B., Freitas, T., Larsen, R. W., Piatibratov, M., Sivozhelezov, V., Yamamoto, A., et al. (2001). Globin-coupled sensors: A class of heme-containing sensors in Archaea and Bacteria. *Proceedings of the National Academy of Sciences of the United States of America*, *98*, 9353–9358.

Hou, S. B., Larsen, R. W., Boudko, D., Riley, C. W., Karatan, E., Zimmer, M., et al. (2000). Myoglobin-like aerotaxis transducers in Archaea and Bacteria. *Nature*, *403*, 540–544.

Keilin, D. (1953). Haemoglobin in fungi. Occurrence of haemoglobin in yeast and the supposed stabilization of the oxygenated cytochrome oxidase. *Nature*, *172*, 390–393.

Keilin, D., & Ryley, J. F. (1953). Haemoglobin in protozoa. *Nature*, *172*, 451.

Keilin, D., & Tissieres, A. (1953). Haemoglobin in moulds: *Neurospora crassa* and *Penicillium notatum*. *Nature*, *172*, 393–394.

Oshino, R., Asakura, T., Takio, K., Oshino, N., Chance, B., & Hagihara, B. (1973a). Purification and molecular properties of yeast hemoglobin. *European Journal of Biochemistry*, *39*, 581–590.

Oshino, R., Oshino, N., Chance, B., & Hagihara, B. (1973b). Studies on yeast hemoglobin. The properties of yeast hemoglobin and its physiological function in the cell. *European Journal of Biochemistry*, *35*, 23–33.

Perutz, M. F. (1986). A bacterial haemoglobin. *Nature*, *322*, 405.

Potts, M., Angeloni, S. V., Ebel, R. E., & Bassam, D. (1992). Myoglobin in a cyanobacterium. *Science*, *256*, 1690–1692.

Vasudevan, S. G., Armarego, W. L. F., Shaw, D. C., Lilley, P. E., Dixon, N. E., & Poole, R. K. (1991). Isolation and nucleotide sequence of the *hmp* gene that encodes a haemoglobin-like protein in *Escherichia coli* K-12. *Molecular and General Genetics*, *226*, 49–58.

Wakabayashi, S., Matsubara, H., & Webster, D. A. (1986). Primary sequence of a dimeric haemoglobin from *Vitreoscilla*. *Nature*, *322*, 481–483.

Wu, G., Wainwright, L. M., & Poole, R. K. (2003). Microbial globins. *Advances in Microbial Physiology*, *47*, 255–310.

CHAPTER ONE

Haem-Based Sensors: A Still Growing Old Superfamily

Francesca Germani, Luc Moens, Sylvia Dewilde[1]

Department of Biomedical Sciences, University of Antwerp, Universiteitsplein 1, B-2610 Antwerp, Belgium
[1]Corresponding author: e-mail address: sylvia.dewilde@ua.ac.be

Contents

1. Introduction and Background: Why *Haem-Based* Sensors?		2
2. Functions of the Haem-Based Sensors		3
2.1 Haem-based sensors with aerotactic function		4
2.2 Haem-based sensors with gene-regulating function		7
2.3 Haem-based sensors with enzymatic function		24
2.4 Haem-based sensors with unknown function		24
2.5 Non-haem globin sensors		26
3. The Evolution of the Globins		28
4. Conclusions and Future Perspectives		29
Acknowledgement		30
References		30

Abstract

The haem-based sensors are chimeric multi-domain proteins responsible for the cellular adaptive responses to environmental changes. The signal transduction is mediated by the sensing capability of the haem-binding domain, which transmits a usable signal to the cognate transmitter domain, responsible for providing the adequate answer.

Four major families of haem-based sensors can be recognized, depending on the nature of the haem-binding domain: (i) the haem-binding PAS domain, (ii) the CO-sensitive carbon monoxide oxidation activator, (iii) the haem NO-binding domain, and (iv) the globin-coupled sensors. The functional classification of the haem-binding sensors is based on the activity of the transmitter domain and, traditionally, comprises: (i) sensors with aerotactic function; (ii) sensors with gene-regulating function; and (iii) sensors with unknown function. We have implemented this classification with newly identified proteins, that is, the *Streptomyces avermitilis* and *Frankia* sp. that present a C-terminal-truncated globin fused to an N-terminal cofactor-free monooxygenase, the structural-related class of non-haem globins in *Bacillus subtilis*, *Moorella thermoacetica*, and *Bacillus anthracis*, and a haemerythrin-coupled diguanylate cyclase in *Vibrio cholerae*.

This review summarizes the structures, the functions, and the structure–function relationships known to date on this broad protein family. We also propose unresolved questions and new possible research approaches.

Advances in Microbial Physiology, Volume 63
ISSN 0065-2911
http://dx.doi.org/10.1016/B978-0-12-407693-8.00001-7

ABBREVIATIONS

Bhr-DGC haemerythrin-coupled diguanylate cyclase
c-di-GMP cyclic dimeric - $(3'-5')$-GMP
CooA carbon monoxide oxidation activator
DGC diguanylate cyclase
Dos direct oxygen sensor
FixL nitrogen fixation gene expression regulator
GcHK globin-coupled histidine kinase
GCS globin-coupled sensor
GReg globin-coupled regulator
HemAT haem-based aerotaxis transducer
HemDGC haem-containing diguanylate cyclase
HNOB haem NO-binding domain
mPER2 mammalian Period protein 2
MtR non-haem-coupled regulator from *Moorella thermoacetica*
NPAS2 neuronal PAS domain protein 2
PAS Per-Arnt-Sim domain
PDE phosphodiesterase
PDEA1 haem-containing phosphodiesterase A1
pXO1-118 and pXO2-61 non-haem-coupled sporulation inhibitory proteins from *Bacillus anthracis*
RcoM CO-responsive transcription regulator
RNAP RNA polymerase
RsbR nonhaem-coupled σB regulator
SwMb *Sperm whale* myoglobin
tGCS truncated globin-coupled sensor

1. INTRODUCTION AND BACKGROUND: WHY HAEM-BASED SENSORS?

The haem-based sensors represent a class of chimeric multi-domain proteins in which a haem-binding sensor domain perceives changes in the intracellular or extracellular environment and converts them into a signal that activates or inactivates the fused transmitter domain.

There is evidence to demonstrate how modularity of proteins is fundamental for efficient transduction pathways involved in gene transcription (McAdams & Shapiro, 2003), biochemical reactions (Papin, Reed, & Palsson, 2004), and protein–protein interactions (Reichmann et al., 2005). In this view, the haem-based sensors represent an example of highly specialized modular proteins evolved in nature to provide adequate answers to changing conditions, for example, oscillation of the nature and concentration of environmental gases.

The haem-binding domain is a highly efficient and sensitive sensor. Indeed, the haem-iron atom is chemically very reactive. It can change its redox state and bind a variety of ligands and has been involved in biological processes since life appeared on Earth (Vinogradov & Moens, 2008).

To remain soluble and active, the haem group needs to be surrounded by a hydrophobic pocket that is folded in, among others, the globin fold. Modifications in the redox state, as well as the absence/presence of a ligand, mostly cause the haem-moiety to undergo structural changes. In the case of multi-domain haem-based sensors, these modifications are spread all over the haem-binding structure and represent the signal that is transferred to the transmitter domain.

Four major families of haem-based sensors can be recognized: (i) the haem-binding Per-Arnt-Sim (PAS) domain, (ii) the CO-sensitive CooA, (iii) the haem NO-binding domain (HNOB), and (iv) the globin-coupled sensors (GCSs).

In *Streptomyces avermitilis* and *Frankia* sp., the presence of a chimeric protein that consists of a C-terminal-truncated globin fused to an N-terminal cofactor-free monooxygenase has been reported (Bonamore et al., 2007). Given that these sensors have yet to be officially classified among the haem-based sensors, we propose to divide the GCS family into two subfamilies, that is, 3/3 GCS and 2/2 GCS, and include the *S. avermitilis* and *Frankia* sp. proteins in the latter group.

Alternative sensing chimeric proteins have been identified. They are related to the known haem-binding sensors for structural or functional reasons, but cannot be ascribed in the superfamily classification as they are lacking the haem. For example, globin-coupled proteins in which the sensor domain is a non-haem globin domain have been identified and characterized in *Bacillus subtilis*, *Moorella thermoacetica*, and *Bacillus anthracis* and represent the first members of an intriguing class of sensors (Chen, Lewis, Harris, Yudkin, & Delumeau, 2003; Quin et al., 2012; Stranzl et al., 2011). Similarly, a haemerythrin that contains a di-iron centre but not a haem group, has been shown in *Vibrio cholerae* and regulates the activity of a diguanylate cyclase (DGC) (Schaller, Ali, Klose, & Kurtz, 2012).

Here, we review the most significant literature on haem-based sensors. We summarize what is known about the function of these proteins, their known structures, and the relationship between these two aspects. Along with this, an evolutionary model is also supported.

2. FUNCTIONS OF THE HAEM-BASED SENSORS

Numerous copies of sensor proteins are present in all kingdoms of life. For example, the number of GCSs alone, identified to date in Bacteria and Archea, is 420 (Vinogradov, Tinajero-Trejo, Poole, & Hoogewijs, 2013).

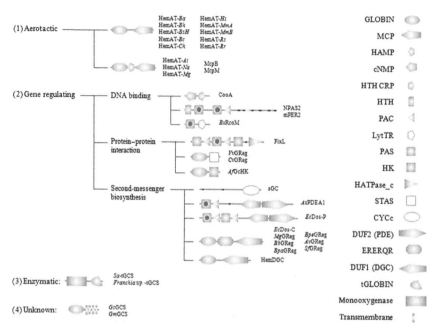

Figure 1.1 Functional classification of the haem-based sensors. The organization of the scheme is made according to the transmitter domain function. The subgroup division is consistent with the classification illustrated in this review. Along with the complete list of the symbols used, the appropriate name of the designated domain is also given. For the haem-binding domains, the presence of the haem group is given as implied. Only in the case of the numerous PAS domain, a small circle inside the graphical symbol indicates the bounded haem group.

Depending on the activity of the transmitter domain, the haem-based sensors can be subdivided into (i) aerotactic sensors, having a methyl-accepting chemotaxis domain (MCP), (ii) gene-regulating sensors, (iii) sensors with an enzymatic function, and (iv) sensors with a yet-uncharacterized function (Fig. 1.1). The haem-based sensors that regulate gene expression can exercise their function via (a) direct DNA binding, (b) protein–protein interaction, or (c) biosynthesis of a second messenger.

2.1. Haem-based sensors with aerotactic function

The aerotactic sensors are responsible for sensing the intracellular O_2 concentration and delivering the signal to the downstream transduction cascade. The result is that the flagella motor is influenced, thus the swimming behaviour changes. The organism moves along the gradient towards (aerophilicity or positive aerotaxis) or away from (aerophobicity or negative aerotaxis) the environmental area with higher O_2 presence.

The haem-based aerotaxis transducer (HemAT) are the only representatives of this class.

2.1.1 The haem-based aerotaxis transducer

Several years ago, Hou and coworkers discovered the first GCSs, the haem-based aerotaxis transducers (HemATs) in *Halobacterium salinarum* (HemAT-*Hs*), *B. subtilis* (HemAT-*Bs*), and *Bacillus halodurans* (HemAT-*Bh*) (Hou, Belisle, et al., 2001; Hou et al., 2000).

Recently, cellular stoichiometry experiments performed on the chemotaxis signalling proteins of *B. subtilis* have been published (Cannistraro, Glekas, Rao, & Ordal, 2011). From quantitative immunoblot studies, it is clear that HemAT is the most expressed chemoreceptor in *B. subtilis*, counting $19,000 \pm 3900$ copies per cell. Although HemAT-*Bs* lacks a transmembrane domain, investigation of cellular localization using fluorescent fusion protein shows clustering of this sensor at the cell poles together with other transmembrane chemoreceptors (Cannistraro et al., 2011).

Interestingly, next to the well-established aerotactic function, the characterization of the twin-arginine protein translocation (Tat) pathway of *B. subtilis*, responsible for the transport of folded proteins to the extra-cytoplasm, highlighted an alternative role for HemAT-*Bs*. Monteferrante and colleagues published early this year proofs of HemAT participation in the PhoD secretion under phosphate starvation conditions. Intriguingly, HemAT colocalizes also with CsbC, a putative pentose transporter with 12 predicted transmembrane passes (Monteferrante et al., 2013).

The protein sequence analysis of *H. salinarum*, *B. subtilis*, and *B. halodurans* HemAT reveals an N-terminal globin domain (residues 1–184 in HemAT-*Hs*, 1–175 in HemAT-*Bs*, and 1–184 in HemAT-*Bh*) with limited homology to the *Sperm whale* myoglobin (*Sw*Mb) and a coupled C-terminal MCP domain (residues 222–489 in HemAT-*Hs*, 198–432 in HemAT-*Bs*, and 222–439 in HemAT-*Bh*) that shares 30% identity with the *E. coli* serine receptor Tsr (Hou, Belisle, et al., 2001; Hou et al., 2000).

Deletion and overexpression of *HemAT* genes influence the movement of these organisms in response to an O_2 gradient, conclusively proving the engagement of HemAT in the aerophobic response in *H. salinarum* and in the aerophilic response in *B. subtilis* (Hou et al., 2000).

As typical for haem–proteins, HemAT reversibly binds diatomic gases (e.g. O_2, CO, NO) (Hou, Freitas, et al., 2001), and the O_2 affinity of HemAT-*Bs* is calculated to be similar to the one of *Sw*Mb, namely, $K_d = 719$ nM (where *Sw*Mb $K_d = 882$ nM) (Aono et al., 2002).

Different techniques were used to collect complementary structural data, proving that HemAT is capable of discriminating among gaseous ligands. Indeed, it ensures an unambiguous response that occurs after signal transduction exclusively with O_2 as haem distal ligand.

A network of hydrogen bonds that involves the haem-bound O_2 and Thr95, and His86 and haem 6-propionate causing conformational changes is essential for the transduction of the signal from the sensor domain to the signalling domain (El-Mashtoly, Gu, et al., 2008; Yoshimura et al., 2006). Moreover, changes at the proximal side of the haem group, with the formation of a hydrogen bond between Tyr133 and the axial His123, are also linked to signal transduction upon O_2 ligand binding (Yoshida, Ishikawa, Aono, & Mizutani, 2012; Yoshimura, Yoshioka, Mizutani, & Aono, 2007).

Using CO as ligand, the presence of an additional binding site at the proximal side of the haem group has been shown. In this respect, Tyr133 assumes a critical role in gating a tunnel that connects the two haem cavities, proximal and distal (Pinakoulaki et al., 2006). This system of tunnel and binding cavities suggests the possibility of fluctuating movements of the ligand between the two environments that might be crucial for determining O_2 affinities and, therefore, the dynamics of the aerotactic response.

After these residue-specific observations, structural changes involving helical movements have been proved (El-Mashtoly, Gu, et al., 2008; El-Mashtoly et al., 2012). Indeed, spectroscopic evidence is available on the central role of the B- and G-helices displacement, specifically triggered by O_2 binding, which leads to signal transduction (El-Mashtoly, Gu, et al., 2008; El-Mashtoly et al., 2012).

The crystal structures of HemAT-*Bs* in the liganded and unliganded forms have been determined and reveal a homo–dimeric organization of the sensor (Zhang & Phillips, 2003). The dimerization interface involves the helices G and H, which form an antiparallel four-helical bundle, part of the Z-helix, and the B–C corner of the globin domain (Fig. 1.2). It has also been noted that the asymmetry of the HemAT-*Bs* structure increases, going from the liganded to the unliganded form. Along with this observation, the authors propose a signalling transduction model in which the unliganded state is energetically unfavourable. When O_2 diffuses from the environment into the cytoplasm, one of the two subunits of the dimer

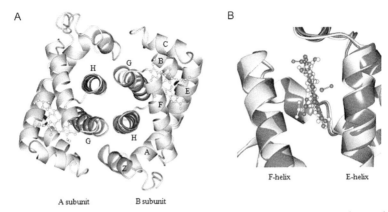

Figure 1.2 HemAT-*Bs* crystal structures. (A) The dimerization interface of the unliganded HemAT-*Bs* homo-dimer is reported (PDB: 1OR6). The contact area is wide and involves helices G and H in an antiparallel four-helical bundle, part of the Z-helix, and the BC corner. (B) Superposition of HemAT-*Bs* unliganded (light grey) and liganded (dark grey, PDB: 1OR4) structures. The presence of the ligand (CN⁻) causes conformational modifications in the haem group and in the haem-pocket. These changes are essential for signal transduction. The haem group and the ligand are depicted with ball-and-stick representation.

binds one gas molecule and passes into the energetically favourable liganded state, undergoing structural rearrangements. These structural modifications further decrease O_2 affinity of the second molecule of the dimer. As the organism moves along the gradient, the O_2 concentration increases progressively until it saturates the second subunit (Zhang & Phillips, 2003).

2.2. Haem-based sensors with gene-regulating function

The regulation of gene expression occurs via (i) DNA direct binding, for example, carbon monoxide oxidation activator (CooA) (Aono, Nakajima, Saito, & Okada, 1996; Aono, Takasaki, Unno, Kamiya, & Nakajima, 1999; He, Shelver, Kerby, & Roberts, 1996; Lanzilotta et al., 2000; Rajeev et al., 2012; Roberts, Thorsteinsson, Kerby, Lanzilotta, & Poulos, 2001; Shelver, Kerby, He, & Roberts, 1997) and neuronal PAS domain protein 2 (NPAS2) (Dioum et al., 2002; Gekakis et al., 1998; Hogenesch, Gu, Jain, & Bradfield, 1998; Reick, Garcia, Dudley, & McKnight, 2001); (ii) protein–protein interaction as in transcription factors and regulators, for example, nitrogen fixation gene expression regulator (FixL), *Vv*GReg, and *Cv*GReg (David et al., 1988; Freitas, Hou, & Alam, 2003; Freitas, Saito, Hou, & Alam, 2005; Gilles-Gonzalez, Ditta, & Helinski, 1991;

Gilles-Gonzalez & Gonzalez, 1993; Iniesta, Hillson, & Shapiro, 2010; Sousa, Tuckerman, Gondim, Gonzalez, & Gilles-Gonzalez, 2013; Virts, Stanfield, Helinski, & Ditta, 1988); and (iii) modulation of the local concentration of the second messenger cyclic-di-GMP by the concerted activity of DGCs and phosphodiesterases (PDEs), for example, haem-containing phosphodiesterase A1 (PDEA1), DOS, and GCS proteins (Chang et al., 2001; Delgado-Nixon, Gonzalez, & Gilles-Gonzalez, 2000; Sasakura et al., 2002; Tal et al., 1998; Weinhouse et al., 1997; Zhao, Brandish, Ballou, & Marletta, 1999).

Each one of these families has been studied during the past years and data are available on their activity pathways and their protein structures.

2.2.1 Regulation of gene expression via protein–DNA interaction

The first mechanism that can influence the expression of selected genes is the direct interaction between the transcription regulator and the gene promoter. When a certain group of genes is regulated, a univocal recognition between the protein and the DNA sequence occurs. The haem-based sensors that regulate gene expression are activated, or inactivated, upon binding of a given diatomic gas. Therefore, the nature and the concentration of the gas is the signal that triggers gene-transcription.

2.2.1.1 The carbon monoxide oxidation activator

The CooA is a transcription regulator first reported in *Rhodospirillum rubrum* (*Rr*CooA) whose action is specifically activated upon binding of CO (Shelver, Kerby, He, & Roberts, 1995; Shelver et al., 1997; Uchida et al., 1998). In the presence of CO, it activates the expression of enzymes involved in the CO oxidation pathway by direct binding to specific DNA regions (Aono et al., 1996; He et al., 1996).

Evolutionary studies on the CooA coding sequence identified homology with the cyclic-AMP receptor protein (CRP) family, in particular with CRP (28% identity and 51% similarity) and FNR (18% identity and 45% similarity) from *Escherichia coli* (Shelver et al., 1995).

In vivo and *in vitro* studies show that CooA alone is sufficient to activate the RNA polymerase (RNAP) and to start the DNA transcription process; no other factors are required. The interaction between CooA and the RNAP happens at the level of the C-terminal domain of the α subunit of the polymerase (He et al., 1999) and is mediated by Glu167, a key residue in the E-helix of CooA (Youn, Thorsteinsson, Conrad, Kerby, & Roberts, 2005).

The CooA F-helix is responsible for the direct DNA binding. On the helix–turn–helix motif (EF-region), three residues have been identified as responsible for the site-specific recognition of DNA: Arg177, Gln178, and Ser181. Their side chains form hydrogen bonds with specific DNA pairs on the CooA-dependent promoter binding site (Aono et al., 1999).

With spectroscopic and mutagenesis techniques, His77 has been identified as critical for conformational changes upon CO binding. These structural changes occur at the haem-binding domain level and are subsequently communicated to the DNA-binding domain, thus allowing the sequence-specific interaction with DNA (Shelver et al., 1999; Uchida et al., 2000; Vogel, Spiro, Shelver, Thorsteinsson, & Roberts, 1999).

The crystal structure of the reduced (Fe^{2+}) RrCooA shows a homo-dimeric form in which the haem-Fe axial ligands are the proximal His77 of the same subunit and the distal Pro2 N-terminal nitrogen from the opposite partner subunit (Fig. 1.3A) (Lanzilotta et al., 2000). Confirming spectroscopic and mutation results, NMR experiments on His77 and Pro2 mutants highlight the binding of CO at the haem distal site in correspondence of Pro2. Therefore, the suggested role for Pro2 is to direct the ligand towards the appropriate binding site (Clark et al., 2004; Yamamoto et al., 2001).

The arrangement of the homo-dimer in the crystal displays two symmetric haem-binding domains coupled via their C-helices to two asymmetric DNA-binding domains (Fig. 1.3B). A comparison with the more symmetric CRP structure suggests that the binding of CO to the haem would cause rearrangements in the C-helix and, consequently, a more symmetric arrangement of the DNA-binding domain (Akiyama, Fujisawa, Ishimori, Morishima, & Aono, 2004; Kerby, Youn, Thorsteinsson, & Roberts, 2003; Lanzilotta et al., 2000). In agreement with this, mutations of residues at the dimer interface, and in particular, the mutation of Gly117 in the C-helix into bulky residues, disrupt the structural modifications that happen upon CO binding, thus, creating inactive CooA (Youn et al., 2001). When CO binds to the haem, structural rearrangements cause a roll-and-slide mechanism in which the C-helix moves towards the haem, thus, forming a wide hydrophobic pocket around the ligand (Yamashita, Hoashi, Tomisugi, Ishikawa, & Uno, 2004).

Kinetics measurements on the RrCooA homo-dimer show positive cooperativity for CO binding (Hill coefficient $n = 1.4$), low CO affinity ($P_{50} = 2.2$ μM, $K_1 = 0.16$ μM^{-1}, $K_2 = 1.3$ μM^{-1}), and slow conformational transitions between an open and a closed form that convert the DNA-binding domains from the inactive to the active state. All these features

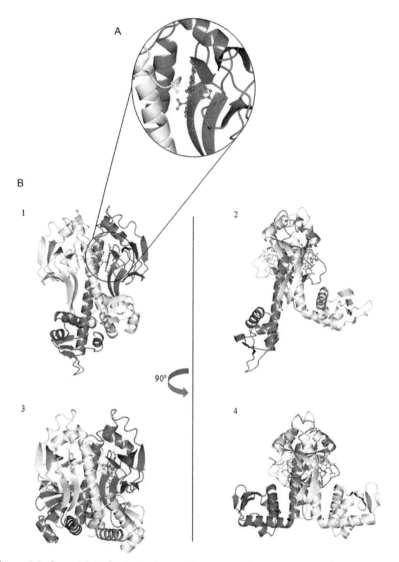

Figure 1.3 Comparison between homo-dimers of unliganded *Rr*CooA (PDB: 1FT9) and imidazole bound *Ch*CooA (PDB: 2FMY) crystal structures. (A) The CooA haem group displays a unique iron axial coordination. The haem proximal ligand is His77, where the distal ligand is Pro2 of the other subunit of the dimer. (B) The frontal and side views of the unliganded *Rr*CooA (1 and 2, respectively) are compared with the frontal and side views of the imidazole bound *Ch*CooA (3 and 4, respectively). When no ligand is coordinated to the haem-iron atom, the homo-dimer displays an asymmetric conformation (2). On the other hand, upon ligand binding, structural rearrangements cause movements of the DNA-binding domain, and the homo-dimer acquires a more symmetric structure (4).

ensure the activation of *Rr*CooA only when CO intracellular concentration occurs at a high (μM) level for sufficiently long time (longer than 1 min), and the activation is efficient over a narrow range of CO concentrations (Puranik et al., 2004).

As expected, the CO-binding kinetics is influenced by the presence of the DNA-binding domain and, even more, when DNA is bound. It has been proposed that the second domain helps to form a hydrophobic trap for CO, whereas the DNA binding rigidifies the whole structure, thus optimizing the CO-dependent signal (Benabbas, Karunakaran, Youn, Poulos, & Champion, 2012).

A *Rr*CooA homologue (29% identity) has been identified in the thermophilic CO-oxidizing bacterium named *Carboxydothermus hydrogenoformans* (*Ch*CooA) (Inagaki et al., 2005), which is activated, next to CO, also upon NO binding (Clark et al., 2006). This unexpected activation is explained by the formation of a six-coordinated Fe(II)–NO adduct, where in *Rr*CooA, NO forms a five-coordinated form (Clark et al., 2006; Reynolds et al., 2000).

The crystal structure of *Ch*CooA bound to exogenous imidazole is known and provides further atomic-resolution information on the molecular mechanisms that drive ligand-binding-dependent activation (Komori, Inagaki, Yoshioka, Aono, & Higuchi, 2007). Upon CO binding, the N-terminal region is displaced and bends, creating a bridge between the two major domains. This rearrangement determines structural modifications in the D-helix, which cause the movement of the DNA-binding domains (Fig. 1.3B). The DNA-binding domains slide closer to the effector-binding domains, thus exposing the F-helix for unambiguous DNA-recognition (Borjigin et al., 2007).

2.2.1.2 The neuronal PAS domain protein 2

The NPAS2 is a dimeric mammalian transcription factor mainly expressed in the forebrain under the control of the retinoic acid-related orphan receptor y (Hogenesch et al., 1998; Takeda, Kang, Angers, & Jetten, 2011; Zhou et al., 1997). It is known that the NPAS2 transcription function is strongly inhibited upon CO binding (Uchida et al., 2005).

Evidence shows that NPAS2 interacts with the brain and muscle Arnt-like protein 1 transcription factor (BMAL1) and, as such, functions as a hetero-dimer (Reick et al., 2001). NPAS2:BMAL1, in particular, is involved in the expression of genes of the circadian rhythm (Dudley et al., 2003; Rutter, Reick, Wu, & McKnight, 2001), and the activation of both factors is dependent on phosphorylation (Dardente, Fortier, Martineau, & Cermakian, 2007).

The DNA-binding region comprises about 100 residues at the N-terminal and is organized in a basic helix–loop–helix motif (bHLH). Two haem-binding PAS domains are present (PAS-A and PAS-B) of about 130 residues each (Dioum et al., 2002). At the C-terminal, there are about 400 additional residues that form a yet-uncharacterized domain (Gu, Hogenesch, & Bradfield, 2000).

The RR spectra of the PAS-B domain in the Fe(III) and Fe(II) form are a mixture of high-spin penta-coordinated and low-spin hexa-coordinated forms. Mutational studies reveal that His335 is the proximal axial ligand (Koudo et al., 2005). In the PAS-B Fe(II)–CO form, there is also a hydrogen bond present between the CO molecule and a distal histidine residue that stabilizes the exogenous ligand in place (Koudo et al., 2005). On the other hand, the dominant species of the PAS-A domain is hexa-coordinated with an equilibrium between the Cys-Fe-His and His-Fe-His forms. In the bHLH-PAS-A construct, this equilibrium is shifted towards the bis-histidine coordination (Uchida, Sagami, Shimizu, Ishimori, & Kitagawa, 2012). The mutation of one of these two histidine axial ligands dramatically decreases the transcriptional capability of NPAS2 (Ishida, Ueha, & Sagami, 2008).

The ligand-binding kinetics for bHLH-PAS-A ($k_{onCO} = 3.3 \times 10^7 \, mol^{-1} \, s^{-1}$), PAS-A ($k_{onCO} < 10^5 \, mol^{-1} \, s^{-1}$), and PAS-B ($k_{onCO} = 7.7 \times 10^5 \, mol^{-1} \, s^{-1}$) have been measured. These differences in the CO-binding affinities highlight the role of the bHLH domain in the stabilization of the haem-binding in NPAS2 (Koudo et al., 2005; Mukaiyama et al., 2006).

The target genes of NPAS2 are numerous. Lack or non-functional mutations in NPAS2 are related to several aberrant states. The best-characterized alterations involve the circadian rhythm (Dudley et al., 2003; Evans et al., 2013; Johansson et al., 2003; Wisor et al., 2008), but during the past years, more and more disorders and diseases have been linked to dysfunctions of NPAS2: for example, problems in the acquisition of specific types of memory (Garcia et al., 2000), autistic disorder (Nicholas et al., 2007), blood pressure (Curtis et al., 2007), Parkinson's disease (Anantharam et al., 2007), winter depression (Partonen et al., 2007) and depression (Lavebratt, Sjoholm, Partonen, Schalling, & Forsell, 2010), thrombogenesis (Westgate et al., 2008), prostate cancer (Zhu et al., 2009), fertility (Kovanen, Saarikoski, Aromaa, Lonnqvist, & Partonen, 2010), and chronic fatigue syndrome (Smith, Fang, Whistler, Unger, & Rajeevan, 2011). The

Ala394Thr polymorphism in NPAS2 has been proposed as a biomarker for non-Hodgkin's lymphoma (Zhu et al., 2007). Some proofs of a correlation between NPAS2 and breast cancer have been reported (Yi et al., 2010; Zhu et al., 2008). Indeed, NPAS2 might have a putative tumour suppressor function (Hoffman, Zheng, Ba, & Zhu, 2008), and its cancer-related transcriptional targets have been identified (Yi, Zheng, Leaderer, Hoffman, & Zhu, 2009). Other research groups, on the contrary, support the hypothesis that it is not possible to incontrovertibly prove any significant direct association between NPAS2 aberrant function and breast cancer risk, as data are still ambiguous (Wang et al., 2011).

2.2.1.3 The mammalian Period protein 2

Recently, another haem-binding protein, homologous to NPAS2 (18% identity and 58% similarity between the PAS domains), has been identified in mouse, namely, mammalian Period protein 2 (mPER2) (Kaasik & Lee, 2004). mPer1 and mPer2 are mammalian Period genes involved in the circadian clock whose expression is controlled by haem. mPER2 and NPAS2 function in concert and their activities are strictly interdependent. It has been shown that when the haem group is synthesized, there is competition between mPER2 and NPAS2 for its binding. It has been proposed that a haem-exchange between them is a critical step to regulate the genes target transcription (Kaasik & Lee, 2004).

mPER2 contains two PAS domains, PAS-A and PAS-B (Hayasaka, Kitanishi, Igarashi, & Shimizu, 2011; Kitanishi et al., 2008). In particular, PAS-B mediates the protein–protein interactions between transcription factors (Hayasaka et al., 2011).

2.2.1.4 The CO-responsive transcription regulator of *Burkholderia xenovorans* (*Bx*RcoM)

Haem-binding transcription factors have been identified also in Bacteria. In the bacterium *B. xenovorans*, two proteins have been isolated, *Bx*RcoM-1 and *Bx*RcoM-2, which function as CO-dependent transcription factors *in vivo* and stimulate the expression of genes for aerobic CO metabolism. Both these proteins show an N-terminal haem-binding PAS domain and a C-terminal LytTR DNA-binding domain (Kerby, Youn, & Roberts, 2008). A unique feature for these proteins is the coordination with a cysteine residue (Cys94) as distal axial ligand (Marvin, Kerby, Youn, Roberts, & Burstyn, 2008; Smith et al., 2012). Nothing can be commented about the architecture of these proteins as structural data have yet to be reported.

2.2.2 Regulation of gene expression via protein–protein interaction
Next to the direct protein–DNA binding, the control of gene expression can occur via the interaction of the haem-based sensor with the polymerase machinery, thus indirectly influencing the transcription.

2.2.2.1 The nitrogen fixation gene expression regulator
In 1987, Ditta and coworkers published evidence that the regulation of nitrogen fixation (*nif*) genes expression levels in *Sinorhizobium meliloti* (formerly known as *Rhizobium meliloti*) increased when environmental O_2 concentration dropped to microaerobic levels (Ditta, Virts, Palomares, & Kim, 1987; Virts et al., 1988). Not much later, a second paper appeared, reporting the first characterization of FixL/FixJ, two proteins responsible for the O_2 dependence of *nif* genes expression (David et al., 1988). Soon after, FixL was identified as an O_2-binding hemoprotein, capable of sensing O_2 with its haem group.

The unliganded form of FixL is catalytically active and transfers the γ-phosphate of one ATP molecule to conserved aspartate residues of the transcription factor FixJ (Gilles-Gonzalez et al., 1991; Gilles-Gonzalez & Gonzalez, 1993). FixJ consists of a phosphorable N-terminal module and a C-terminal transcription activator domain (Batut, Santero, & Kustu, 1991). The FixL-mediated phosphorylation derepresses the DNA-binding domain of FixJ that becomes transcriptionally active (Fig. 1.4) (Gouet et al., 1999).

FixL is a chimeric protein that consists of an optional membrane-anchoring domain, a haem-binding PAS domain, and a histidine kinase (HK) domain. The HK domain, in turn, is divided into an autophosphorylation subdomain, where a histidine is the phospho-accepting residue (H box), and a catalytic subdomain containing the ATP-binding site (Lois, Ditta, & Helinski, 1993; Monson, Weinstein, Ditta, & Helinski, 1992). The activation of the kinase

Figure 1.4 FixL/FixJ activation scheme. The interaction between FixL and FixJ facilitates the initial autophosphorylation of FixL, which transfers the γ-phosphate group from an ATP molecule with the production of one ADP. Subsequently, the transphosphorylation occurs, with the transfer of the phosphate from FixL to FixJ. Phosphorylated FixJ is activated and stimulates the transcription of specific target genes.

domain is driven by the coupled haem-binding PAS domain, but a mutual communication between these two domains is also present. The spin-state of the haem group controls the activation of the kinase, and a phosphate-dependent feedback controls the ligand affinity of the haem (Gilles-Gonzalez, Gonzalez, & Perutz, 1995; Rodgers, Lukat-Rodgers, & Barron, 1996; Sousa et al., 2013). Next to the haem-group spin-state regulation, an alternative inactivation mechanism has been proposed in which an aberrant disulfide bridge is formed in the homo-dimer (Akimoto, Tanaka, Nakamura, Shiro, & Nakamura, 2003) and might also be involved in the ligand discrimination.

Although the expression of *fix* genes has been reported in a number of bacterial strains (Anthamatten & Hennecke, 1991; Anthamatten, Scherb, & Hennecke, 1992; Crosson, McGrath, Stephens, McAdams, & Shapiro, 2005; David et al., 1988; de Philip, Batut, & Boistard, 1990; D'Hooghe, Michiels, & Vanderleyden, 1998; D'Hooghe et al., 1995; Girard et al., 2000; Iniesta et al., 2010; Kaminski & Elmerich, 1991), to date only two of these have been extensively characterized with multiple approaches, which are the transmembrane *S. meliloti* FixL (*Sm*FixL) (Lois et al., 1993) and the soluble *Bradyrhizobium japonicum* FixL (*Bj*FixL) (Anthamatten & Hennecke, 1991).

*Sm*FixL is a 505-residue-long protein that contains the optional trans-membrane domain, organized in four segments. It localizes to the cytoplasmic side of the inner bacterial membrane (Lois et al., 1993). The presence or absence of different domains of the protein influences the ligand-binding kinetics to the haem group. The deletion or retention of the kinase domain causes the rate and the stability of the gas–haem complexes to diminish, suggesting a direct or indirect influence of the second domain on the steric interactions between the iron-bound ligand and the amino acid side chains of the haem-pocket (Gilles-Gonzalez et al., 1994; Miyatake et al., 1999; Rodgers et al., 1996; Rodgers, Lukat-Rodgers, & Tang, 2000).

A closer look at the haem-binding domain reveals the possible role of Ile209 and Ile210 in signal transduction. These two residues are in the neighbourhood of the ligand bound to the haem group and, when O_2 is present, are displaced by steric repulsion. This movement is suggested to be responsible for the conformational changes in the FG-loop, which, in turn, influences the kinase structure, thus regulating its activation (Miyatake et al., 2000; Mukai, Nakamura, Nakamura, Iizuka, & Shiro, 2000). Time-resolved Resonance Raman spectra measured on the full length and on the truncated forms of *Sm*FixL during both O_2 and CO dissociation, as well as mutational studies, seem to confirm the role of the FG-loop in signal transduction (Hiruma, Kikuchi, Tanaka, Shiro, & Mizutani, 2007; Reynolds et al., 2009). Another important amino acid for signal transduction is Arg214. This residue is responsible for rigidifying the haem-moiety when a ligand is present. It has been suggested that this loss of flexibility might be important for down regulating the kinase activity (Tanaka, Nakamura, Shiro, & Fujii, 2006).

The second well-characterized FixL comes from the bacterium *B. japonicum* (*Bj*FixL), whose activity is inversely proportional to O_2 deprivation (Sciotti, Chanfon, Hennecke, & Fischer, 2003). As soon as *Bj*FixL was identified, it was clear that its functions were not limited to the *nif* genes expression stimulation in the absence of O_2. In fact, FixLJ mutants of *B. japonicum* were not able to grow anaerobically, thus suggesting an involvement also in anaerobic metabolism (Anthamatten & Hennecke, 1991; Anthamatten et al., 1992), for example, expression of genes for haem biosynthesis and/or nitrate and nitrite respiration (Mesa, Bedmar, Chanfon, Hennecke, & Fischer, 2003; Nellen-Anthamatten et al., 1998; Page & Guerinot, 1995; Robles, Sanchez, Bonnard, Delgado, & Bedmar, 2006).

*Bj*FixL crystal structure has been determined in the presence of different ligands (CN^- and CO) that help understanding its ligand-binding mechanism. Notably, in the O_2-*Bj*FixL, the hydrogen-bonding networks at the haem distal side (in particular, the ones involving $Fe-O_2$ and the distal residue Arg220-propionate 7) are crucial for the optimal ligand binding and recognition and for the signal transduction via the FG-loop rearrangements (Balland et al., 2006; Gong, Hao, & Chan, 2000; Hao, Isaza, Arndt, Soltis, & Chan, 2002).

In signal transduction, the PAS and the kinase domain act as a unity in which the kinase inactive form is energetically favourable and the liganded haem shifts this equilibrium even further (Ayers & Moffat, 2008; Balland et al., 2005; Dunham et al., 2003). Only two forms are present, the active unliganded form and the inactive O_2-bound form, which can convert one into the other in less than 1 μs (Key, Srajer, Pahl, & Moffat, 2007).

Nonetheless, an interesting effect of O_2 trapping has been reported (Kruglik et al., 2007) in which Arg220 rigidifies the haem group, thus imposing a specific configuration also to the bound O_2. This constraint, in combination with the hydrophobicity and structural properties of the haem distal cavity, creates an O_2 cage that traps the dissociated ligand in place. This finding might be related to the O_2-binding memory hypothesis proposed by Sousa and colleagues in order to explain the non-linear kinase response of this sensor to the haem saturation (Sousa, Tuckerman, Gonzalez, & Gilles-Gonzalez, 2007).

The crystal structure of the PAS domain of *Sm*FixL has been obtained only in the deoxy form (Miyatake et al., 2000), where for the PAS domain of *Bj*FixL several conditions originated well-diffracting crystals (Ayers & Moffat, 2008; Dunham et al., 2003; Gong et al., 2000, 1998; Hao et al., 2002; Key & Moffat, 2005). When comparing the PAS deoxy form of these sensors, striking similarities are noticed. The haem group is accommodated in a hydrophobic pocket formed by a long α-helix (named F) at the proximal side that hosts the haem-coordinating histidine residue, and three antiparallel β-strands (G, H, and I) at the distal side. The proximal and the distal sides are covalently linked by a flexible loop between the F and G segments.

The atomic-resolution structure of *Bj*FixL in the presence of an iron-bound ligand (e.g. CO and CN⁻) at the distal side of the haem does not change significantly the architecture of the pocket, except in the FG-loop, as described (Fig. 1.5). Given the numerous structural affinities between deoxy-*Sm*FixL and deoxy-*Bj*FixL, it is possible to speculate that

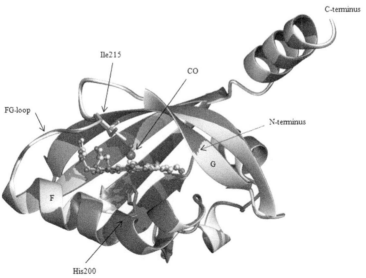

Figure 1.5 Superposition of the *Bj*PAS domain in the unligated (light grey, PDB: 1XJ3) and in the CO-ligated (dark grey: 1XJ2) form. The haem group (represented in ball-and-stick) is hosted in a hydrophobic pocket. The two structures are very similar, the main differences are concentrated in the FG-loop and are probably due to the movement of Ile215 (Ile209 in *Sm*FixL). Indeed, the presence of CO causes steric repulsion to the Ile215 side chain, which adopts a different conformation.

the signal transduction will occur with the same mechanism (Ayers & Moffat, 2008; Hao et al., 2002; Key & Moffat, 2005).

2.2.2.2 The *Vibrio vulnificus* and the *Chromobacterium violaceum* globin-coupled regulators (*Vv*GReg and *Cv*GReg)

To date, only two GCSs have been identified to control gene expression by protein–protein interaction. The first member of this category was identified by Freitas and coworkers in 2003 in the Gram-negative δ-Proteobacterium *V. vulnificus* genome (Freitas et al., 2003), and the second member was identified in 2005 in the δ-Proteobacterium *C. violaceum* (Freitas et al., 2005). These sensors are characterized by the fusion of an N-terminal globin-like domain to a C-terminal STAS domain (sulphate transporter and anti-anti-sigma factor) (Campbell, Westblade, & Darst, 2008). Therefore, the transcription control played by *Vv*GReg and *Cv*GReg involves the indirect stimulation of bacterial RNAP by inactivating the inhibitory anti σ-kinases, similar to what happens in stressosomes of *B. subtilis* (see Section 2.5.1). This causes the liberation of σ factors that direct the specific transcription of target genes or operons.

Only a few crystal structures of STAS domains are known (Marles-Wright et al., 2008; Quin et al., 2012). It is clear that, despite evolutionary distances in the amino acid sequence, all these structures display a conserved fold that comprises four β-strands interspersed among five α-helices.

2.2.2.3 The globin-coupled HK from *Anaeromyxobacter* sp. strain Fw109-5 (*Af*GcHK)

In *Anaeromyxobacter* sp. strain Fw109-5, the first example of a globin domain coupled to an HK has been identified, namely, *Af*GcHK (Kitanishi et al., 2011).

The globin domain of *Af*GcHK displays significant homology to the *Sw*Mb (16% identity and 34% similarity) and to the HemAT-*Bs* (24% identity and 39% similarity). The UV–vis characterization is in-line with the classical spectra of haem-binding globins. RR measurements revealed the presence of a histidine residue (His99) at the proximal side and of a tyrosine residue (Tyr45) at the distal side. Tyr45 is also important for the stabilization of the exogenous ligand.

The HK domain shows one autophosphorylation site (His183) and two phosphorylation sites (Asp52 and Asp169). The presence of a low-spin Fe(III), Fe(II)–O$_2$, or Fe(II)–CO activates the kinase activity, where the high-spin Fe(II) is the inactive form (Kitanishi et al., 2011).

The cognate response regulator, which is activated by *Af*GcHK-mediated phosphorylation, has been identified (*Anae109_2439*). Phosphotransfer reactions have been performed *in vitro* and occur as early as 5 min after ATP addition to the reaction mixture (Kitanishi et al., 2011).

2.2.3 Regulation of gene expression via second messenger biosynthesis

Cyclic dimeric-(3′–5′)-GMP (c-di-GMP) was initially discovered in *Gluconacetobacter xylinus* (previously named *Acetobacter xylinum*) where it regulates cellulose synthesis (Ross et al., 1987). A number of other bacterial functions have been progressively ascribed as controlled by c-di-GMP, that is, biofilm formation, motility, virulence, and the cell cycle (Abel et al., 2011; Bobrov et al., 2011; Duerig et al., 2009; Karaolis et al., 2005; Kulasakara et al., 2006; Ryan et al., 2007, 2010; Tamayo, Pratt, & Camilli, 2007; Tischler & Camilli, 2004), promoting it as a universal bacterial second messenger.

The biosynthesis and the degradation of c-di-GMP are executed by two enzymes, the DGC and the PDE, respectively. Highly conserved motifs can be recognized in these proteins, that is, the GGDEF motif in DGC and the EAL motif in PDE (Ausmees et al., 2001; Matilla, Travieso, Ramos, & Ramos-Gonzalez, 2011; Simm, Morr, Kader, Nimtz, & Romling, 2004).

Most commonly, enzymes with GGDEF and EAL motifs combine the catalytically active domain with a sensor domain. The sensor domain recognizes specific environmental signals, such as the gas concentration, redox state of the electron transport chain, light quorum-sensing, and many others, regulating the synthesis or the degradation of c-di-GMP (Deng et al., 2012; Henry & Crosson, 2011; Ho, Burden, & Hurley, 2000).

2.2.3.1 The direct oxygen sensor from *E. coli* (*Ec*Dos)

The *E. coli* direct O_2 sensor (*Ec*Dos) is expressed via transcription and translation of an operon (*dosCP*). The *dosCP* operon codes for two proteins, one includes a haem-binding PAS domain coupled to a C-terminal PDE domain (*Ec*DosP), the second comprises a globin domain coupled to a DGC domain (*Ec*DosC) (Delgado-Nixon et al., 2000; Tuckerman et al., 2009).

*Ec*DosP is a chimeric protein that shows high-sequence homology between its N-terminal region and the PAS domain of rhizobial FixL (25% identity and 60% similarity) (Blattner et al., 1997) and between its full-length sequence and the *G. xylinus* PDE (30% identity and 50% similarity) (Delgado-Nixon et al., 2000). Spectroscopic analysis on the purified *Ec*DosP identifies a low-spin hexa-coordinated ferric-haem complex with proximal His77 and distal Met95 as axial ligands (Gonzalez et al., 2002; Sasakura et al., 2002). On the other hand, when the CO-bound form is created, Met95 coordinates and stabilizes the ligand in place (Sato et al., 2002; Tanaka, Takahashi, & Shimizu, 2007). Moreover, EcDosP in solution forms concentration-independent homo-dimers (0.01–1 μM), and Met95 is involved also in the intra-dimer interactions (Kobayashi et al., 2010; Lechauve et al., 2009).

Although the gas-binding affinity is cooperative, the affinity of EcDosP for the binding of CO and NO is significantly low ($K_d = 10$ μM) compared to O_2 ($K_d = 13$ μM); therefore, under physiological concentrations, these two gases are negatively selected (Delgado-Nixon et al., 2000; Taguchi et al., 2004).

Several residues are involved in structural modifications upon O_2, CO, and NO binding and, therefore, in signal transduction, that is, Trp53, Asn84, Arg97, Phe113, and Tyr126 (El-Mashtoly, Nakashima, Tanaka, Shimizu, & Kitagawa, 2008; El-Mashtoly, Takahashi, Shimizu, & Kitagawa, 2007; Ito, Araki, et al., 2009; Tanaka & Shimizu, 2008). Structural modifications in the PAS domain are followed by movement in the FG turn, in the G-strand, and in the HI turn, thus activating the enzymatic domain (Ito, Igarashi, & Shimizu, 2009; Park, Suquet, Satterlee, & Kang, 2004).

EcDosC presents an N-terminal globin domain fused to a C-terminal DGC domain. It is likely one of the most expressed DGCs in E. coli (Sommerfeldt et al., 2009) and some of the biofilm-regulating pathways that are under its control have been identified. It regulates the expression of genes encoding for curli fibre subunits (Tagliabue, Maciag, Antoniani, & Landini, 2010) and of the poly-N-acetylglucosamine (PNAG) biosynthetic pgaABCD operon that controls the production of the exopolysaccharide PNAG (Tagliabue, Antoniani, et al., 2010), thus promoting the formation of sessile populations of bacteria enclosed in a self-produced polymeric matrix, namely, the biofilms.

Concerning the characteristics of the globin domain, the proximal and distal ligands of the haem are His98 and Tyr43, respectively. Tyr43, in particular, is important for the O_2 specific recognition and stability (Kitanishi et al., 2010). Leu65, also found at the haem distal side, represents a gate that renders the haem-pocket inaccessible to the environmental water that would render the haem-iron atom more sensitive to oxidation (Nakajima et al., 2012). In conclusion, Tyr43 and Leu65 together cooperate for creating the best conditions for a fully reactive and sensitive globin capable of reversible O_2 binding.

2.2.3.2 The globin-coupled regulator from *Azotobacter vinelandii* (AvGReg)

A. vinelandii globin-coupled regulator (*Av*GReg) is a 472 residues long soluble protein. Hundred and seventy-eight residues at the N-terminus form the globin domain, and 170 residues at the C-terminus form the DGC domain. Until now, the cytosolic form of the globin domain (*Av*GReg178) and an *in vitro* folded full-length form of *Av*GReg have been characterized.

The ligand-binding kinetics have been measured for both *Av*GReg and *Av*GReg178. They show an intriguing k_{off} constant for CO ($4 \, s^{-1}$) that is more than 200 times faster than for *Sw*Mb ($0.019 \, s^{-1}$), whereas the k_{on} is comparable ($1 \times 10^6 \, M^{-1} \, s^{-1}$ for *Av*GReg forms and $0.5 \times 10^6 \, M^{-1} \, s^{-1}$ for *Sw*Mb) (Thijs et al., 2007).

In the study of Thijs et al., the results of extensive UV–vis, RR, and continuous-wavelength EPR spectroscopy characterizations have been shown. Interestingly, the number of haem–iron coordination bonds in *Av*GReg and *Av*GReg178 differs, being six and five, respectively. At first instance, these data suggested that the presence of the DGC domain influences the structural conformation of the globin domain.

We have recently expressed and purified the *in vivo* folded soluble form of *Av*GReg sensor and measured the UV–vis spectra (unpublished data). In this case, the as-expressed *Av*GReg shows penta coordination of the haem-iron atom, as originally reported also for the truncated *Av*GReg178. We therefore can conclude that the presence of the DGC domain does not change the coordination state of the haem group. Most likely, the reported difference in coordination is due to an incorrectly folded form of *Av*GReg.

Thijs et al. (2007) also showed that the haem-pocket conformation in the presence of an external ligand is comparable in both forms, being rather open and without direct His–CO stabilization, therefore making a His-dependent signal transduction unlikely. Significant difference is appreciated in the protein–haem propionates interactions, thus suggesting a central role of these chemical groups in the signalling. These data are in accordance with the presence of the penta coordination in both forms, *Av*GReg and *Av*GReg178.

2.2.3.3 The globin-coupled regulator from *Bordetella pertussis* (BpeGReg)

*Bpe*GReg is the globin–coupled regulator found in *B. pertussis*. It counts for 475 residues, of which 1–155 form the globin domain and 297–475 form the DGC domain. The region comprising residues 156–296 functions as a linker and is supposed to be important for the activity regulation of the DGC domain from the globin domain. The predicted 3D model of the three domains displays an active site for GTP binding and an inhibitory site for c-di-GMP binding, and the molecule is represented as a homo-dimer (Wan et al., 2009).

The capability of synthesizing c-di-GMP upon selective binding of O_2, as well as the dependence of biofilm formation on the BpeGReg activity, has been demonstrated (Wan et al., 2009).

Different states of the globin domain, bound or unbound, drive different levels of DGC activity. When in the Fe(II)–O_2 state, the c-di-GMP synthesis is at the highest level. This level decreases progressively in the presence of Fe(II)–NO, Fe(II)–CO, and Fe(II) (Wan et al., 2009).

The presence of the predicted inhibitory site is confirmed by activity assays performed in the absence and presence of a PDE, which sequesters the newly synthesized c-di-GMP and converts it into pGpG. Without PDE, the reaction catalysed by BpeGReg slows down quickly and stops before the GTP present is completely used. With PDE, the reaction proceeds to completion within minutes (Wan et al., 2009).

Bioinformatic analysis of the linker region sequence identifies a highly conserved residue, His225. Mutations of this residue result in an interruption of the c-di-GMP production, regardless of the haem-group state. On the other hand, the globin domain absorption spectra are not affected by the mutation. These data suggest that His225 is involved in the DCG activation (Wan et al., 2009).

B. pertussis is a human pathogen capable of forming biofilms. The participation of BpeGReg in the biofilm formation pathway has been investigated. The comparison of BpeGReg knock-out and wilde-type shows a much lower biofilm presence in the first case, thus confirming the hypothesis (Wan et al., 2009).

2.2.3.4 The haem-containing diguanylate cyclase from *Desulfotalea psychrophila*

A haem-containing diguanylate cyclase (HemDGC) is expressed in the obligatory anaerobic bacterium *D. psychrophila* as a homo-tetramer (Sawai et al., 2010).

HemDGC displays a globin domain coupled to a DGC and its activity is stimulated upon binding of O_2, probably recognized as a toxic molecule, thus stimulating biofilm formation as a defence mechanism. For this reason, HemDGC is probably involved in the protection of the bacterium against oxidative stress.

RR spectra show that the haem axial ligands are His108 at the proximal side and Tyr55–Gln81 (in presence of O_2) or Gln81 (in presence of CO) at the distal side. It is possible to suggest that the role of Tyr55 in the ligand stabilization is linked to the ligand-discrimination process (Sawai et al., 2010).

2.2.3.5 The haem-containing phosphodiesterase A1 from G. xylinus (AxPDEA1)

The c-di-GMP degradation and, as a consequence, the cellulose production in G. xylinus are controlled by the phosphodiesterase 1 (AxPDEA1) (Chang et al., 2001). AxPDEA1 displays a haem-binding PAS domain that controls the activity of the fused PDE domain depending on environmental O_2 concentration. The deoxy-haem is the active form that leads to the hydrolysis of the c-di-GMP phosphodiester bonds with the formation of linear pGpG (Chang et al., 2001). On the other hand, the formation of c-di-GMP is catalysed by AxDCG2 (Qi, Rao, Luo, & Liang, 2009).

Like AxPDEA1, AxDCG2 is a di-domain protein that comprises a sensing PAS domain and an enzymatic DGC domain. But in this case, the PAS domain binds an FAD cofactor non-covalently and is sensitive to the intracellular redox state and/or O_2 concentration, as the PDE function is activated when FAD is oxidized (Qi et al., 2009).

2.2.3.6 The haemerythrin-coupled diguanylate cyclase in V. cholerae (VcBhr-DGC)

Haemerythrins form a class of non-haem di-iron O_2-binding proteins that display typical sequence and fold and are found only in invertebrates (Vanin et al., 2006). The presence of a chimeric protein has been recently reported in V. cholerae, which consists of an N-terminal bacterial haemerythrin domain fused to a C-terminal DGC domain (VcBhr-DGC) (Schaller et al., 2012).

The haemerythrin domain of VcBhr-DGC binds two non-haem iron atoms and cycles between a di-ferric and a di-ferrous form. The di-ferrous form shows high autoxidation rate when exposed to air, thus being stable only under anaerobic conditions. Moreover, when the haemerythrin is in the di-ferrous state, the activity of the DGC domain is enhanced 10-fold, compared to the inactive di-ferric form (Schaller et al., 2012).

Taking into account that V. cholerae forms biofilms and that the biofilm formation is induced under microaerobic conditions, it can be inferred that VcBhr-DGC is involved in this pathway as an environmental-signal transducing molecule.

2.2.3.7 The haem NO-binding sensors

The bacterial ancestor of the animal soluble guanylyl cyclases (sGC) is represented by the bacterial HNOB sensors (Iyer, Anantharaman, & Aravind,

2003). These proteins show an α-helical fold and bind the haem group, either covalently to a His residue or non-covalently when the His is not present.

The haem-binding domain can be fused to an MCP or be a single-domain protein. HNOBs are encoded by operons that contain genes for the HNOB-associated (HNOBA) protein, another component of the signalling system (Iyer et al., 2003).

Therefore, HNOBs sense intracellular NO presence and provide answers by activating MCPs, PDEs, or DGCs (Gilles-Gonzalez & Gonzalez, 2005; Iyer et al., 2003).

2.3. Haem-based sensors with enzymatic function

A new class of haem-based sensors is represented by chimeric proteins in which the transmitter domain is not directly involved in any of the previously described activities. These sensors mediate different enzymatic reactions in response to a stimulus.

Although to date this class only counts for two members, we believe that other molecules might be present in nature that still have to be characterized.

2.3.1 The truncated globin-coupled sensors

Similarity searches identified the existence of truncated globin-coupled sensors (tGCSs) in S. avermitilis and Frankia sp. that are coupled to antibiotic biosynthesis monooxygenases (Bonamore et al., 2007). These sensors differ from the known GCS for the presence of the 2/2 truncated globin (trHb) in place of the classical 3/3 globin domain, and for the position of this domain at the C-terminus of the sensor.

The preferred form in which the tGCSs are found in solution is homo-dimeric. The sequence alignment, the spectroscopic investigation, and the ligand-binding properties highlight resemblance to the group II trHb. Although the link between the trHb domain and the quinones redox activity of the monooxygenase domain remains unclear, the haem group remains in the reduced O_2-bound state during the catalysis (Bonamore et al., 2007). Further experiments are needed to unveil the physiological role of these sensors.

2.4. Haem-based sensors with unknown function

The role of some haem-based sensors is not yet known. Similarity approaches have been used to predict the function of their transmitter domain, but functional studies, for example, knock-out as well as over-expressing strains, are needed to fully understand the biological role that these molecules play in the bacterial organism.

2.4.1 The GCS from Geobacter sulfurreducens and Geobacter metallireducens

All the GCSs mentioned thus far are soluble proteins. Exceptions to this rule are the GCSs identified in two Gram-negative δ-Proteobacteria, *G. sulfurreducens* (*Gs*GCS) and *G. metallireducens* (*Gm*GCS). *G. metallireducens* is an obligate anaerobe and the first organism found to oxidize organic compounds using iron oxide as the final electron acceptor (Lovley et al., 1993). *G. sulfurreducens* displays the same energy-producing pathway, but, on the other hand, it has been found not to be an obligate anaerobe; therefore, it is also able to grow with O_2 as terminal electron acceptor (Lin, Coppi, & Lovley, 2004; Methe et al., 2003). With respect to this O_2 tolerance, the engagement of the identified transmembrane GCS is yet to be proved.

Both *Gs*GCS and *Gm*GCS have been predicted to display the classical globin-like-sensing domain coupled to a transmembrane transmitter domain with yet unknown function (Freitas et al., 2003). A putative function as S-nitrosoglutathione reductase has been ascribed to both transmitter domains (Freitas et al., 2005). Therefore, *Gs*GCS and *Gm*GCS might be involved in regulating intracytoplasmatic NO levels and in protecting the organism against nitrosative stress. On the other hand, it has been reported that *Gs*GCS expression levels are higher when the bacterium is grown under Fe(III)-reducing conditions, suggesting that it might also have a role in an optimal Fe(III) reduction (Aklujkar et al., 2013; Ding et al., 2006). None-theless, experimental evidences of the real function are yet to be shown.

Very few structural data are available for these GCSs. So far, only the globin domain of *Gs*GCS (GSGCS[162]) has been investigated in this direction (Desmet et al., 2010; Pesce et al., 2009). In the reported crystal structure, *Gs*GCS[162] displays a modified 3/3 globin fold with an extra Z-α-helix preceding the N-terminal A-helix and with the complete loss of the D-helix (Pesce et al., 2009). The molecules are disposed forming homo-dimers at the G–H interface, as previously reported also for HemAT-*Bs* (Zhang & Phillips, 2003) and *Methanosarcina acetivorans* protoglobin (*Ma*PGB) (Freitas et al., 2004), but it is yet to be proved whether this quaternary assembly plays any role in signal transduction.

*Gs*GCS[162] is the first example of bis-His haem coordination that involves proximal His(93)F8 and distal His(66)E11 as axial ligands (Desmet

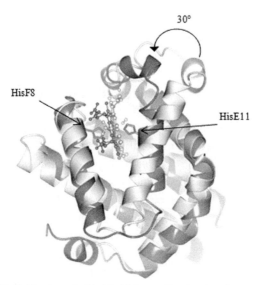

Figure 1.6 *Gs*GCS globin-domain bis-histidyl coordination involves proximal HisF8 and distal HisE11 (dark grey, PDB: 2W31). Such an unprecedented coordination causes the E-helix in a 30° anticlockwise rotation when compared to *Sw*Mb (light grey, PDB: 1MBN), as indicated with the black arrow in the figure.

et al., 2010; Pesce et al., 2009). Such a bis–histidyl coordination is unprecedented, as it usually involves His residues at topological position F8 and E7 at proximal and distal positions, respectively. This special coordination forces the E-helix in a 30° anticlockwise rotation (relative to *Sw*Mb) (Fig. 1.6). This rotation results in an orthogonal orientation of the E-helix to the C-helix and almost parallel to the haem group, thus reshaping the haem distal cavity (Pesce et al., 2009).

2.5. Non-haem globin sensors

The discovery of non-haem globins confirms the high dynamicity and adaptability of this protein fold. Non-haem chimeric globins represent an emerging sensor group that might disclose a big number of variations and specializations.

2.5.1 The B. subtilis σB regulator

B. subtilis is capable of reacting against environmental stress via the stressosome. The *B. subtilis* stressosome is a 1.5-MDa complex that transduces environmental signals, thus mediating the stress response of the bacterium that results in the upregulation of over 125 genes. Three proteins, RsbR, RsbS, and RsbT, constitute the stressosome complex that presents a pseudo-icosahedral core with sensory extensions provided by the N-terminal domain of RsbR (Chen et al., 2003). As such, RsbR comprises an N-terminal sensor domain and a C-terminal STAS domain that interacts with the RNAP σ^B subunit inducing gene transcription.

In 2005, Murray and colleagues published the crystal structure of the RsbR N-terminal domain (N-RsbR). Unexpectedly, N-RsbR displays the classical globin fold architecture, though lacking helices C and D as many known haem-binding globins do (Murray, Delumeau, & Lewis, 2005). The HemAT globin domain was found to be the structural match for N-RsbR and the relative protein sequence alignment revealed 13.5% of identity. Intriguingly, of the universally conserved residues in globins, none is present in RsbR, including the proximal histidine (replaced by an alanine). Again, the putative haem-pocket space is occupied by the helices E and F, brought much closer together than in known globins (Murray et al., 2005).

The signal transduction from the N- to the C-terminal domain is suggested to be mediated by a conserved 13-residue linker region. As N-RsbR is homo-dimeric in the soluble and in the crystallized forms (Murray et al., 2005), the two corresponding linker regions are also close together and interact via a conserved interface (Marles-Wright et al., 2008). Mutations introduced in the linker region at the conserved positions in order to disrupt its dimerization capability, thought to inactivate the stressosome, resulted in 30-fold increased stress-mediated response in unstressed cells. On the other hand, mutations in the non-conserved positions decreased the output in stressed and non-stressed cells (Gaidenko, Bie, Baldwin, & Price, 2012). These findings confirm the involvement of the linker region in the stressosome outcome, but weaken the hypothesis that the non-haem globin domain is responsible for stress signal sensing. Therefore, another hypothesis is that the N-terminal non-haem globin domain serves only as a steady-state regulator of the stressosome, and that the real stress sensor is somewhere downstream of the protein sequence. It is clear that a non-conventional signal transduction, yet to be characterized, is present (Gaidenko et al., 2012).

Some important aspects still remain to be unravelled. We do not know yet whether this non-haem globin domain is the functional form, as a weakly bound cofactor might be lost during purification. In parallel, evidence is available neither on how the stress signal is sensed nor on how it is communicated to the stressosome.

2.5.2 The non-haem-coupled regulator from Moorella thermoacetica and the non-haem-coupled sporulation inhibitory proteins from Bacillus anthracis

The *M. thermoacetica* stressosome, which regulates the biosynthesis of c-di-GMP, has been recently characterized and comprises two components, non-haem-coupled regulator from *M. thermoacetica* (MtR) and MtS. MtR N-terminal domain has been identified as structurally homologous to *B. subtilis* RsbR despite low sequence identity (12%), as such being a non-haem globin (Quin et al., 2012). As for *B. subtilis*, the stress response mechanism in *M. thermoacetica* remains unknown.

Non-haem-coupled sporulation inhibitory proteins from *B. anthracis* (pXO1-118 and pXO2-61) are two sporulation inhibitory proteins identified in *B. anthracis* and are non-haem globins (Stranzl et al., 2011). Instead of the haem group, they bind fatty acids in a hydrophobic tunnel and a hydrophilic chamber. It has been suggested that this kind of non-haem globin senses changes in intracellular fatty acid composition, chloride concentration, and/or pH, thus modulating the sporulation rate of the organism (Stranzl et al., 2011).

3. THE EVOLUTION OF THE GLOBINS

Extensive analyses of the sequences of globins and putative globins have been carried out regularly in the past years in parallel with the constant growth of new genome annotation. Hypotheses on the evolution of single domain and chimeric globins, as well as on the development of functional specializations, have always been proposed and debated (Blank & Burmester, 2012; Freitas et al., 2004; Vinogradov et al., 2007).

As globins are found in all kingdoms of life, it seems to be logical that the ancestor of all globins was present in the first organisms populating the Earth, the Last Universal Common Ancestor (LUCA) some 4000 Myr (Freitas et al., 2004; Moens et al., 1996; Vinogradov et al., 2007). Although the environmental oxygen amount was very low at that time (<0.0008 atm), high local concentration might have been present, thus being lethal for anaerobic organisms and causing oxidative stress. Under these conditions, the evolution of globins started and it has been postulated that their initial role was NO-detoxification (Poole, 2005; Poole & Hughes, 2000). Only at a later stage, they evolved into the multi-domain sensors as we know them to date.

Phylogenetic analyses show that the globin sequences appear to be distributed into three separate lineages, (i) the 3/3 globins of plants and metazoans, the single-domain globins, and the flavohemoglobins; (ii) the 3/3 globins of bacteria (GCSs) and the protoglobins; and (iii) the 2/2 globins. Given that all the three globin subgroups occur only in Bacteria, it has been proposed that globins initially appeared and evolved in these organisms (Vinogradov et al., 2005, 2006). Subsequently, via lateral gene transfers (LGTs), they spread in Archaea and Eukarya, starting the evolutionary process that resulted in the globins we are studying these days (Moens et al., 1996; Vinogradov et al., 2007).

Concerning the formation of chimeric globins, that is, the GCSs, the evolutionary process is schematically divided into three steps: (1) after the emergence of the first 3/3 single-domain and soluble globin ancestor in LUCA, the 2/2 globins evolved and diverged; (2) then alternative domains fused to the 3/3 globins, as such producing the first chimeric GCSs; (3) as last, two sorts of LGTs took place, one between Bacteria and the eukaryote precursor(s), and the other between Bacteria and Archaea (Vinogradov et al., 2007).

It can be hypothesized that the evolution of the haem-based sensors happened similarly. That is, the initial emergence of the simple single domain proteins was followed by their covalent connection and, therefore, the formation of chimeric functional sensors with precise functions.

4. CONCLUSIONS AND FUTURE PERSPECTIVES

The haem-based sensor superfamily comprises numerous chimeric proteins present in Bacteria, Archaea, and Eukarya. Supposedly, they evolved as single domain proteins, which fused to each other afterwards. The haem-based sensors' evolutionary driving force is the need of the organisms and cells to link intra- and extracellular changes with intracellular response. As such, they contain a sensor domain that detects a specific signal and, via structural rearrangements, activates or inactivates the cognate transmitter domain. This mechanism provides a fine tuning of the biochemical reactions, resulting in adapted intracellular modifications.

The haem-based sensors can be classified on the basis of the architecture of the haem-binding domain or on the function of the transmitter domain. In this review, we support the functional classification; therefore we divided the sensors into (i) aerotactic; (ii) gene regulating via protein–DNA interaction, protein–protein interaction, or second messenger biosynthesis; (iii) enzymatic function; and (iv) unknown function. All these subgroups emphasize the extreme flexibility of the haem-binding domains. Indeed, they all rely on the haem–group reactivity (i.e. the binding of a ligand or the change in the redox state of the iron) and translate them in a usable signal for the transmitter domain.

Some of these sensors are better characterized (e.g. HemAT and FixL) both structurally and functionally, where some others are not, due to protein instability, technical complexities, or recent identification (e.g. *Af*GcHK, *Bpe*GReg, and tGCSs).

We have also reported a recently identified class of chimeric proteins that do not bind the haem. These non-haem-binding sensors are an excellent example of the inventiveness of nature and of the plasticity of the globin domain. Despite the absence of the haem group that makes it impossible to sense gases or transfer electrons, they still function as sensors. However, it is not yet clear what the signal is and how it is received.

The scientific community is probably still far from having an exhaustive idea of the whole complexity of the haem-based sensor superfamily, considering that new members are identified regularly. The bioinformatic tools, coupled to the increasing number of sequenced genomes, provide a powerful means for predicting more and more molecules. Physical and bio-physical techniques are also extremely efficient for structural–functional investigation. Nonetheless, we feel that a bigger effort should be made in order to unravel the biological roles of these exciting molecules. Too little is known about the pathways they control and, especially, about the receivers of their messages.

ACKNOWLEDGEMENT

FG is a Ph.D. fellow of the Fund for Scientific Research (FWO). The support of the University of Antwerp (GOA BOF UA 2011-2014) is acknowledged.

REFERENCES

Abel, S., Chien, P., Wassmann, P., Schirmer, T., Kaever, V., Laub, M. T., et al. (2011). Reg-ulatory cohesion of cell cycle and cell differentiation through interlinked phosphoryla-tion and second messenger networks. *Molecular Cell, 43*(4), 550–560. http://dx.doi.org/10.1016/j.molcel.2011.07.018 [Research Support, N.I.H., Extramural Research Support, Non-U.S. Gov't].

Akimoto, S., Tanaka, A., Nakamura, K., Shiro, Y., & Nakamura, H. (2003). O2-specific regulation of the ferrous heme-based sensor kinase FixL from Sinorhizobium meliloti and its aberrant inactivation in the ferric form. *Biochemical and Biophysical Research Com-munications, 304*(1), 136–142 [Research Support, Non-U.S. Gov't].

Akiyama, S., Fujisawa, T., Ishimori, K., Morishima, I., & Aono, S. (2004). Activation mech-anisms of transcriptional regulator CooA revealed by small-angle X-ray scattering. *Journal of Molecular Biology, 341*(3), 651–668. http://dx.doi.org/10.1016/j.jmb.2004.06.040 [Research Support, Non-U.S. Gov't].

Aklujkar, M., Coppi, M. V., Leang, C., Kim, B. C., Chavan, M. A., Perpetua, L. A., et al. (2013). Proteins involved in electron transfer to Fe(III) and Mn(IV) oxides by Geobacter sulfurreducens and Geobacter uraniireducens. *Microbiology, 159*(Pt 3), 515–535. http://dx.doi.org/10.1099/mic.0.064089-0.

Anantharam, V., Lehrmann, E., Kanthasamy, A., Yang, Y., Banerjee, P., Becker, K. G., et al. (2007). Microarray analysis of oxidative stress regulated genes in mesencephalic dopaminergic neuronal cells: Relevance to oxidative damage in Parkinson's disease. *Neurochemistry International, 50*(6), 834–847. http://dx.doi.org/10.1016/j.neuint.2007.02.003 [Research Support, N.I.H., Extramural Research Support, N.I.H., Intramural].

Anthamatten, D., & Hennecke, H. (1991). The regulatory status of the fixL- and fixJ-like genes in Bradyrhizobium japonicum may be different from that in Rhizobium meliloti. *Molecular and General Genetics, 225*(1), 38–48 [Research Support, Non-U.S. Gov't].

Anthamatten, D., Scherb, B., & Hennecke, H. (1992). Characterization of a fixLJ-regulated Bradyrhizobium japonicum gene sharing similarity with the Escherichia coli fnr and Rhizobium meliloti fixK genes. *Journal of Bacteriology, 174*(7), 2111–2120 [Comparative Study. Research Support, Non-U.S. Gov't].

Aono, S., Kato, T., Matsuki, M., Nakajima, H., Ohta, T., Uchida, T., et al. (2002). Resonance Raman and ligand binding studies of the oxygen-sensing signal transducer protein HemAT from Bacillus subtilis. *Journal of Biological Chemistry, 277*(16), 13528–13538. http://dx.doi.org/10.1074/jbc.M112256200 [Research Support, Non-U.S. Gov't].

Aono, S., Nakajima, H., Saito, K., & Okada, M. (1996). A novel heme protein that acts as a carbon monoxide-dependent transcriptional activator in Rhodospirillum rubrum. *Biochemical and Biophysical Research Communications, 228*(3), 752–756. http://dx.doi.org/10.1006/bbrc.1996.1727 [Research Support, Non-U.S. Gov't].

Aono, S., Takasaki, H., Unno, H., Kamiya, T., & Nakajima, H. (1999). Recognition of target DNA and transcription activation by the CO-sensing transcriptional activator CooA. *Biochemical and Biophysical Research Communications, 261*(2), 270–275. http://dx.doi.org/10.1006/bbrc.1999.1046 [Research Support, Non-U.S. Gov't].

Ausmees, N., Mayer, R., Weinhouse, H., Volman, G., Amikam, D., Benziman, M., et al. (2001). Genetic data indicate that proteins containing the GGDEF domain possess diguanylate cyclase activity. *FEMS Microbiology Letters, 204*(1), 163–167 [Research Support, Non-U.S. Gov't].

Ayers, R. A., & Moffat, K. (2008). Changes in quaternary structure in the signaling mechanisms of PAS domains. *Biochemistry, 47*(46), 12078–12086. http://dx.doi.org/10.1021/bi801254c [Research Support, N.I.H., Extramural Research Support, U.S. Gov't, Non-P.H.S.].

Balland, V., Bouzhir-Sima, L., Anxolabehere-Mallart, E., Boussac, A., Vos, M. H., Liebl, U., et al. (2006). Functional implications of the propionate 7-arginine 220 interaction in the FixLH oxygen sensor from Bradyrhizobium japonicum. *Biochemistry, 45*(7), 2072–2084. http://dx.doi.org/10.1021/bi051696h [Research Support, Non-U.S. Gov't].

Balland, V., Bouzhir-Sima, L., Kiger, L., Marden, M. C., Vos, M. H., Liebl, U., et al. (2005). Role of arginine 220 in the oxygen sensor FixL from Bradyrhizobium japonicum. *Journal of Biological Chemistry, 280*(15), 15279–15288. http://dx.doi.org/10.1074/jbc.M413928200 [Research Support, Non-U.S. Gov't].

Batut, J., Santero, E., & Kustu, S. (1991). In vitro activity of the nitrogen fixation regulatory protein FIXJ from Rhizobium meliloti. *Journal of Bacteriology, 173*(18), 5914–5917 [Research Support, Non-U.S. Gov't Research Support, U.S. Gov't, P.H.S.].

Benabbas, A., Karunakaran, V., Youn, H., Poulos, T. L., & Champion, P. M. (2012). Effect of DNA binding on geminate CO recombination kinetics in CO-sensing transcription factor CooA. *Journal of Biological Chemistry, 287*(26), 21729–21740. http://dx.doi.org/10.1074/jbc.M112.345090 [Research Support, N.I.H., Extramural Research Support, U.S. Gov't, Non-P.H.S.].

Blank, M., & Burmester, T. (2012). Widespread occurrence of N-terminal acylation in animal globins and possible origin of respiratory globins from a membrane-bound ancestor. *Molecular and Biological Evolution, 29*(11), 3553–3561. http://dx.doi.org/10.1093/molbev/mss164.

Blattner, F. R., Plunkett, G., 3rd., Bloch, C. A., Perna, N. T., Burland, V., Riley, M., et al. (1997). The complete genome sequence of Escherichia coli K-12. *Science, 277*(5331), 1453–1462 [Research Support, Non-U.S. Gov't Research Support, U.S. Gov't, P.H.S.].

Bobrov, A. G., Kirillina, O., Ryjenkov, D. A., Waters, C. M., Price, P. A., Fetherston, J. D., et al. (2011). Systematic analysis of cyclic di-GMP signalling enzymes and their role in biofilm formation and virulence in Yersinia pestis. *Molecular Microbiology, 79*(2), 533–551. http://dx.doi.org/10.1111/j.1365-2958.2010.07470.x [Research Support, N.I.H., Extramural Research Support, U.S. Gov't, Non-P.H.S.].

Bonamore, A., Attili, A., Arenghi, F., Catacchio, B., Chiancone, E., Morea, V., et al. (2007). A novel chimera: The "truncated hemoglobin-antibiotic monooxygenase" from Streptomyces avermitilis. *Gene, 398*(1–2), 52–61. http://dx.doi.org/10.1016/j.gene.2007.01.038 [Research Support, Non-U.S. Gov't].

Borjigin, M., Li, H., Lanz, N. D., Kerby, R. L., Roberts, G. P., & Poulos, T. L. (2007). Structure-based hypothesis on the activation of the CO-sensing transcription factor CooA. *Acta Crystallographica. Section D, Biological Crystallography, 63*(Pt 3), 282–287. http://dx.doi.org/10.1107/S0907444906051638 [Research Support, N.I.H., Extramural Research Support, Non-U.S. Gov't Research Support, U.S. Gov't, Non-P.H.S.].

Campbell, E. A., Westblade, L. F., & Darst, S. A. (2008). Regulation of bacterial RNA polymerase sigma factor activity: A structural perspective. *Current Opinion in Microbiology, 11*(2), 121–127. http://dx.doi.org/10.1016/j.mib.2008.02.016 [Research Support, N.I.H., Extramural Review].

Cannistraro, V. J., Glekas, G. D., Rao, C. V., & Ordal, G. W. (2011). Cellular stoichiometry of the chemotaxis proteins in Bacillus subtilis. *Journal of Bacteriology, 193*(13), 3220–3227. http://dx.doi.org/10.1128/JB.01255-10 [Comparative Study Research Support, N.I.H., Extramural].

Chang, A. L., Tuckerman, J. R., Gonzalez, G., Mayer, R., Weinhouse, H., Volman, G., et al. (2001). Phosphodiesterase A1, a regulator of cellulose synthesis in Acetobacter xylinum, is a heme-based sensor. *Biochemistry, 40*(12), 3420–3426 [Research Support, Non-U.S. Gov't Research Support, U.S. Gov't, Non-P.H.S. Research Support, U.S. Gov't, P.H.S.].

Chen, C. C., Lewis, R. J., Harris, R., Yudkin, M. D., & Delumeau, O. (2003). A supramolecular complex in the environmental stress signalling pathway of Bacillus subtilis. *Molecular Microbiology, 49*(6), 1657–1669 [Research Support, Non-U.S. Gov't].

Clark, R. W., Lanz, N. D., Lee, A. J., Kerby, R. L., Roberts, G. P., & Burstyn, J. N. (2006). Unexpected NO-dependent DNA binding by the CooA homolog from Carboxydothermus hydrogenoformans. *Proceedings of the National Academy of Sciences of the United States of America, 103*(4), 891–896. http://dx.doi.org/10.1073/pnas.0505919103 [Research Support, N.I.H., Extramural].

Clark, R. W., Youn, H., Parks, R. B., Cherney, M. M., Roberts, G. P., & Burstyn, J. N. (2004). Investigation of the role of the N-terminal proline, the distal heme ligand in the CO sensor CooA. *Biochemistry, 43*(44), 14149–14160. http://dx.doi.org/10.1021/bi0487948 [Research Support, U.S. Gov't, P.H.S.].

Crosson, S., McGrath, P. T., Stephens, C., McAdams, H. H., & Shapiro, L. (2005). Conserved modular design of an oxygen sensory/signaling network with species-specific output. *Proceedings of the National Academy of Sciences of the United States of America, 102*(22), 8018–8023. http://dx.doi.org/10.1073/pnas.0503022102 [Comparative Study

Research Support, N.I.H., Extramural Research Support, U.S. Gov't, Non-P.H.S. Research Support, U.S. Gov't, P.H.S.].

Curtis, A. M., Cheng, Y., Kapoor, S., Reilly, D., Price, T. S., & Fitzgerald, G. A. (2007). Circadian variation of blood pressure and the vascular response to asynchronous stress. *Proceedings of the National Academy of Sciences of the United States of America, 104*(9), 3450–3455. http://dx.doi.org/10.1073/pnas.0611680104 [Comparative Study Research Support, N.I.H., Extramural].

Dardente, H., Fortier, E. E., Martineau, V., & Cermakian, N. (2007). Cryptochromes impair phosphorylation of transcriptional activators in the clock: A general mechanism for circadian repression. *Biochemical Journal, 402*(3), 525–536. http://dx.doi.org/10.1042/BJ20060827 [Research Support, Non-U.S. Gov't].

David, M., Daveran, M. L., Batut, J., Dedieu, A., Domergue, O., Ghai, J., et al. (1988). Cascade regulation of nif gene expression in Rhizobium meliloti. *Cell, 54*(5), 671–683 [Research Support, Non-U.S. Gov't].

Delgado-Nixon, V. M., Gonzalez, G., & Gilles-Gonzalez, M. A. (2000). Dos, a heme-binding PAS protein from Escherichia coli, is a direct oxygen sensor. *Biochemistry, 39*(10), 2685–2691 [Research Support, U.S. Gov't, Non-P.H.S.].

Deng, Y., Schmid, N., Wang, C., Wang, J., Pessi, G., Wu, D., et al. (2012). Cis-2-dodecenoic acid receptor RpfR links quorum-sensing signal perception with regulation of virulence through cyclic dimeric guanosine monophosphate turnover. *Proceedings of the National Academy of Sciences of the United States of America, 109*(38), 15479–15484. http://dx.doi.org/10.1073/pnas.1205037109 [Research Support, Non-U.S. Gov't].

de Philip, P., Batut, J., & Boistard, P. (1990). Rhizobium meliloti Fix L is an oxygen sensor and regulates R. meliloti nifA and fixK genes differently in Escherichia coli. *Journal of Bacteriology, 172*(8), 4255–4262 [Research Support, Non-U.S. Gov't].

Desmet, F., Thijs, L., El Mkami, H., Dewilde, S., Moens, L., Smith, G., et al. (2010). The heme pocket of the globin domain of the globin-coupled sensor of Geobacter sulfurreducens—An EPR study. *Journal of Inorganic Biochemistry, 104*(10), 1022–1028. http://dx.doi.org/10.1016/j.jinorgbio.2010.05.009.

D'Hooghe, I., Michiels, J., & Vanderleyden, J. (1998). The Rhizobium etli FixL protein differs in structure from other known FixL proteins. *Molecular and General Genetics, 257*(5), 576–580 [Research Support, Non-U.S. Gov't].

D'Hooghe, I., Michiels, J., Vlassak, K., Verreth, C., Waelkens, F., & Vanderleyden, J. (1995). Structural and functional analysis of the fixLJ genes of Rhizobium leguminosarum biovar phaseoli CNPAF512. *Molecular and General Genetics, 249*(1), 117–126 [Comparative Study Research Support, Non-U.S. Gov't].

Ding, Y. H., Hixson, K. K., Giometti, C. S., Stanley, A., Esteve-Nunez, A., Khare, T., et al. (2006). The proteome of dissimilatory metal-reducing microorganism Geobacter sulfurreducens under various growth conditions. *Biochimica et Biophysica Acta, 1764*(7), 1198–1206. http://dx.doi.org/10.1016/j.bbapap.2006.04.017 [Research Support, Non-U.S. Gov't Research Support, U.S. Gov't, Non-P.H.S.].

Dioum, E. M., Rutter, J., Tuckerman, J. R., Gonzalez, G., Gilles-Gonzalez, M. A., & McKnight, S. L. (2002). NPAS2: A gas-responsive transcription factor. *Science, 298*(5602), 2385–2387. http://dx.doi.org/10.1126/science.1078456 [Research Support, Non-U.S. Gov't Research Support, U.S. Gov't, P.H.S.].

Ditta, G., Virts, E., Palomares, A., & Kim, C. H. (1987). The nifA gene of Rhizobium meliloti is oxygen regulated. *Journal of Bacteriology, 169*(7), 3217–3223 [Research Support, U.S. Gov't, Non-P.H.S.].

Dudley, C. A., Erbel-Sieler, C., Estill, S. J., Reick, M., Franken, P., Pitts, S., et al. (2003). Altered patterns of sleep and behavioral adaptability in NPAS2-deficient mice. *Science,*

301(5631), 379–383. http://dx.doi.org/10.1126/science.1082795 [Research Support, U.S. Gov't, P.H.S.].

Duerig, A., Abel, S., Folcher, M., Nicollier, M., Schwede, T., Amiot, N., et al. (2009). Second messenger-mediated spatiotemporal control of protein degradation regulates bacterial cell cycle progression. *Genes & Development, 23*(1), 93–104. http://dx.doi.org/10.1101/gad.502409 [Research Support, Non-U.S. Gov't].

Dunham, C. M., Dioum, E. M., Tuckerman, J. R., Gonzalez, G., Scott, W. G., & Gilles-Gonzalez, M. A. (2003). A distal arginine in oxygen-sensing heme-PAS domains is essential to ligand binding, signal transduction, and structure. *Biochemistry, 42*(25), 7701–7708. http://dx.doi.org/10.1021/bi0343370 [Research Support, Non-U.S. Gov't Research Support, U.S. Gov't, Non-P.H.S. Research Support, U.S. Gov't, P.H.S.].

El-Mashtoly, S. F., Gu, Y., Yoshimura, H., Yoshioka, S., Aono, S., & Kitagawa, T. (2008). Protein conformation changes of HemAT-Bs upon ligand binding probed by ultraviolet resonance Raman spectroscopy. *Journal of Biological Chemistry, 283*(11), 6942–6949. http://dx.doi.org/10.1074/jbc.M709209200 [Research Support, Non-U.S. Gov't].

El-Mashtoly, S. F., Kubo, M., Gu, Y., Sawai, H., Nakashima, S., Ogura, T., et al. (2012). Site-specific protein dynamics in communication pathway from sensor to signaling domain of oxygen sensor protein, HemAT-Bs: Time-resolved Ultraviolet Resonance Raman Study. *Journal of Biological Chemistry, 287*(24), 19973–19984. http://dx.doi.org/10.1074/jbc.M112.357855 [Research Support, Non-U.S. Gov't].

El-Mashtoly, S. F., Nakashima, S., Tanaka, A., Shimizu, T., & Kitagawa, T. (2008). Roles of Arg-97 and Phe-113 in regulation of distal ligand binding to heme in the sensor domain of Ec DOS protein. Resonance Raman and mutation study. *Journal of Biological Chemistry, 283*(27), 19000–19010. http://dx.doi.org/10.1074/jbc.M801262200 [Research Support, Non-U.S. Gov't].

El-Mashtoly, S. F., Takahashi, H., Shimizu, T., & Kitagawa, T. (2007). Ultraviolet resonance Raman evidence for utilization of the heme 6-propionate hydrogen-bond network in signal transmission from heme to protein in Ec DOS protein. *Journal of the American Chemical Society, 129*(12), 3556–3563. http://dx.doi.org/10.1021/ja0669777 [Research Support, Non-U.S. Gov't].

Evans, D. S., Parimi, N., Nievergelt, C. M., Blackwell, T., Redline, S., Ancoli-Israel, S., et al. (2013). Common genetic variants in ARNTL and NPAS2 and at chromosome 12p13 are associated with objectively measured sleep traits in the elderly. *Sleep, 36*(3), 431–446. http://dx.doi.org/10.5665/sleep.2466.

Freitas, T. A., Hou, S., & Alam, M. (2003). The diversity of globin-coupled sensors. *FEBS Letters, 552*(2–3), 99–104 [Research Support, Non-U.S. Gov't Research Support, U.S. Gov't, Non-P.H.S. Review].

Freitas, T. A., Hou, S., Dioum, E. M., Saito, J. A., Newhouse, J., Gonzalez, G., et al. (2004). Ancestral hemoglobins in Archaea. *Proceedings of the National Academy of Sciences of the United States of America, 101*(17), 6675–6680. http://dx.doi.org/10.1073/pnas.0308657101 [Research Support, Non-U.S. Gov't Research Support, U.S. Gov't, Non-P.H.S. Research Support, U.S. Gov't, P.H.S.].

Freitas, T. A., Saito, J. A., Hou, S., & Alam, M. (2005). Globin-coupled sensors, protoglobins, and the last universal common ancestor. *Journal of Inorganic Biochemistry, 99*(1), 23–33. http://dx.doi.org/10.1016/j.jinorgbio.2004.10.024 [Research Support, Non-U.S. Gov't Research Support, U.S. Gov't, Non-P.H.S. Review].

Gaidenko, T. A., Bie, X., Baldwin, E. P., & Price, C. W. (2012). Two surfaces of a conserved interdomain linker differentially affect output from the RST sensing module of the Bacillus subtilis stressosome. *Journal of Bacteriology, 194*(15), 3913–3921. http://dx.doi.org/10.1128/JB.00583-12 [Research Support, N.I.H., Extramural].

Garcia, J. A., Zhang, D., Estill, S. J., Michnoff, C., Rutter, J., Reick, M., et al. (2000). Impaired cued and contextual memory in NPAS2-deficient mice. *Science, 288*(5474),

2226–2230 [Research Support, Non-U.S. Gov't Research Support, U.S. Gov't, P.H.S.].

Gekakis, N., Staknis, D., Nguyen, H. B., Davis, F. C., Wilsbacher, L. D., King, D. P., et al. (1998). Role of the CLOCK protein in the mammalian circadian mechanism. *Science*, *280*(5369), 1564–1569 [Research Support, Non-U.S. Gov't Research Support, U.S. Gov't, Non-P.H.S. Research Support, U.S. Gov't, P.H.S.].

Gilles-Gonzalez, M. A., Ditta, G. S., & Helinski, D. R. (1991). A haemoprotein with kinase activity encoded by the oxygen sensor of Rhizobium meliloti. *Nature*, *350*(6314), 170–172. http://dx.doi.org/10.1038/350170a0 [Research Support, U.S. Gov't, Non-P.H.S. Research Support, U.S. Gov't, P.H.S.].

Gilles-Gonzalez, M. A., & Gonzalez, G. (1993). Regulation of the kinase activity of heme protein FixL from the two-component system FixL/FixJ of Rhizobium meliloti. *Journal of Biological Chemistry*, *268*(22), 16293–16297 [Research Support, Non-U.S. Gov't Research Support, U.S. Gov't, P.H.S.].

Gilles-Gonzalez, M. A., & Gonzalez, G. (2005). Heme-based sensors: Defining characteristics, recent developments, and regulatory hypotheses. *Journal of Inorganic Biochemistry*, *99*(1), 1–22. http://dx.doi.org/10.1016/j.jinorgbio.2004.11.006 [Research Support, N.I.H., Extramural Research Support, U.S. Gov't, Non-P.H.S. Research Support, U.S. Gov't, P.H.S. Review].

Gilles-Gonzalez, M. A., Gonzalez, G., & Perutz, M. F. (1995). Kinase activity of oxygen sensor FixL depends on the spin state of its heme iron. *Biochemistry*, *34*(1), 232–236 [Research Support, Non-U.S. Gov't Research Support, U.S. Gov't, P.H.S.].

Gilles-Gonzalez, M. A., Gonzalez, G., Perutz, M. F., Kiger, L., Marden, M. C., & Poyart, C. (1994). Heme-based sensors, exemplified by the kinase FixL, are a new class of heme protein with distinctive ligand binding and autoxidation. *Biochemistry*, *33*(26), 8067–8073 [Comparative Study Research Support, Non-U.S. Gov't Research Support, U.S. Gov't, P.H.S.].

Girard, L., Brom, S., Davalos, A., Lopez, O., Soberon, M., & Romero, D. (2000). Differential regulation of fixN-reiterated genes in Rhizobium etli by a novel fixL–fixK cascade. *Molecular Plant–Microbe Interactions*, *13*(12), 1283–1292. http://dx.doi.org/10.1094/MPMI.2000.13.12.1283 [Research Support, Non-U.S. Gov't].

Gong, W., Hao, B., & Chan, M. K. (2000). New mechanistic insights from structural studies of the oxygen-sensing domain of Bradyrhizobium japonicum FixL. *Biochemistry*, *39*(14), 3955–3962 [Research Support, Non-U.S. Gov't Research Support, U.S. Gov't, P.H.S.].

Gong, W., Hao, B., Mansy, S. S., Gonzalez, G., Gilles-Gonzalez, M. A., & Chan, M. K. (1998). Structure of a biological oxygen sensor: A new mechanism for heme-driven signal transduction. *Proceedings of the National Academy of Sciences of the United States of America*, *95*(26), 15177–15182 [Research Support, Non-U.S. Gov't Research Support, U.S. Gov't, Non-P.H.S. Research Support, U.S. Gov't, P.H.S.].

Gonzalez, G., Dioum, E. M., Bertolucci, C. M., Tomita, T., Ikeda-Saito, M., Cheesman, M. R., et al. (2002). Nature of the displaceable heme-axial residue in the EcDos protein, a heme-based sensor from Escherichia coli. *Biochemistry*, *41*(26), 8414–8421 [Research Support, Non-U.S. Gov't Research Support, U.S. Gov't, P.H.S.].

Gouet, P., Fabry, B., Guillet, V., Birck, C., Mourey, L., Kahn, D., et al. (1999). Structural transitions in the FixJ receiver domain. *Structure*, *7*(12), 1517–1526 [Research Support, Non-U.S. Gov't].

Gu, Y. Z., Hogenesch, J. B., & Bradfield, C. A. (2000). The PAS superfamily: Sensors of environmental and developmental signals. *Annual Review of Pharmacology and Toxicology*, *40*, 519–561. http://dx.doi.org/10.1146/annurev.pharmtox.40.1.519 [Research Support, Non-U.S. Gov't Research Support, U.S. Gov't, P.H.S. Review].

Hao, B., Isaza, C., Arndt, J., Soltis, M., & Chan, M. K. (2002). Structure-based mechanism of O2 sensing and ligand discrimination by the FixL heme domain of Bradyrhizobium japonicum. *Biochemistry*, *41*(43), 12952–12958 [Research Support, Non-U.S. Gov't Research Support, U.S. Gov't, Non-P.H.S. Research Support, U.S. Gov't, P.H.S.].

Hayasaka, K., Kitanishi, K., Igarashi, J., & Shimizu, T. (2011). Heme-binding characteristics of the isolated PAS-B domain of mouse Per2, a transcriptional regulatory factor associated with circadian rhythms. *Biochimica et Biophysica Acta*, *1814*(2), 326–333. http://dx. doi.org/10.1016/j.bbapap.2010.09.007 [In Vitro Research Support, Non-U.S. Gov't].

He, Y., Gaal, T., Karls, R., Donohue, T. J., Gourse, R. L., & Roberts, G. P. (1999). Transcription activation by CooA, the CO-sensing factor from Rhodospirillum rubrum. The interaction between CooA and the C-terminal domain of the alpha subunit of RNA polymerase. *Journal of Biological Chemistry*, *274*(16), 10840–10845 [Research Support, Non-U.S. Gov't Research Support, U.S. Gov't, P.H.S.].

He, Y., Shelver, D., Kerby, R. L., & Roberts, G. P. (1996). Characterization of a CO-responsive transcriptional activator from Rhodospirillum rubrum. *Journal of Biological Chemistry*, *271*(1), 120–123 [Research Support, Non-U.S. Gov't Research Support, U.S. Gov't, Non-P.H.S. Research Support, U.S. Gov't, P.H.S.].

Henry, J. T., & Crosson, S. (2011). Ligand-binding PAS domains in a genomic, cellular, and structural context. *Annual Review of Microbiology*, *65*, 261–286. http://dx.doi.org/ 10.1146/annurev-micro-121809-151631 [Review].

Hiruma, Y., Kikuchi, A., Tanaka, A., Shiro, Y., & Mizutani, Y. (2007). Resonance Raman observation of the structural dynamics of FixL on signal transduction and ligand discrimination. *Biochemistry*, *46*(20), 6086–6096. http://dx.doi.org/10.1021/bi062083n [Comparative Study Research Support, Non-U.S. Gov't].

Ho, Y. S., Burden, L. M., & Hurley, J. H. (2000). Structure of the GAF domain, a ubiquitous signaling motif and a new class of cyclic GMP receptor. *EMBO Journal*, *19*(20), 5288–5299. http://dx.doi.org/10.1093/emboj/19.20.5288.

Hoffman, A. E., Zheng, T., Ba, Y., & Zhu, Y. (2008). The circadian gene NPAS2, a putative tumor suppressor, is involved in DNA damage response. *Molecular Cancer Research*, *6*(9), 1461–1468. http://dx.doi.org/10.1158/1541-7786.MCR-07-2094 [Research Support, N.I.H., Extramural].

Hogenesch, J. B., Gu, Y. Z., Jain, S., & Bradfield, C. A. (1998). The basic-helix-loop-helix-PAS orphan MOP3 forms transcriptionally active complexes with circadian and hypoxia factors. *Proceedings of the National Academy of Sciences of the United States of America*, *95*(10), 5474–5479 [Research Support, Non-U.S. Gov't Research Support, U.S. Gov't, P.H.S.].

Hou, S., Belisle, C., Lam, S., Piatibratov, M., Sivozhelezov, V., Takami, H., et al. (2001). A globin-coupled oxygen sensor from the facultatively alkaliphilic Bacillus halodurans C-125. *Extremophiles*, *5*(5), 351–354 [Research Support, U.S. Gov't, Non-P.H.S.].

Hou, S., Freitas, T., Larsen, R. W., Piatibratov, M., Sivozhelezov, V., Yamamoto, A., et al. (2001). Globin-coupled sensors: A class of heme-containing sensors in Archaea and Bacteria. *Proceedings of the National Academy of Sciences of the United States of America*, *98*(16), 9353–9358. http://dx.doi.org/10.1073/pnas.161185598 [Research Support, Non-U.S. Gov't Research Support, U.S. Gov't, Non-P.H.S.].

Hou, S., Larsen, R. W., Boudko, D., Riley, C. W., Karatan, E., Zimmer, M., et al. (2000). Myoglobin-like aerotaxis transducers in Archaea and Bacteria. *Nature*, *403*(6769), 540–544. http://dx.doi.org/10.1038/35000570 [Research Support, U.S. Gov't, Non-P.H.S.].

Inagaki, S., Masuda, C., Akaishi, T., Nakajima, H., Yoshioka, S., Ohta, T., et al. (2005). Spectroscopic and redox properties of a CooA homologue from Carboxydothermus hydrogenoformans. *Journal of Biological Chemistry*, *280*(5), 3269–3274. http://dx.doi. org/10.1074/jbc.M409884200 [Research Support, Non-U.S. Gov't].

Iniesta, A. A., Hillson, N. J., & Shapiro, L. (2010). Cell pole-specific activation of a critical bacterial cell cycle kinase. *Proceedings of the National Academy of Sciences of the United States of America, 107*(15), 7012–7017. http://dx.doi.org/10.1073/pnas.1001767107 [Research Support, N.I.H., Extramural Research Support, Non-U.S. Gov't Research Support, U.S. Gov't, Non-P.H.S.].

Ishida, M., Ueha, T., & Sagami, I. (2008). Effects of mutations in the heme domain on the transcriptional activity and DNA-binding activity of NPAS2. *Biochemical and Biophysical Research Communications, 368*(2), 292–297. http://dx.doi.org/10.1016/j.bbrc.2008.01.053 [Research Support, Non-U.S. Gov't].

Ito, S., Araki, Y., Tanaka, A., Igarashi, J., Wada, T., & Shimizu, T. (2009). Role of Phe113 at the distal side of the heme domain of an oxygen-sensor (Ec DOS) in the characterization of the heme environment. *Journal of Inorganic Biochemistry, 103*(7), 989–996. http://dx.doi.org/10.1016/j.jinorgbio.2009.04.009 [Research Support, Non-U.S. Gov't].

Ito, S., Igarashi, J., & Shimizu, T. (2009). The FG loop of a heme-based gas sensor enzyme, Ec DOS, functions in heme binding, autoxidation and catalysis. *Journal of Inorganic Biochemistry, 103*(10), 1380–1385. http://dx.doi.org/10.1016/j.jinorgbio.2009.07.012 [Research Support, Non-U.S. Gov't].

Iyer, L. M., Anantharaman, V., & Aravind, L. (2003). Ancient conserved domains shared by animal soluble guanylyl cyclases and bacterial signaling proteins. *BMC Genomics, 4*(1), 5.

Johansson, C., Willeit, M., Smedh, C., Ekholm, J., Paunio, T., Kieseppa, T., et al. (2003). Circadian clock-related polymorphisms in seasonal affective disorder and their relevance to diurnal preference. *Neuropsychopharmacology, 28*(4), 734–739. http://dx.doi.org/10.1038/sj.npp.1300121 [Comparative Study Research Support, Non-U.S. Gov't].

Kaasik, K., & Lee, C. C. (2004). Reciprocal regulation of haem biosynthesis and the circadian clock in mammals. *Nature, 430*(6998), 467–471. http://dx.doi.org/10.1038/nature02724 [Research Support, U.S. Gov't, P.H.S.].

Kaminski, P. A., & Elmerich, C. (1991). Involvement of fixLJ in the regulation of nitrogen fixation in Azorhizobium caulinodans. *Molecular Microbiology, 5*(3), 665–673.

Karaolis, D. K., Rashid, M. H., Chythanya, R., Luo, W., Hyodo, M., & Hayakawa, Y. (2005). c-di-GMP (3'-5'-cyclic diguanylic acid) inhibits Staphylococcus aureus cell–cell interactions and biofilm formation. *Antimicrobial Agents and Chemotherapy, 49*(3), 1029–1038. http://dx.doi.org/10.1128/AAC.49.3.1029-1038.2005.

Kerby, R. L., Youn, H., & Roberts, G. P. (2008). RcoM: A new single-component transcriptional regulator of CO metabolism in bacteria. *Journal of Bacteriology, 190*(9), 3336–3343. http://dx.doi.org/10.1128/JB.00033-08 [Research Support, N.I.H., Extramural Research Support, Non-U.S. Gov't].

Kerby, R. L., Youn, H., Thorsteinsson, M. V., & Roberts, G. P. (2003). Repositioning about the dimer interface of the transcription regulator CooA: A major signal transduction pathway between the effector and DNA-binding domains. *Journal of Molecular Biology, 325*(4), 809–823 [Research Support, Non-U.S. Gov't Research Support, U.S. Gov't, P.H.S.].

Key, J., & Moffat, K. (2005). Crystal structures of deoxy and CO-bound bjFixLH reveal details of ligand recognition and signaling. *Biochemistry, 44*(12), 4627–4635. http://dx.doi.org/10.1021/bi047942r [Research Support, N.I.H., Extramural Research Support, U.S. Gov't, P.H.S.].

Key, J., Srajer, V., Pahl, R., & Moffat, K. (2007). Time-resolved crystallographic studies of the heme domain of the oxygen sensor FixL: Structural dynamics of ligand rebinding and their relation to signal transduction. *Biochemistry, 46*(16), 4706–4715. http://dx.doi.org/10.1021/bi700043c [Research Support, N.I.H., Extramural].

Kitanishi, K., Igarashi, J., Hayasaka, K., Hikage, N., Saiful, I., Yamauchi, S., et al. (2008). Heme-binding characteristics of the isolated PAS-A domain of mouse Per2, a transcriptional regulatory factor associated with circadian rhythms. *Biochemistry, 47*(23), 6157–6168. http://dx.doi.org/10.1021/bi7023892.

Kitanishi, K., Kobayashi, K., Kawamura, Y., Ishigami, I., Ogura, T., Nakajima, K., et al. (2010). Important roles of Tyr43 at the putative heme distal side in the oxygen recognition and stability of the Fe(II)-O2 complex of YddV, a globin-coupled heme-based oxygen sensor diguanylate cyclase. *Biochemistry, 49*(49), 10381–10393. http://dx.doi.org/10.1021/bi100733q [Comparative Study Research Support, Non-U.S. Gov't].

Kitanishi, K., Kobayashi, K., Uchida, T., Ishimori, K., Igarashi, J., & Shimizu, T. (2011). Identification and functional and spectral characterization of a globin-coupled histidine kinase from Anaeromyxobacter sp. Fw109-5. *Journal of Biological Chemistry, 286*(41), 35522–35534. http://dx.doi.org/10.1074/jbc.M111.274811 [Research Support, Non-U.S. Gov't].

Kobayashi, K., Tanaka, A., Takahashi, H., Igarashi, J., Ishitsuka, Y., Yokota, N., et al. (2010). Catalysis and oxygen binding of Ec DOS: A haem-based oxygen-sensor enzyme from *Escherichia coli*. *Journal of Biochemistry, 148*(6), 693–703. http://dx.doi.org/10.1093/jb/mvq103 [Research Support, Non-U.S. Gov't].

Komori, H., Inagaki, S., Yoshioka, S., Aono, S., & Higuchi, Y. (2007). Crystal structure of CO-sensing transcription activator CooA bound to exogenous ligand imidazole. *Journal of Molecular Biology, 367*(3), 864–871. http://dx.doi.org/10.1016/j.jmb.2007.01.043 [Research Support, Non-U.S. Gov't].

Koudo, R., Kurokawa, H., Sato, E., Igarashi, J., Uchida, T., Sagami, I., et al. (2005). Spectroscopic characterization of the isolated heme-bound PAS-B domain of neuronal PAS domain protein 2 associated with circadian rhythms. *FEBS Journal, 272*(16), 4153–4162. http://dx.doi.org/10.1111/j.1742-4658.2005.04828.x [Research Support, Non-U.S. Gov't].

Kovanen, L., Saarikoski, S. T., Aromaa, A., Lonnqvist, J., & Partonen, T. (2010). ARNTL (BMAL1) and NPAS2 gene variants contribute to fertility and seasonality. *PLoS One, 5*(4), e10007. http://dx.doi.org/10.1371/journal.pone.0010007 [Research Support, Non-U.S. Gov't].

Kruglik, S. G., Jasaitis, A., Hola, K., Yamashita, T., Liebl, U., Martin, J. L., et al. (2007). Subpicosecond oxygen trapping in the heme pocket of the oxygen sensor FixL observed by time-resolved resonance Raman spectroscopy. *Proceedings of the National Academy of Sciences of the United States of America, 104*(18), 7408–7413. http://dx.doi.org/10.1073/pnas.0700445104 [Research Support, Non-U.S. Gov't].

Kulasakara, H., Lee, V., Brencic, A., Liberati, N., Urbach, J., Miyata, S., et al. (2006). Analysis of Pseudomonas aeruginosa diguanylate cyclases and phosphodiesterases reveals a role for bis-(3'-5')-cyclic-GMP in virulence. *Proceedings of the National Academy of Sciences of the United States of America, 103*(8), 2839–2844. http://dx.doi.org/10.1073/pnas.0511090103 [Research Support, N.I.H., Extramural Research Support, Non-U.S. Gov't].

Lanzilotta, W. N., Schuller, D. J., Thorsteinsson, M. V., Kerby, R. L., Roberts, G. P., & Poulos, T. L. (2000). Structure of the CO sensing transcription activator CooA. *Natural Structural Biology, 7*(10), 876–880. http://dx.doi.org/10.1038/82820 [Research Support, Non-U.S. Gov't Research Support, U.S. Gov't, Non-P.H.S. Research Support, U.S. Gov't, P.H.S.].

Lavebratt, C., Sjoholm, L. K., Partonen, T., Schalling, M., & Forsell, Y. (2010). PER2 variantion is associated with depression vulnerability. *American Journal of Medical Genetics. Part B, Neuropsychiatric Genetics, 153B*(2), 570–581. http://dx.doi.org/10.1002/ajmg.b.31021 [Research Support, Non-U.S. Gov't].

Lechauve, C., Bouzhir-Sima, L., Yamashita, T., Marden, M. C., Vos, M. H., Liebl, U., et al. (2009). Heme ligand binding properties and intradimer interactions in the full-length sensor protein dos from Escherichia coli and its isolated heme domain. *Journal of*

Biological Chemistry, *284*(52), 36146–36159. http://dx.doi.org/10.1074/jbc.M109. 066811 [Research Support, Non-U.S. Gov't].

Lin, W. C., Coppi, M. V., & Lovley, D. R. (2004). Geobacter sulfurreducens can grow with oxygen as a terminal electron acceptor. *Applied and Environmental Microbiology*, *70*(4), 2525–2528 [Research Support, Non-U.S. Gov't].

Lois, A. F., Ditta, G. S., & Helinski, D. R. (1993). The oxygen sensor FixL of Rhizobium meliloti is a membrane protein containing four possible transmembrane segments. *Journal of Bacteriology*, *175*(4), 1103–1109 [Comparative Study Research Support, Non-U.S. Gov't Research Support, U.S. Gov't, Non-P.H.S. Research Support, U.S. Gov't, P.H.S.].

Lovley, D. R., Giovannoni, S. J., White, D. C., Champine, J. E., Phillips, E. J., Gorby, Y. A., et al. (1993). Geobacter metallireducens gen. nov. sp. nov., a microorganism capable of coupling the complete oxidation of organic compounds to the reduction of iron and other metals. *Archives of Microbiology*, *159*(4), 336–344 [Research Support, U.S. Gov't, Non-P.H.S.].

Marles-Wright, J., Grant, T., Delumeau, O., van Duinen, G., Firbank, S. J., Lewis, P. J., et al. (2008). Molecular architecture of the "stressosome," a signal integration and transduction hub. *Science*, *322*(5898), 92–96. http://dx.doi.org/10.1126/science.1159572 [Research Support, Non-U.S. Gov't].

Marvin, K. A., Kerby, R. L., Youn, H., Roberts, G. P., & Burstyn, J. N. (2008). The transcription regulator RcoM-2 from Burkholderia xenovorans is a cysteine-ligated hemoprotein that undergoes a redox-mediated ligand switch. *Biochemistry*, *47*(34), 9016–9028. http://dx.doi.org/10.1021/bi800486x [Research Support, N.I.H., Extramural].

Matilla, M. A., Travieso, M. L., Ramos, J. L., & Ramos-Gonzalez, M. I. (2011). Cyclic diguanylate turnover mediated by the sole GGDEF/EAL response regulator in Pseudomonas putida: Its role in the rhizosphere and an analysis of its target processes. *Environmental Microbiology*, *13*(7), 1745–1766. http://dx.doi.org/10.1111/j.1462-2920.2011.02499.x [Research Support, Non-U.S. Gov't].

McAdams, H. H., & Shapiro, L. (2003). A bacterial cell-cycle regulatory network operating in time and space. *Science*, *301*(5641), 1874–1877. http://dx.doi.org/10.1126/science.1087694 [Research Support, U.S. Gov't, Non-P.H.S. Research Support, U.S. Gov't, P.H.S. Review].

Mesa, S., Bedmar, E. J., Chanfon, A., Hennecke, H., & Fischer, H. M. (2003). Bradyrhizobium japonicum NnrR, a denitrification regulator, expands the FixLJ-FixK2 regulatory cascade. *Journal of Bacteriology*, *185*(13), 3978–3982 [Research Support, Non-U.S. Gov't].

Methe, B. A., Nelson, K. E., Eisen, J. A., Paulsen, I. T., Nelson, W., Heidelberg, J. F., et al. (2003). Genome of Geobacter sulfurreducens: Metal reduction in subsurface environments. *Science*, *302*(5652), 1967–1969. http://dx.doi.org/10.1126/science.1088727 [Research Support, U.S. Gov't, Non-P.H.S.].

Miyatake, H., Mukai, M., Adachi, S., Nakamura, H., Tamura, K., Iizuka, T., et al. (1999). Iron coordination structures of oxygen sensor FixL characterized by Fe K-edge extended x-ray absorption fine structure and resonance raman spectroscopy. *Journal of Biological Chemistry*, *274*(33), 23176–23184.

Miyatake, H., Mukai, M., Park, S. Y., Adachi, S., Tamura, K., Nakamura, H., et al. (2000). Sensory mechanism of oxygen sensor FixL from Rhizobium meliloti: Crystallographic, mutagenesis and resonance Raman spectroscopic studies. *Journal of Molecular Biology*, *301*(2), 415–431. http://dx.doi.org/10.1006/jmbi.2000.3954 [Research Support, Non-U.S. Gov't].

Moens, L., Vanfleteren, J., Van de Peer, Y., Peeters, K., Kapp, O., Czeluzniak, J., et al. (1996). Globins in nonvertebrate species: Dispersal by horizontal gene transfer and evolution of the structure-function relationships. *Molecular and Biological Evolution*, *13*(2),

324–333 [Comparative Study Research Support, Non-U.S. Gov't Research Support, U.S. Gov't, Non-P.H.S. Research Support, U.S. Gov't, P.H.S.].

Monson, E. K., Weinstein, M., Ditta, G. S., & Helinski, D. R. (1992). The FixL protein of Rhizobium meliloti can be separated into a heme-binding oxygen-sensing domain and a functional C-terminal kinase domain. *Proceedings of the National Academy of Sciences of the United States of America, 89*(10), 4280–4284 [Research Support, U.S. Gov't, Non-P.H.S. Research Support, U.S. Gov't, P.H.S.].

Monteferrante, C. G., Mackichan, C., Marchadier, E., Prejean, M. V., Carballido-Lopez, R., & van Dijl, J. M. (2013). Mapping the twin-arginine protein translocation network of Bacillus subtilis. *Proteomics, 13*(5), 800–811. http://dx.doi.org/10.1002/pmic.201200416.

Mukai, M., Nakamura, K., Nakamura, H., Iizuka, T., & Shiro, Y. (2000). Roles of Ile209 and Ile210 on the heme pocket structure and regulation of histidine kinase activity of oxygen sensor FixL from Rhizobium meliloti. *Biochemistry, 39*(45), 13810–13816 [Research Support, Non-U.S. Gov't].

Mukaiyama, Y., Uchida, T., Sato, E., Sasaki, A., Sato, Y., Igarashi, J., et al. (2006). Spectroscopic and DNA-binding characterization of the isolated heme-bound basic helix-loop-helix-PAS-A domain of neuronal PAS protein 2 (NPAS2), a transcription activator protein associated with circadian rhythms. *FEBS Journal, 273*(11), 2528–2539. http://dx.doi.org/10.1111/j.1742-4658.2006.05259.x [Research Support, Non-U.S. Gov't].

Murray, J. W., Delumeau, O., & Lewis, R. J. (2005). Structure of a nonheme globin in environmental stress signaling. *Proceedings of the National Academy of Sciences of the United States of America, 102*(48), 17320–17325. http://dx.doi.org/10.1073/pnas.0506599102 [Comparative Study Research Support, Non-U.S. Gov't].

Nakajima, K., Kitanishi, K., Kobayashi, K., Kobayashi, N., Igarashi, J., & Shimizu, T. (2012). Leu65 in the heme distal side is critical for the stability of the Fe(II)-O2 complex of YddV, a globin-coupled oxygen sensor diguanylate cyclase. *Journal of Inorganic Biochemistry, 108*, 163–170. http://dx.doi.org/10.1016/j.jinorgbio.2011.09.019 [Research Support, Non-U.S. Gov't].

Nellen-Anthamatten, D., Rossi, P., Preisig, O., Kullik, I., Babst, M., Fischer, H. M., et al. (1998). Bradyrhizobium japonicum FixK2, a crucial distributor in the FixLJ-dependent regulatory cascade for control of genes inducible by low oxygen levels. *Journal of Bacteriology, 180*(19), 5251–5255 [Research Support, Non-U.S. Gov't].

Nicholas, B., Rudrasingham, V., Nash, S., Kirov, G., Owen, M. J., & Wimpory, D. C. (2007). Association of Per1 and Npas2 with autistic disorder: Support for the clock genes/social timing hypothesis. *Molecular Psychiatry, 12*(6), 581–592. http://dx.doi.org/10.1038/sj.mp.4001953 [Research Support, N.I.H., Extramural Research Support, Non-U.S. Gov't].

Page, K. M., & Guerinot, M. L. (1995). Oxygen control of the Bradyrhizobium japonicum hemA gene. *Journal of Bacteriology, 177*(14), 3979–3984 [Research Support, U.S. Gov't, Non-P.H.S.].

Papin, J. A., Reed, J. L., & Palsson, B. O. (2004). Hierarchical thinking in network biology: The unbiased modularization of biochemical networks. *Trends in Biochemical Sciences, 29*(12), 641–647. http://dx.doi.org/10.1016/j.tibs.2004.10.001 [Research Support, Non-U.S. Gov't Review].

Park, H., Suquet, C., Satterlee, J. D., & Kang, C. (2004). Insights into signal transduction involving PAS domain oxygen-sensing heme proteins from the X-ray crystal structure of Escherichia coli Dos heme domain (Ec DosH). *Biochemistry, 43*(10), 2738–2746. http://dx.doi.org/10.1021/bi035980p [Comparative Study Research Support, Non-U.S. Gov't Research Support, U.S. Gov't, P.H.S.].

Partonen, T., Treutlein, J., Alpman, A., Frank, J., Johansson, C., Depner, M., et al. (2007). Three circadian clock genes Per2, Arntl, and Npas2 contribute to winter depression.

Annals of Medicine, 39(3), 229–238. http://dx.doi.org/10.1080/07853890701278795 [Research Support, Non-U.S. Gov't].

Pesce, A., Thijs, L., Nardini, M., Desmet, F., Sisinni, L., Gourlay, L., et al. (2009). HisE11 and HisF8 provide bis-histidyl heme hexa-coordination in the globin domain of Geobacter sulfurreducens globin-coupled sensor. *Journal of Molecular Biology, 386*(1), 246–260. http://dx.doi.org/10.1016/j.jmb.2008.12.023 [Research Support, Non-U.S. Gov't].

Pinakoulaki, E., Yoshimura, H., Daskalakis, V., Yoshioka, S., Aono, S., & Varotsis, C. (2006). Two ligand-binding sites in the O2-sensing signal transducer HemAT: Implications for ligand recognition/discrimination and signaling. *Proceedings of the National Academy of Sciences of the United States of America, 103*(40), 14796–14801. http://dx.doi.org/10.1073/pnas.0604248103 [Research Support, Non-U.S. Gov't].

Poole, R. K. (2005). Nitric oxide and nitrosative stress tolerance in bacteria. *Biochemical Society Transactions, 33*(Pt 1), 176–180. http://dx.doi.org/10.1042/BST0330176 [Research Support, Non-U.S. Gov't Review].

Poole, R. K., & Hughes, M. N. (2000). New functions for the ancient globin family: Bacterial responses to nitric oxide and nitrosative stress. *Molecular Microbiology, 36*(4), 775–783 [Research Support, Non-U.S. Gov't Review].

Puranik, M., Nielsen, S. B., Youn, H., Hvitved, A. N., Bourassa, J. L., Case, M. A., et al. (2004). Dynamics of carbon monoxide binding to CooA. *Journal of Biological Chemistry, 279*(20), 21096–21108. http://dx.doi.org/10.1074/jbc.M400613200 [Research Support, Non-U.S. Gov't Research Support, U.S. Gov't, Non-P.H.S. Research Support, U.S. Gov't, P.H.S.].

Qi, Y., Rao, F., Luo, Z., & Liang, Z. X. (2009). A flavin cofactor-binding PAS domain regulates c-di-GMP synthesis in AxDGC2 from Acetobacter xylinum. *Biochemistry, 48*(43), 10275–10285. http://dx.doi.org/10.1021/bi901121w [Research Support, Non-U.S. Gov't].

Quin, M. B., Berrisford, J. M., Newman, J. A., Basle, A., Lewis, R. J., & Marles-Wright, J. (2012). The bacterial stressosome: A modular system that has been adapted to control secondary messenger signaling. *Structure, 20*(2), 350–363. http://dx.doi.org/10.1016/j.str.2012.01.003 [Research Support, Non-U.S. Gov't].

Rajeev, L., Hillesland, K. L., Zane, G. M., Zhou, A., Joachimiak, M. P., He, Z., et al. (2012). Deletion of the Desulfovibrio vulgaris carbon monoxide sensor invokes global changes in transcription. *Journal of Bacteriology, 194*(21), 5783–5793. http://dx.doi.org/10.1128/JB.00749-12 [Research Support, U.S. Gov't, Non-P.H.S.].

Reichmann, D., Rahat, O., Albeck, S., Meged, R., Dym, O., & Schreiber, G. (2005). The modular architecture of protein-protein binding interfaces. *Proceedings of the National Academy of Sciences of the United States of America, 102*(1), 57–62. http://dx.doi.org/10.1073/pnas.0407280102 [Research Support, Non-U.S. Gov't].

Reick, M., Garcia, J. A., Dudley, C., & McKnight, S. L. (2001). NPAS2: An analog of clock operative in the mammalian forebrain. *Science, 293*(5529), 506–509. http://dx.doi.org/10.1126/science.1060699 [Research Support, Non-U.S. Gov't Research Support, U.S. Gov't, P.H.S.].

Reynolds, M. F., Ackley, L., Blizman, A., Lutz, Z., Manoff, D., Miles, M., et al. (2009). Role of conserved F(alpha)-helix residues in the native fold and stability of the kinase-inhibited oxy state of the oxygen-sensing FixL protein from Sinorhizobium meliloti. *Archives of Biochemistry and Biophysics, 485*(2), 150–159. http://dx.doi.org/10.1016/j.abb.2009.02.011 [Research Support, N.I.H., Extramural Research Support, Non-U.S. Gov't Research Support, U.S. Gov't, Non-P.H.S.].

Reynolds, M. F., Parks, R. B., Burstyn, J. N., Shelver, D., Thorsteinsson, M. V., Kerby, R. L., et al. (2000). Electronic absorption, EPR, and resonance raman spectroscopy of CooA, a CO-sensing transcription activator from R. rubrum, reveals a five-coordinate NO-heme. *Biochemistry*, *39*(2), 388–396 [Comparative Study Research Support, U.S. Gov't, P.H.S.].

Roberts, G. P., Thorsteinsson, M. V., Kerby, R. L., Lanzilotta, W. N., & Poulos, T. (2001). CooA: A heme-containing regulatory protein that serves as a specific sensor of both carbon monoxide and redox state. *Progress in Nucleic Acid Research and Molecular Biology*, *67*, 35–63 [Research Support, Non-U.S. Gov't Research Support, U.S. Gov't, Non-P.H.S. Research Support, U.S. Gov't, P.H.S. Review].

Robles, E. F., Sanchez, C., Bonnard, N., Delgado, M. J., & Bedmar, E. J. (2006). The Bradyrhizobium japonicum napEDABC genes are controlled by the FixLJ-FixK(2)-NnrR regulatory cascade. *Biochemical Society Transactions*, *34*(Pt 1), 108–110. http://dx.doi.org/10.1042/BST0340108 [Research Support, Non-U.S. Gov't].

Rodgers, K. R., Lukat-Rodgers, G. S., & Barron, J. A. (1996). Structural basis for ligand discrimination and response initiation in the heme-based oxygen sensor FixL. *Biochemistry*, *35*(29), 9539–9548. http://dx.doi.org/10.1021/bi9530853 [Research Support, Non-U.S. Gov't Research Support, U.S. Gov't, Non-P.H.S.].

Rodgers, K. R., Lukat-Rodgers, G. S., & Tang, L. (2000). Nitrosyl adducts of FixL as probes of heme environment. *Journal of Biological Inorganic Chemistry*, *5*(5), 642–654 [Research Support, Non-U.S. Gov't Research Support, U.S. Gov't, Non-P.H.S. Research Support, U.S. Gov't, P.H.S.].

Ross, P., Weinhouse, H., Aloni, Y., Michaeli, D., Weinberger-Ohana, P., Mayer, R., et al. (1987). Regulation of cellulose synthesis in Acetobacter xylinum by cyclic diguanylic acid. *Nature*, *325*(6101), 279–281.

Rutter, J., Reick, M., Wu, L. C., & McKnight, S. L. (2001). Regulation of clock and NPAS2 DNA binding by the redox state of NAD cofactors. *Science*, *293*(5529), 510–514. http://dx.doi.org/10.1126/science.1060698 [Research Support, Non-U.S. Gov't Research Support, U.S. Gov't, P.H.S.].

Ryan, R. P., Fouhy, Y., Lucey, J. F., Jiang, B. L., He, Y. Q., Feng, J. X., et al. (2007). Cyclic di-GMP signalling in the virulence and environmental adaptation of Xanthomonas campestris. *Molecular Microbiology*, *63*(2), 429–442. http://dx.doi.org/10.1111/j.1365-2958.2006.05531.x [Research Support, Non-U.S. Gov't].

Ryan, R. P., McCarthy, Y., Andrade, M., Farah, C. S., Armitage, J. P., & Dow, J. M. (2010). Cell-cell signal-dependent dynamic interactions between HD-GYP and GGDEF domain proteins mediate virulence in Xanthomonas campestris. *Proceedings of the National Academy of Sciences of the United States of America*, *107*(13), 5989–5994. http://dx.doi.org/10.1073/pnas.0912839107 [Research Support, Non-U.S. Gov't].

Sasakura, Y., Hirata, S., Sugiyama, S., Suzuki, S., Taguchi, S., Watanabe, M., et al. (2002). Characterization of a direct oxygen sensor heme protein from Escherichia coli. Effects of the heme redox states and mutations at the heme-binding site on catalysis and structure. *Journal of Biological Chemistry*, *277*(26), 23821–23827. http://dx.doi.org/10.1074/jbc.M202738200.

Sato, A., Sasakura, Y., Sugiyama, S., Sagami, I., Shimizu, T., Mizutani, Y., et al. (2002). Stationary and time-resolved resonance Raman spectra of His77 and Met95 mutants of the isolated heme domain of a direct oxygen sensor from Escherichia coli. *Journal of Biological Chemistry*, *277*(36), 32650–32658. http://dx.doi.org/10.1074/jbc.M20455 9200 [Research Support, Non-U.S. Gov't].

Sawai, H., Yoshioka, S., Uchida, T., Hyodo, M., Hayakawa, Y., Ishimori, K., et al. (2010). Molecular oxygen regulates the enzymatic activity of a heme-containing diguanylate cyclase (HemDGC) for the synthesis of cyclic di-GMP. *Biochimica et Biophysica Acta*, *1804*(1), 166–172. http://dx.doi.org/10.1016/j.bbapap.2009.09.028 [Research Support, Non-U.S. Gov't].

Schaller, R. A., Ali, S. K., Klose, K. E., & Kurtz, D. M., Jr. (2012). A bacterial hemerythrin domain regulates the activity of a Vibrio cholerae diguanylate cyclase. *Biochemistry*,

51(43), 8563–8570. http://dx.doi.org/10.1021/bi3011797 [Research Support, N.I.H., Extramural].

Sciotti, M. A., Chanfon, A., Hennecke, H., & Fischer, H. M. (2003). Disparate oxygen responsiveness of two regulatory cascades that control expression of symbiotic genes in Bradyrhizobium japonicum. *Journal of Bacteriology, 185*(18), 5639–5642 [Research Support, Non-U.S. Gov't].

Shelver, D., Kerby, R. L., He, Y., & Roberts, G. P. (1995). Carbon monoxide-induced activation of gene expression in Rhodospirillum rubrum requires the product of cooA, a member of the cyclic AMP receptor protein family of transcriptional regulators. *Journal of Bacteriology, 177*(8), 2157–2163 [Comparative Study Research Support, Non-U.S. Gov't Research Support, U.S. Gov't, Non-P.H.S. Research Support, U.S. Gov't, P.H.S.].

Shelver, D., Kerby, R. L., He, Y., & Roberts, G. P. (1997). CooA, a CO-sensing transcription factor from Rhodospirillum rubrum, is a CO-binding heme protein. *Proceedings of the National Academy of Sciences of the United States of America, 94*(21), 11216–11220 [Research Support, Non-U.S. Gov't Research Support, U.S. Gov't, P.H.S.].

Shelver, D., Thorsteinsson, M. V., Kerby, R. L., Chung, S. Y., Roberts, G. P., Reynolds, M. F., et al. (1999). Identification of two important heme site residues (cysteine 75 and histidine 77) in CooA, the CO-sensing transcription factor of Rhodospirillum rubrum. *Biochemistry, 38*(9), 2669–2678. http://dx.doi.org/10.1021/bi982658j [Research Support, Non-U.S. Gov't Research Support, U.S. Gov't, P.H.S.].

Simm, R., Morr, M., Kader, A., Nimtz, M., & Romling, U. (2004). GGDEF and EAL domains inversely regulate cyclic di-GMP levels and transition from sessility to motility. *Molecular Microbiology, 53*(4), 1123–1134. http://dx.doi.org/10.1111/j.1365-2958.2004.04206.x [Research Support, Non-U.S. Gov't].

Smith, A. K., Fang, H., Whistler, T., Unger, E. R., & Rajeevan, M. S. (2011). Convergent genomic studies identify association of GRIK2 and NPAS2 with chronic fatigue syndrome. *Neuropsychobiology, 64*(4), 183–194. http://dx.doi.org/10.1159/000326692 [Research Support, U.S. Gov't, Non-P.H.S. Research Support, U.S. Gov't, P.H.S.].

Smith, A. T., Marvin, K. A., Freeman, K. M., Kerby, R. L., Roberts, G. P., & Burstyn, J. N. (2012). Identification of Cys94 as the distal ligand to the Fe(III) heme in the transcriptional regulator RcoM-2 from Burkholderia xenovorans. *Journal of Biological Inorganic Chemistry, 17*(7), 1071–1082. http://dx.doi.org/10.1007/s00775-012-0920-1 [Research Support, N.I.H., Extramural].

Sommerfeldt, N., Possling, A., Becker, G., Pesavento, C., Tschowri, N., & Hengge, R. (2009). Gene expression patterns and differential input into curli fimbriae regulation of all GGDEF/EAL domain proteins in Escherichia coli. *Microbiology, 155*(Pt 4), 1318–1331. http://dx.doi.org/10.1099/mic.0.024257-0 [Research Support, Non-U. S. Gov't].

Sousa, E. H., Tuckerman, J. R., Gondim, A. C., Gonzalez, G., & Gilles-Gonzalez, M. A. (2013). Signal transduction and phosphoryl transfer by a FixL hybrid kinase with low oxygen affinity: Importance of the vicinal PAS domain and receiver aspartate. *Biochemistry, 52*(3), 456–465. http://dx.doi.org/10.1021/bi300991r [Research Support, Non-U.S. Gov't].

Sousa, E. H., Tuckerman, J. R., Gonzalez, G., & Gilles-Gonzalez, M. A. (2007). A memory of oxygen binding explains the dose response of the heme-based sensor FixL. *Biochemistry, 46*(21), 6249–6257. http://dx.doi.org/10.1021/bi7003334 [Research Support, Non-U.S. Gov't Research Support, U.S. Gov't, Non-P.H.S.].

Stranzl, G. R., Santelli, E., Bankston, L. A., La Clair, C., Bobkov, A., Schwarzenbacher, R., et al. (2011). Structural insights into inhibition of Bacillus anthracis sporulation by a novel class of non-heme globin sensor domains. *Journal of Biological Chemistry, 286*(10), 8448–8458. http://dx.doi.org/10.1074/jbc.M110.207126 [Research Support, N.I.H., Extramural Research Support, Non-U.S. Gov't Research Support, U.S. Gov't, Non-P.H.S.].

Tagliabue, L., Antoniani, D., Maciag, A., Bocci, P., Raffaelli, N., & Landini, P. (2010). The diguanylate cyclase YddV controls production of the exopolysaccharide poly-N-acetylglucosamine (PNAG) through regulation of the PNAG biosynthetic pgaABCD operon. *Microbiology*, *156*(Pt 10), 2901–2911. http://dx.doi.org/10.1099/mic.0.041350-0 [Research Support, Non-U.S. Gov't].

Tagliabue, L., Maciag, A., Antoniani, D., & Landini, P. (2010). The yddV-dos operon controls biofilm formation through the regulation of genes encoding curli fibers' subunits in aerobically growing Escherichia coli. *FEMS Immunology and Medical Microbiology*, *59*(3), 477–484. http://dx.doi.org/10.1111/j.1574-695X.2010.00702.x.

Taguchi, S., Matsui, T., Igarashi, J., Sasakura, Y., Araki, Y., Ito, O., et al. (2004). Binding of oxygen and carbon monoxide to a heme-regulated phosphodiesterase from Escherichia coli. Kinetics and infrared spectra of the full-length wild-type enzyme, isolated PAS domain, and Met-95 mutants. *Journal of Biological Chemistry*, *279*(5), 3340–3347. http://dx.doi.org/10.1074/jbc.M301013200.

Takeda, Y., Kang, H. S., Angers, M., & Jetten, A. M. (2011). Retinoic acid-related orphan receptor gamma directly regulates neuronal PAS domain protein 2 transcription in vivo. *Nucleic Acids Research*, *39*(11), 4769–4782. http://dx.doi.org/10.1093/nar/gkq1335 [Research Support, N.I.H., Extramural Research Support, N.I.H., Intramural].

Tal, R., Wong, H. C., Calhoon, R., Gelfand, D., Fear, A. L., Volman, G., et al. (1998). Three cdg operons control cellular turnover of cyclic di-GMP in Acetobacter xylinum: Genetic organization and occurrence of conserved domains in isoenzymes. *Journal of Bacteriology*, *180*(17), 4416–4425 [Research Support, Non-U.S. Gov't].

Tamayo, R., Pratt, J. T., & Camilli, A. (2007). Roles of cyclic diguanylate in the regulation of bacterial pathogenesis. *Annual Review of Microbiology*, *61*, 131–148. http://dx.doi.org/10.1146/annurev.micro.61.080706.093426 [Research Support, N.I.H., Extramural Review].

Tanaka, A., Nakamura, H., Shiro, Y., & Fujii, H. (2006). Roles of the heme distal residues of FixL in O2 sensing: A single convergent structure of the heme moiety is relevant to the downregulation of kinase activity. *Biochemistry*, *45*(8), 2515–2523. http://dx.doi.org/10.1021/bi051989a [Research Support, Non-U.S. Gov't].

Tanaka, A., & Shimizu, T. (2008). Ligand binding to the Fe(III)-protoporphyrin IX complex of phosphodiesterase from Escherichia coli (Ec DOS) markedly enhances catalysis of cyclic di-GMP: Roles of Met95, Arg97, and Phe113 of the putative heme distal side in catalytic regulation and ligand binding. *Biochemistry*, *47*(50), 13438–13446. http://dx.doi.org/10.1021/bi8012017 [Comparative Study Research Support, Non-U.S. Gov't].

Tanaka, A., Takahashi, H., & Shimizu, T. (2007). Critical role of the heme axial ligand, Met95, in locking catalysis of the phosphodiesterase from Escherichia coli (Ec DOS) toward Cyclic diGMP. *Journal of Biological Chemistry*, *282*(29), 21301–21307. http://dx.doi.org/10.1074/jbc.M701920200 [Research Support, Non-U.S. Gov't].

Thijs, L., Vinck, E., Bolli, A., Trandafir, F., Wan, X., Hoogewijs, D., et al. (2007). Characterization of a globin-coupled oxygen sensor with a gene-regulating function. *Journal of Biological Chemistry*, *282*(52), 37325–37340. http://dx.doi.org/10.1074/jbc.M705541200 [Research Support, Non-U.S. Gov't Research Support, U.S. Gov't, Non-P.H.S.].

Tischler, A. D., & Camilli, A. (2004). Cyclic diguanylate (c-di-GMP) regulates Vibrio cholerae biofilm formation. *Molecular Microbiology*, *53*(3), 857–869. http://dx.doi.org/10.1111/j.1365-2958.2004.04155.x [Research Support, U.S. Gov't, Non-P.H.S. Research Support, U.S. Gov't, P.H.S.].

Tuckerman, J. R., Gonzalez, G., Sousa, E. H., Wan, X., Saito, J. A., Alam, M., et al. (2009). An oxygen-sensing diguanylate cyclase and phosphodiesterase couple for c-di-GMP control. *Biochemistry*, *48*(41), 9764–9774. http://dx.doi.org/10.1021/bi901409g [Research Support, Non-U.S. Gov't Research Support, U.S. Gov't, Non-P.H.S.].

Uchida, T., Ishikawa, H., Ishimori, K., Morishima, I., Nakajima, H., Aono, S., et al. (2000). Identification of histidine 77 as the axial heme ligand of carbonmonoxy CooA by picosecond time-resolved resonance Raman spectroscopy. *Biochemistry*, *39*(42), 12747–12752 [Research Support, Non-U.S. Gov't].

Uchida, T., Ishikawa, H., Takahashi, S., Ishimori, K., Morishima, I., Ohkubo, K., et al. (1998). Heme environmental structure of CooA is modulated by the target DNA binding. Evidence from resonance Raman spectroscopy and CO rebinding kinetics. *Journal of Biological Chemistry*, *273*(32), 19988–19992 [Research Support, Non-U.S. Gov't].

Uchida, T., Sagami, I., Shimizu, T., Ishimori, K., & Kitagawa, T. (2012). Effects of the bHLH domain on axial coordination of heme in the PAS-A domain of neuronal PAS domain protein 2 (NPAS2): Conversion from His119/Cys170 coordination to His119/His171 coordination. *Journal of Inorganic Biochemistry*, *108*, 188–195. http://dx.doi.org/10.1016/j.jinorgbio.2011.12.005 [Research Support, Non-U.S. Gov't].

Uchida, T., Sato, E., Sato, A., Sagami, I., Shimizu, T., & Kitagawa, T. (2005). CO-dependent activity-controlling mechanism of heme-containing CO-sensor protein, neuronal PAS domain protein 2. *Journal of Biological Chemistry*, *280*(22), 21358–21368. http://dx.doi.org/10.1074/jbc.M412350200 [Research Support, Non-U.S. Gov't].

Vanin, S., Negrisolo, E., Bailly, X., Bubacco, L., Beltramini, M., & Salvato, B. (2006). Molecular evolution and phylogeny of sipunculan hemerythrins. *Journal of Molecular Evolution*, *62*(1), 32–41. http://dx.doi.org/10.1007/s00239-004-0296-0.

Vinogradov, S. N., Hoogewijs, D., Bailly, X., Arredondo-Peter, R., Gough, J., Dewilde, S., et al. (2006). A phylogenomic profile of globins. *BMC Evolutionary Biology*, *6*, 31. http://dx.doi.org/10.1186/1471-2148-6-31 [Comparative Study Research Support, Non-U.S. Gov't].

Vinogradov, S. N., Hoogewijs, D., Bailly, X., Arredondo-Peter, R., Guertin, M., Gough, J., et al. (2005). Three globin lineages belonging to two structural classes in genomes from the three kingdoms of life. *Proceedings of the National Academy of Sciences of the United States of America*, *102*(32), 11385–11389. http://dx.doi.org/10.1073/pnas.0502103102 [Comparative Study Research Support, Non-U.S. Gov't].

Vinogradov, S. N., Hoogewijs, D., Bailly, X., Mizuguchi, K., Dewilde, S., Moens, L., et al. (2007). A model of globin evolution. *Gene*, *398*(1–2), 132–142. http://dx.doi.org/10.1016/j.gene.2007.02.041 [Research Support, Non-U.S. Gov't].

Vinogradov, S. N., & Moens, L. (2008). Diversity of globin function: Enzymatic, transport, storage, and sensing. *Journal of Biological Chemistry*, *283*(14), 8773–8777. http://dx.doi.org/10.1074/jbc.R700029200 [Review].

Vinogradov, S. N., Tinajero-Trejo, M., Poole, R. K., & Hoogewijs, D. (2013). Bacterial and archaeal globins—A revised perspective. *Biochimica et Biophysica Acta* http://dx.doi.org/10.1016/j.bbapap.2013.03.021.

Virts, E. L., Stanfield, S. W., Helinski, D. R., & Ditta, G. S. (1988). Common regulatory elements control symbiotic and microaerobic induction of nifA in Rhizobium meliloti. *Proceedings of the National Academy of Sciences of the United States of America*, *85*(9), 3062–3065 [Research Support, U.S. Gov't, Non-P.H.S. Research Support, U.S. Gov't, P.H.S.].

Vogel, K. M., Spiro, T. G., Shelver, D., Thorsteinsson, M. V., & Roberts, G. P. (1999). Resonance Raman evidence for a novel charge relay activation mechanism of the CO-dependent heme protein transcription factor CooA. *Biochemistry*, *38*(9), 2679–2687. http://dx.doi.org/10.1021/bi982375r [Research Support, U.S. Gov't, P.H.S.].

Wan, X., Tuckerman, J. R., Saito, J. A., Freitas, T. A., Newhouse, J. S., Denery, J. R., et al. (2009). Globins synthesize the second messenger bis-(3'-5')-cyclic diguanosine mono-phosphate in bacteria. *Journal of Molecular Biology*, *388*(2), 262–270. http://dx.doi.org/10.1016/j.jmb.2009.03.015 [Research Support, N.I.H., Intramural Research Support, Non-U.S. Gov't Research Support, U.S. Gov't, Non-P.H.S.].

Wang, F., Hu, Z., Yang, R., Tang, J., Liu, Y., Hemminki, K., et al. (2011). A variant affect-ing miRNAs binding in the circadian gene Neuronal PAS domain protein 2 (NPAS2) is not associated with breast cancer risk. *Breast Cancer Research and Treatment*, *127*(3), 769–775. http://dx.doi.org/10.1007/s10549-010-1157-8 [Research Support, Non-U. S. Gov't].

Weinhouse, H., Sapir, S., Amikam, D., Shilo, Y., Volman, G., Ohana, P., et al. (1997). c-di-GMP-binding protein, a new factor regulating cellulose synthesis in Acetobacter xylinum. *FEBS Letters*, *416*(2), 207–211 [Research Support, Non-U.S. Gov't].

Westgate, E. J., Cheng, Y., Reilly, D. F., Price, T. S., Walisser, J. A., Bradfield, C. A., et al. (2008). Genetic components of the circadian clock regulate thrombogenesis in vivo. *Circulation*, *117*(16), 2087–2095. http://dx.doi.org/10.1161/CIRCULATIONAHA. 107.739227 [Research Support, N.I.H., Extramural Research Support, Non-U.S. Gov't].

Wisor, J. P., Pasumarthi, R. K., Gerashchenko, D., Thompson, C. L., Pathak, S., Sancar, A., et al. (2008). Sleep deprivation effects on circadian clock gene expression in the cerebral cortex parallel electroencephalographic differences among mouse strains. *Journal of Neu-roscience*, *28*(28), 7193–7201. http://dx.doi.org/10.1523/JNEUROSCI.1150-08.2008 [Research Support, N.I.H., Extramural Research Support, U.S. Gov't, Non-P.H.S.].

Yamamoto, K., Ishikawa, H., Takahashi, S., Ishimori, K., Morishima, I., Nakajima, H., et al. (2001). Binding of CO at the Pro2 side is crucial for the activation of CO-sensing tran-scriptional activator CooA. (1)H NMR spectroscopic studies. *Journal of Biological Chem-istry*, *276*(15), 11473–11476. http://dx.doi.org/10.1074/jbc.C100047200 [Research Support, Non-U.S. Gov't].

Yamashita, T., Hoashi, Y., Tomisugi, Y., Ishikawa, Y., & Uno, T. (2004). The C-helix in CooA rolls upon CO binding to ferrous heme. *Journal of Biological Chemistry*, *279*(45), 47320–47325. http://dx.doi.org/10.1074/jbc.M407766200 [Research Support, Non-U.S. Gov't].

Yi, C., Mu, L., de la Longrais, I. A., Sochirca, O., Arisio, R., Yu, H., et al. (2010). The circadian gene NPAS2 is a novel prognostic biomarker for breast cancer. *Breast Cancer Research and Treatment*, *120*(3), 663–669. http://dx.doi.org/10.1007/s10549-009-0484-0 [Research Support, N.I.H., Extramural].

Yi, C. H., Zheng, T., Leaderer, D., Hoffman, A., & Zhu, Y. (2009). Cancer-related tran-scriptional targets of the circadian gene NPAS2 identified by genome-wide ChIP-on-chip analysis. *Cancer Letters*, *284*(2), 149–156. http://dx.doi.org/10.1016/j.canlet. 2009.04.017 [Research Support, N.I.H., Extramural].

Yoshida, Y., Ishikawa, H., Aono, S., & Mizutani, Y. (2012). Structural dynamics of proximal heme pocket in HemAT-Bs associated with oxygen dissociation. *Biochimica et Biophysica Acta*, *1824*(7), 866–872. http://dx.doi.org/10.1016/j.bbapap.2012.04.007 [Research Support, Non-U.S. Gov't].

Yoshimura, H., Yoshioka, S., Kobayashi, K., Ohta, T., Uchida, T., Kubo, M., et al. (2006). Specific hydrogen-bonding networks responsible for selective O2 sensing of the oxygen sensor protein HemAT from Bacillus subtilis. *Biochemistry*, *45*(27), 8301–8307. http://dx. doi.org/10.1021/bi060315c [Research Support, Non-U.S. Gov't].

Yoshimura, H., Yoshioka, S., Mizutani, Y., & Aono, S. (2007). The formation of hydrogen bond in the proximal heme pocket of HemAT-Bs upon ligand binding. *Biochemical and Biophysical Research Communications*, *357*(4), 1053–1057. http://dx.doi.org/10.1016/j.bbrc.2007.04.041 [Research Support, Non-U.S. Gov't].

Youn, H., Kerby, R. L., Thorsteinsson, M. V., Conrad, M., Staples, C. R., Serate, J., et al. (2001). The heme pocket afforded by Gly117 is crucial for proper heme ligation and activity of CooA. *Journal of Biological Chemistry, 276*(45), 41603–41610. http://dx.doi.org/10.1074/jbc.M106165200 [Research Support, Non-U.S. Gov't Research Support, U.S. Gov't, P.H.S.].

Youn, H., Thorsteinsson, M. V., Conrad, M., Kerby, R. L., & Roberts, G. P. (2005). Dual roles of an E-helix residue, Glu167, in the transcriptional activator function of CooA. *Journal of Bacteriology, 187*(8), 2573–2581. http://dx.doi.org/10.1128/JB.187.8.2573-2581.2005 [Research Support, Non-U.S. Gov't].

Zhang, W., & Phillips, G. N., Jr. (2003). Structure of the oxygen sensor in Bacillus subtilis: Signal transduction of chemotaxis by control of symmetry. *Structure, 11*(9), 1097–1110 [Comparative Study Research Support, Non-U.S. Gov't].

Zhao, Y., Brandish, P. E., Ballou, D. P., & Marletta, M. A. (1999). A molecular basis for nitric oxide sensing by soluble guanylate cyclase. *Proceedings of the National Academy of Sciences of the United States of America, 96*(26), 14753–14758 [Research Support, Non-U.S. Gov't Research Support, U.S. Gov't, P.H.S.].

Zhou, Y. D., Barnard, M., Tian, H., Li, X., Ring, H. Z., Francke, U., et al. (1997). Molecular characterization of two mammalian bHLH-PAS domain proteins selectively expressed in the central nervous system. *Proceedings of the National Academy of Sciences of the United States of America, 94*(2), 713–718 [Research Support, Non-U.S. Gov't Research Support, U.S. Gov't, P.H.S.].

Zhu, Y., Leaderer, D., Guss, C., Brown, H. N., Zhang, Y., Boyle, P., et al. (2007). Ala394Thr polymorphism in the clock gene NPAS2: A circadian modifier for the risk of non-Hodgkin's lymphoma. *International Journal of Cancer, 120*(2), 432–435. http://dx.doi.org/10.1002/ijc.22321 [Research Support, N.I.H., Extramural Research Support, Non-U.S. Gov't].

Zhu, Y., Stevens, R. G., Hoffman, A. E., Fitzgerald, L. M., Kwon, E. M., Ostrander, E. A., et al. (2009). Testing the circadian gene hypothesis in prostate cancer: A population-based case-control study. *Cancer Research, 69*(24), 9315–9322. http://dx.doi.org/10.1158/0008-5472.CAN-09-0648 [Research Support, N.I.H., Extramural Research Support, Non-U.S. Gov't].

Zhu, Y., Stevens, R. G., Leaderer, D., Hoffman, A., Holford, T., Zhang, Y., et al. (2008). Non-synonymous polymorphisms in the circadian gene NPAS2 and breast cancer risk. *Breast Cancer Research and Treatment, 107*(3), 421–425. http://dx.doi.org/10.1007/s10549-007-9565-0 [Research Support, N.I.H., Extramural].

CHAPTER TWO

The Diversity of 2/2 (Truncated) Globins

Alessandra Pesce[*], Martino Bolognesi[†,‡], Marco Nardini[†,1]

[*]Department of Physics, University of Genova, Genova, Italy
[†]Department of Biosciences, University of Milano, Milano, Italy
[‡]CIMAINA and CNR Institute of Biophysics, Milano, Italy
[1]Corresponding author: e-mail address: marco.nardini@unimi.it

Contents

1. Introduction	50
2. Fold and Fold Variation in 2/2Hb Groups I, II, III	52
3. The Haem Environment	54
4. Tunnels and Cavities Through 2/2Hb Protein Matrix	60
5. Proposed Functions for the 2/2Hb Family	66
6. Conclusions	71
References	72

Abstract

Small size globins that have been defined as 'truncated haemoglobins' or as '2/2 haemoglobins' have increasingly been discovered in microorganisms since the early 1990s. Analysis of amino acid sequences allowed to distinguish three groups that collect proteins with specific and common structural properties. All three groups display 3D structures that are based on four main α-helices, which are a subset of the conventional eight-helices globin fold. Specific features, such as the presence of protein matrix tunnels that are held to promote diffusion of functional ligands to/from the haem, distinguish members of the three groups. Haem distal sites vary for their accessibility, local structures, polarity, and ligand stabilization mechanisms, suggesting functional roles that are related to O_2/NO chemistry. In a few cases, such activities have been proven in vitro and in vivo through deletion mutants. The issue of 2/2 haemoglobin varied biological functions throughout the three groups remains however fully open.

ABBREVIATIONS

2/2Hb 2-on-2 globin
Hb haemoglobin
Mb myoglobin
(non-)vertebrate Hbs vertebrate and non-vertebrate Hbs
amino acid residues have been labelled using their three-letter codes and the topological site they occupy within the globin fold

Advances in Microbial Physiology, Volume 63
ISSN 0065-2911
http://dx.doi.org/10.1016/B978-0-12-407693-8.00002-9

49

1. INTRODUCTION

The globin family has long been known from studies of vertebrate myoglobin (Mb) and haemoglobins (Hbs), which are haemoproteins typically composed of about 150 amino acids. In the past 30 years, the Hb superfamily has been enriched by the discovery of Hbs and related haemoproteins in virtually all kingdoms of life; among these non-symbiotic Hbs in plants and symbiotic Hbs in plants other than legumes, chimaeric flavoHbs comprised of an N-terminal globin linked to a FAD reductase domain in bacteria and yeasts, neuroglobins and cytoglobins in vertebrates, globin-coupled sensors and protoglobins in eubacteria and in archaea, and globins that fall in the 110–130 amino acid range (per haem), which have been called 'truncated Hbs', in protozoa and in bacteria. Such large variety of Hb types suggests an ancient origin for their genes and stresses the concept that Hbs/globins cover functions that stretch well beyond that of simple oxygen carriers.

In this review, we focus our attention on the truncated Hb family based on the extensive sequence and crystallographic investigations performed in various laboratories starting from the early 2000s. The term 'truncated Hb' was first introduced to refer to the size of these small globins. However, structural considerations underline the fact that the tertiary structure of these proteins results from careful editing of the classical globin fold through an evolutionary/engineering process that affects the whole polypeptide chain, rather than just acting through simple truncation of the N- and C-terminal ends. For this reason, the historical term 'truncated Hb' was recently replaced by '2/2Hbs' (read 2-on-2 Hbs) in relation to specific features of their folds (see below).

The 2/2Hbs are small oxygen-binding haemoproteins, identified in bacteria, higher plants, and in certain unicellular eukaryotes, building a clear separate cluster within the haemoglobin superfamily (Nardini, Pesce, Milani, & Bolognesi, 2007; Vinogradov, Tinajero-Trejo, Poole, & Hoogewijs, 2013; Vuletich & Lecomte, 2006; Wittenberg, Bolognesi, Wittenberg, & Guertin, 2002). 2/2Hbs display amino acid sequences that are 20–40 residues shorter than (non-)vertebrate Hbs, to which they are loosely related by sequence similarity (sequence identity to vertebrate Hbs falls well below 20%). Based on amino acid sequence analysis, three 2/2Hbs phylogenetic groups (groups I, II, and III, whose members are designated by the N, O, and P suffixes, respectively) were recognized, proteins being orthologous within each group and paralogous across the groups

(Vuletich & Lecomte, 2006). Group I and group II can be further separated into two and four subgroups, respectively, whereas group III displays a high level of overall sequence conservation. Despite the conserved small size of 2/2Hbs, sequence identity among proteins from the different groups is low ($\leq$20% overall identity), but may be higher than 80% within a given group. Phylogenetic analyses further suggest an evolutionary scenario where group II *HbO* gene is the ancestral gene, and group I and group III genes are the result of duplication and transfer events (Vuletich & Lecomte, 2006).

In some cases, 2/2Hbs from more than one group can coexist in the same organism, indicating a diversification of their functions. In general, members of the 2/2Hb family are monomeric or dimeric proteins characterized by medium to very high oxygen affinities, with cases of ligand-binding cooperativity (Couture, Yeh, et al., 1999). Some of the organisms hosting 2/2Hbs are aggressive pathogenic bacteria; others perform photosynthesis, fix nitrogen, or may display distinctive metabolic capabilities. Although very little is known about their role *in vivo*, possible functions of 2/2Hbs that are consistent with the observed biophysical properties include long-term ligand or substrate storage, nitric oxide (NO) detoxification, O_2/NO sensing, redox reactions, and O_2 delivery under hypoxic conditions (Nardini et al., 2007; Vuletich & Lecomte, 2006; Wittenberg et al., 2002).

So far, a number of three-dimensional structures belonging to all three groups have been solved by X-ray crystallography and NMR methods, thus providing a clear picture of the structural features specific for each group: six structures from group I 2/2HbNs (from *Chlamydomonas eugametos* (Pesce et al., 2000), *Paramecium caudatum* (Pesce et al., 2000), *Mycobacterium tuberculosis* (Milani et al., 2001), *Synechocystis* sp. (*Synechocystis* 6803: Falzone, Vu, Scott, & Lecomte, 2002; Hoy, Kundu, Trent, Ramaswamy, & Hargrove, 2004; Trent, Kundu, Hoy, & Hargrove, 2004; *Synechococcus* 7002: Scott et al., 2010), and *Tetrahymena pyriformis* (Igarashi, Kobayashi, & Matsuoka, 2011)), five structures of group II 2/2HbOs (from *Mycobacterium tuberculosis* (Milani et al., 2003), *Bacillus subtilis* (Giangiacomo, Ilari, Boffi, Morea, & Chiancone, 2005), *Thermobifida fusca* (Bonamore et al., 2005) *Geobacillus stearothermophilus* (Ilari et al., 2007), and *Agrobacterium tumefaciens* (Pesce et al., 2011)), and one structure from group III 2/2HbPs (from *Campylobacter jejuni* (Nardini et al., 2006)). Additionally, the NMR method was applied to characterize the haem ligand binding site of 2/2HbP from *Helicobacter hepaticus* in solution (Nothnagel, Winer, Vuletich, Pond, & Lecomte, 2011).

2. FOLD AND FOLD VARIATION IN 2/2Hb GROUPS I, II, III

The globin fold of 2/2Hb (Fig. 2.1) has been described as consisting of a simplified version of the 'classical' globin fold (a 3-on-3 α-helical sandwich; Perutz, 1979) typical of sperm whale Mb. The topology of the 2/2Hb fold is characterized by a 2-on-2 α-helical sandwich based on four α-helices, corresponding to the B-, E-, G-, and H-helices of the classical globin fold (Nardini et al., 2007; Pesce et al., 2000). The helix pairs B/E and G/H are arranged each in antiparallel fashion and assembled in a sort of α-helical bundle which surrounds and protects the haem group from the solvent. Although the G- and H-helices generally match the globin fold

Sperm whale myoglobin (PDB-code 1A6M)

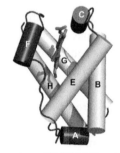

P. caudatum 2/2HbN (PDB-code 1DLW)

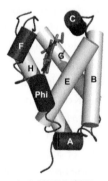

M. tuberculosis 2/2HbO (PDB-code 1NGK)

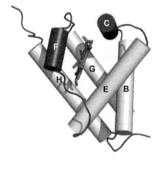

C. jejuni 2/2HbP (PDB-code 2IG3)

Figure 2.1 Comparative view of the classical 3/3 globin fold (sperm whale myoglobin) with the 2/2 globin fold in groups I (HbN), II (HbO), and III (HbP). Helices are shown as cylinders and labelled. The helices structurally conserved within 3/3 and 2/2 folds are shown in grey. The haem is shown in stick representation. (For colour version of this figure, the reader is referred to the online version of this chapter.)

topology, they may be much shorter or bent, as compared to the secondary structure elements in sperm whale Mb. The most striking differences between the 2-on-2 and the 3-on-3 globin folds are (i) the drastically short-ened A-helix (completely deleted in group III 2/2HbP), (ii) the absence of the D-helix, (iii) the presence of a long polypeptide segment (pre-F) in extended conformation, and (iv) a variable-length F-helix (reduced to a one-turn-helix in group I and III 2/2Hbs) that properly supports the haem-coordinated proximal HisF8 residue (Pesce et al., 2000; Milani et al., 2003; Nardini et al., 2006) (Fig. 2.1). Structural differences are evident also on the haem distal side, where the 2/2 fold CD–D region differs in length relative to (non-)vertebrate globins.

Structural superposition of 2/2Hbs of known structure highlights a general good conservation of the α-helical scaffold among the three groups, with the overall fold of group III 2/2HbP equally diverging in its C_α trace from group I and group II 2/2Hbs. Interesting group-specific structural variability/plasticity can be recognized at defined sites of the tertiary structure and correlated to attainment and stabilization of the compact 2/2Hb fold. At the N- and C-termini, the A-helix can be either very short or fully absent (as in group III), while the H-helix is highly variable in length and linearity, being kinked in group I, short in group II, unusually long in group III. Other important structural variations are localized in the core of the protein. For instance, the polypeptide stretch bridging the C- and E-helices is usually trimmed to about three residues in group I 2/2HbNs and group II 2/2HbOs; on the contrary, a 3–7 amino acid insertion is invariantly found in group III 2/2HbPs. Such elongation of the CD region has structural implications on the spanning of the C- and E-helices and on the 3_{10} helical character of helix C. Indeed, in group III 2/2HbP from *C. jejuni*, the C- and E-helices are elongated by one additional turn at their C- and N-termini, respectively, relative to the corresponding helices in group I and group II 2/2Hbs, not affecting, however, the position of the E-helix relative to the haem distal site (Nardini et al., 2006). Variable capping interactions in the CE inter-helical region of group III 2/2HbP, however, suggest that secondary structure boundaries may not be conserved and that C- and E-helix lengths, side chain locations, and haem accessibility may differ across the group (Nothnagel, Winer, et al., 2011). Additionally, in group III 2/2HbP, the C-helix displays a clear α-helical character, whereas it is a 3_{10} helix in group I and group II 2/2Hbs and in (non-)vertebrate globins (Bolognesi, Bordo, Rizzi, Tarricone, & Ascenzi, 1997). Despite the group-specific structural variations, in all 2/2Hbs, the CD region and the E-, F-, and G-helices build the protein crevice hosting the haem, which is shielded from the solvent and stabilized by well-conserved polar/electro-static interactions involving the porphyrin propionates.

A stable and properly structured haem crevice in the context of such a short polypeptide chain has been correlated to the presence of three glycine motifs, conserved among sequences of group I 2/2HbNs and group II 2/2HbOs. Such Gly motifs are located at the AB and EF inter-helical corners, and just before the short F-helix; they are thought to provide the protein backbone flexibility needed to stabilize the short A-helix in a conformation locked onto the B- and E-helices, and to support the pre-F segment in building the haem pocket (Milani et al., 2001; Pesce et al., 2000). A similar stabilization, however, cannot be achieved in group III 2/2HbPs, where the AB Gly–Gly motif is absent, due to full deletion of the A-helix. As a consequence, the 2/2HbP amino-terminus cannot face the BE inter-helical region, and the protein residues preceding the B-helix extend towards the GH region; such conformation is opposite to that found in group I and II 2/2Hbs. Similarly, a clear Gly–Gly motif cannot be recognized in the EF region of group III 2/2HbP, although scattered Gly residues are present in this region of the sequence. Despite the absence of a clear EF Gly–Gly motif, B- and E-helices in group III 2/2HbPs are oriented as in groups I and II, their stabilization being achieved through group-specific hydrophobic contacts at the B/E helical interface.

Other group-specific structural variation is localized inside the haem binding pocket, with group III 2/2HbPs displaying a high degree of similarity in sequence and structure of the distal region to group II 2/2HbOs and, simultaneously, sharing a proximal side-extended EF region typically found in group I 2/2HbNs. Structural differences and group-specific residues at the haem distal and proximal sites are correlated to different ligand-binding properties of 2/2Hbs.

3. THE HAEM ENVIRONMENT

Despite the trimmed globin fold, the 2/2 helical sandwich provides a minimal scaffold, formed by the E- and G-helices, by the CD region and by the F-helix, that allows efficient incorporation of the haem group within the protein (Fig. 2.1). Besides the HisF8–Fe coordination bond, in all 2/2Hbs, the haem is stabilized by a network of van der Waals contacts with hydro-phobic residues at the conserved topological positions C6, C7, CD1, E14, F4, FG3, G8, and H11. Other protein–haem interactions may arise from residues located in regions of the 2/2 fold that vary in the three groups, such as the CD and FG segments and the amino-terminal part of the H-helix. Further stabilizing interactions are provided by hydrogen bonds between the haem and polar residues, involving Thr/Tyr at sites E2, E5, and EF6, and by salt bridges involving haem propionates and Arg/Lys residues located at position E10 in all 2/2Hbs, at position F2 in group I 2/2HbNs, and at position F7 in group II 2/2HbOs (where F7 is invariantly Arg). It should be noted that LysF7, despite being involved in electrostatic stabilization of the haem also in group III *C. jejuni* 2/2HbP, is not conserved in other group III globins (Nardini et al., 2006). Further salt bridge interactions may also derive from sequence-specific substitutions in the surroundings of the haem propionates, as in the case of *M. tuberculosis* 2/2HbN ArgE6, and 2/2HbO ArgEF10. The crystallographic studies on group I 2/2HbNs and group II 2/2HbOs have shown that haem isomerism (insertion of a fraction of the haem groups into the globin structure in an inverted orientation) may be present (Milani et al., 2005).

The conformation of the Fe-coordinated proximal HisF8 is typical of an unstrained imidazole ring, with the imidazole plane lying in a staggered azimuthal orientation relative to the haem pyrrole nitrogen atoms, thus facilitating haem in-plane location of the iron atom, and supporting fast oxygen association (Bolognesi et al., 1997; Wittenberg et al., 2002) and electron donation to the bound distal ligand.

The 2/2Hb haem distal site cavity, hosting the exogenous ligands, is characterized by unusual residues as compared to (non-)vertebrate globins. It should be noted that in all 2/2Hbs, the E-helix falls close to the haem distal face due to the 'pulling action' of the shortened CD region, thus causing side chain crowding of the distal site residues at topological positions B10, CD1, E7, E11, E14, E15, and G8. Among these, group-specific selections of residues display polar character and allow the formation of networks of hydrogen bonds functional to the stabilization of the diatomic haem ligand, or implicated in the rebinding of dissociated ligands (Samuni et al., 2003). Distal site polarity is expected to favour oxygen chemistry in the haem crevice, as in peroxidases (Hiner, Raven, Thorneley, García-Cánovas, & Rodríguez-López, 2002).

There have been a number of experimental studies devoted to the spectroscopic and structural characterization of the binding of several diatomic ligands (CO, O_2, NO, and cyanide) to the 2/2HbN in *M. tuberculosis* (Couture, Yeh, et al., 1999; Milani, Ouellet, et al., 2004; Ouellet, Milani, Couture, Bolognesi, & Guertin, 2006; Ouellet et al.,

2008; Yeh, Couture, Ouellet, Guertin, & Rousseau, 2000); these indicate that the ligand binding is largely controlled by a pair of interacting amino acids (GlnE11 and Tyr10) in the haem distal site that participate in hydrogen bonding with the haem-bound diatomic O_2 ligand. Indeed, in 2/2HbN from *M. tuberculosis*, a direct hydrogen bond occurs between TyrB10 side chain and the ligand (O_2 or cyanide, in the ferrous or ferric haem states, respectively), stabilized by GlnE11 that interacts with TyrB10 (Couture, Yeh, et al., 1999; Milani et al., 2001; Milani, Ouellet, et al., 2004; Yeh et al., 2000). It has been shown that the main barrier to ligand binding to deoxy *M. tuberculosis* 2/2HbN is the displacement of a distal cavity water molecule, which is mainly stabilized by residue TyrB10, but not coordinated to the haem iron. As observed in the TyrB10/GlnE11 apolar mutants (TyrB10Phe/Val and GlnE11Val/Ala, respectively), once this kinetic barrier is lowered, CO and O_2 binding is very fast with rates approaching $1 - 2 \times 10^9 \, M^{-1}s^{-1}$. These large values almost certainly represent the upper limit for ligand binding to a haem protein and also indicate that the iron atom in 2/2HbN is highly reactive (Ouellet et al., 2008). In *P. caudatum* 2/2HbN and in *C. eugametos* 2/2HbN, residue TyrB10, buried in the inner part of the haem pocket, is properly oriented through hydrogen bonds towards residues GlnE7 and Thr/GlnE11, to provide stabilization of the haem-bound distal ligand (Pesce et al., 2000). In *T. pyriformis* 2/2HbN, TyrB10 and GlnE7 are hydrogen bonded to the haem-bound O_2 molecule. Furthermore, TyrB10 is hydrogen bonded to GlnE7 and GlnE11 residues. Mutation of these residues results in fast O_2 dissociation and autoxidation (Igarashi et al., 2011). In all cases, the strongly conserved TyrB10 plays a pivotal role in ligand stabilization through a direct hydrogen bond to the haem ligand. In general, when in group I a hydrogen bonding residue is present at B10, a Gln is located at E7 or E11, or at both these sites, likely completing the distal site hydrogen-bonded network. On the contrary, when a side chain devoid of hydrogen-bonding capabilities is (rarely) hosted at B10, then large hydrophobic residues are coupled at the E7 and E11 sites (Nardini et al., 2007; Vuletich & Lecomte, 2006).

In group I 2/2Hbs, an example of *bis*-histidine hexacoordination has been reported for group I 2/2HbN from the cyanobacterium *Synechococcus* sp. strain PCC 7002 and PCC 6803 (involving the proximal/distal residues HisF8 and HisE10, respectively), where binding of an exogenous ligand to the haem requires the dissociation of the Fe-coordinated HisE10 from the haem and a large conformational change of the B- and E-helices (Couture et al., 2000; Falzone et al., 2002; Hoy et al., 2004; Scott et al., 2002, 2010; Trent et al., 2004; Vu, Nothnagel, Vuletich, Falzone, & Lecomte, 2004).

The hexacoordinate *Synechocystis* sp. 2/2HbN shows also a unique haem–protein covalent interaction between HisH16 and the 2-vinyl group of the haem. This post-translational modification prevents haem loss and has the potential to modulate the reactivity of the haem group (Couture et al., 2000; Falzone et al., 2002; Hoy et al., 2004; Scott et al., 2002, 2010). In particular, for *Synechocystis* 6803 2/2HbN, it has been shown that the post-translational modification has little effect on the protein structure, perturbing the backbone dynamics only modestly, and that the specificity and rate of the cross-linking reaction depended critically on the nature of the sixth ligand to the haem iron (Nothnagel, Love, & Lecomte, 2009; Nothnagel, Preimesberger, et al., 2011; Pond, Majumdar, & Lecomte, 2012).

Although the endogenous haem hexacoordination is not a prominent trend for 2/2Hbs, it has been also observed, under specific conditions, in other 2/2Hbs. For instance, the *C. eugametos* 2/2HbN may also display a six-coordinate haem-Fe atom in its ferric state, while the ferrous derivative displays a five-coordinate high spin haem-Fe atom at neutral pH, and a six-coordinate low spin species at alkaline pH where the sixth ligand to the haem-Fe atom is held to be either TyrB10 or to be stabilized by this residue (Couture, Das, et al., 1999; Couture & Guertin, 1996). Haem hexacoordination has also been observed in the ferrous derivative of group II *M. leprae* 2/2HbO at neutral pH (Visca, Fabozzi, Petrucca, et al., 2002).

In group II 2/2HbO, specific residue substitutions characterize the distal site environment relative to group I 2/2HbN. In this group, TyrB10 is strictly conserved, and so is TrpG8. The five available structures (Bonamore et al., 2005; Giangiacomo et al., 2005; Ilari et al., 2007; Milani et al., 2003; Pesce et al., 2011) show these residues to be located at the haem distal site, but not necessarily involved in direct ligand binding. Several group II 2/2HbOs display a polar residue, His or Tyr, at the topological position CD1, which in globins was thought to harbour a strictly conserved Phe (Kapp et al., 1995) whose role was to fasten the haem in its binding site (Ptitsyn & Ting, 1999).

In *M. tuberculosis* 2/2HbO, TyrCD1 is the residue responsible for hydrogen bonding to the diatomic ligand (Milani et al., 2003). Further ligand-stabilizing interactions are provided by TrpG8, whose indole NE1 atom is hydrogen bonded to the haem-bound ligand and to TyrCD1 OH (Boechi et al., 2008; Guallar, Lu, Borrelli, Egawa, & Yeh, 2009; Milani

et al., 2003; Ouellet et al., 2003, 2007). The crystal structure of
M. tuberculosis 2/2HbO has also shown that the simultaneous presence
of Tyr residues at the B10 and CD1 sites may trigger the formation of a
covalent (iso–dityrosine like) bond between the two side chains (Milani
et al., 2003), whose functional role is yet unclear.

A similar network of interaction has also been described in *T. fusca*
(Bonamore et al., 2005) and in *B. subtilis* (Boechi, Mañez, Luque,
Martì, & Estrin, 2010; Feis et al., 2008; Giangiacomo et al., 2005) 2/2HbOs,
where TyrB10, Phe/TyrCD1, and TrpG8 are mainly involved in a
hydrogen-bonding network, thus stabilizing the exogenous ligands. In
T. fusca 2/2HbO, the carbonyl oxygen of the acetate ion ligand is stabilized
by hydrogen bonds with residues TyrCD1 and TrpG8 (Bonamore et al.,
2005). When instead a Phe residue is present at the CD1 position, as in
B. subtilis 2/2HbO, TyrB10 assumes the role of hydrogen bond donor
for the interactions with the exogenous ligand (Giangiacomo et al.,
2005). GlnE11, TrpG8, and ThrE7 complete the polar distal frame with
GlnE11 side chain and the TrpG8 indolic nitrogen atom at hydrogen bonding distance to the bound ligand (Giangiacomo et al., 2005). The
G. stearothermophilus 2/2HbO haem pocket displays a hydrogen bonding
network involving TyrB10 and TrpG8 residues similar to *B. subtilis*
2/2HbO (Ilari et al., 2007). Interestingly in *M. tuberculosis* 2/2HbO, when
TyrCD1 is mutated to Phe, is TyrB10 the hydrogen bonding residue for the
haem-bound ligand (Ouellet et al., 2003), thus mimicking in *M. tuberculosis*
2/2HbO what has been observed in *B. subtilis* and *G. stearothermophilus*
2/2HbOs (Giangiacomo et al., 2005; Ilari et al., 2007).

Recently, *B. subtilis* and *T. fusca*, group II 2/2HbOs have been demonstrated to be able to bind CO in the ferrous state (Droghetti et al., 2010),
and sulphide (Nicoletti et al., 2010) or fluoride (Nicoletti et al., 2011) in
the ferric state. The architecture of the distal cavities of *B. subtilis* and
T. fusca 2/2HbOs can be compared with those of the few reported examples
of sulphide-binding haem proteins (Rizzi, Wittenberg, Coda, Ascenzi,
& Bolognesi, 1996). Molecular dynamics simulation indicates that only
TrpG8 residue contributes to the sulphide stabilization through
direct hydrogen-bonding interaction, thus accounting for the relatively
high affinity for sulphide in these proteins (Nicoletti et al., 2010).

A. tumefaciens 2/2HbO is the first example of a structure where the
topological position CD1 is occupied by His. Here, the haem distal site is
characterized by the presence of a highly intertwined hydrogen-bonding
network, involving residues TyrB10, HisCD1, SerE7, TrpG8, and three

water molecules. In particular, the haem-coordinated water molecule is directly hydrogen bonded to a distal site intervening water molecule and to TrpG8 (Pesce et al., 2011). Thus, overall, the group II-conserved TrpG8 seems to be the residue playing a crucial/pivotal role in stabilizing the ligand and modulating its escape rate out of the distal pocket (Bonamore et al., 2005; Giangiacomo et al., 2005; Ilari et al., 2007; Milani et al., 2003; Pesce et al., 2011).

2/2HbOs also tend to contain a small residue at the E7 site, typically Ala, Ser, or Thr. A small residue at the E7 position may suggest the presence of an E7 route entry system to facilitate the accessibility of diatomic ligands to the 2/2HbO haem distal site (see below).

Group III sequences (all bacterial) display the largest extent of conservation (Wittenberg et al., 2002), all containing PheB9, TyrB10, PheCD1, HisE7, PheE14, TrpE15, and TrpG8. Such a large number of strictly conserved residues near the haem group suggest a narrow range of chemical properties and group III 2/2HbPs form a single homogeneous class. It should be noted, however, that recent sequence analysis based on an expanded and corrected bacterial genome database containing 181 group III 2/2HbPs in eight *phyla* showed group III to be less homogeneous than originally thought and raised the possibility that diverse chemical behaviours may be exhibited by its members (Nothnagel, Winer, et al., 2011).

In group III *C. jejuni* 2/2HbP, the haem distal pocket residues TyrB10, PheCD1, HisE7, IleE11, PheE14, and TrpG8 surround the haem-bound ligand (Nardini et al., 2006). Similar residues are also conserved in the 2/2HbP from *Helicobacter hepaticus* (Nothnagel, Winer, et al., 2011). Among these, only HisE7 is specific (and fully conserved) in group III (Vuletich & Lecomte, 2006), although the hydrophobic character of the haem pocket distal side is a highly conserved feature of group III globins. Contrary to group II, group III 2/2HbPs display Phe (or hydrophobic) residue at position CD1 and a hydrophobic residue at site E11. In *C. jejuni* 2/2HbP, the only hydrogen-bonding residues involved in ligand stabilization are TyrB10 and TrpG8, while no ligand-stabilizing interactions may be provided by residues at CD1 and E11 positions, nor by HisE7. Mutagenesis and molecular dynamics studies revealed that in the wild-type protein, the main residue responsible for oxygen stabilization is TyrB10. Bound oxygen is further stabilized by a hydrogen bond from either TrpG8 or HisE7, depending on the orientation of the Fe–O–O moiety, the hydrogen bond to TrpG8 being stronger than to HisE7. Most importantly, the coexistence of multiple conformations for the residues in the distal cavity, each characterized by a distinct pattern of hydrogen-bonding interaction, creates differences in the local polarity and affects the stabilization of the haem-bound ligand, the behaviour of each residue being affected by the other residues. Therefore, the oxygen-binding affinity (which in *C. jejuni* 2/2HbP is very high due to a low dissociation rate constant of 0.0041 s^{-1}) is the result of a cooperative property (Arroyo Mañez et al., 2011; Lu, Egawa, Wainwright, Poole, & Yeh, 2007). Furthermore, in the crystal structure of the *C. jejuni* 2/2HbP–cyanide complex, residue HisE7 occurs in two distinct conformations, corresponding to side-chain orientations that point towards the solvent or towards the haem distal site, respectively. Alternative position for the strictly conserved HisE7 suggests a ligand-gating mechanism similar to that described in *P. caudatum* 2/2HbN (Das et al., 2000; Nardini et al., 2006).

4. TUNNELS AND CAVITIES THROUGH 2/2Hb PROTEIN MATRIX

Despite the high-density packing of residues in the protein core, inner cavities or tunnels are often found in the protein matrix. Although such residue packing 'defects' may hamper the thermodynamic stability of a folded protein, their presence offers an evolutionary, possibly functional, advantage to the hosting protein. For instance, in enzymes they may provide preferred paths or intramolecular docking stations for the diffusion of substrates and products (Milani et al., 2003; Raushel, Thoden, & Holden, 2003; Weeks, Lund, & Raushel, 2006). In globins, the haem site is often buried inside the protein chain, which prevents direct contact with solvent. Therefore, the ligand has to trace its way to the haem by traversing the globin helical fold. The migration pathways are commonly believed to result from thermal fluctuations of the protein molecular structure, and the ligand access sites are located in (what are held to be) evolutionarily optimized well-defined regions of proteins that can be identified with systematic experimental and computational efforts (Brunori et al., 1999).

The analysis of the three-dimensional structures of 2/2Hbs have shown that the group I 2/2HbN fold is characterized by the presence of interconnected protein matrix apolar cavities, or a continuous tunnel, which connect the protein surface to an inner region merging with the haem distal site (Milani, Pesce, et al., 2004; Pesce, Milani, Nardini, & Bolognesi, 2008). Such peculiar feature may be related to the orientation of the CD–D region that forces positioning of the E-helix close to the haem distal face, thus preventing ligand access to the distal site cavity through the classical E7.

The protein matrix tunnel, linking the protein surface to the haem distal site, appears to be conserved in group I 2/2Hbs, with the exception of the hexacoordinated *Synechocystis* sp. 2/2HbN, likely because of the conformational transitions required in this protein distal site region to achieve haem hexacoordination (Hoy et al., 2004). An alternative haem distal site access through the exposed 8-methyl edge of the haem group and near the propionates has been proposed (Falzone et al., 2002; Scott et al., 2010).

In agreement with the availability of cavities in the protein matrix, it has been shown that at least three group I 2/2Hbs, from *C. eugametos*, *M. tuberculosis*, and *P. caudatum*, can bind Xe atoms in the crystalline state. The Xe atoms map experimentally at multiple sites and with comparable topology within the tunnel/cavity path in these 2/2Hbs (Milani, Pesce, et al., 2004). In particular, in *C. eugametos* 2/2HbN and in *M. tuberculosis* 2/2HbN, the tunnel is composed of two roughly orthogonal branches converging at the haem distal site from two distinct protein surface access sites. On one hand, a 20 Å long tunnel branch connects the protein region nestled between the AB and GH hinges to the haem distal site. On the other, a short tunnel branch of about 8 Å connects an opening in the protein structure between G- and H-helices to the haem (Fig. 2.2). In *P. caudatum* 2/2HbN, the haem site is connected to the solvent region by a three-cavity system, topologically distributed along the tunnel's long branch described above (Milani, Pesce, et al., 2004). A similar, but more open, tunnel system is also present in *T. pyriformis* 2/2HbN, where the exit of the short branch differs slightly in orientation relative to *M. tuberculosis* 2/2HbN. The tunnel volume of *T. pyriformis* 2/2HbN is about 380 $Å^3$, which is similar to that of *C. eugametos* 2/2HbN (400 $Å^3$), but larger than that of *M. tuberculosis* and *P. caudatum* HbNs (265 $Å^3$ and 180 $Å^3$, respectively) (Igarashi et al., 2011). Although protein cavity volumes vary among 2/2HbNs, these values are not correlated with O_2 association rate constants (Couture, Das, et al., 1999; Couture, Yeh, et al., 1999; Das et al., 2000; Igarashi et al., 2011; Ouellet et al., 2008).

Residues lining the tunnel branches are hydrophobic and are substantially conserved throughout group I (Vuletich & Lecomte, 2006). PheE15, a well-conserved residue, adopts two conformations in *M. tuberculosis* 2/2HbN (Fig. 2.2). In one, PheE15 benzene side chain blocks the longer channel of the tunnel path (the so-called closed state) and in the other it does not (the open state) (Milani et al., 2001; Milani, Pesce, et al., 2004). *M. tuberculosis* 2/2HbN is endowed with a potent nitric oxide dioxygenase activity which allows it to relieve nitrosative stress and enhance *in vivo*

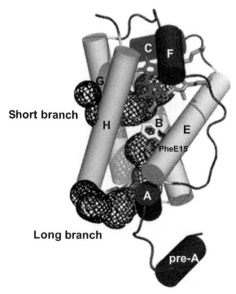

Figure 2.2 The protein matrix tunnel (short and long branches) observed in group I *M. tuberculosis* 2/2HbN. The tunnel surface, defined by a 1.4 Å radius probe, is portrayed as a mesh. Residue PheE15, causing the main restriction in the diameter of the long branch tunnel, is shown in the close conformation in stick representation and labelled. Helices are shown as cylinders and labelled. The helices structurally conserved within 3/3 and 2/2 folds are shown in grey. The haem is shown in stick representation. (For colour version of this figure, the reader is referred to the online version of this chapter.)

survival of its host through the rapid oxidation of NO to harmless nitrate (Couture, Yeh, et al., 1999; Ouellet et al., 2002; Pathania, Navani, Gardner, Gardner, & Dikshit, 2002). Migration of O_2 and NO to the *M. tuberculosis* 2/2HbN distal haem cavity is driven by a dual-path mechanism. In fact, by long molecular dynamics simulations (0.1 ms), it has been shown that in deoxy 2/2HbN, PheE15 adopts the closed conformation and hence the O_2 ligand enters the protein *via* the short channel. In the case of oxygenated 2/2HbN, the PheE15 prefers the open conformation, thus facilitating entrance of the second ligand (NO) *via* the long tunnel branch (Bidon-Chanal et al., 2006; Bidon-Chanal, Martì, Estrin, & Luque, 2007). Recent mutagenesis studies also support this view on the diffusion of small diatomic ligands through the *M. tuberculosis* 2/2HbN protein matrix tunnel system, pointing out the delicate structural balance imposed by the PheE15 gate, which not only regulates ligand migration but also contributes to avoid the collapse of helices B and E, thus preserving the ligand accessibility along the tunnel long branch (Oliveira et al., 2012).

In *T. pyriformis* 2/2HbN, the hydrophobic residues Leu11, Ala20, and Leu94 form a gate delimiting the outer solvent space from the inner region, whereas Leu54 and Leu90 are located at the connecting gaps. Leu54 corresponds to the gating residue PheE15 of *M. tuberculosis* 2/2HbN and may contribute significantly to molecular processes that sustain diffusion of diatomic ligands to the haem (Igarashi et al., 2011).

Strikingly, the *M. tuberculosis* 2/2HbN nitric-oxide dioxygenase activity has been correlated with the presence of the *M. tuberculosis* 2/2HbN-specific pre-A helix (Fig. 2.2). Deletion of pre-A region from the *M. tuberculosis* 2/2HbN drastically reduces its ability to scavenge nitric oxide, whereas its insertion at the N-terminus of pre-A lacking 2/2HbN of *M. smegmatis* improved its nitric-oxide dioxygenase activity. Molecular dynamics simulations show that the excision of the pre-A motif results in distinct changes in the protein dynamics, which cause the gate of the tunnel long branch to be trapped into a closed conformation, thus impeding migration of diatomic ligands towards the haem distal site (Lama et al., 2009).

Other molecular dynamics studies have proposed a different view of NO access to the haem distal site in *M. tuberculosis* 2/2HbN, whereby the ligand molecule preferentially would enter the protein matrix through the tunnel short branch, and once inside the protein NO diffuses through a series of cavities corresponding to experimental Xenon-binding pockets (Daigle, Rousseau, Guertin, & Lagüe, 2009). NO diffusion along the tunnel long branch was found to be hindered by the PheE15 side-chain obstruction, countering the dual-path mechanism proposed previously (Bidon-Chanal et al., 2006). Moreover, NO entering the tunnel long branch would preferentially bypass the PheE15 barrier by means of an additional tunnel located between the E- and H-helices (Daigle et al., 2009). The presence of other ligand migration pathways through the *M. tuberculosis* 2/2HbN matrix, potentially competing with the short and the long tunnel branches, was also proposed for paths located between E- and H-helices or between the C- and F-helices. Both of these paths are surrounded by at least one polar residue and are expected to be the preferred escape channel for removing the products of NO detoxification reaction (such as the nitrate anion) from the protein matrix (Martì et al., 2008; Mishra & Meuwly, 2010).

Contrary to group I, group II 2/2Hbs do not show an evident tunnel/cavity system connecting the protein surface to the haem distal pocket. The protein matrix tunnel observed in 2/2HbN, is dramatically restricted in 2/2HbO, where different relative orientations of the G- and H-helices, and increased side-chain volumes at topological sites B1, B5, G8, G9, and

G12, mostly fill the protein matrix tunnel space (Bonamore et al., 2005; Giangiacomo et al., 2005; Ilari et al., 2007; Milani et al., 2003; Pesce et al., 2011). In particular, in 2/2HbOs, the bulky side-chain of the conserved TrpG8 obstructs the short tunnel branch and the deeper part of the distal site pocket typical of 2/2HbNs. The 2/2HbN long tunnel branch retains only two cavities in 2/2HbOs, both fully shielded from solvent contact (Milani et al., 2003). Interestingly, in *M. tuberculosis* 2/2HbO, residue TrpG8 is also responsible for blocking the region corresponding to the 2/2HbN long tunnel and therefore key for ligand entry (Boechi et al., 2008; Ouellet et al., 2007). The restriction of the cavities within the protein matrix becomes extreme in *T. fusca* 2/2HbO, where no internal cavities are detected, due to substitutions with larger residues relative to other 2/2HbOs, or by conformational differences of conserved or similar size residues (Bonamore et al., 2005).

The substantial absence of a protein matrix tunnel system is mirrored by the conserved presence of a small distal site E7 residue in group II 2/2HbOs (Vuletich & Lecomte, 2006), which does not hinder entrance to the haem distal cavity. Therefore, in 2/2HbOs, diatomic ligands (such as O_2, CO, and NO) may preferably access the haem distal site through an E7 route. Molecular dynamics simulations, however, showed that once the protein is oxygenated, both the E7 route and the path corresponding to the *M. tuberculosis* 2/2HbN tunnel long branch can contribute to ligand entry, because they present similar barriers. This mechanism differs from the case of 2/2HbN, in which each ligand has been proposed to migrate through a separate pathway (Bidon-Chanal et al., 2006). The change in the free energy barrier for the long tunnel is due to the TrpG8 interaction with the haem-bound O_2. The short-tunnel E7 barrier does not change significantly upon oxygenation; consequently, the overall barrier presented by the short-tunnel E7 is similar in the oxygenated and deoxygenated states of the protein. This fact is consistent with the experimental kinetic constants for ligand migration. The results highlight the importance of TrpG8 in regulating ligand migration in 2/2HbO, since not only is it responsible for the high barrier observed in the long tunnel, but it also blocks the short tunnel branch displayed by group I 2/2HbNs. Furthermore, TrpG8 is important in anchoring TyrCD1 and LeuE11 side chains, thereby allowing the stabilization of the haem-bound ligand *via* hydrogen bonds donated from TrpG8 and TyrCD1. Following its dissociation, the ligand can migrate between three temporary docking sites, which are modulated by the conformational rearrangements of the side chains of several critical distal amino acids, including the TrpG8, LeuE11, TyrCD1, and AlaE7. The initial migration of the ligand within the distal pocket leads to its rebinding to the haem iron atom or to its escape into the solvent *via* a hydrophobic tunnel that coincides with the internal cavities found in the crystallographic structure of 2/2HbO (Guallar et al., 2009). This would indicate that the presence of the conserved residues in group II and group III, but not in group I, is responsible for the significantly different migration patterns in 2/2HbO and 2/2HbN.

Within group II 2/2HbOs, the protein from *B. subtilis* is a peculiar case, as the presence of a Thr residue at position E7 typically blocks the E7 path and the X-ray crystallographic structure does not exhibit a clear tunnel for ligand migration (Giangiacomo et al., 2005). However, O_2 association rate constant k_{on} is higher than that found for *M. tuberculosis* 2/2HbO, and similar to that of *M. tuberculosis* 2/2HbN (Couture, Yeh, et al., 1999; Giangiacomo et al., 2005; Pathania, Navani, Rajamohan, & Dikshit, 2002). The structural and the kinetic data have been reconciled by classical molecular dynamics simulations of the oxy, carboxy, and deoxy proteins which showed that GlnE11 presents an alternate conformation, giving rise to a wide ligand migration tunnel, topologically related to the long tunnel branch found in group I 2/2HbNs. In *B. subtilis* 2/2HbO, residue TrpG8 does not block the tunnel, as generally assumed by inspection of the crystal structure, due to a rearrangement in the distal site involving GlnE11, and the tunnel is open due to the lack of the bulky PheE15, the tunnel gating residue in *M. tuberculosis* 2/2HbN. On the other hand, the results for the CO and O_2 bound protein show that GlnE11 is directly involved in the stabilization of the coordinated ligand, playing a similar role as TyrB10 and TrpG8 in other 2/2Hbs (Boechi et al., 2010). These results underline once more the plasticity and redundancy of several residues within the globin fold that account for the varied ligand-binding kinetics observed.

Analysis of group III *C. jejuni* 2/2HbP structure shows no evident protein matrix tunnel/cavity system, mostly due to the peculiar backbone conformation of the pre-B helix residues, and to bulky side-chain substitutions (conserved among members of group III) at residues that define the tunnel/cavities walls in group I and II 2/2Hbs (Nardini et al., 2006). Since HisE7 (conserved in group III) adopts two alternate conformations ('open' and 'closed') in *C. jejuni* 2/2HbP, E7 haem-distal-site gating has been proposed to play a functional role for ligand diffusion to the haem, in the absence of a protein matrix tunnel/cavity system (Nardini et al., 2006).

5. PROPOSED FUNCTIONS FOR THE 2/2Hb FAMILY

Although the number of the deposited 2/2Hb sequences has grown rapidly over the past years, limited functional information is currently available for these proteins. Examples of proposed functions, consistent with observed biophysical properties, include nitric oxide detoxification, protection from reactive oxygen and nitrogen species, dioxygen scavenging, and recently sulphide binding (Hill et al., 1996; Nicoletti et al., 2010; Ouellet et al., 2002; Parrilli et al., 2010; Scott et al., 2010; Vinogradov & Moens, 2008). However, the diversity of physiological and environmental contexts in which 2/2Hbs are found suggests that additional enzymatic activities and insights into haem chemistry are yet to be discovered. For example, group I 2/2HbN of the unicellular green alga *C. eugametos* is induced in response to active photosynthesis and is localized partly along the chloroplast thylakoid membranes (Couture & Guertin, 1996). Group I 2/2HbN from the ciliated protozoa *P. caudatum* may supply O_2 to the mitochondria (Wittenberg et al., 2002). Moreover, group I 2/2HbN from the *Nostoc* sp. cyanobacterium is thought to protect the nitrogen–fixation apparatus from oxidative damage through O_2 scavenging (Hill et al., 1996).

Most of the functional analyses have been reported for 2/2Hbs from mycobacterial species, in particular, group I 2/2HbN in *M. tuberculosis*, *M. bovis*, *M. smegmatis,* and *M. avium*, group II 2/2HbO in *M. leprae* and all the above species, while group III 2/2HbP in *M. avium* only (Vinogradov et al., 2006; Vuletich & Lecomte, 2006; Wittenberg et al., 2002). The regression in content of 2/2Hb paralogues from *M. avium* (three 2/2Hbs), through *M. tuberculosis* (two 2/2Hbs), to *M. leprae* (one 2/2Hb) has been proposed to reflect an adaptation from saprophytic lifestyle to obligate intracellular parasitism, which paralleled the loss of functions provided by 2/2HbN and 2/2HbP. These results are consistent with the general notion that a 2/2HbO-like globin provided the ancestor structure from which 2/2HbNs and 2/2HbPs, as well as the classical 3-on-3 structural fold, originated (Nakajima, Álvarez-Salgado, Kikuchi, & Arredondo-Peter, 2005; Vinogradov et al., 2006; Vuletich & Lecomte, 2006).

Mycobacterial 2/2Hbs have been mostly implicated in scavenging of reactive nitrogen species. During infection, mycobacteria have to face the toxic effects of reactive nitrogen species, primarily NO, produced by activated macrophages expressing inducible NO synthase (Cooper, Adams, Dalton, Appelberg, & Ehlers, 2002; MacMicking et al., 1997; Nathan &

Shiloh, 2000; Ohno et al., 2003; Schnappinger, Schoolnik, & Ehrt, 2006; Visca, Fabozzi, Milani, Bolognesi, & Ascenzi, 2002). The distinct features of the haem active site structure of NO-responsive mycobacterial 2/2Hbs and their ligand-binding properties (Milani et al., 2001, 2003; Milani, Pesce, et al., 2004; Milani et al., 2005; Visca, Fabozzi, Milani, et al., 2002), combined with co-occurrence of multiple 2/2Hb classes in individual mycobacterial species and their temporal expression patterns *in vivo* (Fabozzi, Ascenzi, Renzi, & Visca, 2006; Ouellet et al., 2002, 2003), suggest that these globins play different physiological functions (Ascenzi, Bolognesi, Milani, Guertin, & Visca, 2007). For instance, *M. tuberculosis* 2/2HbN is endowed with a potent nitric-oxide dioxygenase activity and has been found to relieve nitrosative stress (Couture, Yeh, et al., 1999; Pathania, Navani, Gardner, et al., 2002) enhancing *in vivo* survival of a heterologous host, *Salmonella enterica typhimurium*, within the macrophages (Pawaria et al., 2007). These findings strongly support the NO scavenging and detoxification roles of 2/2HbN, which may be vital for *in vivo* survival and pathogenicity of *M. tuberculosis*. Similarly, a *M. bovis* mutant lacking 2/2HbN does not oxidize NO to NO_3^- and shows decreased respiration upon exposure to NO (Ouellet et al., 2002). Although to a lesser extent, a similar protective effect was also reported for *M. smegmatis* 2/2HbN in the homologous system (Lama, Pawaria, & Dikshit, 2006).

Detoxification of NO to nitrate is the hypothetical physiological function proposed also for *T. pyriformis* 2/2HbN. Based on oxygen affinity measurements, it has been estimated that *T. pyriformis* 2/2HbN within the cell would be maintained in the Fe(II)–O_2 form, indicating that *T. pyriformis* 2/2HbN does not function as an oxygen transporter. In addition, nitrosative stress mediated by sodium nitroprusside inhibits glyceraldehyde 3-phosphate dehydrogenase activity in *T. pyriformis* (Fourrat, Iddar, Valverde, Serrano, & Soukri, 2007). Therefore, *T. pyriformis* must have acquired a mechanism that senses and protects against nitrosative stress conditions, such as NO exposure (Igarashi et al., 2011).

Other widely studied members of the group I 2/2HbNs are cyanobacterial globins. 2/2HbN from the cyanobacterium *Synechococcus* sp. strain PCC 7002 and PCC 6803 have been characterized structurally and biochemically, focusing mainly on their two unusual structural properties: the bis-histidyl coordination of the haem iron in the absence of an exogenous ligand, and the post-translational covalent attachment of the haem to the globin by modification of the 2-vinyl substituent (Falzone et al., 2002; Lecomte et al., 2004; Lecomte, Vu, & Falzone, 2005; Pond et al., 2012; Scott et al., 2002, 2010; Trent et al., 2004; Vu, Vuletich, Kuriakose, Falzone, & Lecomte, 2004; Vuletich, Falzone, & Lecomte, 2006). The function of *Synecocistis* 2/2HbNs and their relationship to the metabolism of dioxygen, nitric oxide, or various reactive nitrogen and oxygen species are still largely unknown, although comparison to similar 2/2 haemoglobins suggests that reversible dioxygen binding is not its main activity. Recently, *in vitro* and *in vivo* experiments on cyanobacterium *Synechococcus* sp. strain PCC 7002 showed that its transcription profiles indicate that the protein is not strongly regulated under any of a large number of growth conditions and that the gene is probably constitutively expressed. High levels of nitrate, used as the sole source of nitrogen, and exposure to nitric oxide were tolerated better by the wild-type strain than by a 2/2HbN null mutant, whereas overproduction of protein in the null mutant background restored the wild-type growth. The cellular contents of reactive oxygen/nitrogen species were elevated in the null mutant under all conditions and were highest under NO challenge or in the presence of high nitrate concentrations. A peroxidase assay showed that purified 2/2HbN does not possess significant hydrogen peroxidase activity. Taken together, all these evidences suggested for 2/2HbN from cyanobacterium *Synechococcus* sp. strain PCC 7002 a protection role from reactive nitrogen species which cells could encounter naturally during growth on nitrate or under denitrifying conditions (Scott et al., 2010).

The physiological role of *M. tuberculosis* 2/2HbO has been primarily related to O_2 metabolism. 2/2HbO was hypothesized to be endowed with O_2 uptake or delivery properties during mycobacterial hypoxia and latency (Liu, He, & Chang, 2004; Pathania, Navani, Gardner, et al., 2002). This hypothesis is in apparent contrast with the low O_2 association and dissociation rates reported for 2/2HbO (Ouellet et al., 2003), and with its constitutive expression under aerobic conditions during the whole growth cycle of *M. bovis* (Mukai, Savard, Ouellet, Guertin, & Yeh, 2002; Pathania, Navani, Rajamohan, et al., 2002). 2/2HbO (II)–O_2 could still be able to sustain bacterial aerobic respiration by scavenging NO or other reactive species that would block the respiratory chain. In this context, the high stability of 2/2HbO(II)–O_2 would secure the reaction with NO even at very low O_2 tensions, as those that may exist in infected or necrotic tissue (Fabozzi et al., 2006). Interestingly, *M. leprae* 2/2HbO has been proposed to be involved in both H_2O_2 and NO scavenging, protecting from nitrosative and oxidative stress, and sustaining mycobacterial respiration (Ascenzi, De Marinis, Coletta, & Visca, 2008). Under anaerobic and highly oxidative conditions, as in the macrophagic environment where *M. leprae* is faced with H_2O_2 (Ascenzi, Bolognesi, & Visca, 2007; Ascenzi & Visca, 2008; Visca, Fabozzi, Milani, et al., 2002), the rapid formation of 2/2HbO—Fe (IV)=O occurs, which in turn facilitates NO scavenging, leading to the formation of haem-Fe(III) and NO_2^-. In turn, NO acts as antioxidant of 2/2HbO—Fe(IV)=O, which could be responsible for the oxidative damage of the mycobacterium. This reaction does not require partner redox enzymes, since the haem–protein oscillates between the haem-Fe(III) and haem-Fe(IV)=O forms, being helped by NO in maintaining an efficient H_2O_2 reduction rate. In this framework, it can be understood why *M. leprae* 2/2HbO—Fe(III) does not require a reductase system(s), which indeed has not been identified yet in this elusive mycobacterium (Ascenzi, Bolognesi, & Visca, 2007; Ascenzi & Visca, 2008). The catalytic parameters for NO scavenging by haem-Fe(II)—O_2 and haem-Fe(IV)=O are similar and high enough, suggesting that both reactions could take place *in vivo* (Ascenzi et al., 2008; Ascenzi & Visca, 2008).

The H_2O_2-induced *M. leprae* 2/2HbO—Fe(IV)=O formation could be relevant for *M. leprae* survival *in vivo* in the presence not only of NO and NO_2^- but also of peroxynitrite (Ascenzi, De Marinis, Visca, Ciaccio, & Coletta, 2009). The formation of peroxynitrite can in fact result from a secondary reaction of NO and the superoxide radical, which is concomitantly produced by activated macrophages. Then, peroxynitrite could rapidly react with CO_2 at the site of inflammation leading to the formation of strong oxidant and nitrating species (Ascenzi, Bocedi, et al., 2006; Goldstein, Lind, & Merényi, 2005). Peroxynitrite detoxification by *M. leprae* 2/2HbO has been shown to be rapid; therefore, 2/2HbO might be an important contributor to such function (Ascenzi, Milani, & Visca, 2006). Furthermore, as reported for NO and NO_2^- (Ascenzi et al., 2008), peroxynitrite acts as an antioxidant preventing the *M. leprae* 2/2HbO—Fe(IV)=O-mediated oxidation of mycobacterial (macro)molecules such as membrane lipids (i.e. lipid peroxidation) (Ascenzi et al., 2009).

The defence mechanisms against reactive oxygen and nitrogen species represent important components in the evolutionary adaptations, particularly under extreme environmental conditions. In this framework, *in vivo* and *in vitro* experiments have been performed in order to understand the roles of group II 2/2HbO from the Antarctic bacterium *Pseudoalteromonas haloplanktis* TAC125 (encoded by the *PSHAa0030* gene) in NO detoxification mechanisms (Coppola et al., 2013). The presence of multiple genes encoding 2/2Hbs and a flavohaemoglobin in this bacterium strongly suggests that these proteins fulfil important physiological roles, perhaps associated to the peculiar features of the Antarctic habitat (Giordano et al., 2007). Inactivation of the *PSHAa0030* gene renders the mutant bacterial strain sensitive to high O_2 levels, hydrogen peroxide, and nitrosating agents (Parrilli et al., 2010). Furthermore, when the *PSHAa0030* gene was cloned and over-expressed in a flavohaemoglobin-deficient mutant of *E. coli*, unable to metabolize NO, and the resulting strain was analyzed for its growth properties and oxygen uptake in the presence of NO, it was shown that *P. haloplanktis* 2/2HbO indeed protects growth and cellular respiration of the heterologous host from the toxic effects of NO donors. Moreover, the ferric form of *P. haloplanktis* 2/2HbO was shown to catalyze peroxynitrite isomerization *in vitro*, confirming its potential role in scavenging reactive nitrogen species (Coppola et al., 2013).

Recently, group II 2/2HbOs have been also implicated in metabolic pathways involving physiologically relevant sulphur compounds. *B. subtilis* and *T. fusca* group II 2/2HbOs have been shown to bind sulphide with an affinity constant in the sub-micromolar range such that they are partially saturated with sulphide when recombinantly expressed in *E. coli* (Nicoletti et al., 2010). Thus, these proteins have been proposed to play a direct role as sulphide scavenger under high oxygen growth conditions, due to the oxygen-dependent down-regulation of the competing cysteine synthase B (which is instead a very effective sulphide scavenger under low oxygen conditions). Also, a highly oxidative environment would favour the oxidation of the 2/2Hb haem-Fe atom, thus allowing prompt formation of the high affinity ferric sulphide adduct (Nicoletti et al., 2010). Interestingly, it has recently been demonstrated that the gene encoding *B. subtilis* 2/2HbO (as well as most 2/2Hbs from bacilli and staphylococci) is contained within a thiol redox pathway that is implicated in the bacterial response to the thiol oxidative stress (Larsson, Rogstam, & von Wachenfeldt, 2007). In this framework, it has been proposed that *B. subtilis* 2/2HbO could participate (directly or indirectly) in the complex redox pathway of sulphur metabolism in *Bacillus* sp. (Nicoletti et al., 2010).

Among group III 2/2HbPs, the protein from *Campylobacter jejuni* is the most characterized from the structural and biochemical view points (Bolli et al., 2008; Nardini et al., 2006), together with the more recently reported 2/2HbP from *Helicobacter hepaticus* (Nothnagel, Winer, et al., 2011). *C. jejuni*, one of the most important etiological agents of bacterial gastroenteritis worldwide, hosts two Hbs: a single domain globin named Cgb, and a group III 2/2HbP named Ctb. Although both globins are up-regulated by the transcription factor NssR (which regulates expression of a small regulon that includes *cgb* and *ctb*) in response to nitrosative stress (Elvers et al., 2005), only Cgb has been proposed to protect the bacterium against nitrosative stress, likely *via* a NO dioxygenase reaction during the initial stages, followed by a denitrosylase mechanism upon prolonged exposure to NO (Elvers, Wu, Gilberthorpe, Poole, & Park, 2004; Shepherd, Bernhardt, & Poole, 2011). The role of *C. jejuni* 2/2HbP is unclear: it is not directly involved in NO detoxification and it displays an extremely high O_2 affinity, making it unlikely to be an O_2 carrier. Based on the polarity of the haem distal cavity, reminiscent of that found in cytochrome *c* peroxidase, *C. jejuni* 2/2HbP has been proposed to be involved in (pseudo)enzymatic O_2 chemistry (Lu et al., 2007; Nardini et al., 2006; Wainwright, Elvers, Park, & Poole, 2005). Recently, attempts to define a function for *C. jejuni* 2/2HbP have been pursued by examining the effects of a *ctb* mutation on the NO transcriptome and *cgb* gene expression during normoxia and hypoxia. Based on these data, it was proposed that, by binding NO or O_2, *C. jejuni* 2/2HbP dampens the response to NO under hypoxic conditions and limits *cgb* expression, perhaps because Cgb function (i.e. NO detoxification) requires O_2-dependent chemistry (Smith, Shepherd, Monk, Green, & Poole, 2011).

It is worth noticing also that *C. jejuni* 2/2HbP displays the highest affinity as well as the fastest combination and the slowest dissociation rates for cyanide binding within the Hb superfamily. Thus, it was suggested that this 2/2HbP may act as a cyanide scavenger facilitating *C. jejuni* survival (Bolli et al., 2008).

6. CONCLUSIONS

2/2Hbs populate a branch of the Hb superfamily tree, which has been discovered and enriched with data starting in the early 1990s. Our current knowledge on these 'minimal' globins is rather extensive as far as their primary structures are concerned (more than 1000 gene sequences have been identified to date; Vinogradov et al., 2013). Such information allowed to distinguish groups I, II, and III 2/2Hbs and provided insight into their distribution through the evolutionary *phyla*. Indeed, 2/2Hbs have been mostly found among protozoa and bacteria, although their presence in plants and in some lower eukaryotes has also been reported. Despite the relatively contained size of 2/2Hbs in all three groups, several crystal structures have shown that careful editing of the 'classical' globin (i.e. Mb) 3-on-3 α-helical fold, and the introduction of Gly-based motifs (in groups I and II), provides efficient enclosure of the haem in a protein environment that is essentially based on four α-helices (hence the 2/2 helical fold acronym). A key feature emerging from the crystal structures is the presence (at least in groups I and II) of strategic and 2/2Hb-specific protein matrix tunnels or cavities, with conserved topology, which are held to support diffusion of small physiological ligands to/from the haem. The variety of distal site haem cavities, and the properties of the lining amino acids, is substantial and compatible with the display of different (pseudo)enzymatic activities. These are often related to detoxifying mechanisms devised by a pathogen in response to nitrogen and oxygen reactive species produced by the host. Other roles have been considered, although these are mainly hypothetic rather than proved *in vivo*. Notably, based on affinity of kinetic considerations, in just one case an intracellular O_2 transport/delivery role has been considered. The haem distal site in 2/2Hbs (as in Mb and Hb) is suited to bind small ligands, likely diatomic gaseous molecules, such as O_2, NO, and CO. Thus, compared to a classical enzyme acting on larger substrates, exploring 2/2Hb functions *in vivo* appears more complex, despite the extensive knowledge available on haem biochemistry. A further complicating factor is related to the pathogenicity of some 2/2Hb carrying microorganisms, or to the limited knowledge we have on their lifestyles and basic microbiology. After more than 15 years in this field, it is felt that one of the main open challanges is a thorough analysis and description of 2/2Hb *in vivo* functions.

REFERENCES

Arroyo Mañez, P., Lu, C., Boechi, L., Martí, M. A., Shepherd, M., Wilson, J. L., et al. (2011). Role of the distal hydrogen-bonding network in regulating oxygen affinity in the truncated hemoglobin III from *Campylobacter jejuni*. *Biochemistry, 50*, 3946–3956.

Ascenzi, P., Bocedi, A., Bolognesi, M., Fabozzi, G., Milani, M., & Visca, P. (2006). Nitric oxide scavenging by *Mycobacterium leprae* GlbO involves the formation of the ferric heme-bound peroxynitrite intermediate. *Biochemical and Biophysical Research Communications, 339*, 450–456.

Ascenzi, P., Bolognesi, M., Milani, M., Guertin, M., & Visca, P. (2007). Mycobacterial truncated hemoglobins: From genes to functions. *Gene, 398*, 42–51.

Ascenzi, P., Bolognesi, M., & Visca, P. (2007). *NO dissociation represents the rate limiting step for O_2-mediated oxidation of ferrous nitrosylated *Mycobacterium leprae* truncated hemoglobin O. *Biochemical and Biophysical Research Communications, 357*, 809–814.

Ascenzi, P., De Marinis, E., Coletta, M., & Visca, P. (2008). H_2O_2 and •NO scavenging by *Mycobacterium leprae* truncated hemoglobin O. *Biochemical and Biophysical Research Communications, 373*, 197–201.

Ascenzi, P., De Marinis, E., Visca, P., Ciaccio, C., & Coletta, M. (2009). Peroxynitrite detoxification by ferryl *Mycobacterium leprae* truncated hemoglobin O. *Biochemical and Biophysical Research Communications, 380*, 392–396.

Ascenzi, P., Milani, M., & Visca, P. (2006). Peroxynitrite scavenging by ferrous truncated hemoglobin GlbO from *Mycobacterium leprae*. *Biochemical and Biophysical Research Communications*, *351*, 528–533.

Ascenzi, P., & Visca, P. (2008). Scavenging of reactive nitrogen species by mycobacterial truncated hemoglobins. *Methods in Enzymology*, *436*, 317–337.

Bidon-Chanal, A., Martì, M. A., Crespo, A., Milani, M., Orozco, M., Bolognesi, M., et al. (2006). Ligand-induced dynamical regulation of NO conversion in *Mycobacterium tuberculosis* truncated hemoglobin-N. *Proteins*, *64*, 457–464.

Bidon-Chanal, A., Martì, M. A., Estrin, D. A., & Luque, F. J. (2007). Dynamical regulation of ligand migration by a gate-opening molecular switch in truncated hemoglobin-N from *Mycobacterium tuberculosis*. *Journal of the American Chemical Society*, *129*, 6782–6788.

Boechi, L., Mañez, P. A., Luque, F. J., Martì, M. A., & Estrin, D. A. (2010). Unraveling the molecular basis for ligand binding in truncated hemoglobins: The trHbO *Bacillus subtilis* case. *Proteins*, *78*, 962–970.

Boechi, L., Martì, M. A., Milani, M., Bolognesi, M., Luque, F. J., & Estrin, D. A. (2008). Structural determinants of ligand migration in *Mycobacterium tuberculosis* truncated hemoglobin O. *Proteins*, *73*, 372–379.

Bolli, A., Ciaccio, C., Coletta, M., Nardini, M., Bolognesi, M., Pesce, A., et al. (2008). Ferrous *Campylobacter jejuni* truncated hemoglobin P displays an extremely high reactivity for cyanide—A comparative study. *FEBS Journal*, *275*, 303–315.

Bolognesi, M., Bordo, D., Rizzi, M., Tarricone, C., & Ascenzi, P. (1997). Nonvertebrate hemoglobins: Structural bases for reactivity. *Progress in Biophysics and Molecular Biology*, *68*, 29–68.

Bonamore, A., Ilari, A., Giangiacomo, L., Bellelli, A., Morea, V., & Boffi, A. (2005). A novel thermostable hemoglobin from the actinobacterium *Thermobifida fusca*. *FEBS Journal*, *16*, 4189–4201.

Brunori, M., Cutruzzolà, F., Savino, C., Travaglini-Allocatelli, C., Vallone, B., & Gibson, Q. H. (1999). Structural dynamics of ligand diffusion in the protein matrix: A study on a new myoglobin mutant Y(B10) Q(E7) R(E10). *Biophysical Journal*, *76*, 1259–1269.

Cooper, A. M., Adams, L. B., Dalton, D. K., Appelberg, R., & Ehlers, S. (2002). IFN-gamma and NO in mycobacterial disease: New jobs for old hands. *Trends in Microbiology*, *10*, 221–226.

Coppola, D., Giordano, D., Tinajero-Trejo, M., di Prisco, G., Ascenzi, P., Poole, R. K., et al. (2013). Antarctic bacterial haemoglobin and its role in the protection against nitrogen reactive species. *Biochimica et Biophysica Acta – Proteins and Proteomics*, *1834*, 1923–1931.

Couture, M., Das, T. K., Lee, H. C., Peisach, J., Rousseau, D. L., Wittenberg, B. A., et al. (1999). Chlamydomonas chloroplast ferrous hemoglobin. Heme pocket structure and reactions with ligands. *The Journal of Biological Chemistry*, *274*, 6898–6910.

Couture, M., Das, T. K., Savard, P. Y., Ouellet, Y., Wittenberg, J. B., Wittenberg, B. A., et al. (2000). Structural investigations of the hemoglobin of the cyanobacterium *Synechocystis* PCC6803 reveal a unique distal heme pocket. *European Journal of Biochemistry*, *267*, 4770–4780.

Couture, M., & Guertin, M. (1996). Purification and spectroscopic characterization of a recombinant chloroplastic hemoglobin from the green unicellular alga *Chlamydomonas eugametos*. *European Journal of Biochemistry*, *242*, 779–787.

Couture, M., Yeh, S. R., Wittenberg, B. A., Wittenberg, J. B., Ouellet, Y., Rousseau, D. L., et al. (1999). A cooperative oxygen-binding hemoglobin from *Mycobacterium tuberculosis*. *Proceedings of the National Academy of Sciences of the United States of America*, *96*, 11223–11228.

Daigle, R., Rousseau, J. A., Guertin, M., & Lagüe, P. (2009). Theoretical investigations of nitric oxide channeling in *Mycobacterium tuberculosis* truncated hemoglobin N. *Biophysical Journal, 97*, 2967–2977.

Das, T. K., Weber, R. E., Dewilde, S., Wittenberg, J. B., Wittenberg, B. A., Yamauchi, K., et al. (2000). Ligand binding in the ferric and ferrous states of *Paramecium* hemoglobin. *Biochemistry, 39*, 14330–14340.

Droghetti, E., Nicoletti, F. P., Bonamore, A., Boechi, L., Arroyo Mañez, P., Estrin, D. A., et al. (2010). Heme pocket structural properties of a bacterial truncated hemoglobin from *Thermobifida fusca. Biochemistry, 49*, 10394–10402.

Elvers, K. T., Turner, S. M., Wainwright, L. M., Marsden, G., Hinds, J., Cole, J. A., et al. (2005). NssR, a member of the Crp-Fnr superfamily from *Campylobacter jejuni*, regulates a nitrosative stress-responsive regulon that includes both a single-domain and a truncated haemoglobin. *Molecular Microbiology, 57*, 735–750.

Elvers, K. T., Wu, G., Gilberthorpe, N. J., Poole, R. K., & Park, S. F. (2004). Role of an inducible single-domain hemoglobin in mediating resistance to nitric oxide and nitrosative stress in *Campylobacter jejuni* and *Campylobacter coli. Journal of Bacteriology, 186*, 5332–5341.

Fabozzi, G., Ascenzi, P., Renzi, S. D., & Visca, P. (2006). Truncated hemoglobin GlbO from *Mycobacterium leprae* alleviates nitric oxide toxicity. *Microbial Pathogenesis, 40*, 211–220.

Falzone, C. J., Vu, B. C., Scott, N. L., & Lecomte, J. T. (2002). The solution structure of the recombinant hemoglobin from the cyanobacterium *Synechocystis* sp. PCC 6803 in its hemichrome state. *Journal of Molecular Biology, 324*, 1015–1029.

Feis, A., Lapini, A., Catacchio, B., Brogioni, S., Foggi, P., Chiancone, E., et al. (2008). Unusually strong H-bonding to the heme ligand and fast geminate recombination dynamics of the carbon monoxide complex of *Bacillus subtilis* truncated hemoglobin. *Biochemistry, 47*, 902–910.

Fourrat, L., Iddar, A., Valverde, F., Serrano, A., & Soukri, A. (2007). Effects of oxidative and nitrosative stress on *Tetrahymena pyriformis* glyceraldehyde-3-phosphate dehydrogenase. *The Journal of Eukaryotic Microbiology, 54*, 338–346.

Giangiacomo, L., Ilari, A., Boffi, A., Morea, V., & Chiancone, E. (2005). The truncated oxygen-avid hemoglobin from *Bacillus subtilis*: X-ray structure and ligand binding properties. *The Journal of Biological Chemistry, 280*, 9192–9202.

Giordano, D., Parrilli, E., Dettaï, A., Russo, R., Barbiero, G., Marino, G., et al. (2007). The truncated hemoglobins in the Antarctic psychrophilic bacterium *Pseudoalteromonas haloplanktis* TAC125. *Gene, 398*, 69–77.

Goldstein, S., Lind, J., & Merényi, G. (2005). Chemistry of peroxynitrites and peroxynitrates. *Chemical Reviews, 105*, 2457–2470.

Guallar, V., Lu, C., Borrelli, K., Egawa, T., & Yeh, S. R. (2009). Ligand migration in the truncated hemoglobin-II from *Mycobacterium tubercolosis. The Journal of Biological Chemistry, 284*, 3106–3116.

Hill, D. R., Belbin, T. J., Thorsteinsson, M. V., Bassam, D., Brass, S., Ernst, A., et al. (1996). GlbN (cyanoglobin) is a peripheral membrane protein that is restricted to certain *Nostoc* spp. *Journal of Bacteriology, 178*, 6587–6598.

Hiner, A. N., Raven, E. L., Thorneley, R. N., García-Cánovas, F., & Rodríguez-López, J. N. (2002). Mechanisms of compound I formation in heme peroxidases. *Journal of Inorganic Biochemistry, 91*, 27–34.

Hoy, J. A., Kundu, S., Trent, J. T., 3rd., Ramaswamy, S., & Hargrove, M. S. (2004). The crystal structure of *Synechocystis* hemoglobin with a covalent heme linkage. *The Journal of Biological Chemistry, 279*, 16535–16542.

Igarashi, J., Kobayashi, K., & Matsuoka, A. (2011). A hydrogen-bonding network formed by the B10-E7-E11 residues of a truncated hemoglobin from *Tetrahymena pyriformis* is critical for stability of bound oxygen and nitric oxide detoxification. *Journal of Biological Inorganic Chemistry*, *16*, 599–609.

Ilari, A., Kjelgaard, P., von Wachenfeldt, C., Catacchio, B., Chiancone, E., & Boffi, A. (2007). Crystal structure and ligand binding properties of the truncated hemoglobin from *Geobacillus stearothermophilus*. *Archives of Biochemistry and Biophysics*, *457*, 85–94.

Kapp, O. H., Moens, L., Vanfleteren, J., Trotman, C. N., Suzuki, T., & Vinogradov, S. N. (1995). Alignment of 700 globin sequences: Extent of amino acid substitution and its correlation with variation in volume. *Protein Science*, *10*, 2179–2190.

Lama, A., Pawaria, S., Bidon-Chanal, A., Anand, A., Gelpí, J. L., Arya, S., et al. (2009). Role of Pre-A motif in nitric oxide scavenging by truncated hemoglobin, HbN, of *Mycobacterium tuberculosis*. *The Journal of Biological Chemistry*, *284*, 14457–14468.

Lama, A., Pawaria, S., & Dikshit, K. L. (2006). Oxygen binding and NO scavenging properties of truncated hemoglobin, HbN, of *Mycobacterium smegmatis*. *FEBS Letters*, *580*, 4031–4041.

Larsson, J. T., Rogstam, A., & von Wachenfeldt, C. (2007). YjbH is a novel negative effector of the disulfide stress regulator, Spx, in *Bacillus subtilis*. *Molecular Microbiology*, *66*, 669–684.

Lecomte, J. T., Vu, B. C., & Falzone, C. J. (2005). Structural and dynamic properties of *Synechocystis* sp. PCC 6803 Hb revealed by reconstitution with Zn-protoporphyrin IX. *Journal of Inorganic Biochemistry*, *99*, 1585–1592.

Lecomte, J. T., Vuletich, D. A., Vu, B. C., Kuriakose, S. A., Scott, N. L., & Falzone, C. J. (2004). Structural properties of cyanobacterial hemoglobins: The unusual heme-protein cross-link of *Synechocystis* sp. PCC 6803 Hb and *Synechococcus* sp. PCC 7002 Hb. *Micron*, *35*, 71–72.

Liu, C., He, Y., & Chang, Z. (2004). Truncated hemoglobin O of *Mycobacterium tuberculosis*: The oligomeric state change and the interaction with membrane components. *Biochemical and Biophysical Research Communications*, *316*, 1163–1172.

Lu, C., Egawa, T., Wainwright, L. M., Poole, R. K., & Yeh, S. R. (2007). Structural and functional properties of a truncated haemoglobin from a food-borne pathogen *Campylobacter jejuni*. *The Journal of Biological Chemistry*, *282*, 13627–13636.

MacMicking, J. D., North, R. J., LaCourse, R., Mudgett, J. S., Shah, S. K., & Nathan, C. F. (1997). Identification of nitric oxide synthase as a protective locus against tuberculosis. *Proceedings of the National Academy of Sciences of the United States of America*, *94*, 5243–5248.

Martì, M. A., Capece, L., Bidon-Chanal, A., Crespo, A., Guallar, V., Luque, F. J., et al. (2008). Nitric oxide reactivity with globins as investigated through computer simulation. *Methods in Enzymology*, *437*, 477–498.

Milani, M., Ouellet, Y., Ouellet, H., Guertin, M., Boffi, A., Antonini, G., et al. (2004). Cyanide binding to truncated hemoglobins: A crystallographic and kinetic study. *Biochemistry*, *43*, 5213–5221.

Milani, M., Pesce, A., Nardini, M., Ouellet, H., Ouellet, Y., Dewilde, S., et al. (2005). Structural bases for heme binding and diatomic ligand recognition in truncated hemoglobins. *Journal of Inorganic Biochemistry*, *99*, 97–109.

Milani, M., Pesce, A., Ouellet, Y., Ascenzi, P., Guertin, M., & Bolognesi, M. (2001). *Mycobacterium tuberculosis* hemoglobin N displays a protein tunnel suited for O_2 diffusion to the heme. *EMBO Journal*, *20*, 3902–3909.

Milani, M., Pesce, A., Ouellet, Y., Dewilde, S., Friedman, J., Ascenzi, P., et al. (2004). Heme-ligand tunneling in group I truncated hemoglobins. *The Journal of Biological Chemistry*, *279*, 21520–21525.

Milani, M., Savard, P. Y., Ouellet, H., Ascenzi, P., Guertin, M., & Bolognesi, M. (2003). A TyrCD1/TrpG8 hydrogen bond network and a TyrB10TyrCD1 covalent link shape the heme distal site of *Mycobacterium tuberculosis* hemoglobin O. *Proceedings of the National Academy of Sciences of the United States of America*, *100*, 5766–5771.

Mishra, S., & Meuwly, M. (2010). Atomistic simulation of NO dioxygenation in group I truncated hemoglobin. *Journal of the American Chemical Society, 132,* 2968–2982.

Mukai, M., Savard, P. Y., Ouellet, H., Guertin, M., & Yeh, S. R. (2002). Unique ligand-protein interactions in a new truncated hemoglobin from *Mycobacterium tuberculosis. Biochemistry, 41,* 3897–3905.

Nakajima, S., Álvarez-Salgado, E., Kikuchi, T., & Arredondo-Peter, R. (2005). Prediction of folding pathway and kinetics among plant hemoglobins using an average distance map method. *Proteins: Structure, Function, and Bioinformatics, 61,* 500–506.

Nardini, M., Pesce, A., Labarre, M., Richard, C., Bolli, A., Ascenzi, P., et al. (2006). Structural determinants in the group III truncated hemoglobin from *Campylobacter jejuni. The Journal of Biological Chemistry, 281,* 37803–37812.

Nardini, M., Pesce, A., Milani, M., & Bolognesi, M. (2007). Protein fold and structure in the truncated (2/2) globin family. *Gene, 398,* 2–11.

Nathan, C., & Shiloh, M. U. (2000). Reactive oxygen and nitrogen intermediates in the relationship between mammalian hosts and microbial pathogens. *Proceedings of the National Academy of Sciences of the United States of America, 97,* 8841–8848.

Nicoletti, F. P., Comandini, A., Bonamore, A., Boechi, L., Boubeta, F. M., Feis, A., et al. (2010). Sulfide binding properties of truncated hemoglobins. *Biochemistry, 49,* 2269–2278.

Nicoletti, F. P., Droghetti, E., Boechi, L., Bonamore, A., Sciamanna, N., Estrin, D. A., et al. (2011). Fluoride as a probe for H-bonding interactions in the active site of heme proteins: The case of *Thermobifida fusca* hemoglobin. *Journal of the American Chemical Society, 133,* 20970–20980.

Nothnagel, H. J., Love, N., & Lecomte, J. T. (2009). The role of the heme distal ligand in the post-translational modification of *Synechocystis* hemoglobin. *Journal of Inorganic Biochemistry, 103,* 107–116.

Nothnagel, H. J., Preimesberger, M. R., Pond, M. P., Winer, B. Y., Adney, E. M., & Lecomte, J. T. (2011). Chemical reactivity of *Synechococcus* sp. PCC 7002 and *Synechocystis* sp. PCC 6803 hemoglobins: Covalent heme attachment and bishistidine coordination. *Journal of Biological Inorganic Chemistry, 16,* 539–552.

Nothnagel, H. J., Winer, B. Y., Vuletich, D. A., Pond, M. P., & Lecomte, J. T. (2011). Structural properties of 2/2 hemoglobins: The group III protein from *Helicobacter hepaticus. IUBMB Life, 63,* 197–205.

Ohno, H., Zhu, G., Mohan, V. P., Chu, D., Kohno, S., Jacobs, W. R., Jr., et al. (2003). The effects of reactive nitrogen intermediates on gene expression in *Mycobacterium tuberculosis. Cellular Microbiology, 5,* 637–648.

Oliveira, A., Singh, S., Bidon-Chanal, A., Forti, F., Martì, M. A., Boechi, L., et al. (2012). Role of PheE15 gate in ligand entry and nitric oxide detoxification function of *Mycobacterium tuberculosis* truncated hemoglobin N. *PLoS One, 7,* e49291.

Ouellet, Y. H., Daigle, R., Lagüe, P., Dantsker, D., Milani, M., Bolognesi, M., et al. (2008). Ligand binding to truncated hemoglobin N from *Mycobacterium tuberculosis* is strongly modulated by the interplay between the distal heme pocket residues and internal water. *The Journal of Biological Chemistry, 283,* 27270–27278.

Ouellet, H., Juszczak, L., Dantsker, D., Samuni, U., Ouellet, Y. H., Savard, P. Y., et al. (2003). Reactions of *Mycobacterium tuberculosis* truncated hemoglobin O with ligands reveal a novel ligand-inclusive hydrogen bond network. *Biochemistry, 42,* 5764–5774.

Ouellet, Y., Milani, M., Couture, M., Bolognesi, M., & Guertin, M. (2006). Ligand interactions in the distal heme pocket of *Mycobacterium tuberculosis* truncated hemoglobin N: Roles of TyrB10 and GlnE11 residues. *Biochemistry, 45,* 8770–8781.

Ouellet, H., Milani, M., Labarre, M., Bolognesi, M., Couture, M., & Guertin, M. (2007). The roles of Tyr(CD1) and Trp(G8) in *Mycobacterium tuberculosis* truncated hemoglobin

O in ligand binding and on the heme distal site architecture. *Biochemistry, 46,* 11440–11450.

Ouellet, H., Ouellet, Y., Richard, C., Labarre, M., Wittenberg, B., Wittenberg, J., et al. (2002). Truncated hemoglobin HbN protects *Mycobacterium bovis* from nitric oxide. *Proceedings of the National Academy of Sciences of the United States of America, 99,* 5902–5907.

Parrilli, E., Giuliani, M., Giordano, D., Russo, R., Marino, G., Verde, C., et al. (2010). The role of a 2-on-2 haemoglobin in oxidative and nitrosative stress resistance of Antarctic *Pseudoalteromonas haloplanktis* TAC125. *Biochimie, 92,* 1003–1009.

Pathania, R., Navani, N. K., Gardner, A. M., Gardner, P. R., & Dikshit, K. L. (2002). Nitric oxide scavenging and detoxification by the *Mycobacterium tuberculosis* haemoglobin, HbN in *Escherichia coli. Molecular Microbiology, 45,* 1303–1314.

Pathania, R., Navani, N. K., Rajamohan, G., & Dikshit, K. L. (2002). *Mycobacterium tuberculosis* hemoglobin HbO associates with membranes and stimulates cellular respiration of recombinant *Escherichia coli. The Journal of Biological Chemistry, 277,* 15293–15302.

Pawaria, S., Rajamohan, G., Gambhir, V., Lama, A., Varshney, G. C., & Dikshit, K. L. (2007). Intracellular growth and survival of *Salmonella* enterica serovar *Typhimurium* carrying truncated hemoglobins of *Mycobacterium tuberculosis. Microbial Pathogenesis, 42,* 119–128.

Perutz, M. F. (1979). Regulation of oxygen affinity of hemoglobin: Influence of structure of the globin on the heme iron. *Annual Review of Biochemistry, 48,* 327–386.

Pesce, A., Couture, M., Dewilde, S., Guertin, M., Yamauchi, K., Ascenzi, P., et al. (2000). A novel two-over-two alpha-helical sandwich fold is characteristic of the truncated hemoglobin family. *EMBO Journal, 19,* 2424–2434.

Pesce, A., Milani, M., Nardini, M., & Bolognesi, M. (2008). Mapping heme-ligand tunnels in group I truncated(2/2) hemoglobins. *Methods in Enzymology, 436,* 303–315.

Pesce, A., Nardini, M., Labarre, M., Richard, C., Wittenberg, J. B., Wittenberg, B. A., et al. (2011). Structural characterization of a group II 2/2 hemoglobin from the plant pathogen *Agrobacterium tumefaciens. Biochimica et Biophysica Acta – Proteins and Proteomics, 1814,* 810–816.

Pond, M. P., Majumdar, A., & Lecomte, J. T. (2012). Influence of heme post-translational modification and distal ligation on the backbone dynamics of a monomeric hemoglobin. *Biochemistry, 51,* 5733–5747.

Ptitsyn, O. B., & Ting, K. L. (1999). Non-functional conserved residues in globins and their possible role as a folding nucleus. *Journal of Molecular Biology, 291,* 671–682.

Raushel, F. M., Thoden, J. B., & Holden, H. M. (2003). Enzymes with molecular tunnels. *Accounts of Chemical Research, 36,* 539–548.

Rizzi, M., Wittenberg, J. B., Coda, A., Ascenzi, P., & Bolognesi, M. (1996). Structural bases for sulfide recognition in *Lucina pectinata* hemoglobin I. *Journal of Molecular Biology, 258,* 1–5.

Samuni, U., Dankster, D., Ray, A., Wittenberg, J. B., Wittenberg, B. A., Dewilde, S., et al. (2003). Kinetic modulation in carbonmonoxy derivatives of truncated hemoglobins: The role of distal heme pocket residues and extended apolar tunnel. *The Journal of Biological Chemistry, 278,* 27241–27250.

Schnappinger, D., Schoolnik, G. K., & Ehrt, S. (2006). Expression profiling of host pathogen interactions: How *Mycobacterium tuberculosis* and the macrophage adapt to one another. *Microbes and Infection, 8,* 1132–1140.

Scott, N. L., Falzone, C. J., Vuletich, D. A., Zhao, J., Bryant, D. A., & Lecomte, J. T. (2002). Truncated hemoglobin from the cyanobacterium *Synechococcus* sp. PCC 7002: evidence for hexacoordination and covalent adduct formation in the ferric recombinant protein. *Biochemistry, 41,* 6902–6910.

Scott, N. L., Xu, Y., Shen, G., Vuletich, D. A., Falzone, C. J., Li, Z., et al. (2010). Functional and structural characterization of the 2/2 hemoglobin from *Synechococcus* sp. PCC 7002. *Biochemistry, 49,* 7000–7011.

Shepherd, M., Bernhardt, P. V., & Poole, R. K. (2011). Globin-mediated nitric oxide detoxification in the foodborne pathogenic bacterium *Campylobacter jejuni* proceeds via a dioxygenase or denitrosylase mechanism. *Nitric Oxide, 25*, 229–233.

Smith, H. K., Shepherd, M., Monk, C., Green, J., & Poole, R. K. (2011). The NO-responsive hemoglobins of *Campylobacter jejuni*: Concerted responses of two globins to NO and evidence *in vitro* for globin regulation by the transcription factor NssR. *Nitric Oxide, 25*, 234–241.

Trent, J. T., 3rd., Kundu, S., Hoy, J. A., & Hargrove, M. S. (2004). Crystallographic analysis of *synechocystis* cyanoglobin reveals the structural changes accompanying ligand binding in a hexacoordinate hemoglobin. *Journal of Molecular Biology, 341*, 1097–1108.

Vinogradov, S. N., Hoogewijs, D., Bailly, X., Arredondo-Peter, R., Gough, J., Dewilde, S., et al. (2006). A phylogenomic profile of globins. *BMC Evolutionary Biology, 6*, 31–47.

Vinogradov, S. N., & Moens, L. (2008). Diversity of globin function: Enzymatic, transport, storage, and sensing. *The Journal of Biological Chemistry, 283*, 8773–8777.

Vinogradov, S. N., Tinajero-Trejo, M., Poole, R. K., & Hoogewijs, D. (2013). Bacterial and archaeal globins—A revised perspective. *Biochimica et Biophysica Acta – Proteins and Proteomics, 1834*, 1789–1800.

Visca, P., Fabozzi, G., Milani, M., Bolognesi, M., & Ascenzi, P. (2002). Nitric oxide and *Mycobacterium leprae* pathogenicity. *IUBMB Life, 54*, 95–99.

Visca, P., Fabozzi, G., Petrucca, A., Ciaccio, C., Coletta, M., De Sanctis, G., et al. (2002). The truncated hemoglobin from *Mycobacterium leprae*. *Biochemical and Biophysical Research Communications, 294*, 1064–1070.

Vu, B. C., Nothnagel, H. J., Vuletich, D. A., Falzone, C. J., & Lecomte, J. T. J. (2004). Cyanide binding to hexacoordinate cyanobacterial hemoglobins: Hydrogen-bonding network and heme pocket rearrangement in ferric H117A *Synechocystis* hemoglobin. *Biochemistry, 43*, 12622–12633.

Vu, B. C., Vuletich, D. A., Kuriakose, S. A., Falzone, C. J., & Lecomte, J. T. (2004). Characterization of the heme-histidine crosslink in cyanobacterial hemoglobins from *Synechocystis* sp. PCC 6803 and *Synechococcus* sp. PCC 7002. *Journal of Biological Inorganic Chemistry, 9*, 183–194.

Vuletich, D. A., Falzone, C. J., & Lecomte, J. T. (2006). Structural and dynamic repercussions of heme binding and heme-protein cross-linking in *Synechococcus* sp. PCC 7002 hemoglobin. *Biochemistry, 45*, 14075–14084.

Vuletich, D. A., & Lecomte, J. T. (2006). A phylogenetic and structural analysis of truncated hemoglobins. *Journal of Molecular Evolution, 62*, 196–210.

Wainwright, L. M., Elvers, K. T., Park, S. F., & Poole, R. K. (2005). A truncated haemoglobin implicated in oxygen metabolism by the microaerophilic food-borne pathogen *Campylobacter jejuni*. *Microbiology (Reading, England), 151*, 4079–4091.

Weeks, A., Lund, L., & Raushel, F. M. (2006). Tunneling of intermediates in enzyme-catalyzed reactions. *Current Opinion in Chemical Biology, 10*, 465–472.

Wittenberg, J. B., Bolognesi, M., Wittenberg, B. A., & Guertin, M. (2002). Truncated hemoglobins: A new family of hemoglobins widely distributed in bacteria, unicellular eukaryotes, and plants. *The Journal of Biological Chemistry, 277*, 871–874.

Yeh, S. R., Couture, M., Ouellet, Y., Guertin, M., & Rousseau, D. L. (2000). A cooperative oxygen binding hemoglobin from *Mycobacterium tuberculosis*. Stabilization of heme ligands by a distal tyrosine residue. *The Journal of Biological Chemistry, 275*, 1679–1684.

CHAPTER THREE

Protoglobin: Structure and Ligand-Binding Properties

Alessandra Pesce[*], Martino Bolognesi[†,‡], Marco Nardini[†,1]
*Department of Physics, University of Genova, Genova, Italy
†Department of Biosciences, University of Milano, Milano, Italy
‡CIMAINA and CNR Institute of Biophysics, Milano, Italy
[1]Corresponding author: e-mail address: marco.nardini@unimi.it

Contents

1. Introduction   80
2. Overall Structure   82
3. The Two-Tunnel System   85
4. The Haem Environment   87
5. Biochemical and Functional Characterisation   90
6. Conclusions   93
References   94

Abstract

Protoglobin is the first globin identified in Archaea; its biological role is still unknown, although it can bind O_2, CO and NO reversibly *in vitro*. The X-ray structure of *Methanosarcina acetivorans* protoglobin revealed several peculiar structural features. Its tertiary structure can be considered as an expanded version of the canonical globin fold, characterised by the presence of a pre-A helix (named Z) and a 20-residue N-terminal extension. Other unusual trends are a large distortion of the haem moiety, and its complete burial in the protein matrix due to the extended CE and FG loops and the 20-residue N-terminal loop. Access of diatomic ligands to the haem has been proposed to be granted by two tunnels, which are mainly defined by helices B/G (tunnel 1) and B/E (tunnel 2), and whose spatial orientation and topology give rise to an almost orthogonal two-tunnel system unprecedented in other globins. At a quaternary level, protoglobin forms a tight dimer, mostly based on the inter-molecular four-helix bundle built by the G- and H-helices, similar to that found in globin-coupled sensor proteins, which share with protoglobin a common phylogenetic origin. Such unique structural properties, together with an unusually low O_2 dissociation rate and a selectivity ratio for O_2/CO binding that favours O_2 ligation, make protoglobin a peculiar case for gaining insight into structure to function relationships within the globin superfamily. While recent structural and biochemical data have given answers to important questions, the functional issue is still unclear and it is expected to represent the major focus of future investigations.

Advances in Microbial Physiology, Volume 63
ISSN 0065-2911
http://dx.doi.org/10.1016/B978-0-12-407693-8.00003-0

ABBREVIATIONS

2/2Hb truncated globin
ApPgb Pgb from *Aeropyrum pernix*
FHb flavo-haemoglobin
GCS globin-coupled sensor
Hb haemoglobin
MaPgb Pgb from *Methanosarcina acetivorans*
MaPgb* *Methanosarcina acetivorans* protoglobin whose Cys(101)E20 is replaced by Ser; amino acid residues have been labeled using their three-letter codes the sequence numbering (in parentheses), and the topological site they occupy within the globin fold
MaPgb*-**ΔN20** *Methanosarcina acetivorans* protoglobin whose the first 20 amino acids at the N-terminal have been deleted
MaPgb*-**ΔN20Z** *Methanosarcina acetivorans* protoglobin whose 33 N-terminal amino acids have been removed
Mb myoglobin
MCP methyl-accepting chemotaxis protein
Pgb protoglobin
SDgb single-domain globin
SDSgb single-domain sensor globin

1. INTRODUCTION

Much of the diversity within the globin superfamily has been disclosed by progresses achieved in the past 2 decades, which unravelled several novel haemoglobins (Hbs) endowed with distinct structural and functional features. A survey of putative globins in sequenced genomes within all kingdoms of life showed the presence of three globin lineages and two structural families (Vinogradov et al., 2006, 2007; Vinogradov, Tinajero-Trejo, Poole, & Hoogewijs, 2013). One lineage includes the chimeric flavo-haemoglobins (FHbs), comprising an N-terminal globin domain and a C-terminal FAD- and NAD-binding domain, homologous to the ferredoxin–NADP+ reductases and related single-domain globins (SDgbs); a second lineage includes two-domain globin-coupled sensors (GCS), single-domain sensor globins (SDSgbs) and a small number of related single-domain protoglobins (Pgbs) (Freitas, Hou, & Alam, 2003; Freitas et al., 2004; Freitas, Saito, Hou, & Alam, 2005; Hou et al., 2001). All members of these two globin families belong to the same structural class exhibiting the canonical 3/3 globin fold found in myoglobin (Mb), with the haem group surrounded by the A-, B- and E-helices roughly on one side and the F-, G- and H-helices on the other (Bolognesi, Bordo, Rizzi,

Tarricone, & Ascenzi, 1997; Perutz, 1979). The third lineage encompasses the "truncated" globins that are structurally distinct from 3/3 globins. This lineage consists of three subgroups, all sharing an abbreviated single globin domain folded as a 2/2 α-helical sandwich (2/2Hb), characterised by a very short or absent A-helix, a brief CE inter-helical region and most of the F-helix occurring as a loop, with only the B-, E-, G- and H-helices left to surround the haem group (Nardini, Pesce, Milani, & Bolognesi, 2007; Pesce, Bolognesi, & Nardini, 2013; Vuletich & Lecomte, 2006).

Pgbs belong phylogenetically to the same cluster as the GCS group within the globin superfamily. However, while GCSs are chimeric haem-proteins, tentatively classified either as aerotactic or gene regulating, which couple a globin-like sensor domain (usually larger than that of Mb) to a transmitter domain of variable structure and function, Pgbs are single-domain variants devoid of the transmitter domain (Freitas et al., 2003, 2004, 2005; Hou et al., 2001).

Pgb derives its name from a very first phylogenetic analysis of the GCS family, which suggested that each GCS globin domain evolved independently, with its particular signalling partner, and predicted the existence of an ancestor globin, or Pgb, within more primitive organisms. As such, Pgb was first characterised in two Archaea, in the obligate aerobic hyper-thermophile *Aeropyrum pernix* (*Ap*Pgb) and in the strictly anaerobic meth-anogen *Methanosarcina acetivorans* (*Ma*Pgb). At that time the SDSgbs had not been identified yet, and Pgb was postulated to be the single-domain ancestor not only for GCS but also possibly for all contemporary Hbs (Freitas et al., 2004). Such prototypic view of Pgb soon appeared unlikely, based on more extended surveys of putative globins in sequenced genomes: the identification of SDSgbs within the GCS lineage that could explain the emergence of chimeric GCSs, the surprisingly distant relationship between Pgbs and chimeric sensor sequences, and the sequence homogeneity among Pgbs support the idea that Pgbs are a relatively recent group of globins emerged in globin evolution (Vinogradov et al., 2007). Within this evolu-tionary scenario, a model for globin evolution finding its root in a bacterial FHb-like single-domain ancestral globin was proposed (Vinogradov et al., 2007). Thus, the FHb family globins displaying enzymatic functions in the early bacteria are held to have evolved *via* horizontal gene transfer in multi-cellular eukaryotes; such event brought about new properties, includ-ing reversible binding of important diatomic ligands, such as O_2, NO and sulphide, enabling the evolution of transport and storage functions (Vinogradov & Moens, 2008).

Sixteen Pgbs have been identified to date in Archaea (five Pgbs) and Bacteria (Hoogewijs, Dewilde, Vierstraete, Moens, & Vinogradov, 2012). Although the physiological role of Pgb is totally unknown, as is the nature of its physiological ligand(s), over the past few years several progresses have been made in the structural, biochemical and biophysical characterisation of this unusual globin, which are reviewed in the following sections.

2. OVERALL STRUCTURE

To date only one Pgb three-dimensional structure, from *M. acetivorans*, has been reported (Nardini et al., 2008). The protein bears the Cys(101)E20 → Ser mutation that was engineered for crystallisation purposes, and hereafter will be simply termed *Ma*Pgb*.

*Ma*Pgb* is folded as an expanded version of the canonical globin fold (i.e. the Mb fold), consisting of a 3-on-3 α-helical sandwich built by the AGH–BEF helices, with the support of the C- and D-helices (Bolognesi et al., 1997; Perutz, 1979). Contrary to the classical globin fold, however, in *Ma*Pgb*, the D-helix is absent, an additional N-terminal Z-helix precedes the A-helix, and a short H′-helix (one helical turn) falls close to the C-terminus. The Z-helix is further preceded by 20 amino acids held next to the haem propionates by hydrogen bonds that link Pro(7), Gly(8), Tyr(9), Thr(10), Ala(18) and Phe(20) to residues of the E- and F-helices in the protein α-helical core. As a consequence, this N-terminal loop together with other extended loops connecting the C-helix with the E-helix and the F-helix with the G-helix (which all are longer than in classical globins and are conserved in known Pgb amino acid sequences) completely bury the haem within the protein matrix, and the haem propionates are solvent inaccessible (Fig. 3.1A). Such structural feature is very unusual within the globin family 3D structures, where approximately 30% of the haem surface is usually solvent accessible. Thus, ligand diffusion to the haem pocket in *Ma*Pgb* must occur *via* paths different from direct haem access through the E7 residue gate. The proposed alternative ligand diffusion paths in *Ma*Pgb* are located between the B- and G-helices (tunnel 1), and between the B- and E-helices (tunnel 2) (Fig. 3.1B).

The location and extention of the Z-helix in the Pgb fold are reminiscent of the globin domain of the haem-based O_2 sensor responsible for aerotaxis in aerobic *Bacillus subtilis* (Zhang & Phillips, 2003), and of the GCS globin domain from the strictly anaerobic δ-proteobacterium *Geobacter sulfurreducens* (Pesce et al., 2009), with which *Ma*Pgb* shares approximately

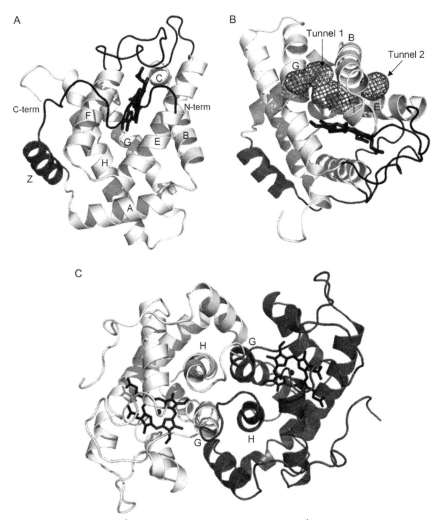

Figure 3.1 The *Ma*Pgb* fold. (A) Tertiary structure of *Ma*Pgb*. Secondary structure elements conforming to the classical 3/3 globin fold are shown in white (labels A through H). Structural elements specific of *Ma*Pgb* are displayed in grey (Z-helix) and in black (N-terminal, CE and FG loops). (B) The N-terminal region, the CE and FG loops bury the haem (black) and prevent access of small ligands to the haem distal cavity, which is connected to the solvent region by tunnel 1 and tunnel 2 (grey mesh). (C) Quaternary structure of *Ma*Pgb*. The two *Ma*Pgb* subunits within the dimer, interacting mainly through the G- and H-helices (helical bundle), are shown in white and grey (haem in black). (For colour version of this figure, the reader is referred to the online version of this chapter.)

10% sequence identity. However, major structural differences are present between $MaPgb^*$ and the globin domain of GCSs at the 20 N-terminal loop, which extend in opposite directions, and at the CE and FG loop regions (where $MaPgb^*$ shows 6- and 12-residue insertions, respectively). These differences have important structural implications for the haem-binding pocket of the GCS globin domain, where the haem is solvent accessible from the propionates edge, whereas specific mutations and local shifts of the B-, E- and G-helices (particularly in *G. sulfurreducens* GCS, which is hexacoordinated) prevent the formation of the apolar tunnels observed in the Pgb fold.

The structural role of the Pgb-specific N-terminal region was analysed by deletion mutants where (i) the first 20 residues are truncated ($MaPgb^*$-ΔN20 mutant) and (ii) the 20 N-terminal residues and the following Z-helix, a protein stretch covering residues 1–33, are omitted ($MaPgb^*$-ΔN20Z mutant) (Ciaccio et al., 2013). A comparative analysis of the $MaPgb^*$-ΔN20 and $MaPgb^*$ crystal structures suggests that the 20 N-terminal residues are not required to attain the correct Pgb fold. On the other hand, deletion of the 20 N-terminal residues partly uncovers the haem cavity, which is instead blocked by such N-terminal segment in the full-length protein. Such engineered haem accessibility may provide an alternative route for haem/ligand exchange in $MaPgb^*$-ΔN20, and, indeed, it appears to facilitate haem access for an exogenous ligand, such as azide, as well as the exchange of solvent molecules between the haem pocket and the bulk solution (Ciaccio et al., 2013). The structural details provided by the crystallographic investigation are also in agreement with CD spectra in the UV region, which suggest that the $MaPgb^*$-ΔN20 mutant maintains a secondary structure content matching that of $MaPgb^*$. Moreover, the CD spectra in the Soret region indicate minor perturbation of the haem cavity in $MaPgb^*$-ΔN20, which could result from solvent exposure or exchange in the haem region uncovered by deletion of the 20 N-terminal residues.

On the contrary, in the $MaPgb^*$-ΔN20Z mutant, a dramatic structural change occurs upon removal of the 33 N-terminal residues, such that the protein appears to collapse, partly bringing about a relevant alteration of the haem distal side, with the likely formation of a hexacoordinated haem, both in the ferric and ferrous species, and direct implications for ligand binding. Indeed, modelling studies indicate that the contemporary absence of the N-terminal loop and of the Z-helix would expose to the solvent a wide hydrophobic surface from the protein core, thus inducing the dramatic structural re-arrangements that are indicated by the spectral changes observed in the UV and Soret regions of the CD spectra (Ciaccio et al., 2013).

At the quaternary structure level, MaPgb* is a dimer both in solution and in the crystal (Abbruzzetti et al., 2012; Nardini et al., 2008). The core of the association interface (>2000 Å^2) is provided mostly by residues belonging to the G- and H-helices, which build an inter-molecular four-helix bundle, with the additional contribution of the Z-helices, and of the BC and FG hinges (Fig. 3.1C). Overall, the quaternary structure of MaPgb* strongly resembles that of the GCS globin domain (Pesce et al., 2009; Zhang & Phillips, 2003).

Molecular dynamics simulations revealed that the MaPgb* dimer is more stable than the monomeric form, although the dissociation of the dimer into separate monomers does not promote a substantial alteration in the overall fold of the protein, but induces local re-arrangements in selected structural elements, for instance, a relevant displacement of the G-helix and, to a lower extent, of the H-helix. Such finding was not unexpected, as both helices are directly involved in defining the dimeric four-helix bundle interface and dissociation of the dimer increases dramatically their exposure to the aqueous solvent, thus promoting conformational relaxation of the monomeric species compared to the dimer. It is worth noting that the molecular dynamics results indicate that access of ligands through the two tunnels is only granted in the dimeric form, as tunnel 1 is primarily closed in the monomeric species. Strikingly, however, when the same analysis is performed for the oxygenated forms of the protein, significant structural re-arrangements are also identified for the B- and E-helices. As both these helical segments delineate the two tunnels in the MaPgb* protein matrix, it was suggested that ligand migration is likely affected not only by the dimeric assembly of the protein but also by the binding of ligands to the haem cavity (Forti et al., 2011).

3. THE TWO-TUNNEL SYSTEM

Access of diatomic ligands, such as O_2, CO and NO, to the fully buried MaPgb* haem is granted by two orthogonal apolar tunnels that reach the haem distal region from entry sites at the B/G (tunnel 1) and B/E (tunnel 2) helix interfaces (Fig. 3.1B). Notably, the topology of such two-tunnel system is a unique feature of Pgb, totally unrelated with the ligand diffusion path through the E7-gate first reported for Mb (Bolognesi et al., 1982), and with the system of internal cavities/tunnels found in other globins (Nardini et al., 2007; Salter et al., 2008).

Tunnel 1 is a straight apolar protein matrix tunnel ($\sim$7 Å in diameter and $\sim$16 Å in length), lined by residues Ile(56)B5, Thr(59)B8, Trp(60)B9, Phe(63)B12, Phe(93)E11, Ile(97)E15, Phe(145)G7, Pro(148)G10, Ile(149) G11, Thr(152)G14 and Met(153)G15. Tunnel 2 ($\sim$5 Å in diameter and $\sim$10 Å in length) is a straight and short opening to the haem distal cavity nestled among residues Tyr(61)B10, Leu(64)B13, Gly(65)B14, Leu(71) C5, Phe(74)CD1 and Leu(86)E4. Both tunnels host one water molecule at their solvent side aperture. In addition, a core cavity of $\sim$75 Å^3 is located between the distal and proximal haem sides; the cavity hosts four mutually hydrogen-bonded water molecules. All the residues lining the protein tunnels and the inner cavity are conserved in known Pgbs, which would be consistent with their implication in diatomic ligand diffusion to and from the haem, multi-ligand storage and/or (pseudo-)enzymatic reactivity. Such functional roles would rely on structural principles that are entirely different from those shown for Xenon cavities in Mb and in truncated 2/2Hbs (Brunori & Gibson, 2001; Milani et al., 2005). Indeed, the structures of ferric MaPgb* in complex with Xenon (in the presence of cyanide or azide as haem Fe-ligand) clearly show a Xenon atom trapped inside tunnel 1, in a hydrophobic cavity efficiently sealed by Trp(60)B9 side chain, which moves into the distal site upon cyanide/azide binding (see below). On the contrary, no Xenon atoms bind at tunnel 2, due to its short length and more hydrophilic nature (Pesce, Tilleman, et al., 2013).

Molecular dynamics simulations showed that while tunnel 2 is always accessible to diatomic ligands in both deoxygenated and oxygenated forms of the protein, the accessibility of tunnel 1 is controlled through the synergistic effect of both the ligation and the oligomerization states of the protein. In particular, steric hindrance between Phe(93)E11 and the haem-bound ligand would alter the structural and dynamical behaviour of the B- and E-helices, thus facilitating the opening of tunnel 1, while dimerisation would affect the spatial organisation of the G-helix, which, in turn, would modify the structure of tunnel 1 (Forti et al., 2011). More controversial is the role of Phe(145)G8. Molecular dynamics simulations and electron paramagnetic resonance experiments have suggested that the accessibility of ligands through tunnel 1 is also regulated by the side chain of Phe(145)G8, which can adopt open and closed conformations (Forti et al., 2011; Van Doorslaer et al., 2012). However, there are no crystallographic evidences of alternate conformations for this residue in all crystal structures of MaPgb*

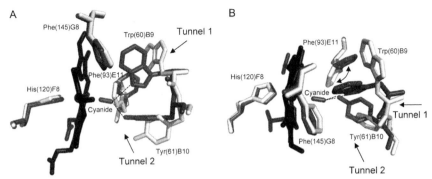

Figure 3.2 The haem distal site of *Ma*Pgb*. Superimposition of the ferric ligand-free *Ma*Pgb* (white) onto the *Ma*Pgb*(III)-cyanide structure (grey). The haem is shown in black. Residues lining the haem distal pocket are indicated and shown in stick representation, together with the proximal His(120)F8 residue. Hydrogen bonds to the haem-Fe (III)-bound cyanide are indicated by dashed lines. The haem distal cavity entrance sites of tunnel 1 and tunnel 2 are indicated by arrows. (A) Side view and (B) top view: rotation of the Phe(93)E11 side chain upon ligand binding is also indicated. (For colour version of this figure, the reader is referred to the online version of this chapter.)

(and mutants) reported so far, where irrespective of the ligation state of the protein and of the open/closed conformation of tunnel 1 (due to the location of Trp(60)B9 side chain) Phe(145)G8 is always in the open conformation (Fig. 3.2).

4. THE HAEM ENVIRONMENT

The Pgb haem cavity is characterised by strong hydrophobicity, both at the proximal and distal sites. Besides the proximal His(120)F8-Fe coordination bond, *Ma*Pgb*–haem interactions include van der Waals contacts with hydrophobic residues at the conserved topological positions: Ile(116)F4, Thr(128), Ile(137), Leu(142)G4, Trp(185)H17 at the haem-proximal side, and Trp(60)B9, Tyr(61)B10, Val(64)B13, Phe(74)CD1, Val(89)E7, Phe(93)E11, Phe(145)G7 and Ile(149)G11 at the haem distal side. In particular, Phe(74)CD1, Phe(93)E11 and Phe(145)G7 provide π–π interactions with three out of four porphyrins pyrrole rings. Further stabilising interactions are provided by hydrogen bonds between the haem propionates and polar residues, involving Tyr(85)E3, Arg(92)E10, Tyr(112) and Arg(119)F7, and interaction with a buried water molecule.

The high-resolution crystal structure of *Ma*Pgb* provides unequivocal evidence of substantial distortion in the porphyrin ring system, likely supported by the sealed nature of the haem cavity. The main out-of-plane contribution to the MaPgb* haem distortion is ruffling (which leaves opposite carbon atoms equally displaced and alternatively above and below the mean porphyrin plane), while the MaPgb* haem in-plane distortion is mainly ascribed to a strong breathing mode, which involves the symmetric compression–expansion of the porphyrin ring (with expansion associated to destabilisation of O_2 binding, whereas the opposite trend is found for compression). Therefore, haem compression due to the restricted MaPgb* haem-binding pocket is expected to lead to a sizable stabilisation of O_2 binding, overcoming the destabilisation due to ruffling, thus resulting in stabilisation of the haem-bound O_2, as compared to the ideal planar reference haem model (Bikiel et al., 2010). This is in agreement with the unusually low in $vitro$ O_2 dissociation rate reported for MaPgb* $(0.092–0.0094 \text{ s}^{-1})$, appearing particularly low considering that the haem-bound O_2 was found not to be stabilised by any hydrogen bond to the protein (Nardini et al., 2008). This unusually low O_2 dissociation rate constant in MaPgb* has been correlated to the large deviations from planarity of the porphyrin system (Bikiel et al., 2010; Nardini et al., 2008). Thus, the O_2 binding behaviour of ferrous MaPgb* suggests a scenario whereby evolutionary events could subtly regulate O_2 affinity by shaping the haem cavity to favour porphyrin out-of-plane distortions in order to decrease ligand affinity, or to compress the porphyrin ring, for instance, by the presence of bulky residues (usually Leu and/or Phe), which would selectively push the pyrrole rings out of the porphyrin plane, to promote the reverse effect (Bikiel et al., 2010).

Very recently, structural studies on ferric MaPgb* (and selected mutants) have further characterised the distal site plasticity and the ligand-binding modes in MaPgb* (Pesce et al., 2011; Pesce, Tilleman, et al., 2013). The first intriguing finding was that ferric MaPgb* can productively bind a plethora of ligands, comprising classical globin ligands such as cyanide, azide and imidazole, and a less common ligand such as nicotinamide. Contrary to what reported for the oxygenated protein, in ferric MaPgb* all ligands are productively stabilised by hydrogen bonds. The prototypical case is represented by the Fe-bound cyanide, which is stabilised by two hydrogen bonds provided by Trp(60)B9 and Tyr(61)B10 side chains. While ligand stabilisation by the OH group of Tyr(61)B10 requires only one side chain rotation relative to the ligand-free ferric MaPgb*, the hydrogen bond provided by the Nε2 atom of Trp(60)B9 requires a complex re-arrangement of the haem distal site cavity (Fig. 3.2). The presently available data identify three residues at key topological positions B9, B10 and E11 that appear to be involved not only in ligand recognition and binding but also in the dynamic regulation of ligand exchange between the solvent and the distal pocket.

Phe(93)E11 appears to play a role in ligand sensing and discrimination. In fact, the Phe(93)E11 side chain rotates by ~120°, relative to ligand-free ferric MaPgb, upon cyanide binding (Fig. 3.2B); a similar side chain rotation occurs upon azide and imidazole binding, but it is somehow impaired in the case of nicotinamide. This ligand-linked conformational change makes room for a consequent "induced" conformational re-arrangement of Trp(60)B9 side chain, which inserts into the distal site and hydrogen bonds the Fe-bound ligand. Such a sequential mechanism is confirmed by the structure of the ferric cyanide-bound Trp(60)B9Ala mutant, where the ligand is coordinated to the haem-Fe atom and stabilised only by the Tyr(61)B10 OH group, the Phe(93)E11 residue being rotated by about 120° relative to ligand-free MaPgb*, even in the absence of Trp(60)B9. The structure of ferric MaPgb* in complex with nicotinamide reveals that despite the 60° rotation of Phe(93)E11 side chain, Trp(60)B9 does not enter the haem distal cavity, thus indicating that there must be a certain rotation of Phe(93)E11 ligand sensor to trigger the conformational change of Trp(60)B9.

Sensitivity response of Phe(93)E11 to the haem-bound ligand is consistently observed also through molecular dynamics simulations, which evidenced how the haem-bound ligand introduces steric hindrance onto the side chain of Phe(93)E11 (Forti et al., 2011). However, the ligand-sensing role of Phe(93)E11 does not necessarily require the presence of an aromatic residue at the E11 topological position (although Phe is conserved in Pgbs), as demonstrated by the structure of the cyanide-bound Phe(93)E11Leu mutant, where the ligand-binding mode and the distal site geometry are essentially identical to those found in the MaPgb*–cyanide structure. Cyanide dissociation rate constants, measured in solution for ferric MaPgb* and selected mutants, are in good agreement with the ligand-stabilisation events and mechanisms described by the crystal structures (Pesce, Tilleman, et al., 2013).

A relevant finding (probably the most important) is that the conformational adaptability shown by MaPgb* haem distal site residues results in coupling between ligand sensing/binding and haem distal site accessibility through the two-tunnel system, as Trp(60)B9 and Tyr(61)B10 side chains line the two separate tunnels. Thus, the comparison of ferric MaPgb* in its liganded and unliganded states identifies two distinct haem distal site arrangements: (i) a closed distal site conformation, whereby residue Trp(60) B9 points towards the Fe–bound ligand and shuts tunnel 1 and (ii) an open distal site conformation, where the Trp(60)B9 side chain points away from the distal pocket, thus keeping tunnel 1 in its open state. In addition, considering that for a ligand such as nicotinamide, the ligation state of the protein does not result in the relocation of the Trp(60)B9 side chain into the haem distal cavity, nor in the transition of tunnel 1 from the open to the closed state, the reactivity of $MaPgb$ appears to be regulated also by the nature of the haem ligand (Pesce, Tilleman, et al., 2013).

Another interesting point is that the insertion of Trp(60)B9 into the haem distal cavity upon ligand binding is coupled to a change in the backbone conformation of the neighbouring 149–154 region, located in the second half of the G-helix, at the subunit interface of the $MaPgb^*$ homodimer. Within the homodimeric $MaPgb^*$ G–H four-helix bundle (Fig. 3.1C), tight packing involves specifically the N-terminal half of the G-helices and the C-terminal half of the H-helices, while the remaining interface regions are mostly solvent exposed, thus allowing the 149–154 region to afford some flexibility to compensate the structural changes transmitted from reshaping of the haem distal site upon ligand binding. Alternatively, the 149–154 region might also be able to influence/modulate the architecture and ligand-binding properties of the haem distal cavity through association of an (unknown) effector molecule or even a partner protein. Therefore, the 149–154 region has been proposed to be a potential allosteric regulation site (Pesce, Tilleman, et al., 2013), although (negative) co-operativity has only been reported for O_2 but not for CO binding (Abbruzzetti et al., 2012; Nardini et al., 2008).

5. BIOCHEMICAL AND FUNCTIONAL CHARACTERISATION

Originally, Pgbs were identified in the genomes of the Actinobacterium *Thermobifida fusca*, the green non-sulphur bacterium *Chloroflexus aurantiacus* and the two Archaea *A. pernix* and *M. acetivorans*. Pgbs from the latter two Archaea were characterised and found to purify in the oxidised state; generation of reduced Pgb proved to require an environment free of O_2, as they autoxidise rapidly, at the rates of 0.0032 and 0.0027 s^{-1}, corresponding to half-lives of 3.6 and 4.3 min for oxy-$MaPgb$ and oxy-$ApPgb$, respectively (Freitas et al., 2004, 2005). Globins with such high oxidation rates and long reduction times are usually not suitable for operation within high O_2 environments. Indeed, the oxygen requirements for these two Archaea are prohibitively small (nanomolar level) or null; in fact, the obligate aerobe *A. pernix* is also a hyperthermophile and only ~25 µM of O_2 can dissolve in water (1 atm pressure) at its optimal growth temperature of 95 °C, and *M. acetivorans* is a strict anaerobe.

For *Ma*Pgb, a possible involvement in either facilitating O_2 detoxification or acting as CO sensor/supplier in methanogenesis has been proposed (Freitas et al., 2003, 2004, 2005; Hou et al., 2001). Very unusually, *Ma*Pgb shows a selectivity ratio for O_2/CO binding that favours O_2 ligation and anti-cooperativity in ligand binding (Abbruzzetti et al., 2012; Nardini et al., 2008). Such property could be related to the fact that *M. acetivorans* takes advantage of acetate, methanol, CO_2 and CO as carbon sources for methanogenesis; methane production occurs simultaneously with the formation of a proton gradient that is essential for energy harvesting (Lessner et al., 2006; Oelgeschläger & Rother, 2008; Rother & Metcalf, 2004). Therefore, the ability to convert CO to methane might indicate that CO is the actual ligand of *Ma*Pgb *in vivo*, supporting the hypothesis of a very ancient origin for such metabolic pathway(s) (Ferry & House, 2006; Oelgeschläger & Rother, 2008).

The possibility for Pgb to play a yet unclear role in CO metabolism of Archaea has driven efforts for the biochemical and kinetic characterisation of CO binding and dissociation, also taking advantage of the stability of the CO complexes, as shown for many other globins in the literature (Brunori & Gibson, 2001). Early rapid mixing experiments have shown that CO (and O_2) binding to *Ma*Pgb* is a biphasic process (Nardini et al., 2008). This feature raises the question whether such heterogeneity arises from the existence of two molecular populations in equilibrium, or whether it is linked to the presence of the reported two-tunnel system. By means of vibrational and time-resolved spectroscopy experiments, it was shown that the protein exists in two distinct conformations displaying different affinities for CO, based on an equilibrium influenced by the ligation state of the protein (Abbruzzetti et al., 2012). Ligation shifts the equilibrium towards a high-affinity species (in the following indicated as *Ma*Pgb^{*r}) characterised by low dissociation rates. In the absence of the ligand, the protein adopts preferentially a lower affinity conformation (in the following indicated as *Ma*Pgb^{*t}), displaying high dissociation rates. Upon ligand dissociation, switching from the high- to the low-affinity conformation occurs on the sub-millisecond time scale. Such conformational change appears to involve only the tertiary structure of each subunit, not extending its effects to the partner subunit of the dimer. However, it cannot be excluded that the quaternary structure modulates protein affinity for CO by affecting the tertiary structure of the monomers. Recent molecular dynamics investigations have shown that dimerisation has profound consequences on the structure and the dynamics of $MaPgb^*$ (Forti et al., 2011). Thus, a mechanism of homotropic allosteric control of affinity for CO appears to be operative in $MaPgb^*$, where changes in rate constants for CO binding to and dissociation from the haem upon ligation/ deligation may bear consequences for a yet unknown reaction, sequentially involving two substrates.

The recent structures of $MaPgb^*$ in complex with different ligands allowed to propose a stereochemical mechanism which could be at the basis of the conformational switching between the $MaPgb^{*r}$ and $MaPgb^{*t}$ states. According to such mechanism, Phe(93)E11 residue could sense the presence of a haem-bound ligand through its steric hindrance. Information on the liganded state would then be transferred to Trp(60)B9, which would re-arrange its side chain enabling a hydrogen bond to the haem-bound ligand and the concomitant closing of tunnel 1. The presence of more than one residue, namely Trp(60)B9 and Tyr(61)B10, involved in hydrogen bond interactions with the haem-bound ligand, and the conformational flexibility of the distal site residues, which can modulate hydrogen bond strength, may explain the observed heterogeneity in resonance Raman stretching bands and in the CO dissociation kinetics. Conversely, the structural bases for the heterogeneous ligand-binding kinetics remain unclear, although the effects exerted by the haem-proximal site residues as well as by other residues along the tunnel pathways may be relevant.

Besides the putative involvement of $MaPgb$ in CO metabolism, very recently, kinetics and thermodynamics of ferric and ferrous $MaPgb$ nitrosylation have shown that addition of NO to the ferric protein leads to the transient formation of the $MaPgb(III)–NO$ complex in equilibrium with $MaPgb(II)–NO^+$. In turn, $MaPgb(II)–NO^+$ is converted to $MaPgb(II)$ by OH^--based catalysis. Then, $MaPgb(II)$ binds NO very rapidly leading to the formation of the $MaPgb(II)–NO$ complex. The rate-limiting step for reductive nitrosylation of $MaPgb(III)$ is represented by the OH^--mediated reduction of $MaPgb(II)–NO^+$ to $MaPgb(II)$ (Ascenzi et al., 2013). These results suggest a potential role of $MaPgb$ in scavenging of reactive nitrogen and oxygen species, which appears pivotal in the physiology of the strictly anaerobic *M. acetivorans*. In fact, multiple functional roles can be envisaged for $MaPgb$, as its scavenging function(s) might co-exist with enzymatic activity(ies) to facilitate the conversion of CO to methane, as an adaptation to different gaseous environments.

Furthermore, it has been demonstrated that *A. pernix* Pgb and the globin domains of GCSs use a common signalling mechanism, which would allow *Ap*Pgb to communicate with heterologous C-terminal transmitter domains to perform signal transduction and gene regulation. Chimeric receptors consisting of the GCS or *Ap*Pgb globin domains and the C-terminal methyl-accepting chemotaxis protein (MCP) signalling domain of the *E. coli* chemotaxis transducer Tsr reversibly bind oxygen and mediate aerotactic responses in *E. coli* (Saito, Wan, Lee, Hou, & Alam, 2008). The observation that single-domain *Ap*Pgb is functionally compatible with the MCP signalling domain for aerotactic signalling is intriguing, since it suggests that, besides sequence and structural similarities, a functional relationship exists between Pgbs and GCSs, and that *Ap*Pgb has an inherent capacity for signal transduction. Thus, the existence of Pgb as a single-domain globin could allow Pgbs for the flexibility to perform multiple functions.

6. CONCLUSIONS

The globin domain has been adopted by nature to host the haem in an increasing number of successful engineering experiments. Pgb is a fascinating example of structural modulation of the classical 3/3 globin fold, which translates into new access routes to the haem, into a unique distortion from planarity of the porphyrin system that affects the protein reactivity versus O_2, and into a specific quaternary assembly shared with the phylogenetic related GCS proteins (Nardini et al., 2008). What makes Pgb unique in the globin panorama is the striking flexibility/adaptability of the haem distal site residues, which allow a structural crosstalk between ligand recognition/binding and haem accessibility through the two-tunnel system, and the coupling of potential reactivity in the haem cavity with ligation (Pesce et al., 2011; Pesce, Tilleman, et al., 2013).

Thus, the accessibility to the haem cavity, at least through tunnel 1, is linked to (and possibly modulated by) the ligation state of the protein through an inter-twinned mechanism of side chain re-arrangements, which involve three conserved residues at the key topological B9, B10 and E11 sites. Moreover, the nature of the haem ligand appears to be capable of driving the distal site architecture, the dynamics of the B-, E- and G-helices and the open/closed states of tunnel 1 through conformational relocation of

Trp(60)B9 side chain. Thus, Pgbs maintain the ligand-stabilisation mechanism based on residue B10, as most invertebrate globins (Bolognesi et al., 1997), but evolved an E7-independent ligand-to-haem path based also on residues at the B9 and E11 topological positions.

On the basis of the above considerations, $MaPgb$ has been suggested to be involved in a ligand-controlled bimolecular process, where loading of the protein with a first ligand would facilitate ligand binding to the haem cavity of the second subunit (Forti et al., 2011). This hypothesis is in keeping with experimental findings derived from kinetic studies of CO binding to $MaPgb^*$ in solution and gels, which show that the heterogeneous ligand-binding kinetics is affected by ligand concentration, and highlight a ligand-dependent equilibrium between two conformational species related to fast and slow CO-rebinding processes in $MaPgb^*$. The putative dual path ligand exchange mechanism (typical of some enzymes) could bear functional implications for a yet undiscovered role in $M.$ $acetivorans$ CO metabolism (Abbruzzetti et al., 2012). At present, the in $vivo$ functional role of Pgb is, indeed, an unanswered and challenging question. The structural flexibility/adaptability of the tertiary structure, together with its ability to form a quaternary assembly similar to GCS proteins, strongly suggests that Pgbs may serve several different biological functions, including protection from nitrosative and oxidative stress, and formation (together with other GCS proteins) of a common signalling mechanism addressing diverse physiological functions, such as aerotactic response. A key test on Pgb function would probably require mutagenesis in $vivo$ and analysis of the mutant phenotype. A significant part of future work on Pgbs is, therefore, expected to derive from microbiology, the results of which should be integrated with other approaches, such as expression profiles, biochemical experiments and structural information.

REFERENCES

Abbruzzetti, S., Tilleman, L., Bruno, S., Viappiani, C., Desmet, F., Van Doorslaer, S., et al. (2012). Ligation tunes protein reactivity in an ancient haemoglobin: Kinetic evidence for an allosteric mechanism in *Methanosarcina acetivorans* protoglobin. *PLoS One, 7*, e33614.

Ascenzi, P., Pesce, A., Nardini, M., Bolognesi, M., Ciaccio, C., Coletta, M., et al. (2013). Reductive nitrosylation of *Methanosarcina acetivorans* protoglobin: A comparative study. *Biochemical and Biophysical Research Communications, 430*, 1301–1305.

Bikiel, D. E., Forti, F., Boechi, L., Nardini, M., Luque, F. J., Martí, M. A., et al. (2010). Role of haem distortion on oxygen affinity in haem proteins: The protoglobin case. *The Journal of Physical Chemistry B, 114*, 8536–8543.

Bolognesi, M., Bordo, D., Rizzi, M., Tarricone, C., & Ascenzi, P. (1997). Nonvertebrate hemoglobins: Structural bases for reactivity. *Progress in Biophysics and Molecular Biology, 68*, 29–68.

Bolognesi, M., Cannillo, E., Ascenzi, P., Giacometti, G. M., Merli, A., & Brunori, M. (1982). Reactivity of ferric *Aplysia* and sperm whale myoglobins towards imidazole. X-ray and binding study. *Journal of Molecular Biology*, *158*, 305–315.

Brunori, M., & Gibson, Q. H. (2001). Cavities packing defects in the structural dynamics of myoglobin. *EMBO Reports*, *2*, 674–679.

Ciaccio, C., Pesce, A., Tundo, G. R., Tilleman, L., Bertolacci, L., Dewilde, S., et al. (2013). Functional and structural roles of the N-terminal extension in *Methanosarcina acetivorans* protoglobin. *Biochimica et Biophysica Acta – Proteins and Proteomics*, *1834*, 1813–1823.

Ferry, J. G., & House, C. H. (2006). The stepwise evolution of early life driven by energy conservation. *Molecular Biology and Evolution*, *23*, 1286–1296.

Forti, F., Boechi, L., Bikiel, D. E., Martì, M. A., Nardini, M., Bolognesi, M., et al. (2011). Ligand migration in *Methanosarcina acetivorans* protoglobin: Effects of ligand binding and dimeric assembly. *The Journal of Physical Chemistry B*, *115*, 13771–13780.

Freitas, T. A. K., Hou, S., & Alam, M. (2003). The diversity of globin-coupled sensors. *FEBS Letters*, *552*, 99–104.

Freitas, T. A. K., Hou, S., Dioum, E. M., Saito, J. A., Newhouse, J., Gonzalez, G., et al. (2004). Ancestral hemoglobins in Archaea. *Proceedings of the National Academy of Sciences of the United States of America*, *101*, 6675–6680.

Freitas, T. A. K., Saito, J. A., Hou, S., & Alam, M. (2005). Globin-coupled sensors, protoglobins, and the last universal common ancestor. *Journal of Inorganic Biochemistry*, *99*, 23–33.

Hoogewijs, D., Dewilde, S., Vierstraete, A., Moens, L., & Vinogradov, S. N. (2012). A phylogenetic analysis of the globins in fungi. *PLoS One*, *7*, e31856.

Hou, S., Freitas, T., Larsen, R. W., Piatibratov, M., Sivozhelezov, V., Yamamoto, A., et al. (2001). Globin-coupled sensors: A class of haem-containing sensors in Archaea and Bacteria. *Proceedings of the National Academy of Sciences of the United States of America*, *98*, 9353–9358.

Lessner, D. J., Li, L., Eejtar, T., Andreev, V. P., Reichlen, M., Hill, K., et al. (2006). An unconventional pathway for reduction of CO_2 to methane in CO-grown *Methanosarcina acetivorans* revealed by proteomics. *Proceedings of the National Academy of Sciences of the United States of America*, *103*, 17921–17926.

Milani, M., Pesce, A., Nardini, M., Ouellet, H., Ouellet, Y., Dewilde, S., et al. (2005). Structural bases for haem binding and diatomic ligand recognition in truncated hemoglobins. *Journal of Inorganic Biochemistry*, *99*, 97–109.

Nardini, M., Pesce, A., Milani, M., & Bolognesi, M. (2007). Protein fold and structure in the truncated (2/2) globin family. *Gene*, *398*, 2–11.

Nardini, M., Pesce, A., Thijs, L., Saito, J. A., Dewilde, S., Alam, M., et al. (2008). Archaeal protoglobin structure indicates new ligand diffusion paths and modulation of haem-reactivity. *EMBO Reports*, *9*, 157–163.

Oelgeschläger, E., & Rother, M. (2008). Carbon monoxide-dependent energy metabolism in anaerobic bacteria and archaea. *Archives of Microbiology*, *190*, 257–269.

Perutz, M. F. (1979). Regulation of oxygen affinity of hemoglobin: Influence of structure of the globin on the haem iron. *Annual Review of Biochemistry*, *48*, 327–386.

Pesce, A., Bolognesi, M., & Nardini, M. (2013). The diversity of 2/2 (truncated) globins. *Advances in Microbial Physiology*, *63*, 49–79.

Pesce, A., Thijs, L., Nardini, M., Desmet, F., Sisinni, L., Gourlay, L., et al. (2009). HisE11 and HisF8 provide bis-histidyl haem hexa-coordination in the globin domain of *Geobacter sulfurreducens* globin-coupled sensor. *Journal of Molecular Biology*, *386*, 246–260.

Pesce, A., Tilleman, L., Dewilde, S., Ascenzi, P., Coletta, M., Ciaccio, C., et al. (2011). Structural heterogeneity and ligand gating in ferric *Methanosarcina acetivorans* protoglobin mutants. *IUBMB Life*, *63*, 287–294.

Pesce, A., Tilleman, L., Donné, J., Aste, E., Ascenzi, P., Ciaccio, C., et al. (2013). Structure and haem-distal site plasticity in *Methanosarcina acetivorans* protoglobin. *PLoS One, 8,* e66144.

Rother, M., & Metcalf, W. W. (2004). Anaerobic growth of *Methanosarcina acetivorans* C2A on carbon monoxide: An unusual way of life for a methanogenic archaeon. *Proceedings of the National Academy of Sciences of the United States of America, 101,* 16929–16934.

Saito, J. A., Wan, X., Lee, K. S., Hou, S., & Alam, M. (2008). Globin-coupled sensors and protoglobins share a common signaling mechanism. *FEBS Letters, 582,* 1840–1846.

Salter, M. D., Nienhaus, K., Nienhaus, G. U., Dewilde, S., Moens, L., Pesce, A., et al. (2008). The apolar channel in *Cerebratulus lacteus* hemoglobin is the route for O_2 entry and exit. *Journal of Biological Chemistry, 283,* 35689–35702.

Van Doorslaer, S., Tilleman, L., Verrept, B., Desmet, F., Maurelli, S., Trandafir, F., et al. (2012). Marked difference in the electronic structure of cyanide-ligated ferric protoglobins and myoglobin due to heme ruffling. *Inorganic Chemistry, 51,* 8834–8841.

Vinogradov, S. N., Hoogewijs, D., Bailly, X., Arredondo-Peter, R., Gough, J., Dewilde, S., et al. (2006). A phylogenomic profile of globins. *BMC Evolutionary Biology, 6,* 31–47.

Vinogradov, S. N., Hoogewijs, D., Bailly, X., Mizuguchi, K., Dewilde, S., Moens, L., et al. (2007). A model of globin evolution. *Gene, 398,* 132–142.

Vinogradov, S. N., & Moens, L. (2008). Diversity of globin function: Enzymatic, transport, storage, and sensing. *Journal of Biological Chemistry, 283,* 8773–8777.

Vinogradov, S. N., Tinajero-Trejo, M., Poole, R. K., & Hoogewijs, D. (2013). Bacterial and archaeal globins—A revised perspective. *Biochimica et Biophysica Acta – Proteins and Proteomics, 1834,* 1789–1800.

Vuletich, D. A., & Lecomte, J. T. (2006). A phylogenetic and structural analysis of truncated hemoglobins. *Journal of Molecular Evolution, 62,* 196–210.

Zhang, W., & Phillips, G. N., Jr. (2003). Structure of the oxygen sensor in *Bacillus subtilis*: Signal transduction of chemotaxis by control of symmetry. *Structure, 11,* 1097–1110.

CHAPTER FOUR

The Globins of *Campylobacter jejuni*

Mariana Tinajero-Trejo[*], Mark Shepherd[†,1]

[*]Department of Molecular Biology and Biotechnology, University of Sheffield, Sheffield, United Kingdom
[†]School of Biosciences, University of Kent, Canterbury, United Kingdom
[1]Corresponding author: e-mail address: m.shepherd@kent.ac.uk

Contents

1.	Overview	99
2.	NO and RNS in Biology	100
3.	Oxygen and Reactive Oxygen Species in Biology	107
4.	Respiratory Metabolism of *Campylobacter*	108
	4.1 Microaerobic respiration	109
	4.2 Respiration under oxygen-limited conditions	109
	4.3 Impact of nitrosative stress upon *Campylobacter* respiration	110
5.	*Campylobacter* Single-Domain Globin, Cgb: Functional and Structural Characterisation	111
	5.1 Functional characterisation	111
	5.2 Cgb reduction: The redox partner conundrum	112
	5.3 Structural characterisation	113
	5.4 Biophysical and mechanistic characterisation	116
6.	*Campylobacter* Truncated Globin, Ctb: Functional and Structural Characterisation	120
	6.1 Functional characterisation	120
	6.2 Structural characterisation	121
	6.3 Biophysical and mechanistic characterisation	123
7.	Control of Globin Expression	123
	7.1 Control of globin expression by NssR	123
	7.2 NssR classification, targets and mechanistic insights	126
	7.3 Structural modelling of NssR	128
	7.4 NrfA of *Campylobacter*	129
	7.5 Induction of Cgb by nitrite and nitrate is independent of NrfA	130
	7.6 NO resistance in oxygen-limited conditions	131
8.	Integrated Response Involving Cgb and ctb	131
9.	Conclusions	132
	Acknowledgements	133
	References	133

Advances in Microbial Physiology, Volume 63
ISSN 0065-2911
http://dx.doi.org/10.1016/B978-0-12-407693-8.00004-2

Abstract

Campylobacter jejuni is a zoonotic Gram-negative bacterial pathogen that is exposed to reactive nitrogen species, such as nitric oxide, from a variety of sources. To combat the toxic effects of this nitrosative stress, *C. jejuni* upregulates a small regulon under the control of the transcriptional activator NssR, which positively regulates the expression of a single-domain globin protein (Cgb) and a truncated globin protein (Ctb). Cgb has previously been shown to detoxify nitric oxide, but the role of Ctb remains contentious. As *C. jejuni* is amenable to genetic manipulation, and its globin proteins are easily expressed and purified, a combination of mutagenesis, complementation, transcriptomics, spectroscopic characterisation and structural analyses has been used to probe the regulation, function and structure of Cgb and Ctb. This ability to study Cgb and Ctb with such a multi-pronged approach is a valuable asset, especially since only a small fraction of known globin proteins have been functionally characterised.

ABBREVIATIONS

CcP cytochrome *c* peroxidase
Cgb *Campylobacter* single-domain haemoglobin
CioAB *Campylobacter* cyanide-insensitive oxidase
Ctb *Campylobacter*-truncated haemoglobin
DMSO dimethyl sulphoxide
FHb flavohaemoglobin
GSNO *S*-nitrosoglutathione
HbN *M. tuberculosis* TrHbI
HbO *M. tuberculosis* TrHbII
Hmp *E. coli* flavohaemoglobin
Mb myoglobin
MFS Miller Fisher syndrome
Nap nitrate reductase
NOC-5 3-(aminopropyl)-1-hydroxy-3-isopropyl-2-oxo-1-triazene
NOC-7 3-(2-hydroxy-1-methyl-2-nitrosohydrazino)-*N*-methyl-1-propanamine
NOC-18 (*Z*)-1-[2-(2-aminoethyl)-*N*-(2-ammonioethyl)amino]diazen-1-ium-1,2-diolate
NOD nitric oxide dioxygenase
NOS NO synthase
NssR *Campylobacter* nitrosative stress–sensing regulator
RNS reactive nitrogen species
ROS reactive oxygen species
SDHb single-domain haemoglobin
SNAP *S*-nitroso-*N*-acetylpenicillamine
SNOs *S*-nitrosothiols
SNP sodium nitroprusside
swMb sperm whale myoglobin
T family truncated Mb-fold family
TMAO trimethylamine-*N*-oxide
TrHb truncated haemoglobin
Vgb *Vitreoscilla* single-domain haemoglobin

1. OVERVIEW

In 1913, McFadyean and Stockman isolated a *Vibrio*-like organism from aborted ovine foetuses. After the creation of the genus *Campylobacter* in 1963 (Butzler, 2004; Skirrow, 2006), Véron and Chateline (1973) reported four species within the genus *Campylobacter*: *C. fetus*, *C. coli*, *C. jejuni* and *C. sputorum*. Currently, the *Campylobacteraceae* family comprises at least 15 species of *Campylobacter* (Debruine, Gevers, & Vandamme, 2008) being found in a range of niches, from free-living environmental organisms to commensals or parasites in humans and domestic animals.

Campylobacter has been recognised as one of the main causes of bacterial gastroenteritis in the developed world (Friedman, Neimann, Wegener, & Tauxe, 2000; Scallan et al., 2011; Simonsen et al., 2011). It is estimated that 9.4 million episodes of foodborne illness arise every year in the United States; 9% of those cases and 15% of the related hospitalisations are caused by *Campylobacter* species (Scallan et al., 2011).

C. jejuni is a commensal in the lower intestine of chickens but a pathogen in humans. Contamination of poultry products during processing constitutes an important route of transmission (Friedman et al., 2000). *C. jejuni* infection produces an inflammatory response (Bakhiet et al., 2004; Jacobs et al., 1998; Koga et al., 2005; Zheng, Meng, Zhao, Singh, & Song, 2008) associated with pathological symptoms. Even though the majority of the symptoms related to *Campylobacteriosis* are self-limited and infection is generally restricted to the intestine, invasion of other tissues, mainly in immunocompromised and elderly patients, leads to significant morbidity and mortality (Allos, 2001; Wassenaar & Blaser, 1999). Moreover, the development of autoimmune diseases such as Guillain–Barré syndrome and Miller Fisher syndrome, that mainly occur as sequelae of gastrointestinal infection (Fisher, 1956; Tam et al., 2007), has been closely related to a preceding infection by *C. jejuni*.

During infection and pathogenesis, *C. jejuni* is exposed to nitric oxide (NO) and a range of other reactive nitrogen species (RNS) (see Sections 2 and 4.3). The ability of the bacterium to detoxify these compounds has been mainly associated with the expression of two haemoglobins (Cgb and Ctb) in response to nitrosative stress.

The most recent bioinformatics survey of sequences such as globins in over 2200 bacterial genomes has provided a complete inventory of globins present in this kingdom (Vinogradov, Tinajero-Trejo, Poole, & Hoogewijs, 2013). This study revealed genes encoding globins in half of the bacterial genomes within members of the three globin families. In addition, a more comprehensive nomenclature for both prokaryotic and eukaryotic globins was proposed: the canonical 3/3 alpha-helical fold (the myoglobin (Mb)-fold) includes two families: the Mb-like family containing both single-domain haemoglobins (SDHbs) and flavohaemoglobins (FHbs) and the sensor globin family enclosing protoglobins and single-domain sensor globins (SDSgb). The truncated Mb-fold family (T family) displays a 2/2 alpha-helical fold (truncated Mb-fold) and contains truncated haemoglobins (TrHb1, TrHb2 and TrHb3) (Vinogradov et al., 2013).

Despite the vast number of globins in bacteria, our understanding of their physiological functions is restricted to only a few examples. Table 4.1 summarises the current understanding of *in vivo* function(s) and regulation of the most extensively studied bacterial globins: the FHb of *Escherichia coli* (Hmp), the single-domain globin of *Vitreoscilla* (Vgb), the truncated globins of *Mycobacterium tuberculosis* (HbN and HbO) and the single-domain (Cgb) and truncated (Ctb) globins of *C. jejuni*. Resistance to nitrosative stress and involvement in oxygen transfer are the most prevalent functions for bacterial haemoglobins (see Table 4.1). However, given the number of bacterial genomes that encode globin-like sequences (~1100) (Vinogradov et al., 2013), the number of globins whereby functions have been elucidated via the robust approach of mutation/complementation is very limited (~15).

C. jejuni has been shown to be a good model for the analysis of globin function. Indeed, (i) construction of mutants for both *cgb* and *ctb* has allowed the inference of functions, (ii) the regulation of globin expression under conditions of nitrosative stress has been extensively studied, (iii) purified globins have been subject to structural and kinetic characterisation and (iv) heterologous expression, as a tool to explore function, has also been exploited. However, many questions remain, pertaining to the physiological function of the globins of *Campylobacter* (especially Ctb), and the potential interaction between Cgb and Ctb is yet to be investigated. This review aims to provide an overview of our current understanding of the globins of *C. jejuni*, including the impact of gene deletions on phenotype, transcriptional regulation and the biochemical/biophysical/structural characterisation of Cgb and Ctb.

2. NO AND RNS IN BIOLOGY

NO is a free radical with an unpaired electron and readily reacts with a variety of biological molecules (Halliwell & Gutteridge, 2007). The

Table 4.1 Selected bacterial haemoglobins: function(s) and regulation

Organism	Family, sub-family and globin name	Crystal structure solved	Function(s) inferred by mutation/complementation	Regulation Regulator(s) involved in natural host	Environmental factors and/or compounds involved in upregulation	Heterologous expression Host	Function(s) inferred	References
Escherichia coli	M FHb Hmp	Yes	O$_2$-dependent NO detoxification to nitrate (alleviation of NO toxicity)	Fnr, MetR, NsrR	Nitrosative stress	*E. coli* WT[a]	Oxidative stress resistance, NO consumption, increased cell growth in microaerophilic conditions	Anjum, Ioannidis, and Poole (1998), Bodenmiller and Spiro (2006), Bollinger, Bailey, and Kallio (2001), Corker and Poole (2003), Cruz-Ramos et al. (2002), Flatley et al. (2005), Frey, Farres, Bollinger, and Kallio (2002), Gardner and Gardner (2002), Gardner, Costantino, and Salzman (1998), Gardner, Gardner, Martin, and Salzman (1998), Hausladen, Gow, and Stamler (1998), Hernandez-Urzua et al. (2003), Ilari, Ceci, Ferrari, Rossi, and Chiancone (2002), Justino, Vicente,

Continued

Table 4.1 Selected bacterial haemoglobins: function(s) and regulation—cont'd

| Organism | Family, sub-family and globin name | Crystal structure solved | Function(s) inferred by mutation/ complementation | Regulation | | Heterologous expression | | References |
				Regulator(s) involved in natural host	Environmental factors and/or compounds involved in upregulation	Host	Function(s) inferred	
Vitreoscilla stercoraria	M SDHb Vgb	Yes	No	Unidentified	Upregulation in oxygen-limited conditions in the native host and in *E. coli*,	*E. coli* WT[a], *Bacillus subtilis* WT[a]	Oxygen transfer, NO scavenging	Teixeira, and Saraiva (2005), Membrillo-Hernandez et al. (1999), Membrillo-Hernández, Coopamah, Channa, Hughes, and Poole (1998), Membrillo-Hernandez, Kim, Cook, and Poole (1997), Mukhopadhyay and Schellhorn (1994), Poole et al. (1996), Stevanin et al. (2000), Svensson et al. (2010) Bollinger and Kallio (2007), Bolognesi et al. (1999), Dikshit, Dikshit, and Webster (1990), Dikshit, Spaulding, Braun, and Webster (1989),

					carbon-limited conditions in E. coli			Dikshit and Webster (1988), Dikshit, Dikshit, Liu, and Webster (1992), Dikshit et al. (2002), Kallio and Bailey (1996), Kallio, Tsai et al. (2007), Kallio, Tsai and Bailey (1996), Khosla and Bailey (1988, 1989), Khosla, Curtis, Bydalek, Swartz, and Bailey (1990), Khosla, Curtis, DeModena, Rinas, and Bailey (1990), Ramandeep et al. (2001), Tarricone, Galizzi, Coda, Ascenzi, and Bolognesi (1997), Tsai, Kallio, and Bailey (1995), Webster and Hackett (1966), Yang, Webster, and Stark (2005)
Campylobacter jejuni	M SDHb Cgb	Yes	NO and RNS detoxification	NssR	Nitrosative stress	E. coli WT[a], E. coli hmp	NO scavenging	Avila-Ramirez et al. (2013), Bollinger et al. (2001), Elvers et al. (2005), Elvers, Wu, Gilberthorpe, Poole, and Park (2004), Frey et al. (2002), Monk, Pearson, Mulholland,

Continued

Table 4.1 Selected bacterial haemoglobins: function(s) and regulation—cont'd

Organism	Family, sub-family and globin name	Crystal structure solved	Function(s) inferred by mutation/complementation	Regulation		Heterologous expression		References
				Regulator(s) involved in natural host	Environmental factors and/or compounds involved in upregulation	Host	Function(s) inferred	
								Smith, and Poole (2008), Pittman et al. (2007), Shepherd et al. (2010), Wainwright, Elvers, Park, and Poole (2005)
	TrHb TrHbIII Ctb	Yes	Oxygen metabolism	NssR	Constitutively expressed at low levels Nitrosative stress	E. coli hmp	NO scavenging	Avila-Ramirez et al. (2013), Elvers et al. (2005), Monk et al. (2008), Nardini et al. (2006), Smith, Shepherd, Monk, Green, and Poole (2011), Wainwright et al. (2005)
Mycobacterium tuberculosis	TrHb TrHbI HbN	Yes	No	Unidentified	Stationary phase, nitrite, SNP, hypoxia, intracellular growth in	E. coli hmp, M. smegmatis WT[a], Salmonella serovar	NO scavenging	Bidon-Chanal et al. (2006), Couture et al. (1999), Joseph, Madhavilatha, Kumar, and Mundayoor (2012), Lama et al. (2009), Milani, Ouellet, et al.

				macrophages, early response to oxidative and nitrosative stress	Typhimurium *hmp*		(2004), Milani et al. (2001), Milani, Pesce, et al. (2004), Ouellet, Milani, Couture, Bolognesi, and Guertin (2006), Pathania, Navani, Gardner, and Dikshit (2002), Pawaria, Lama, Raje, and Dikshit (2008), Pawaria et al. (2007), Savard et al. (2011)
TrHb TrHbII HbO	Yes	Unidentified	Unidentified	Constitutively expressed at low levels Nitrite, H_2O_2, hypoxia, intracellular growth in macrophages	*E. coli* WT[a], *E. coli cyoB*, *M. smegmatis* WT[a], *Salmonella* serovar Typhimurium *hmp*	Oxygen transfer, increased cell growth	Joseph et al. (2012), Liu, He, and Chang (2004), Milani et al. (2003), Ouellet et al. (2007), Pathania, Navani, Rajamohan, and Dikshit (2002), Pawaria et al. (2008, 2007)

[a]Wild type.

biological chemistry of NO and related molecules is complex; NO in biological environments reacts with several targets and generates a large amount of species that in turn interact with other molecules (for an overview, see Lehnert & Scheidt, 2009). For instance, the intracellular NO toxicity mechanisms are related to the oxidation of NO and the production of various poisonous substances such as the nitrosating nitrosonium ion (NO^+), nitrite (NO_2^-) and peroxynitrite ($ONOO^-$) (Poole & Hughes, 2000). Peroxynitrite arises from the reaction of NO with superoxide (O_2^-) (Hughes, 1999), and in living cells reacts with carbon dioxide to produce the adduct ($ONOOCO_2^-$). This product is broken down via two mechanisms, the first one producing carbon dioxide (CO_2) and nitrate (NO_3^-) and the other evolving nitrogen dioxide (NO_2) and the carbonate radical ion (reviewed in Bowman, McLean, Poole, & Fukuto, 2011; Poole & Hughes, 2000).

Although NO is a toxic molecule, it has important functions in biological systems, mainly in signalling and defence mechanisms. Generation of NO by endothelial cells produces relaxation of vascular smooth muscle, in part via the activation of the guanylate cyclase (Murad, 1986). NO is produced by NO synthases (NOSs) in various cell types by the oxidation of L-arginine to L-citrulline and NO in an NADPH- and O_2-dependent manner (Stuehr, 1999). NOSs were firstly reported in mammals where three isoforms were identified, two constitutively expressed, endothelial and neuronal, and an inducible NO synthase able to produce high levels of NO in response to infection (reviewed in Alderton, Cooper, & Knowles, 2001; Lowenstein & Padalko, 2004).

Elevated concentrations of NO generated by the immune system cause inhibition of key bacterial enzymes such as terminal oxidases (Stevanin et al., 2000) and enzymes with iron–sulphur (Fe–S) centres (Gardner, Costantino, Szabo, & Salzman, 1997). NO produced by the host can diffuse easily across the bacterial membrane where it reacts with haems (Hausladen, Gow, & Stamler, 2001), Fe–S clusters (Cruz-Ramos et al., 2002) and thiols (Hess, Matsumoto, Kim, Marshall, & Stamler, 2005). Besides these toxic effects, NO is also a modulator of protein function in bacteria through the S-nitrosylation of specific cysteine thiols: the presence of NO and other RNS produces nitrosative stress, eliciting adaptive responses such as the expression of genes related to NO and RNS tolerance and detoxification (Avila-Ramirez et al., 2013; Flatley et al., 2005; Monk et al., 2008; Moore, Nakano, Wang, Ye, & Helmann, 2004; Mukhopadhyay, Zheng, Bedzyk, LaRossa, & Storz, 2004; Pullan et al., 2007; Richardson, Dunman, & Fang, 2006 and many others).

Exposure to endogenously produced NO has been proposed, particularly in microorganisms that use nitrite as an electron acceptor under anaerobic conditions. Indeed, enteric bacteria, such as *E. coli*, produce low concentrations of intracellular NO when nitrite is reduced to ammonia (Corker & Poole, 2003). Moreover, several bacteria possess NOS enzymes (reviewed by Bowman et al., 2011); bacterial NO production in Gram positives has been proposed to confer resistance to antibiotics through chemical modification and alleviation of oxidative stress (Gusarov, Shatalin, Starodubtseva, & Nudler, 2009), although the physiological role for NO production in Gram negatives is poorly understood.

Resistance to NO and RNS in bacteria has been related to the presence of haemoglobins (see Table 4.1) (Poole, 2005). However, other proteins have also been associated with tolerance to nitrosative stress. For instance, while the FHb Hmp represents the main mechanism for NO detoxification under aerobic conditions in *E. coli* (Gardner, Costantino, et al., 1998), the NO sensor NorR positively regulates the expression of the flavorubredoxin protein (NorV) and its reductase partner (NorW) during NO exposure under anaerobic conditions, and this system mediates the reduction of NO to nitrous oxide (N_2O) (Gardner & Gardner, 2002; Hutchings, Mandhana, & Spiro, 2002).

3. OXYGEN AND REACTIVE OXYGEN SPECIES IN BIOLOGY

The limited tolerance of microorganisms for oxygen is well known. Anaerobes and microaerophiles are unable to grow in air-saturated environments and committed aerobes experience deleterious or even lethal effects in hypoxic conditions. The stability of the oxygen molecule and its relatively weak capacity to accept electrons make it an inefficient oxidant of organic molecules such as amino acids and nucleic acids. However, oxygen readily interacts with organic radicals and transitions metals. Whereas oxygen is a relatively weak electron acceptor, superoxide (O_2^-), hydrogen peroxide (H_2O_2) and the hydroxide ion (OH^-) are much stronger oxidants (Imlay, 2003).

The term oxidative stress has been defined as 'a disturbance in the pro-oxidant–anti-oxidant balance in favour of the former, leading to potential damage' (Sies, 1991). The rate of O_2^- and H_2O_2 formation dictates the level of oxidative stress experienced by a microorganism (Imlay, 2003). Partially reduced oxygen species form via the reduction of oxygen by electron transfer enzymes, especially the ubiquitous flavoenzymes (Massey et al., 1969), generating reactive oxygen species (ROS) (O_2^-, H_2O_2) (Fridovich, 1999; Imlay, 2008). High levels of ROS cause irreversible damage such as the oxidation of Fe–S proteins and DNA (Farr, D'Ari, & Touati, 1986; Farr & Kogoma, 1991; Imlay, 2008; Jang & Imlay, 2007). Oxidation of DNA is a consequence of a very rapid reaction between H_2O_2 and transition metals such as ferrous iron to form hydroxyl radicals (via the Fenton reaction) that in turn attack both base and sugar molecules producing irreversible damage to the double strand (Henle et al., 1999; Hutchinson, 1985).

During bacterial infection, high concentrations of O_2^- are produced within the phagocyte mainly via the NADPH oxidase complex (Babior, 1999; Miller & Britigan, 1997), which plays a major antibacterial role (Huang & Brumell, 2009; Mastroeni et al., 2000). In humans, chronic granulomatous disease, developed as a consequence of a genetically defective production of O_2^-, causes recurrent life-threatening bacterial and fungal infections (van den Berg et al., 2009; Winkelstein et al., 2000).

Oxidative stress triggers specific microbial defence responses. The upregulation of the SoxR(S) and OxyR regulons in the presence of ROS (Aussel et al., 2011; Greenberg, Monach, Chou, Josephy, & Demple, 1990; Tsaneva & Weiss, 1990; Zheng et al., 2001) is followed by the induction of a number of protective proteins, including superoxide dismutase (SOD) and catalase. *E. coli* strains that lack SOD grow poorly under aerobic but not anaerobic conditions due to the accumulation of O_2^- (Carlioz & Touati, 1986), and the accumulation of H_2O_2 in a catalase/peroxidase mutant (Jang & Imlay, 2007; Park, You, & Imlay, 2005), establishing the preponderant role for these enzymes as ROS scavengers in bacteria (Imlay, 2008).

4. RESPIRATORY METABOLISM OF *CAMPYLOBACTER*

Since *Campylobacter* possesses an incomplete glycolytic pathway and lacks fermentative metabolism (Parkhill et al., 2000), oxidative phosphorylation provides the major route for energy production. Nevertheless, the bioenergetics and stress responses of this microorganism remain poorly understood. *C. jejuni* NCTC 11168 (Parkhill et al., 2000) encodes a highly complex branched electron transport system including two terminal oxidases, and a variety of terminal reductases involved in both microaerobic and anaerobic respiration (Sellars, Hall, & Kelly, 2002; Smith, Finel, Korolik, & Mendz, 2000). However, the presence of an oxygen-dependent ribonucleotide reductase (class II) that is essential for DNA synthesis prevents growth of *C. jejuni* under strictly anaerobic conditions (Sellars et al., 2002). It is, therefore, unsurprising that when oxygen transfer rates are manipulated via variation of media volumes in batch culture (Pirt, 1985), there is a clear correlation between oxygen availability and growth of *C. jejuni* (Wainwright et al., 2005).

4.1. Microaerobic respiration

C. jejuni has two terminal oxidases, a cytochrome *bd*-type quinol oxidase and a *cb*-type cytochrome *c* oxidase (Fouts et al., 2005; Hofreuter et al., 2006; Parkhill et al., 2000). However, spectral signals produced by high-spin haems *b* and *d*, characteristic of these oxidases, are absent in *C. jejuni* cells. Expression of the *cydAB* operon has been linked to formate respiration and survival during growth under atmospheres of 5% oxygen. Since resistance to cyanide is normally associated with CydAB complexes, which is not the case with the *bd*-type oxidase of *C. jejuni*, this oxidase was renamed CioAB (cyanide-insensitive oxidase). Expression of *cioAB* increases at higher oxygen tensions, and CioAB shows a relatively low affinity for oxygen ($K_m = 0.8$ μM) and a V_{max} of >20 nmol mg^{-1} s^{-1}. The *cb*-type cytochrome *c* oxidase, a cyanide-sensitive complex encode by *ccoNOQP*, plays a major role in respiration under microaerobic conditions, shows a higher oxygen affinity ($K_m = 0.04$ μM) and a V_{max} of 6–9 nmol mg^{-1} s^{-1} and might be essential for viability (Jackson et al., 2007). The presence of two terminal oxidases with different affinities for oxygen may represent an adaptation to changing oxygen tensions; CioAB is suitable under microaerobic conditions, and the *cb*-type oxidase allows aerobic respiration during conditions of extreme oxygen limitation.

4.2. Respiration under oxygen-limited conditions

The ability of *C. jejuni* to use a range of alternative respiratory substrates such as fumarate, nitrate, nitrite, trimethylamine-*N*-oxide (TMAO), dimethyl sulphoxide (DMSO), mesaconate, crotonate, sulphite, lactate and hydrogen peroxide has been demonstrated (Guccione et al., 2010; Mead, 1989; Myers & Kelly, 2005; Pittman et al., 2007; Saengkerdsub, Kim, Anderson, Nisbet, & Ricke, 2006; Sellars et al., 2002; Weerakoon, Borden, Goodson, Grimes, & Olson, 2009; Weingarten, Taveirne, & Olson, 2009). Even though anaerobic conditions represent a stress condition for *C. jejuni*, where growth has proved to be severely impaired, micro-aerobic cultures with extremely restricted oxygen transfer (oxygen-limited conditions) are able to grow in media supplemented with fumarate, nitrate, nitrite, TMAO or DMSO, indicating alternative pathways for electron acceptor-dependent energy conservation (Sellars et al., 2002). Variability of oxygen availability during host colonisation can be expected, although the behaviour of *Campylobacter* in the presence of different electron acceptors under oxygen-limited conditions *in vivo* remains unexplored.

4.3. Impact of nitrosative stress upon *Campylobacter* respiration

Campylobacter is exposed to nitrosative stresses from a variety of sources from different environmental niches, in addition to NO arising from the action of NOS (Section 2). Production of NO other than the specific host defence response is likely to come from other nitrogenous species, such as nitrite on the skin (Suschek, Schewe, Sies, & Kroncke, 2006) and in the oral cavity (Rausch-Fan & Matejka, 2001). Indeed, reaction of dietary nitrite with stomach acid produces NO, and reduction of dietary nitrate to nitrite (Olin et al., 2001) by oral microflora exacerbates this process. Furthermore, nitrates are used as a preservative on meat, further increasing the exposure of *Campylobacter* to sources of nitrosative stress.

Like most bacteria, aerobic respiration of *Campylobacter* is inhibited by NO. However, *Campylobacter* has a range of respiratory complexes involved in anaerobic respiration that can process sources of nitrosative stress: *C. jejuni* NCTC 11168 possesses both periplasmic nitrite reductase (Nrf) and nitrate reductase (Nap) (Pittman & Kelly, 2005; Sellars et al., 2002). NrfA is a pentahaem cytochrome *c* Nrf, playing a role as the terminal enzyme in the dissimilatory reduction of nitrite to ammonia (Pittman & Kelly, 2005; Sellars et al., 2002). The *nap* operon in *C. jejuni* is composed of *napAGHBLD*, and the periplasmic machinery consists of two subunits, NapA and NapB. NapA ($\sim$90 kDa) is the catalytic subunit that reduces nitrate to nitrite and contains a bis-molybdenum guanosine dinucleoside cofactor and a [4Fe–4S] group, and NapB ($\sim$16 kDa) is a di-haem *c*-type cytochrome (Butler et al., 2001). Even though this operon lacks *napC*, a gene encoding a tetra-haem cytochrome that couples nitrate reduction to quinol oxidation in *E. coli* (Brondijk, Nilavongse, Filenko, Richardson, & Cole, 2004), *C. jejuni* does encode a putative *napC* gene that is probably related to the Nrf system (reviewed by Pittman & Kelly, 2005).

5. *CAMPYLOBACTER* SINGLE-DOMAIN GLOBIN, CGB: FUNCTIONAL AND STRUCTURAL CHARACTERISATION

5.1. Functional characterisation

The presence of a globin-like sequence in *C. jejuni* NCTC 11168 (Parkhill et al., 2000) (Cj1586) resembling that of the Vgb was reported for the first time in 2001 (Bollinger et al., 2001). Since then, important progress has contributed towards the characterisation of the physiological role, structural characteristics and regulation of this haemoglobin, now known as Cgb (for *Campylobacter* globin). Cgb is a member of the Mb-like haemoglobins, belonging to the single-domain haemoglobin sub-family (SDHb) (Vinogradov et al., 2013). Cgb, composed of 140 residues (16.1 kDa), shares 42% amino acid identity with the Vgb. Although Cgb lacks the reductase domain present in the FHbs, it shares a high level of sequence homology with the globin domains of FHbs from *Bacillus subtilis*, *Salmonella enterica* serovar Typhimurium and *E. coli* (39%, 34% and 33%, respectively) (Elvers et al., 2004).

The first evidence for the involvement of Cgb in resistance against nitrosative stress came from a study aimed to measure the protection provided by the heterologous expression of several bacterial haemoglobins in *E. coli* (Frey et al., 2002). Cultures of Cgb-expressing cells showed a significant improvement in growth compared to the parental strain in the presence of sodium nitroprusside (SNP), a clinically relevant NO donor (Miller & Megson, 2007; Wang et al., 2002). It is now well known that resistance against NO and nitrosative stress agents is linked to the presence of Cgb in *Campylobacter*. There are several studies that support this: (i) growth of *C. jejuni* in the presence of the nitrosative agent *S*-nitrosoglutathione (GSNO) is impaired by mutation of *cgb* (Avila-Ramirez et al., 2013; Elvers et al., 2004); (ii) tolerance to GSNO, SNP and NO is significantly diminished in strains lacking *cgb* (Elvers et al., 2004; Wainwright et al., 2005); (iii) Cgb consumes NO and protects aerobic respiration of *C. jejuni* and *E. coli* from NO-mediated respiratory inhibition (Avila-Ramirez et al., 2013; Elvers et al., 2004; Monk et al., 2008); (iv) Colorectal adenocarcinoma cells infected with a *cgb*-defective strain accumulate a higher concentration of NO than cells infected with the parent strain or uninfected cells (Elvers et al., 2004) and (v) Cgb is a member of a small regulon positively controlled by NssR under nitrosative stress conditions (see Section 7.1), and the expression of *cgb* is triggered by GSNO, SNP,

NOC-18, spermine NONOate, nitrate and nitrite (Elvers et al., 2005; Pittman et al., 2007).

Little work has been performed on the reaction intermediates of the Cgb-mediated detoxification of NO. However, a close homolog in *E. coli*, the FHb Hmp, is an extensively studied microbial NO detoxification mechanism converting NO and O_2 to the harmless nitrate ion via a dioxygenase (NO dioxygenase, NOD) (Gardner et al., 2006, 2000; Gardner, Gardner, Martin, & Salzman, 1998) or denitrosylase reaction (Hausladen et al., 1998, 2001). Biophysical studies of the Cgb haem pocket support the conversion of NO to nitrate (Lu, Mukai, et al., 2007) via the following general reaction:

$$CgbFe(II) + O_2 + NO \rightarrow CgbFe(III) + NO_3^- \qquad (4.1)$$

If Cgb catalyses the conversion of NO to nitrate, there are two conditions that must be fulfilled *in vivo*; oxygen availability and the existence of an efficient reductase system for the regeneration of the ferrous haem cofactor (Fe(II)) following oxidation by NO. The correlation between oxygen availability and resistance to nitrosative stress has indeed been demonstrated in *Campylobacter*, and resistance to NO and GSNO under microaerobic or oxygen-limited conditions shows important differences. For instance, cultures pre-treated with GSNO at higher rates of oxygen diffusion show that respiration and growth are better protected from NO inhibition, and consumption of NO is also more efficient (Avila-Ramirez et al., 2013; Monk et al., 2008). Since the expression of Cgb occurs maximally in the presence of oxygen, albeit microaerobic conditions (Avila-Ramirez et al., 2013; Elvers et al., 2005, 2004; Monk et al., 2008; Wainwright et al., 2005), an O_2-dependent NO detoxification mechanism for Cgb seems plausible. However, nitrate production by Cgb has not yet been demonstrated either *in vitro* or *in vivo*. Indeed, overexpression of Cgb in *E. coli* failed to increase the NO consumption activity measured in soluble cell extracts, and there were no differences in nitrate production compared to the control (Frey et al., 2002), although it is possible that the presence of the FHb Hmp in the system could be masking the effect of Cgb.

5.2. Cgb reduction: The redox partner conundrum

The conversion of NO and O_2 to NO_3^- by single-domain globin proteins involves the oxidation of the haem cofactor, which must be reduced for subsequent enzymatic turnovers. Hence, the existence of partner reductase proteins for single-domain globins has been speculated upon for many years. The use of electrons from the respiratory chain for the reduction of Cgb has arisen as a possibility. Indeed, the ability of Vgb and other bacterial haemoglobins to associate with the cytoplasmic membrane (Park, Kim, Howard, Stark, & Webster, 2002) and their involvement in oxygen transfer (Dikshit et al., 1992) support this idea.

FHbs circumvent the need for a separate redox partner by encoding a reductase domain. The reduction of the ferric haem (Fe(III)) takes place via electron transfer from the reductase domain (or FNR, ferredoxin-NADP reductase-like domain) to the N-terminal haem domain in an NAD(P)H-dependent reaction via a non-covalently bound FAD (Gardner, Costantino, et al., 1998; Hausladen et al., 1998; Hernandez-Urzua et al., 2003). Since there are important differences between Cgb and FHbs, it has been suggested that Cgb may not interact with a protein homologous to the reductase domain of FHbs. Indeed, the residue Lys-84 conserved in the globin domain of FHbs is responsible for the formation of a salt bridge between the domains (Ermler, Siddiqui, Cramm, & Friedrich, 1995), yet is absent in Cgb.

A lactate dehydrogenase enzyme, encoded by a gene (*cj1585c*) adjacent to *cgb*, has been suggested as a candidate for a redox partner for Cgb (Thomas et al., 2010). The spectroscopic characteristics of this protein make it a good candidate for the Cgb electron donor, and *cj1585c* is also upregulated by NO (Avila-Ramirez et al., 2013). However, the upregulation of *cj1585c* in response to NO occurs only in oxygen-limited conditions, where *cgb* induction is diminished (Avila-Ramirez et al., 2013) (see Section 7). Hence, the involvement of Cj1585c in an oxygen-dependent detoxification mechanism seems rather unlikely.

5.3. Structural characterisation

At least three classes of bacterial globin are recognised, namely, the FHbs, the single-domain 'myoglobin-like' globins and the truncated globins: for a recent review, see Poole and Shepherd (2010). FHbs are distinguished by the presence of an N-terminal globin domain (a 3-over-3 α-helical fold similar to Mb) with an additional C-terminal domain with binding sites for FAD and NAD(P)H. Single-domain globins also exhibit a 3-over-3 α-helical fold similar to Mb but have no separate C-terminal domain. All globin subunits consist of 6–8 α-helical segments that fold around a haem group, which is coordinated to a histidine residue via a central iron atom. Indeed, the structure of Cgb conforms to this general globin fold (Fig. 4.1).

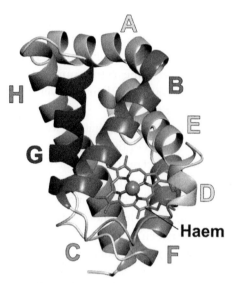

Figure 4.1 Backbone topology of *Campylobacter* globin, Cgb. The figure depicts the 3-over-3 α-helical fold of Cgb with the haem cofactor (PDB id = 2WY4; Shepherd et al., 2010). Helices/regions are labelled according to conventional globin nomenclature. (For colour version of this figure, the reader is referred to the online version of this chapter.)

The structure of cyanide–bound Cgb was solved by X-ray crystallography to a resolution of 1.35 Å (Shepherd et al., 2010) and was found to adopt a classic 3-over-3 α-helical globin fold (Fig. 4.1). The helices constituting the globin fold are labelled A–H in sequence order, according to standard globin nomenclature, and the amino acids within each helix are also numbered sequentially. The C and D regions adopt 3_{10}- and α-helical conformations, respectively, and the ligand-binding (distal) pockets of Cgb are constructed from the B-, E- and part of the G-helices (Fig. 4.1). Structural overlays (Fig. 4.2) indicate considerable structural homology with Vgb (RMSD = 1.30 Å, 110 residues), the globin domain of Hmp (RMSD = 1.64 Å, 134 residues) and sperm whale myoglobin (swMb) (RMSD = 1.83 Å, 116 residues). Whereas Vgb is a dimer (Tarricone et al., 1997) and Hmp has an FAD-binding reductase domain (Ilari, Bonamore, et al., 2002), Cgb was purified and crystallised as a monomer with a single globin domain.

The identity of the amino acids in the B10 and E7 positions (i.e. the 10th residue on the B-helix and the 7th residue on the E-helix) is known to be important for modulating ligand binding. In mammalian globins, the E7 position is almost invariably occupied by a histidine. The HisE7 of Mb

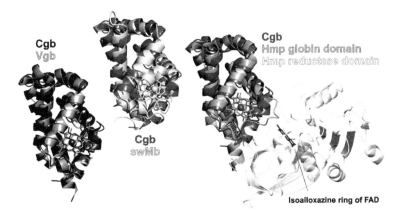

Figure 4.2 Structural overlays of Cgb. Overlays of the Cgb backbone with Vgb from *Vitreoscilla stercoraria* (PDB id = 1VHB; Tarricone et al., 1997) (A), sperm whale myoglobin (PDB id = 2JHO; Arcovito et al., 2007) (B), and the N-terminal domain of Hmp from *E. coli* (PDB id = 1GVH; Ilari, Bonamore, Farina, Johnson, & Boffi, 2002) (C). (For interpretation of the references to colour in this figure legend, the reader is referred to the online version of this chapter.)

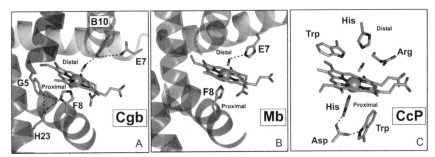

Figure 4.3 The haem-binding cleft of Cgb (PDB id = 2WY4; Shepherd et al., 2010), sperm whale myoglobin (PDB id = 2JHO; Arcovito et al., 2007) and yeast cytochrome *c* peroxidase (PDB id = 2PCC; Pelletier & Kraut, 1992). Structures contain either a cyanide or sulphenate ion in the distal pocket. Only side chains that interact with the proximal or distal ligands are shown. (For colour version of this figure, the reader is referred to the online version of this chapter.)

stabilises the haem–bound dioxygen by H-bonding to it (Fig. 4.3), whereas the B10 position is typically occupied by hydrophobic residues. In contrast to mammalian globins such as Mb, the distal residues of peroxidases and oxidases are much more polar, for example, the B10 and E7 residues in Cgb (an NOD) are occupied by tyrosine and glutamine (Fig. 4.3A and B), and the GlnE7 residue is further stabilised by a hydrogen-bonding network involving Lysine 46 in the D-region (not shown). As a result of this stabilising hydrogen-bonding network, Cgb exhibits higher ligand affinities compared to mammalian globins such as Mb. This is evidenced by the dissociation constants for oxygen binding being 6 nM and 0.86 μM for Cgb (Lu, Mukai, et al., 2007) and swMb (Gibson, Olson, McKinnie, & Rohlfs, 1986), respectively. The presence of such a hydrogen-bonding network is not unique to Cgb, as both Hmp and Vgb also possess tyrosine and glutamine residues at the B10 and E7 positions, respectively, although neither Hmp nor Vgb structures exhibit hydrogen-bonding interactions between these residues (Shepherd et al., 2010). This is possibly because the region surrounding the E7 residue is disordered in the Vgb structures and is interacting with the reductase domain in Hmp. In addition, the organisation of hydrogen-bonding residues in Cgb is reminiscent of cytochrome c peroxidase (CcP; Fig. 4.3C), which is unsurprising since the NOD mechanism of Cgb has been proposed to proceed via a peroxidase-like mechanism that requires a positively polar distal environment (Lu, Mukai, et al., 2007).

In addition to the hydrogen-bonding network found in the distal pocket of Cgb, another network of hydrogen bonds is found in the proximal pocket, where the haem cofactor is coordinated to the F8 histidine (Fig. 4.3A). The 'catalytic triad' of H23Glu, G5Tyr and F8His form a stable network of hydrogen bonds, resulting in a 90° rotation in the imidazole ring of the axial F8His ligand, similar to that of CcP (Fig. 4.3C). This contrasts to the F8His residue of Mb (Fig. 4.3B), which does not interact with other amino acid side chains (Fig. 4.3B). The presence of a hydrogen-bonding network in the proximal pocket is proposed to impose imidazolate (negatively charged) character upon the F8His residue, which is also implicated in the peroxidase-like mechanism of Cgb catalysis (Lu, Mukai, et al., 2007).

5.4. Biophysical and mechanistic characterisation

The structure of Cgb is consistent with an imidazolate character of the axial F8His and a positively polar distal environment, both necessary for the well-characterised 'push–pull' model for peroxidase-like enzymes (Poulos, 1996). Such a mechanism was proposed for O—O bond cleavage of the proposed peroxynitrite intermediate (Eq. 4.2) of Hmp during an NOD reaction (Mukai, Mills, Poole, & Yeh, 2001) and is depicted in Fig. 4.4.

$$Fe^{2+}-O-O+\bullet N=O \rightarrow \left[Fe^{3+}-O-O-NO^{-}\right] \rightarrow Fe^{3+}+NO_{3}^{-} \quad (4.2)$$

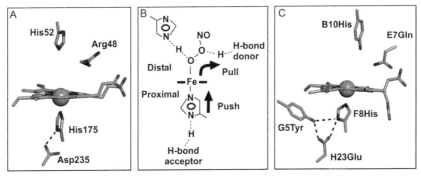

Figure 4.4 The 'push–pull' model for the activation of the peroxynitrite-bound intermediate of Cgb. (A) The active site of yeast cytochrome *c* peroxidase (PDB id = 2PCC; Pelletier & Kraut, 1992). (B) The 'push–pull' model for the cleavage of O—O bonds in peroxidases. (C) The haem-binding cleft of Cgb (PDB id = 2WY4; Shepherd et al., 2010). (For colour version of this figure, the reader is referred to the online version of this chapter.)

Evidence for the imidazolate character of the axial F8His ligand was provided by resonance Raman spectroscopy, where the iron-histidine-stretching mode was found to be 251 cm^{-1} (Lu, Egawa, Wainwright, Poole, & Yeh, 2007), an unusually high frequency for a globin protein. This spectral feature was dependent upon pH, suggesting that protonation of the H23Glu residue perturbs the imidazolate character of the F8His residue. The requirement for this proximal hydrogen-bonding network for the imidazolate character of the F8His residue was later confirmed by mutagenesis, whereby substitution of either H23Glu or G5Tyr decreased the iron-histidine-stretching frequency to 225 cm^{-1} (Shepherd et al., 2010), close to that for the neutral F8His of Mb. Intriguingly, mutation of the G5Tyr residue to Phe also resulted in the partial formation of a six coordinate haem species, which was attributed to the axial coordination of the haem iron by the B10Tyr in the distal pocket, highlighting the potential influence of the proximal hydrogen-bonding network upon ligand binding. To further demonstrate the effect of the proximal residues upon ligand binding, the E134A mutant (H23) was shown to have a higher *on* rate for negatively charged CN$^-$ ions: mutation of the H23Glu will diminish the imidazolate character (negative charge) of the F8His residue, raising the formal charge (positive) of the haem iron.

The positively polar distal environment was characterised through the analysis of a CO-bound derivative via a resonance Raman approach. CO has been widely used to probe the electronic environments of haem cofactors, largely because of the influence of the distal and proximal charge balance upon the vibrational spectroscopy of the iron—CO bond. Essentially, the Fe–C–O moiety can exist in one of two extreme resonance structures, depending upon the extent to which the d_π orbital of the iron donates electron density to the π^* orbital of CO shown in Eq. (4.3):

$$\text{His} - \text{Fe}^{\delta-} - \text{C} \equiv \text{O}^{\delta+}(\text{I}) \rightarrow \text{His} - \text{Fe} = \text{C} = \text{O}(\text{II}) \qquad (4.3)$$

Generally, a positive polar environment destabilises form (I) and promotes the 'backbonding' interaction, leading to a stronger Fe—CO bond and a weaker C—O bond. Hence, the frequency of the v_{Fe-CO} mode is typically inversely correlated with that of the v_{C-O} mode in a linear fashion (Egawa & Yeh, 2005; Rousseau, Li, Couture, & Yeh, 2005; Spiro & Wasbotten, 2005). Form II is referred to as closed, where the haem-bound CO is stabilised by B10Tyr and E7Gln, whereas form I is known as the open conformation that lacks these hydrogen-bonding interactions. In the case of Cgb, both resonance conformers were observed: CO was proposed to be stabilised by a hydrogen bond from the Tyr-28 (B10)/Gln-52 (E7) pair (closed), or the ligand remained bound without a hydrogen bond (open). Globins that can exist in both these resonance forms, such as Cgb, Hmp and HbN from *M. tuberculosis*, have all been implicated in protecting the respective organisms from nitrosative stress (Couture et al., 1999; Frey & Kallio, 2005; Mukai et al., 2001).

Further evidence for the peroxidase-like character of Cgb was investigated through testing the reactivity towards peroxides. Surprisingly, Cgb exhibited lower reactivity towards hydrogen peroxide compared to Mb, although rapid decomposition of the organic peroxide *m*-chloroperbenzoic acid by Cgb demonstrates its potential function as an organic peroxidase (Cgb was fivefold faster than Mb). In addition, peroxidase-like haemoproteins with imidazolate axial ligands typically have lower redox midpoints than haem proteins with neutral axial ligands. For example, horseradish peroxidase has a midpoint potential of -250 mV (Yamada, Makino, & Yamazaki, 1975), compared to 58 mV for Mb (Rayner, Stocker, Lay, & Witting, 2004). Unsurprisingly, Cgb has a low redox midpoint of -134 mV at pH 7 (Shepherd et al., 2010), reflecting the negative charge imposed upon the axial histidine by the proximal hydrogen-bonding network. Indeed, mutation of the H23Glu residue increased the redox midpoint to -110 mV, reflecting coordination of the haem by a more neutral axial histidine.

Taken together, the structural and biophysical data are consistent with a peroxidase such as 'push–pull' mechanism (Fig. 4.4) that facilitates the isomerisation of the proposed peroxynitrite intermediate (Eq. 4.2) in a dioxygenase reaction. However, as mentioned in Section 6.1, the reaction mechanism may also proceed via a denitrosylase route whereby NO binds to the ferrous haem first (Fig. 4.5). Indeed, this has been shown to occur *in vitro* with Hmp of *E. coli* (Hausladen et al., 2001), where the unliganded ferrous haem binds preferentially to NO ($K_d = 8$ pM) compared to oxygen ($K_d = 12$ nM). However, a denitrosylase reaction involves the formation of a nitrosyl-haem complex, which requires the cleavage of an N—O bond rather than an O—O linkage, a less thermodynamically favourable process. Given that Cgb is able to bind NO in both the ferrous and ferric forms (Shepherd, Bernhardt, & Poole, 2011), whereas only ferrous Cgb can bind oxygen, the relative availability of NO and oxygen and the redox poise of the cell will influence the route via which Cgb can detoxify NO.

An alternative mechanism has been investigated for Cgb, involving the formation of an oxoferryl (Fe(IV)=O) species (Shepherd et al., 2011). Since *Campylobacter* will be exposed to both hydrogen peroxide and NO during colonisation of the host, it was proposed that ferric Cgb could react with hydrogen peroxide to form an oxoferryl species, as is the case for horseradish peroxidase. The oxoferryl species would subsequently react with NO to produce nitrite and nitrate. However, although forward rate constants for the formation of the oxoferryl species were comparable to other globins, they were orders of magnitude slower than *bona fide* peroxidases such as horseradish peroxidase. Furthermore, reaction of NO with oxoferryl Cgb was extremely slow, which rules out this mechanism as a viable route for NO detoxification *in vivo*.

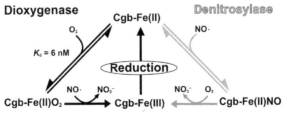

Figure 4.5 Proposed denitrosylase and dioxygenase routes in the catalytic cycle of Cgb. (For colour version of this figure, the reader is referred to the online version of this chapter.)

6. *CAMPYLOBACTER* TRUNCATED GLOBIN, CTB: FUNCTIONAL AND STRUCTURAL CHARACTERISATION

6.1. Functional characterisation

In 2004, Elvers et al. identified a second haemoglobin-like protein (Cj0465c) in addition to the single-domain globin Cgb. This locus encodes a truncated globin designated Ctb (*Campylobacter*-truncated globin) that is constitutively expressed at low levels, and under nitrosative stress conditions (GSNO and *S*-nitroso-*N*-acetylpenicillamine (SNAP)), expression is increased in an NssR-dependent manner (see Section 7) (Wainwright et al., 2005).

Ctb is a member of the truncated haemoglobin family (T family), a distinctive phylogenetic group characterised by the 2/2 Mb-fold with members in archaea, eukaryotes and bacteria (Vinogradov et al., 2013; Wittenberg, Bolognesi, Wittenberg, & Guertin, 2002). More specifically, Ctb is classified within the poorly explored group III of the TrHbIII (Pesce et al., 2000; Vinogradov et al., 2005). Although Ctb has been studied for several years (Bolli et al., 2008; Lu, Egawa, et al., 2007; Nardini et al., 2006; Wainwright, Wang, Park, Yeh, & Poole, 2006), its physiological function remains unclear. Attempts to elucidate the role of this enigmatic globin include testing its ability to improve cellular growth in microaerophilic environments and its capacity to confer cellular protection against toxic oxygen tensions and nitrosative stress conditions.

The ability of Vgb to improve microaerobic growth in *E. coli* and other bacterial and eukaryotic species has been extensively documented (reviewed by Frey, Shepherd, Jokipii-Lukkari, Haggman, & Kallio, 2011). Analysis of a *C. jejuni* strain that lacks *ctb* revealed a slower growth rate during the stationary phase in microaerobic conditions compared to the wild-type parental strain (Wainwright et al., 2005), suggesting a role for Ctb in oxygen transfer. However, growth under oxygen-limited conditions failed to show these differences. The influence of Ctb upon oxygen consumption is equally interesting: (i) respiration rates of the *ctb* mutant were 50% compared to wild type; (ii) the K_M values for oxygen among the *ctb*, *cgb* and *cgb/ctb* mutants and the wild type are all comparable, but the V_{max} determined by the deoxygenation of oxy-leghaemoglobin (Contreras et al., 1999; D'mello et al., 1994, 1995, 1996; Smith, Hill, & Anthony, 1990) is greater for the *ctb* mutant than the others (Wainwright et al., 2005) and (iii) the oxygen consumption rates of the *ctb*-lacking strain measured *in vivo* decreases below

~1 µM but increases in the range between 1 µM and air saturation concentrations (Wainwright et al., 2005). This data suggest that Ctb could be a regulator of cellular oxygen consumption: given that high external oxygen tensions are toxic and consumption of O_2 by Ctb might offer protection during microaerophilic growth.

The studies discussed earlier suggest a possible role for Ctb in promoting microaerobic growth and moderating respiration in *C. jejuni*. Nonetheless, understanding the physiological relevance of these observations is not straightforward. For instance, the *ctb* gene is upregulated under nitrosative stress conditions, yet expression is not induced by variations in oxygen concentration or oxidative stress (Wainwright et al., 2005). Indeed, induction of Ctb expression by GSNO and SNAP was demonstrated (by Western blotting), but paraquat or peroxides did not influence Ctb production. It is possible that the differences in expression patterns between Cgb and Ctb have physiological relevance; the expression of Cgb is strictly dependent upon NssR (see Section 7.1) and, as far as we know, only occurs under nitrosative stress conditions (Elvers et al., 2005; Pittman et al., 2007). However, Ctb is constitutively expressed at low levels and induced under nitrosative stress conditions (Wainwright et al., 2005), raising the question 'is the constitutive level of Ctb enough to play a significant role in regulation of intracellular oxygen tensions?' Furthermore, are higher levels of this globin needed to play the same, or perhaps a different role, under nitrosative stress conditions? Nonetheless, Ctb fails to offer growth protection in the presence of NO or other RNS in *C. jejuni* (Wainwright et al., 2005). When *ctb* was expressed under the control of an arabinose-inducible promoter in an *E. coli hmp* mutant, aerobic respiration was protected from NO inhibition compared to cells carrying an empty vector. Interestingly, cells expressing Ctb consumed NO at similar rates under aerobic and anaerobic conditions, while the controls accumulated a significant concentration of NO (Avila-Ramirez et al., 2013), perhaps suggesting a secondary role in NO detoxification in *Campylobacter*.

6.2. Structural characterisation

The structure of dimeric cyanide-bound Ctb was solved via X-ray crystallography to a resolution of 2.15 Å (Nardini et al., 2006) and was the first class III-truncated globin to be structurally characterised. This revealed a 2-over-2 α-helical fold, as for all truncated globins characterised to date (Fig. 4.6A). Previous structures of type I and II globins revealed conserved Gly-Gly sequence motifs located at the AB inter-helical hinge and

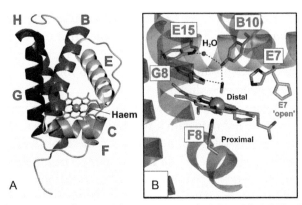

Figure 4.6 The structure of *Campylobacter*-truncated globin, Ctb. (A) The figure depicts the 2-over-2 α-helical fold of Ctb (Nardini et al., 2006) (PDB id = 2IG3). (B) The haem-binding cleft of Ctb, highlighting the hydrogen-bonding network in the distal pocket. The E7His residue was crystallised in 'closed' and 'open' conformations. (For colour version of this figure, the reader is referred to the online version of this chapter.)

C-terminal to the E-helix, which were considered to be essential for the 2-over-2-fold. However, these were absent in Ctb, further distinguishing Ctb from other classes of truncated globin.

Figure 4.6B depicts the distal- and proximal-binding pockets of Ctb with a cyanide ion bound in the active site. Whereas class I and class II-truncated globins have been shown to possess cavity/tunnel systems for ligand migration to/from the distal pocket, these features were absent in the Ctb structure. However, the crystal structure revealed the E7His residue in two alternative conformations, 'open' and 'closed' (Fig. 4.6B), which implicated E7His in a potential gating mechanism for ligand entry/exit, as is the case for Mb (Nardini et al., 2006). The distal pocket also contains conserved B10Tyr, G8Trp and E15Trp residues, all participating in a hydrogen-bonding network with an active site water molecule (Fig. 4.6B). The proximal pocket of Ctb is reminiscent of that of Mb (Fig. 4.3B). As for Mb, the F8His residue that provides axial coordination to the haem iron does not participate in hydrogen bonding to other amino acid side chains, although like Mb does hydrogen bond to the main chain carbonyl of an adjacent residue (Pro68 for Ctb). The F8His is shielded from the solvent by the F7Lys residue, which is electrostatically coupled to the haem propionates, whereas class I- and class II-truncated globins have solvent-accessible F8His residues.

Whereas the proximal histidyl coordination of the Ctb haem resembles that of Mb (Fig. 4.3B), the polar distal pocket of Ctb resembles that of CcP

(Fig. 4.3C). However, deletion of *ctb* does not induce sensitivity to a variety of peroxides (Wainwright et al., 2005), indicating an alternative function for Ctb other than peroxide decomposition.

6.3. Biophysical and mechanistic characterisation

As for Cgb, resonance Raman spectroscopy with CO as a structural probe (see Section 5.4) was used to confirm that Ctb also has a positively polar distal pocket (Lu, Egawa, et al., 2007). In addition, this approach was used to probe the role of B10Tyr and E7His residues in ligand stabilisation: these residues were proposed to contribute to the high affinity for oxygen exhibited by Ctb ($K_d = 5$ nM). Substitution of these residues for non-hydrogen-bonding amino acids combined with oxygen association/dissociation measurements and Raman spectroscopy yielded data consistent with the G8Trp residue providing the major contribution towards ligand stabilisation. Indeed, this was later confirmed by combining similar resonance Raman techniques with molecular dynamics to conclude that, when oxygen is bound, Ctb may exist in one of two conformations whereby both conformers rely upon the G8Trp for stabilising interactions (Arroyo Manez et al., 2011). In line with this important role for the G8Trp residue, mutation to Phe resulted in a large increase in the oxygen dissociation rate.

7. CONTROL OF GLOBIN EXPRESSION

7.1. Control of globin expression by NssR

The inducibility of the *cgb* gene was first described in cells of *C. coli* carrying a vector containing a reporter gene (*astA*) under control of the *cgb* promoter (Hendrixson & DiRita, 2003). Induction of Cgb expression by GSNO and SPN, but not by methyl viologen, was reported. This result and immuno-blotting tests using Cgb polyclonal antibodies were consistent with Cgb expression under nitrosative stress conditions (Elvers et al., 2004).

Screening of the *C. jejuni* genome (Parkhill et al., 2000) revealed the presence of three potential proteins that could sense NO and induce the expression of Cgb (Elvers et al., 2004); Fur, an iron sensor containing a Fe^{2+} cofactor that, in *C. jejuni*, is associated with iron acquisition, flagellar biogenesis and non-iron ion transport (Butcher, Sarvan, Brunzelle, Couture, & Stintzi, 2012), PerR, a Fe^{2+}-containing metalloregulator involved in peroxide stress responses (Butcher et al., 2012; Mongkolsuk & Helmann, 2002) and Cj0466, belonging to the Crp-Fnr superfamily of transcription regulators (Korner, Sofia, & Zumft, 2003). Mutants of either *fur* or *perR* exposed to GSNO maintained the Cgb expression pattern found in the parental strain. However, a *fur* mutant was more sensitive to GSNO (Elvers et al., 2005): chemical interaction between exogenous RNS and endogenously generated ROS was suggested. Indeed, hypersensitivity to nitrosative stress in the *E. coli fur* mutant has been related to the de-repression of the iron assimilation system, consequently generating oxidative stress (Mukhopadhyay et al., 2004). More recently, the insensitivity of a *C. jejuni fur* mutant to inhibition by NO was reported, and a role for the Fur-regulated genes in protection against nitrosative stress, related to augmentation of iron acquisition for repairing of damaged Fe–S and haem proteins, was suggested (Monk et al., 2008).

Mutation of the *cj0466* locus abolished GSNO-mediated Cgb expression; since Cgb expression was apparently dependent upon the product of Cj0466 under nitrosative stress conditions, the protein was designated NssR (nitrosative stress sensing regulator) (Elvers et al., 2005). Nonetheless, an increased sensitivity of the *nssR* mutant to methyl viologen that, in *Campylobacter*, is related to superoxide production (Purdy, Cawthraw, Dickinson, Newell, & Park, 1999) suggests an additional role for this regulator.

The transcriptional response to nitrosative stress, obtained by comparison of microarray data from microaerobic batch cultures of wild-type *C. jejuni* in the absence and presence of GSNO, showed up regulation of eight genes, encoding Cgb (*cj1586*), Ctb (*cj0565c*), four probable integral membrane proteins (*cj0830*, *cj0851c*, *cj0313* and *cj0430*), a probable peptide ABC transport system permease protein (*cj1582c*) and a hypothetical protein with unknown function (*cj0761*) (Elvers et al., 2005). However, microarray data comparing GSNO-treated cultures of the wild type and the *nssR* mutant defined the scope of the NssR-dependent response; *cgb*, *ctb*, *cj0761* and *cj0830* were upregulated, and this was also confirmed by RT-PCR (Elvers et al., 2005). A more detailed analysis of transcriptional changes upon addition of GSNO performed in continuous cultures showed transcriptional up regulation ($\geq$ 2-fold increased) of 97 genes from a total of 1632 genes (Monk et al., 2008). Upregulation of the NssR regulon was confirmed; *cgb* (320-fold), *ctb* (63.8-fold), Cj0761 (49.7-fold) and Cj0830 (12.3-fold) and the presence of Cgb and Ctb in cultures treated with GSNO were demonstrated by proteomic analysis (Monk et al., 2008). Interestingly, *nssR* was also modestly upregulated (2.2-fold), suggesting the existence of a regulatory mechanism affecting the expression of NssR under conditions of nitrosative stress. Other genes upregulated under this condition were *cj0757*, *cj0758* and *cj0759*, homologs to *hrcA*, *grpE* and *dnaK* that are involved in the heat-shock response (Parkhill et al., 2000). In addition, other upregulated genes included *cj0311*, encoding Ctc that is involved in general stress response (Volkert, Loewen, Switala, Crowley, & Conley, 1994), *trxA* and *trxB* (encoding a thioredoxin and its reductase) with a role in oxidative stress tolerance, and 9 genes from a group of 18 genes involved in iron transport that show transcriptional changes under low iron conditions (Holmes et al., 2005). Certainly, the effect of nitrosative stress in the upregulation of iron acquisition has been demonstrated in a number of other bacteria (Hernandez-Urzua et al., 2007; Moore et al., 2004; Mukhopadhyay et al., 2004; Richardson et al., 2006). The binding of NO to iron-bound Fur maintains the de-repression of genes under control of this transcriptional repressor.

Differences in the chemistry of S-nitrosothiols (SNOs) such as GSNO (a nitrosative agent) and NOCs (NO donors) and their biological interactions have been recently reviewed (Bowman et al., 2011). Indeed, SNOs and their derivative species play biologically relevant roles far beyond the simple release of NO (Hess et al., 2005). Due to the moderate stability of GSNO, this compound is widely used in bacterial growth experiments, although for the purpose of studying the biological effects of NO it is not ideal. For instance, transfer of NO^+ from GSNO to membrane thiols is proposed in *Bacillus* species (Morris & Hansen, 1981), but other studies suggest that toxicity is related to active transport: in *E. coli*, S-nitroso-L-cysteinylglycine, an SNO-derived nitrosated dipeptide, is transported inwards (via the Dpp-encoded dipeptide permease), resulting in intracellular transnitrosation reactions (Jarboe, Hyduke, Tran, Chou, & Liao, 2008; Laver et al., 2013). Interestingly, comparison of the transcriptional responses of *C. jejuni* exposed to GSNO (Monk et al., 2008) or a combination of NOCs (NOC-5 and NOC-7) shows common features. The Cgb and Ctb globins, heat-shock proteins and regulators seem to be similarly affected by either NO or GSNO (Smith et al., 2011). In contrast, transcriptional responses of *E. coli* in continuous culture in the presence of GSNO or NOCs revealed some similarities but several important differences (Pullan et al., 2007).

NO is released from GSNO under biological conditions (Singh, Hogg, Joseph, & Kalyanaraman, 1996), albeit at concentrations much lower than the GSNO concentration: 500 µM GSNO releases less than 5 µM NO (Jarboe et al., 2008). It is therefore possible that the transcriptional profile of *C. jejuni* in the presence of GSNO is a direct result of NO release. Indeed, it has been suggested that, in *E. coli*, the upregulation of *hmp* via the transcriptional regulator NsrR results from the submicromolar levels of NO released from GSNO. In the aforementioned transcriptomic studies of *C. jejuni*, cultures were exposed to 250 µM GSNO (Monk et al., 2008) or 10 µM NOC-5 plus 10 µM NOC-7 (Smith et al., 2011). It is unlikely that a concentration of NO under 2.5 µM (putatively released from GSNO) was responsible for the majority of the transcriptional changes, although given the extremely high affinity of bacterial globins for NO, it seems plausible that micromolar concentrations of NO should induce their expression.

Remarkable differences in the transcriptional response of *C. jejuni* to NO elicited by variations in oxygen availability have been recently reported (Avila-Ramirez et al., 2013). Genes induced by NOCs in microaerobic conditions and oxygen-limited conditions are mutually exclusive. In oxygen-limited conditions, only 11 genes show upregulation, and members of the NssR regulon were absent. Notably, the *cj1585c* gene encoding a lactate dehydrogenase, located next to *cgb*, was marginally upregulated under oxygen-limited conditions but not in microaerobic cultures. This protein has been proposed as the Cgb redox partner, although since *cj1585c* induction does not coincide with the expression of Cgb (the absence of Cgb in oxygen-limited cultures under conditions of nitrosative stress has been proved by immunoblotting), this now seems unlikely. This finding raises a number of new questions related to the ability of this microorganism to survive in the host, where variations in oxygen concentrations are expected.

7.2. NssR classification, targets and mechanistic insights

NssR belongs to branch E of the Crp-Fnr superfamily (Korner et al., 2003; Matsui, Tomita, & Kanai, 2013) and is the only member of this branch to be involved in NO regulation. Some members of the Crp-Fnr family do not directly interact with the signal molecule but with an independent sensor system that increases the concentration of the regulator (Fischer, 1994). However, since the expression of NssR is only very modestly increased under nitrosative stress conditions (Elvers et al., 2005), this suggests a role for NssR as an NO sensor and regulator of the response.

In the *C. jejuni* genome, the −35 regions for σ^{70}-controlled promoters are not conserved (Petersen, Larsen, Ussery, On, & Krogh, 2003). Analysis of promoters for *cgb*, *ctb*, *cj0761*, *cj0830* and *nssR* showed a typical −10 motif and an Fnr-like binding motif upstream of the −10 region (Elvers et al., 2005). The consensus sequence (TTAAC-N_4-GTAA) is similar to the suggested recognition sequences for the regulator of virulence gene expression PrfA in *Listeria monocytogenes* (TTAACA-N_2-TGTTAA) (Korner et al., 2003) and for the NO-sensing regulator Nnr from *Paraccocus pantotrophus* (TTAAC-N4-GTCAA) (Saunders, Ferguson, & Baker, 2000). Thus, the architecture of the four genes regulated by NssR is reminiscent of a class II Fnr-dependent promoter (Guest, Green, Irvine, & Spiro, 1996). The specificity of the NssR recognition for the Fnr-like motif in the *cgb* gene (TTAACacaaGTCAA) was demonstrated by comparing the wild-type and modified sequence (CTAACacaaGTCAG) in transcriptional fusions to *lacZ* (Wosten, Boeve, Koot, van Nuenen, & van der Zeijst, 1998): modification of the motif prevented NssR-mediated transcriptional activation (Elvers et al., 2005).

The *nssR* and the *ctb* genes are divergently transcribed sharing a putative NssR-binding sequence in their promoter regions, implicating NssR in the regulation of both *ctb* and *nssR* expression (Elvers et al., 2005). Certainly, the modest upregulation of *nssR* in continuous cultures exposed to GSNO (Monk et al., 2008) might be due to autoregulation. NssR binds to a specific *ctb* promoter region (−32 bp) in the absence of GSNO ($K_{d(app)}$ ~50 nM) (Smith et al., 2011), in agreement with the low but constitutive expression of Ctb in the absence of nitrosative conditions (Elvers et al., 2005). However, the presence of GSNO or NOCs in gel shift experiments did not show an increase in the affinity of NssR for the *ctb* promoter (Smith et al., 2011). This finding has significant implications since it is likely that NssR differs from other well-characterised regulators within the Crp/Fnr family that display enhanced DNA binding in the presence of their related signal molecules compared to the absence of the signal ($K_{d(app)}$ in the nanomolar range and in the micromolar range, respectively) (Green, Scott, & Guest, 2001). In this context, it seems plausible that NssR binds permanently to the promoters of genes that are members of the regulon and that conformational changes induce the binding of the transcriptional machinery upon exposure to NO. Nitrosylation of the sole cysteine or nitration of one of the several tyrosine residues present in the structure of NssR has been suggested as possible mechanisms (Smith et al., 2011). Nitration implies the production of peroxynitrite, a product of the reaction between NO and superoxide. Since superoxide production has been demonstrated for other bacterial haemoglobins (Membrillo-Hernandez, Ioannidis, & Poole, 1996), production of this compound by the oxyferrous form of Ctb seems plausible. However, it has been shown in a strain lacking *ctb* that Cgb expression is elevated under oxygen-limited conditions (Smith et al., 2011), conditions that are inconsistent with superoxide production.

Even though the exact mechanism for NssR sensing remains elusive, it has been suggested that iron is required for the interaction of NssR with NO or for the repair of NssR following interaction with NO. Indeed, in other members of the Fnr family, the presence of a 4Fe–4S group has been implicated in the NO response (Korner et al., 2003), although NssR lacks the cysteine signature for binding the cluster (Monk et al., 2008).

7.3. Structural modelling of NssR

To facilitate the development of hypotheses for the NssR mechanism, structural modelling was performed using the online RaptorX server (Kallberg et al., 2012; Peng & Xu, 2011). This highlights structural homology between NssR and transcriptional regulators that bind cyclic nucleotides, haem and α-ketoglutarate; of special interest are catabolite activator protein from *Thermus thermophilus* (top hit, 2EV0), CooA from *Carboxydothermus hydrogenoformans* (binds haem, 2FMY) and NtcA from *Synechococcus elongatus* (a global nitrogen regulator, 2XGX). Although RaptorX predicts structural homology between NssR- and cAMP-binding proteins (Fig. 4.7A), the most obvious candidate to occupy the ligand-binding cleft is haem, given the requirement of iron for NssR activity *in vivo* (Monk et al., 2008) and the involvement of this cofactor in NO- and CO-sensing by the bacterial transcription factors DNR (Giardina et al., 2008) and CooA (Nakajima et al., 2001). Indeed, a histidine residue, a common haem ligand, is predicted in the vicinity of the ligand-binding cleft (Fig. 4.7B). Given that NssR only

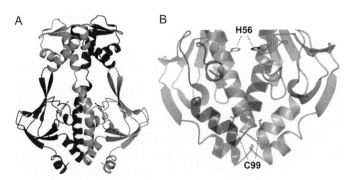

Figure 4.7 Structural modelling of the transcriptional regulator NssR. (A) Catabolite activator protein (PDB id = 4EV0) was used as a structural template for NssR. cAMP is shown bound to each monomer. (B) The C99 and H56 side chains are shown in the apoprotein dimer. (For colour version of this figure, the reader is referred to the online version of this chapter.)

has one cysteine residue (Cys99) and one histidine (His56) residue, it is unlikely that a complex Fe–S cluster is present, although a simple metal cofactor is possible. An intriguing idea is that the formation of an inter-molecular disulphide bond between the Cys99 residues of each monomer (Fig. 4.7B) can modulate transcriptional activation, especially since NssR activity is abolished under anoxic conditions (Avila-Ramirez et al., 2013) (oxygen influences cellular redox poise). Future investigations into the cofactor/ligand-binding capabilities of NssR and the potential for redox modulation of intermolecular disulphides may contribute towards the elu-cidation of the molecular switch that initiates the response of *Campylobacter* to nitrosative stress.

7.4. NrfA of *Campylobacter*

NrfA is a pentahaem cytochrome *c* nitrite reductase that catalyses the dissim-ilatory reduction of nitrite to ammonia (Pittman & Kelly, 2005; Sellars et al., 2002). A wide range of bacteria encode NrfA homologs, and the ability of these proteins to reduce both nitrite and NO has been demonstrated *in vitro* (Bamford et al., 2002; Costa et al., 1990). In *E. coli*, NrfA catalyses the reduc-tion of NO to ammonia under anaerobic conditions (Costa et al., 1990; Poock, Leach, Moir, Cole, & Richardson, 2002; van Wonderen, Burlat, Richardson, Cheesman, & Butt, 2008), and it has been suggested that NrfA could play a role in anaerobic NO detoxification together with the flavohaemoglobin Hmp and the flavorubredoxin NorV (Poock et al., 2002).

Pittman et al. explored the role of the constitutively expressed NrfA in the protection of *C. jejuni* against nitrosative stress by testing a range of mutants affected in key genes of the *nap* and *nrf* operons. Nitrite is produced from nitrate in stoichiometric quantities in oxygen-limited cultures of the *C. jejuni* parental strain, although an *napA*-lacking strain failed to produce nitrite: this is consis-tent with NapA being the sole Nap in *C. jejuni*. Likewise, consumption of nitrite under the same conditions was dependent upon NrfA and the partner electron donor NrfH. Under microaerobic conditions, an *nrfA* mutant was hypersensitive to NO donors (spermine NONOate, GSNO and SNAP), aerobic respiration was severely inhibited by NO and NO consumption was diminished compared to the parental strain (Pittman et al., 2007). This suggests that NrfA can offer a significant protection from nitrosative stress.

The level of protection attributable to NrfA and Cgb has not been directly compared. Growth of an *nrfA* mutant in the presence of nitrite was slower than the isogenic wild type, whereas growth of *cgb* or *nssR*

mutants under the same conditions displayed a more severe growth inhibition (Pittman et al., 2007). For these reasons, it has been suggested that NrfA
might provide a constitutive defence against NO and RNS allowing
Campylobacter to survive before Cgb induction is triggered (Pittman et al.,
2007). Interestingly, under oxygen-limited conditions (where NssR fails
to upregulate Cgb), loss of *nrfA* does not elicit sensitivity to GSNO
(Avila-Ramirez et al., 2013). This would seem to suggest that neither
Cgb nor NrfA play a role in resisting nitrosative stress under conditions
of oxygen limitation.

7.5. Induction of Cgb by nitrite and nitrate is independent of NrfA

Reactive nitrogen compounds such as NO and N_2O are produced during
growth with nitrite and nitrate as electron donors (Corker & Poole, 2003;
Weiss, 2006), and Hmp expression is induced by exposure to these compounds in *E. coli* (Poole et al., 1996). Given that Cgb is also induced by
nitrite and nitrate (Pittman et al., 2007), a similar mechanism of transcriptional activation is proposed for this globin, although the production of
NO or its reactive derivatives in *Campylobacter* remains undemonstrated.
The level of Cgb induction by nitrite and nitrate was similar in wild-type
and the *nrfA* strains. In contrast, in an *napA* mutant, Cgb expression occurred
only in the presence of nitrite implicating the reduction of nitrite as the
actual source of nitrosative stress (NO production perhaps) (Pittman
et al., 2007). However, since nitrite still induces Cgb in an *nrfA*-lacking
strain, this suggests that NrfA is not the source of nitrosative stress that
induces Cgb expression in *C. jejuni*. Furthermore, similar levels of nitrite-
mediated respiratory inhibition in the wild-type and the *nrfA* mutant suggest
an NrfA-independent mechanism for the generation of NO in this
microaerophilic microorganism.

The putative source of NO as a by-product of the nitrite reduction in
Campylobacter remains elusive. However, it is well known that globin proteins can produce NO from nitrite, which leads to an attractive hypothesis:
since Ctb is also upregulated in response to nitrosative stress, this truncated
globin may convert nitrite to NO, which is subsequently converted to
nitrate by Cgb. However, this is a speculative mechanism and remains to
be experimentally verified. Alternatively, it is possible that the molecular
switch for NssR activation could involve the binding of nitrite or nitrate
to NssR (Smith et al., 2011), rather than a direct interaction of the

transcriptional regulator with NO. Similarly, this remains a speculative hypothesis and will require experimental verification.

7.6. NO resistance in oxygen-limited conditions

An adaptive response to nitrosative stress that is independent of NssR, Cgb or NrfA has been suggested under oxygen-limited conditions (Avila-Ramirez et al., 2013). Under microaerobic conditions, NssR-dependent Cgb induction in cultures pre-treated with GSNO protects *C. jejuni* respiration from NO inhibition (Avila-Ramirez et al., 2013; Elvers et al., 2004). However, under conditions of oxygen limitation, GSNO pre-treatment produces a more modest, yet significant, decrease in NO-mediated inhibition of respiration (Avila-Ramirez et al., 2013) compared to untreated cells. As this protection proved to be independent of NssR, an inducible NO detoxification system, unrelated to either members of the NssR regulon or NrfA, seems to be playing a role. This putative system remains to be characterised, yet may be of importance during *Campylobacter* infection, where oxygen tension is known to be extremely low.

8. INTEGRATED RESPONSE INVOLVING CGB AND CTB

Direct interplay between haemoglobins co-existing in the same microorganism has not been reported. However, given that *cgb* and *ctb* are members of the same regulon, their upregulation is linked to nitrosative stress (via NssR) and considering the role of Ctb under these conditions is not obvious (Elvers et al., 2005), an integrated response involving Cgb and Ctb is possible.

Elevated expression of Cgb in a *ctb* mutant was recently reported under oxygen-limited conditions (Smith et al., 2011). This finding lead to speculation that Ctb modulates the availability of NO and, as a consequence, controls NssR activity. This phenomenon was observed under microaerobic conditions, where the haem cofactor of Ctb exists in an oxyferrous form (Wainwright et al., 2006), and the binding of NO to the ferrous iron of Ctb is therefore likely to be hindered. This suggests that when nitrosative stress is encountered under microaerobic conditions, NssR is activated by the available NO. However, when the oxygen concentration is limited, the unliganded ferrous haem of Ctb binds NO and consequently prevents the induction of Cgb (Smith et al., 2011). The failure of NO-exposed cells to reveal the NssR regulon under oxygen-limited conditions, demonstrated by transcriptomic analysis (Avila-Ramirez et al., 2013), can also be explained by this hypothesis.

Several studies support the idea of interplay between the *C. jejuni* globins: (i) diminished viability in cultures of a *cgb* mutant (Elvers et al., 2004) is reverted by mutation of the second globin, being comparable to the parent strain or a *ctb* mutant (Wainwright et al., 2005); (ii) a double mutant (*cgb*, *ctb*) is less sensitive to GSNO than a *cgb* mutant, suggesting a partial suppression of the *cgb* phenotype by the absence of the second globin (Avila-Ramirez et al., 2013); (iii) a *ctb* mutant shows a slower growth profile compared to the parent strain during the stationary phase in microaerobic cultures, whereas growth rates of the wild-type strain and a *cgb*, *ctb* double mutant are similar (Wainwright et al., 2005).

9. CONCLUSIONS

Adaptations of *C. jejuni* to nitrosative stress conditions seem to be linked to oxygen availability. Under microaerobic conditions, NO and RNS trigger the induction of the NssR-dependent regulon that includes the expression of the Cgb and Ctb globins, whereas in oxygen-limited environments an inducible NssR-independent system offers protection. Even though microaerobic resistance to NO and nitrosative stress agents is associated with the presence of Cgb, the reaction intermediates during the Cgb-mediated detoxification of NO are yet to be experimentally identified. However, biophysical studies of the Cgb haem pocket support the conversion of NO to nitrate. Indeed, much is known about the influence of hydrogen-bonding networks upon the ligand-binding properties, and how the haem environment is well suited to the conversion of NO to nitrate via a dioxygenase or denitrosylase mechanism. However, these mechanisms involve the oxidation of the haem cofactor, and a reductase system for the regeneration of the ferrous haem has not yet been identified, although the possibility remains that Cgb is reduced by endogenous small molecules (e.g. glutathione).

The truncated globin Ctb is constitutively expressed at low levels and is overexpressed under nitrosative stress conditions in an NssR-dependent manner. However, Ctb fails to offer protection against NO or RNS. A possible role for Ctb in promoting microaerobic growth and moderating respiration in *C. jejuni* has been suggested, and biophysical and mechanistic characterisation is consistent with the involvement of this globin in peroxidase-like chemistry. Additional analyses will be required to confidently assign a functional role for this class III-truncated globin.

Given that *cgb* and *ctb* are members of the same regulon and their upregulation is linked to nitrosative stress (via NssR), an integrated response involving Cgb and Ctb has been suggested. Unravelling the molecular mechanisms, controlling the cofactor/ligand-binding capabilities of NssR may contribute towards the elucidation of the molecular switch that initiates the response of *Campylobacter* to nitrosative stress under aerobic conditions. However, mechanisms that confer NO tolerance during oxygen limitation remain to be identified. Given the host of biochemical, biophysical and genetic tools available to study the globins of *Campylobacter*, these evasive mechanistic details will inevitably reveal themselves upon interrogation via a multi-pronged approach.

ACKNOWLEDGEMENTS

The authors wish to thank the Consejo Nacional de Ciencia y Tecnologia (Mexico) for aid through Grant number 99171 and Consejo Estatal de Ciencia, Tecnología e Innovación de Michoacán through Grant number 007 (Mariana Tinajero-Trejo).

REFERENCES

Alderton, W. K., Cooper, C. E., & Knowles, R. G. (2001). Nitric oxide synthases: Structure, function and inhibition. *The Biochemical Journal*, *357*, 593–615.

Allos, B. M. (2001). *Campylobacter jejuni* infections: Update on emerging issues and trends. *Clinical Infectious Diseases*, *32*, 1201–1206.

Anjum, M. F., Ioannidis, N., & Poole, R. K. (1998). Response of the NAD(P)H-oxidising flavohaemoglobin (Hmp) to prolonged oxidative stress and implications for its physiological role in *Escherichia coli*. *FEMS Microbiology Letters*, *166*, 219–223.

Arcovito, A., Benfatto, M., Cianci, M., Hasnain, S. S., Nienhaus, K., Nienhaus, G. U., et al. (2007). X-ray structure analysis of a metalloprotein with enhanced active-site resolution using in situ X-ray absorption near edge structure spectroscopy. *Proceedings of the National Academy of Sciences of the United States of America*, *104*, 6211–6216.

Arroyo Manez, P., Lu, C., Boechi, L., Marti, M. A., Shepherd, M., Wilson, J. L., et al. (2011). Role of the distal hydrogen-bonding network in regulating oxygen affinity in the truncated hemoglobin III from *Campylobacter jejuni*. *Biochemistry*, *50*, 3946–3956.

Aussel, L., Zhao, W., Hebrard, M., Guilhon, A. A., Viala, J. P., Henri, S., et al. (2011). *Salmonella* detoxifying enzymes are sufficient to cope with the host oxidative burst. *Molecular Microbiology*, *80*, 628–640.

Avila-Ramirez, C., Tinajero-Trejo, M., Davidge, K. S., Monk, C. E., Kelly, D. J., & Poole, R. K. (2013). Do globins in microaerophilic *Campylobacter jejuni* confer nitrosative stress tolerance under oxygen limitation? *Antioxidants & Redox Signaling*, *18*, 424–431.

Babior, B. M. (1999). NADPH oxidase: An update. *Blood*, *93*, 1464–1476.

Bakhiet, M., Al-Salloom, F. S., Qareiballa, A., Bindayna, K., Farid, I., & Botta, G. A. (2004). Induction of alpha and beta chemokines by intestinal epithelial cells stimulated with *Campylobacter jejuni*. *The Journal of Infection, 48,* 236–244.

Bamford, V. A., Angove, H. C., Seward, H. E., Thomson, A. J., Cole, J. A., Butt, J. N., et al. (2002). Structure and spectroscopy of the periplasmic cytochrome *c* nitrite reductase from *Escherichia coli*. *Biochemistry, 41,* 2921–2931.

Bidon-Chanal, A., Marti, M. A., Crespo, A., Milani, M., Orozco, M., Bolognesi, M., et al. (2006). Ligand-induced dynamical regulation of NO conversion in *Mycobacterium tuberculosis* truncated hemoglobin-N. *Proteins, 64,* 457–464.

Bodenmiller, D. M., & Spiro, S. (2006). The *yjeB* (*nsrR*) gene of *Escherichia coli* encodes a nitric oxide-sensitive transcriptional regulator. *Journal of Bacteriology, 188,* 874–881.

Bolli, A., Ciaccio, C., Coletta, M., Nardini, M., Bolognesi, M., Pesce, A., et al. (2008). Ferrous *Campylobacter jejuni* truncated hemoglobin P displays an extremely high reactivity for cyanide—A comparative study. *The FEBS Journal, 275,* 633–645.

Bollinger, C. J., Bailey, J. E., & Kallio, P. T. (2001). Novel hemoglobins to enhance micro-aerobic growth and substrate utilization in *Escherichia coli*. *Biotechnology Progress, 17,* 798–808.

Bollinger, C. J., & Kallio, P. T. (2007). Impact of the small RNA RyhB on growth, physiology and heterologous protein expression in *Escherichia coli*. *FEMS Microbiology Letters, 275,* 221–228.

Bolognesi, M., Boffi, A., Coletta, M., Mozzarelli, A., Pesce, A., Tarricone, C., et al. (1999). Anticooperative ligand binding properties of recombinant ferric *Vitreoscilla* homodimeric hemoglobin: A thermodynamic, kinetic and X-ray crystallographic study. *Journal of Molecular Biology, 291,* 637–650.

Bowman, L. A. H., McLean, S., Poole, R. K., & Fukuto, J. (2011). The diversity of microbial responses to nitric oxide and agents of nitrosative stress: Close cousins but not identical twins. *Advances in Microbial Physiology, 59,* 135–219.

Brondijk, T. H., Nilavongse, A., Filenko, N., Richardson, D. J., & Cole, J. A. (2004). NapGH components of the periplasmic nitrate reductase of *Escherichia coli* K-12: Location, topology and physiological roles in quinol oxidation and redox balancing. *The Biochemical Journal, 379,* 47–55.

Butcher, J., Sarvan, S., Brunzelle, J. S., Couture, J. F., & Stintzi, A. (2012). Structure and regulon of Campylobacter jejuni ferric uptake regulator Fur define apo-Fur regulation. *Proceedings of the National Academy of Sciences of the United States of America, 109,* 10047–10052.

Butler, C. S., Ferguson, S. J., Berks, B. C., Thomson, A. J., Cheesman, M. R., & Richardson, D. J. (2001). Assignment of haem ligands and detection of electronic absorption bands of molybdenum in the di-haem periplasmic nitrate reductase of *Paracoccus pantotrophus*. *FEBS Letters, 500,* 71–74.

Butzler, J. P. (2004). *Campylobacter*, from obscurity to celebrity. *Clinical Microbiology and Infection, 10,* 868–876.

Carlioz, A., & Touati, D. (1986). Isolation of superoxide dismutase mutants in *Escherichia coli*: Is superoxide dismutase necessary for aerobic life? *The EMBO Journal, 5,* 623–630.

Contreras, M. L., Escamilla, J. E., DelArenal, I. P., Davila, J. R., D'mello, R., & Poole, R. K. (1999). An unusual cytochrome *o'*-type cytochrome *c* oxidase in a *Bacillus cereus* cytochrome a_3 mutant has a very high affinity for oxygen. *Microbiology, 145,* 1563–1573.

Corker, H., & Poole, R. K. (2003). Nitric oxide formation by *Escherichia coli*. Dependence on nitrite reductase, the NO-sensing regulator Fnr, and flavohemoglobin Hmp. *The Journal of Biological Chemistry, 278,* 31584–31592.

Costa, C., Macedo, A., Moura, I., Moura, J. J., Le Gall, J., Berlier, Y., et al. (1990). Regulation of the hexaheme nitrite/nitric oxide reductase of *Desulfovibrio desulfuricans*, Wolinella succinogenes and Escherichia coli. A mass spectrometric study. *FEBS Letters, 276,* 67–70.

Couture, M., Yeh, S. R., Wittenberg, B. A., Wittenberg, J. B., Ouellet, Y., Rousseau, D. L., et al. (1999). A cooperative oxygen-binding hemoglobin from *Mycobacterium tuberculosis*. *Proceedings of the National Academy of Sciences of the United States of America, 96*, 11223–11228.

Cruz-Ramos, H., Crack, J., Wu, G., Hughes, M. N., Scott, C., Thomson, A. J., et al. (2002). NO sensing by FNR: Regulation of the *Escherichia coli* NO-detoxifying flavohaemoglobin, Hmp. *The EMBO Journal, 21*, 3235–3244.

Debruine, L., Gevers, D., & Vandamme, P. (2008). Taxonomy of the family *Campylobacteraceae*. In I. Nachamkin, C. M. Szymanski, & M. J. Blaser (Eds.), *Campylobacter* (3th ed., pp. 1–25). Washington, DC: ASM Press.

Dikshit, R. P., Dikshit, K. L., Liu, Y. X., & Webster, D. A. (1992). The bacterial hemoglobin from *Vitreoscilla* can support the aerobic growth of *Escherichia coli* lacking terminal oxidases. *Archives of Biochemistry and Biophysics, 293*, 241–245.

Dikshit, K. L., Dikshit, R. P., & Webster, D. A. (1990). Study of *Vitreoscilla* globin (*vgb*) gene expression and promoter activity in *E. coli* through transcriptional fusion. *Nucleic Acids Research, 18*, 4149–4155.

Dikshit, K. L., Spaulding, D., Braun, A., & Webster, D. A. (1989). Oxygen inhibition of globin gene transcription and bacterial haemoglobin synthesis in *Vitreoscilla*. *Journal of General Microbiology, 135*, 2601–2609.

Dikshit, K. L., & Webster, D. A. (1988). Cloning, characterization and expression of the bacterial globin gene from *Vitreoscilla* in *Escherichia coli*. *Gene, 70*, 377–386.

D'mello, R., Hill, S., & Poole, R. K. (1994). Determination of the oxygen affinities of terminal oxidases in *Azotobacter vinelandii* using the deoxygenation of oxyleghaemoglobin and oxymyoglobin: Cytochrome *bd* is a low-affinity oxidase. *Microbiology, 140*, 1395–1402.

D'mello, R., Hill, S., & Poole, R. K. (1995). The oxygen affinity of cytochrome *bo'* in *Escherichia coli* determined by the deoxygenation of oxyleghemoglobin and oxymyoglobin: K_m values for oxygen are in the submicromolar range. *Journal of Bacteriology, 177*, 867–870.

D'mello, R., Hill, S., & Poole, R. K. (1996). The cytochrome *bd* quinol oxidase in *Escherichia coli* has an extremely high oxygen affinity and two oxygen-binding haems: Implications for regulation of activity *in vivo* by oxygen inhibition. *Microbiology, 142*, 755–763.

Egawa, T., & Yeh, S. R. (2005). Structural and functional properties of hemoglobins from unicellular organisms as revealed by resonance Raman spectroscopy. *Journal of Inorganic Biochemistry, 99*, 72–96.

Elvers, K. T., Turner, S. M., Wainwright, L. M., Marsden, G., Hinds, J., Cole, J. A., et al. (2005). NssR, a member of the Crp-Fnr superfamily from *Campylobacter jejuni*, regulates a nitrosative stress-responsive regulon that includes both a single-domain and a truncated haemoglobin. *Molecular Microbiology, 57*, 735–750.

Elvers, K. T., Wu, G., Gilberthorpe, N. J., Poole, R. K., & Park, S. F. (2004). Role of an inducible single-domain hemoglobin in mediating resistance to nitric oxide and nitrosative stress in *Campylobacter jejuni* and *Campylobacter coli*. *Journal of Bacteriology, 186*, 5332–5341.

Ermler, U., Siddiqui, R. A., Cramm, R., & Friedrich, B. (1995). Crystal structure of the flavohemoglobin from *Alcaligenes eutrophus* at 1.75 A resolution. *The EMBO Journal, 14*, 6067–6077.

Farr, S. B., D'Ari, R., & Touati, D. (1986). Oxygen-dependent mutagenesis in *Escherichia coli* lacking superoxide dismutase. *Proceedings of the National Academy of Sciences of the United States of America, 83*, 8268–8272.

Farr, S. B., & Kogoma, T. (1991). Oxidative stress responses in *Escherichia coli* and *Salmonella typhimurium*. *Microbiological Reviews, 55*, 561–585.

Fischer, H. M. (1994). Genetic regulation of nitrogen fixation in rhizobia. *Microbiological Reviews, 58*, 352–386.

Fisher, M. (1956). An unusual variant of acute idiopathic polyneuritis (syndrome of ophthalmoplegia, ataxia and areflexia). *The New England Journal of Medicine, 255*, 57–65.

Flatley, J., Barrett, J., Pullan, S. T., Hughes, M. N., Green, J., & Poole, R. K. (2005). Transcriptional responses of *Escherichia coli* to *S*-nitrosoglutathione under defined chemostat conditions reveal major changes in methionine biosynthesis. *The Journal of Biological Chemistry, 280*, 10065–10072.

Fouts, D. E., Mongodin, E. F., Mandrell, R. E., Miller, W. G., Rasko, D. A., Ravel, J., et al. (2005). Major structural differences and novel potential virulence mechanisms from the genomes of multiple *Campylobacter* species. *PLoS Biology, 3*, e15.

Frey, A. D., Farres, J., Bollinger, C. J., & Kallio, P. T. (2002). Bacterial hemoglobins and flavohemoglobins for alleviation of nitrosative stress in *Escherichia coli*. *Applied and Environmental Microbiology, 68*, 4835–4840.

Frey, A. D., & Kallio, P. T. (2005). Nitric oxide detoxification—A new era for bacterial globins in biotechnology? *Trends in Biotechnology, 23*, 69–73.

Frey, A. D., Shepherd, M., Jokipii-Lukkari, S., Haggman, H., & Kallio, P. T. (2011). The single-domain globin of *Vitreoscilla*: Augmentation of aerobic metabolism for biotechnological applications. In R. K. Poole (Ed.), *Advances in microbial physiology, Vol. 58*, (pp. 81–139). London: Academic Press.

Fridovich, I. (1999). Fundamental aspects of reactive oxygen species, or what's the matter with oxygen? *Annals of the New York Academy of Sciences, 893*, 13–18.

Friedman, C. R., Neimann, J., Wegener, H. C., & Tauxe, R. V. (2000). Epidemiology of *Campylobacter jejuni* infections in the United States and other industrialized nations. In I. Nachamkin & M. J. Blaser (Eds.), *Campylobacter* (pp. 121–138). Washington, DC: ASM Press.

Gardner, P. R., Costantino, G., & Salzman, A. L. (1998). Constitutive and adaptive detoxification of nitric oxide in *Escherichia coli*. Role of nitric-oxide dioxygenase in the protection of aconitase. *The Journal of Biological Chemistry, 273*, 26528–26533.

Gardner, P. R., Costantino, G., Szabo, C., & Salzman, A. L. (1997). Nitric oxide sensitivity of the aconitases. *The Journal of Biological Chemistry, 272*, 25071–25076.

Gardner, A. M., & Gardner, P. R. (2002). Flavohemoglobin detoxifies nitric oxide in aerobic, but not anaerobic, Escherichia coli. Evidence for a novel inducible anaerobic nitric oxide-scavenging activity. *The Journal of Biological Chemistry, 277*, 8166–8171.

Gardner, P. R., Gardner, A. M., Brashear, W. T., Suzuki, T., Hvitved, A. N., Setchell, K. D., et al. (2006). Hemoglobins dioxygenate nitric oxide with high fidelity. *Journal of Inorganic Biochemistry, 100*, 542–550.

Gardner, P. R., Gardner, A. M., Martin, L. A., Dou, Y., Li, T., Olson, J. S., et al. (2000). Nitric-oxide dioxygenase activity and function of flavohemoglobins. sensitivity to nitric oxide and carbon monoxide inhibition. *The Journal of Biological Chemistry, 275*, 31581–31587.

Gardner, P. R., Gardner, A. M., Martin, L. A., & Salzman, A. L. (1998). Nitric oxide dioxygenase: An enzymic function for flavohemoglobin. *Proceedings of the National Academy of Sciences of the United States of America, 95*, 10378–10383.

Giardina, G., Rinaldo, S., Johnson, K. A., Di Matteo, A., Brunori, M., & Cutruzzola, F. (2008). NO sensing in *Pseudomonas aeruginosa*: Structure of the transcriptional regulator DNR. *Journal of Molecular Biology, 378*, 1002–1015.

Gibson, Q. H., Olson, J. S., McKinnie, R. E., & Rohlfs, R. J. (1986). A kinetic description of ligand binding to sperm whale myoglobin. *The Journal of Biological Chemistry, 261*, 10228–10239.

Green, J., Scott, C., & Guest, J. R. (2001). Functional versatility in the CRP-FNR superfamily of transcription factors: FNR and FLP. In R. K. Poole (Ed.), *Advances in microbial physiology, Vol. 44*, (pp. 1–34). London: Academic Press.

Greenberg, J. T., Monach, P., Chou, J. H., Josephy, P. D., & Demple, B. (1990). Positive control of a global antioxidant defense regulon activated by superoxide-generating agents in *Escherichia coli*. *Proceedings of the National Academy of Sciences of the United States of America*, 87, 6181–6185.

Guccione, E., Hitchcock, A., Hall, S. J., Mulholland, F., Shearer, N., van Vliet, A. H., et al. (2010). Reduction of fumarate, mesaconate and crotonate by Mfr, a novel oxygen-regulated periplasmic reductase in *Campylobacter jejuni*. *Environmental Microbiology*, 12, 576–591.

Guest, J. R., Green, J., Irvine, A. S., & Spiro, S. (1996). The FNR modulon and FNR-regulated gene expression. In E. C. C. Lin & A. S. Lynch (Eds.), *Regulation of gene expression in* Escherichia Coli (pp. 317–342). Georgetown: R.G. Landes Co.

Gusarov, I., Shatalin, K., Starodubtseva, M., & Nudler, E. (2009). Endogenous nitric oxide protects bacteria against a wide spectrum of antibiotics. *Science*, 325, 1380–1384.

Halliwell, B., & Gutteridge, J. M. (2007). *Free radicals in biology and medicine*. Oxford: Oxford University Press.

Hausladen, A., Gow, A. J., & Stamler, J. S. (1998). Nitrosative stress: Metabolic pathway involving the flavohemoglobin. *Proceedings of the National Academy of Sciences of the United States of America*, 95, 14100–14105.

Hausladen, A., Gow, A., & Stamler, J. S. (2001). Flavohemoglobin denitrosylase catalyzes the reaction of a nitroxyl equivalent with molecular oxygen. *Proceedings of the National Academy of Sciences of the United States of America*, 98, 10108–10112.

Hendrixson, D. R., & DiRita, V. J. (2003). Transcription of sigma54-dependent but not sigma28-dependent flagellar genes in *Campylobacter jejuni* is associated with formation of the flagellar secretory apparatus. *Molecular Microbiology*, 50, 687–702.

Henle, E. S., Han, Z., Tang, N., Rai, P., Luo, Y., & Linn, S. (1999). Sequence-specific DNA cleavage by Fe2+-mediated fenton reactions has possible biological implications. *The Journal of Biological Chemistry*, 274, 962–971.

Hernandez-Urzua, E., Mills, C. E., White, G. P., Contreras-Zentella, M. L., Escamilla, E., Vasudevan, S. G., et al. (2003). Flavohemoglobin Hmp, but not its individual domains, confers protection from respiratory inhibition by nitric oxide in *Escherichia coli*. *The Journal of Biological Chemistry*, 278, 34975–34982.

Hernandez-Urzua, E., Zamorano-Sanchez, D. S., Ponce-Coria, J., Morett, E., Grogan, S., Poole, R. K., et al. (2007). Multiple regulators of the Flavohaemoglobin (*hmp*) gene of *Salmonella enterica* serovar Typhimurium include RamA, a transcriptional regulator conferring the multidrug resistance phenotype. *Archives of Microbiology*, 187, 67–77.

Hess, D. T., Matsumoto, A., Kim, S.-O., Marshall, H. E., & Stamler, J. S. (2005). Protein S-nitrosylation: Purview and parameters. *Nature Reviews. Molecular Cell Biology*, 6, 150–166.

Hofreuter, D., Tsai, J., Watson, R. O., Novik, V., Altman, B., Benitez, M., et al. (2006). Unique features of a highly pathogenic *Campylobacter jejuni* strain. *Infection and Immunity*, 74, 4694–4707.

Holmes, K., Mulholland, F., Pearson, B. M., Pin, C., McNicholl-Kennedy, J., Ketley, J. M., et al. (2005). *Campylobacter jejuni* gene expression in response to iron limitation and the role of Fur. *Microbiology*, 151, 243–257.

Huang, J., & Brumell, J. H. (2009). NADPH oxidases contribute to autophagy regulation. *Autophagy*, 5, 887–889.

Hughes, M. N. (1999). Relationships between nitric oxide, nitroxyl ion, nitrosonium cation and peroxynitrite. *Biochimica et Biophysica Acta Bioenergetics*, 1411, 263–272.

Hutchings, M. I., Mandhana, N., & Spiro, S. (2002). The NorR protein of *Escherichia coli* activates expression of the flavorubredoxin gene *norV* in response to reactive nitrogen species. *Journal of Bacteriology*, 184, 4640–4643.

Hutchinson, F. (1985). Chemical changes induced in DNA by ionizing radiation. *Progress in Nucleic Acid Research and Molecular Biology*, 32, 115–154.

Ilari, A., Bonamore, A., Farina, A., Johnson, K. A., & Boffi, A. (2002). The X-ray structure of ferric *Escherichia coli* flavohemoglobin reveals an unexpected geometry of the distal heme pocket. *The Journal of Biological Chemistry, 277*, 23725–23732.

Ilari, A., Ceci, P., Ferrari, D., Rossi, G. L., & Chiancone, E. (2002). Iron incorporation into *Escherichia coli* Dps gives rise to a ferritin-like microcrystalline core. *The Journal of Biological Chemistry, 277*, 37619–37623.

Imlay, J. A. (2003). Pathways of oxidative damage. *Annual Review of Microbiology, 57*, 395–418.

Imlay, J. A. (2008). Cellular defenses against superoxide and hydrogen peroxide. *Annual Review of Biochemistry, 77*, 755–776.

Jackson, R. J., Elvers, K. T., Lee, L. J., Gidley, M. D., Wainwright, L. M., Lightfoot, J., et al. (2007). Oxygen reactivity of both respiratory oxidases in *Campylobacter jejuni*: The *cydAB* genes encode a cyanide-resistant, low-affinity oxidase that is not of the cytochrome *bd* type. *Journal of Bacteriology, 189*, 1604–1615.

Jacobs, B. C., Rothbarth, P. H., van der Meche, F. G., Herbrink, P., Schmitz, P. I., de Klerk, M. A., et al. (1998). The spectrum of antecedent infections in Guillain-Barre syndrome: A case-control study. *Neurology, 51*, 1110–1115.

Jang, S., & Imlay, J. A. (2007). Micromolar intracellular hydrogen peroxide disrupts metabolism by damaging iron-sulfur enzymes. *The Journal of Biological Chemistry, 282*, 929–937.

Jarboe, L. R., Hyduke, D. R., Tran, L. M., Chou, K. J., & Liao, J. C. (2008). Determination of the *Escherichia coli* S-nitrosoglutathione response network using integrated biochemical and systems analysis. *The Journal of Biological Chemistry, 283*, 5148–5157.

Joseph, S. V., Madhavilatha, G. K., Kumar, R. A., & Mundayoor, S. (2012). Comparative analysis of mycobacterial truncated hemoglobin promoters and the groEL2 promoter in free-living and intracellular mycobacteria. *Applied and Environmental Microbiology, 78*, 6499–6506.

Justino, M. C., Vicente, J. B., Teixeira, M., & Saraiva, L. M. (2005). New genes implicated in the protection of anaerobically grown *Escherichia coli* against nitric oxide. *The Journal of Biological Chemistry, 280*, 2636–2643.

Kallberg, M., Wang, H., Wang, S., Peng, J., Wang, Z., Lu, H., et al. (2012). Template-based protein structure modeling using the RaptorX web server. *Nature Protocols, 7*, 1511–1522.

Kallio, P. T., & Bailey, J. E. (1996). Intracellular expression of *Vitreoscilla* hemoglobin (VHb) enhances total protein secretion and improves the production of alpha-amylase and neutral protease in *Bacillus subtilis*. *Biotechnology Progress, 12*, 31–39.

Kallio, P. T., Heidrich, J., Koskenkorva, T., Bollinger, C. J. T., Farrés, J., & Frey, A. D. (2007). Analysis of novel hemoglobins during microaerobic growth of HMP-negative *Escherichia coli*. *Enzyme and Microbial Technology, 40*, 329–336.

Kallio, P. T., Tsai, P. S., & Bailey, J. E. (1996). Expression of *Vitreoscilla* hemoglobin is superior to horse heart myoglobin or yeast flavohemoglobin expression for enhancing *Escherichia coli* growth in a microaerobic bioreactor. *Biotechnology Progress, 12*, 751–757.

Khosla, C., & Bailey, J. E. (1988). Heterologous expression of a bacterial haemoglobin improves the growth properties of recombinant *Escherichia coli*. *Nature, 331*, 633–635.

Khosla, C., & Bailey, J. E. (1989). Characterization of the oxygen-dependent promoter of the *Vitreoscilla* hemoglobin gene in *Escherichia coli*. *Journal of Bacteriology, 171*, 5995–6004.

Khosla, C., Curtis, J. E., Bydalek, P., Swartz, J. R., & Bailey, J. E. (1990). Expression of recombinant proteins in *Escherichia coli* using an oxygen-responsive promoter. *Biotechnology (N. Y), 8*, 554–558.

Khosla, C., Curtis, J. E., DeModena, J., Rinas, U., & Bailey, J. E. (1990). Expression of intracellular hemoglobin improves protein synthesis in oxygen-limited *Escherichia coli*. *Biotechnology (N. Y)*, *8*, 849–853.

Koga, M., Gilbert, M., Li, J., Koike, S., Takahashi, M., Furukawa, K., et al. (2005). Antecedent infections in Fisher syndrome: A common pathogenesis of molecular mimicry. *Neurology*, *64*, 1605–1611.

Korner, H., Sofia, H. J., & Zumft, W. G. (2003). Phylogeny of the bacterial superfamily of Crp-Fnr transcription regulators: Exploiting the metabolic spectrum by controlling alternative gene programs. *FEMS Microbiology Reviews*, *27*, 559–592.

Lama, A., Pawaria, S., Bidon-Chanal, A., Anand, A., Gelpi, J. L., Arya, S., et al. (2009). Role of Pre-A motif in nitric oxide scavenging by truncated hemoglobin, HbN, of *Mycobacterium tuberculosis*. *The Journal of Biological Chemistry*, *284*, 14457–14468.

Laver, J. R., McLean, S., Bowman, L. A., Harrison, L. J., Read, R. C., & Poole, R. K. (2013). Nitrosothiols in bacterial pathogens and pathogenesis. *Antioxidants & Redox Signaling*, *18*, 309–322.

Lehnert, N., & Scheidt, W. R. (2009). Preface for the inorganic chemistry forum: The coordination chemistry of nitric oxide and its significance for metabolism, signaling, and toxicity in biology. *Inorganic Chemistry*, *49*, 6223–6225.

Liu, C., He, Y., & Chang, Z. (2004). Truncated hemoglobin *o* of *Mycobacterium tuberculosis*: The oligomeric state change and the interaction with membrane components. *Biochemical and Biophysical Research Communications*, *316*, 1163–1172.

Lowenstein, C. J., & Padalko, E. (2004). INOS (NOS2) at a glance. *Journal of Cell Science*, *117*, 2865–2867.

Lu, C., Egawa, T., Wainwright, L. M., Poole, R. K., & Yeh, S. R. (2007). Structural and functional properties of a truncated hemoglobin from a food-borne pathogen *Campylobacter jejuni*. *The Journal of Biological Chemistry*, *282*, 13627–13636.

Lu, C., Mukai, M., Lin, Y., Wu, G., Poole, R. K., & Yeh, S.-R. (2007). Structural and functional properties of a single-domain hemoglobin from the food-borne pathogen *Campylobacter jejuni*. *The Journal of Biological Chemistry*, *282*, 25917–25928.

Massey, V., Strickland, S., Mayhew, S. G., Howell, L. G., Engel, P. C., Matthews, R. G., et al. (1969). The production of superoxide anion radicals in the reaction of reduced flavins and flavoproteins with molecular oxygen. *Biochemical and Biophysical Research Communications*, *36*, 891–897.

Mastroeni, P., Vazquez-Torres, A., Fang, F. C., Xu, Y., Khan, S., Hormaeche, C. E., et al. (2000). Antimicrobial actions of the NADPH phagocyte oxidase and inducible nitric oxide synthase in experimental salmonellosis. II. Effects on microbial proliferation and host survival in vivo. *The Journal of Experimental Medicine*, *192*, 237–248.

Matsui, M., Tomita, M., & Kanai, A. (2013). Comprehensive computational analysis of bacterial CRP/FNR superfamily and its target motifs reveals stepwise evolution of transcriptional networks. *Genome Biology and Evolution*, *5*, 267–282.

Mead, G. C. (1989). Microbes of the avian cecum: Types present and substrates utilized. *The Journal of Experimental Zoology. Supplement*, *3*, 48–54.

Membrillo-Hernandez, J., Coopamah, M. D., Anjum, M. F., Stevanin, T. M., Kelly, A., Hughes, M. N., et al. (1999). The flavohemoglobin of *Escherichia coli* confers resistance to a nitrosating agent, a "Nitric oxide Releaser," and paraquat and is essential for transcriptional responses to oxidative stress. *The Journal of Biological Chemistry*, *274*, 748–754.

Membrillo-Hernández, J., Coopamah, M. D., Channa, A., Hughes, M. N., & Poole, R. K. (1998). A novel mechanism for upregulation of the *Escherichia coli* K-12 *hmp* (flavohaemoglobin) gene by the 'NO releaser', S-nitrosoglutathione: Nitrosation of homocysteine and modulation of MetR binding to the *glyA-hmp* intergenic region. *Molecular Microbiology*, *29*, 1101–1112.

Membrillo-Hernandez, J., Ioannidis, N., & Poole, R. K. (1996). The flavohaemoglobin (HMP) of *Escherichia coli* generates superoxide in vitro and causes oxidative stress in vivo. *FEBS Letters, 382*, 141–144.

Membrillo-Hernandez, J., Kim, S. O., Cook, G. M., & Poole, R. K. (1997). Paraquat regulation of *hmp* (flavohemoglobin) gene expression in *Escherichia coli* K-12 is SoxRS independent but modulated by sigma S. *Journal of Bacteriology, 179*, 3164–3170.

Milani, M., Ouellet, Y., Ouellet, H., Guertin, M., Boffi, A., Antonini, G., et al. (2004). Cyanide binding to truncated hemoglobins: A crystallographic and kinetic study. *Biochemistry, 43*, 5213–5221.

Milani, M., Pesce, A., Ouellet, Y., Ascenzi, P., Guertin, M., & Bolognesi, M. (2001). *Mycobacterium tuberculosis* hemoglobin N displays a protein tunnel suited for O2 diffusion to the heme. *The EMBO Journal, 20*, 3902–3909.

Milani, M., Pesce, A., Ouellet, Y., Dewilde, S., Friedman, J., Ascenzi, P., et al. (2004). Heme-ligand tunneling in group I truncated hemoglobins. *The Journal of Biological Chemistry, 279*, 21520–21525.

Milani, M., Savard, P. Y., Ouellet, H., Ascenzi, P., Guertin, M., & Bolognesi, M. (2003). A TyrCD1/TrpG8 hydrogen bond network and a TyrB10TyrCD1 covalent link shape the heme distal site of *Mycobacterium tuberculosis* hemoglobin O. *Proceedings of the National Academy of Sciences of the United States of America, 100*, 5766–5771.

Miller, R. A., & Britigan, B. E. (1997). Role of oxidants in microbial pathophysiology. *Clinical Microbiology Reviews, 10*, 1–18.

Miller, M. R., & Megson, I. L. (2007). Recent developments in nitric oxide donor drugs. *British Journal of Pharmacology, 151*, 305–321.

Mongkolsuk, S., & Helmann, J. D. (2002). Regulation of inducible peroxide stress responses. *Molecular Microbiology, 45*, 9–15.

Monk, C. E., Pearson, B. M., Mulholland, F., Smith, H. K., & Poole, R. K. (2008). Oxygen- and NssR-dependent globin expression and enhanced iron acquisition in the response of *Campylobacter* to nitrosative stress. *The Journal of Biological Chemistry, 283*, 28413–28425.

Moore, C. M., Nakano, M. M., Wang, T., Ye, R. W., & Helmann, J. D. (2004). Response of *Bacillus subtilis* to nitric oxide and the nitrosating agent sodium nitroprusside. *Journal of Bacteriology, 186*, 4655–4664.

Morris, S. L., & Hansen, J. N. (1981). Inhibition of *Bacillus cereus* spore outgrowth by covalent modification of a sulfhydryl group by nitrosothiol and iodoacetate. *Journal of Bacteriology, 148*, 465–471.

Mukai, M., Mills, C. E., Poole, R. K., & Yeh, S. R. (2001). Flavohemoglobin, a globin with a peroxidase-like catalytic site. *The Journal of Biological Chemistry, 276*, 7272–7277.

Mukhopadhyay, S., & Schellhorn, H. E. (1994). Induction of *Escherichia coli* hydroperoxidase I by acetate and other weak acids. *Journal of Bacteriology, 176*, 2300–2307.

Mukhopadhyay, P., Zheng, M., Bedzyk, L. A., LaRossa, R. A., & Storz, G. (2004). Prominent roles of the NorR and Fur regulators in the *Escherichia coli* transcriptional response to reactive nitrogen species. *Proceedings of the National Academy of Sciences of the United States of America, 101*, 745–750.

Murad, F. (1986). Cyclic guanosine monophosphate as a mediator of vasodilation. *The Journal of Clinical Investigation, 78*, 1–5.

Myers, J. D., & Kelly, D. J. (2005). A sulphite respiration system in the chemoheterotrophic human pathogen *Campylobacter jejuni. Microbiology, 151*, 233–242.

Nakajima, H., Honma, Y., Tawara, T., Kato, T., Park, S. Y., Miyatake, H., et al. (2001). Redox properties and coordination structure of the heme in the CO-sensing transcriptional activator CooA. *The Journal of Biological Chemistry, 276*, 7055–7061.

Nardini, M., Pesce, A., Labarre, M., Richard, C., Bolli, A., Ascenzi, P., et al. (2006). Structural determinants in the group III truncated hemoglobin from *Campylobacter jejuni*. *The Journal of Biological Chemistry*, *281*, 37803–37812.

Olin, A. C., Aldenbratt, A., Ekman, A., Ljungkvist, G., Jungersten, L., Alving, K., et al. (2001). Increased nitric oxide in exhaled air after intake of a nitrate-rich meal. *Respiratory Medicine*, *95*, 153–158.

Ouellet, Y., Milani, M., Couture, M., Bolognesi, M., & Guertin, M. (2006). Ligand interactions in the distal heme pocket of *Mycobacterium tuberculosis* truncated hemoglobin N: Roles of TyrB10 and GlnE11 residues. *Biochemistry*, *45*, 8770–8781.

Ouellet, H., Milani, M., LaBarre, M., Bolognesi, M., Couture, M., & Guertin, M. (2007). The roles of Tyr(CD1) and Trp(G8) in *Mycobacterium tuberculosis* truncated hemoglobin O in ligand binding and on the heme distal site architecture. *Biochemistry*, *46*, 11440–11450.

Park, K. W., Kim, K. J., Howard, A. J., Stark, B. C., & Webster, D. A. (2002). *Vitreoscilla* hemoglobin binds to subunit I of cytochrome *bo* ubiquinol oxidases. *The Journal of Biological Chemistry*, *277*, 33334–33337.

Park, S., You, X., & Imlay, J. A. (2005). Substantial DNA damage from submicromolar intracellular hydrogen peroxide detected in Hpx-mutants of Escherichia coli. *Proceedings of the National Academy of Sciences of the United States of America*, *102*, 9317–9322.

Parkhill, J., Wren, B. W., Mungall, K., Ketley, J. M., Churcher, C., Basham, D., et al. (2000). The genome sequence of the food-borne pathogen *Campylobacter jejuni* reveals hypervariable sequences. *Nature*, *403*, 665–668.

Pathania, R., Navani, N. K., Gardner, A. M., Gardner, P. R., & Dikshit, K. L. (2002). Nitric oxide scavenging and detoxification by the *Mycobacterium tuberculosis* haemoglobin, HbN in *Escherichia coli*. *Molecular Microbiology*, *45*, 1303–1314.

Pathania, R., Navani, N. K., Rajamohan, G., & Dikshit, K. L. (2002). *Mycobacterium tuberculosis* hemoglobin HbO associates with membranes and stimulates cellular respiration of recombinant *Escherichia coli*. *The Journal of Biological Chemistry*, *277*, 15293–15302.

Pawaria, S., Lama, A., Raje, M., & Dikshit, K. L. (2008). Responses of *Mycobacterium tuberculosis* hemoglobin promoters to in vitro and in vivo growth conditions. *Applied and Environmental Microbiology*, *74*, 3512–3522.

Pawaria, S., Rajamohan, G., Gambhir, V., Lama, A., Varshney, G. C., & Dikshit, K. L. (2007). Intracellular growth and survival of *Salmonella enterica* serovar Typhimurium carrying truncated hemoglobins of *Mycobacterium tuberculosis*. *Microbial Pathogenesis*, *42*, 119–128.

Pelletier, H., & Kraut, J. (1992). Crystal structure of a complex between electron transfer partners, cytochrome *c* peroxidase and cytochrome *c*. *Science*, *258*, 1748–1755.

Peng, J., & Xu, J. (2011). RaptorX: Exploiting structure information for protein alignment by statistical inference. *Proteins*, *79*(Suppl. 10), 161–171.

Pesce, A., Couture, M., Dewilde, S., Guertin, M., Yamauchi, K., Ascenzi, P., et al. (2000). A novel two-over-two alpha-helical sandwich fold is characteristic of the truncated hemoglobin family. *The EMBO Journal*, *19*, 2424–2434.

Petersen, L., Larsen, T. S., Ussery, D. W., On, S. L., & Krogh, A. (2003). RpoD promoters in *Campylobacter jejuni* exhibit a strong periodic signal instead of a -35 box. *Journal of Molecular Biology*, *326*, 1361–1372.

Pirt, S. J. (1985). *Principles of microbe and cell cultivation*. Oxford: Blackwell Scientific Publications.

Pittman, M. S., Elvers, K. T., Lee, L., Jones, M. A., Poole, R. K., Park, S. F., et al. (2007). Growth of *Campylobacter jejuni* on nitrate and nitrite: Electron transport to NapA and NrfA via NrfH and distinct roles for NrfA and the globin Cgb in protection against nitrosative stress. *Molecular Microbiology*, *63*, 575–590.

Pittman, M. S., & Kelly, D. J. (2005). Electron transport through nitrate and nitrite reductases in *Campylobacter jejuni*. *Biochemical Society Transactions*, *33*, 190–192.

Poock, S. R., Leach, E. R., Moir, J. W., Cole, J. A., & Richardson, D. J. (2002). Respiratory detoxification of nitric oxide by the cytochrome *c* nitrite reductase of *Escherichia coli*. *The Journal of Biological Chemistry*, *277*, 23664–23669.

Poole, R. K. (2005). Nitric oxide and nitrosative stress tolerance in bacteria. *Biochemical Society Transactions*, *33*, 176–180.

Poole, R. K., Anjum, M. F., Membrillo-Hernández, J., Kim, S. O., Hughes, M. N., & Stewart, V. (1996). Nitric oxide, nitrite, and Fnr regulation of *hmp* (flavohemoglobin) gene expression in *Escherichia coli* K-12. *Journal of Bacteriology*, *178*, 5487–5492.

Poole, R. K., & Hughes, M. N. (2000). New functions for the ancient globin family: Bacterial responses to nitric oxide and nitrosative stress. *Molecular Microbiology*, *36*, 775–783.

Poole, R. K., & Shepherd, M. (2010). Bacterial Globins. In G. C. K. Roberts (Ed.), *Encyclopedia of biophysics*. Springer-Verlag Berlin Heidelberg.

Poulos, T. L. (1996). Ligands and electrons and haem proteins. *Nature Structural Biology*, *3*, 401–403.

Pullan, S. T., Gidley, M. D., Jones, R. A., Barrett, J., Stevanin, T. A., Read, R. C., et al. (2007). Nitric oxide in chemostat-cultured *Escherichia coli* is sensed by Fnr and other global regulators: Unaltered methionine biosynthesis indicates lack of S-nitrosation. *Journal of Bacteriology*, *189*, 1845–1855.

Purdy, D., Cawthraw, S., Dickinson, J. H., Newell, D. G., & Park, S. F. (1999). Generation of a superoxide dismutase (SOD)-deficient mutant of *Campylobacter coli*: Evidence for the significance of SOD in *Campylobacter* survival and colonization. *Applied and Environmental Microbiology*, *65*, 2540–2546.

Ramandeep, Hwang, K. W., Raje, M., Kim, K. J., Stark, B. C., Dikshit, K. L., et al. (2001). *Vitreoscilla* hemoglobin. Intracellular localization and binding to membranes. *The Journal of Biological Chemistry*, *276*, 24781–24789.

Rausch-Fan, X., & Matejka, M. (2001). From plaque formation to periodontal disease, is there a role for nitric oxide? *European Journal of Clinical Investigation*, *31*, 833–835.

Rayner, B. S., Stocker, R., Lay, P. A., & Witting, P. K. (2004). Regio- and stereo-chemical oxidation of linoleic acid by human myoglobin and hydrogen peroxide: Tyr(103) affects rate and product distribution. *The Biochemical Journal*, *381*, 365–372.

Richardson, A. R., Dunman, P. M., & Fang, F. C. (2006). The nitrosative stress response of *Staphylococcus aureus* is required for resistance to innate immunity. *Molecular Microbiology*, *61*, 927–939.

Rousseau, D. L., Li, D., Couture, M., & Yeh, S. R. (2005). Ligand-protein interactions in nitric oxide synthase. *Journal of Inorganic Biochemistry*, *99*, 306–323.

Saengkerdsub, S., Kim, W. K., Anderson, R. C., Nisbet, D. J., & Ricke, S. C. (2006). Effects of nitro compounds and feedstuffs on in vitro methane production in chicken cecal contents and rumen fluid. *Anaerobe*, *12*, 85–92.

Saunders, N. F., Ferguson, S. J., & Baker, S. C. (2000). Transcriptional analysis of the *nirS* gene, encoding cytochrome *cd1* nitrite reductase, of *Paracoccus pantotrophus* LMD 92.63. *Microbiology*, *146*(Pt 2), 509–516.

Savard, P. Y., Daigle, R., Morin, S., Sebilo, A., Meindre, F., Lague, P., et al. (2011). Structure and dynamics of *Mycobacterium tuberculosis* truncated hemoglobin N: Insights from NMR spectroscopy and molecular dynamics simulations. *Biochemistry*, *50*, 11121–11130.

Scallan, E., Hoekstra, R. M., Angulo, F. J., Tauxe, R. V., Widdowson, M. A., Roy, S. L., et al. (2011). Foodborne illness acquired in the United States—Major pathogens. *Emerging Infectious Diseases*, *17*, 7–15.

Sellars, M. J., Hall, S. J., & Kelly, D. J. (2002). Growth of *Campylobacter jejuni* supported by respiration of fumarate, nitrate, nitrite, trimethylamine-N-oxide, or dimethyl sulfoxide requires oxygen. *Journal of Bacteriology, 184,* 4187–4196.

Shepherd, M., Barynin, V., Lu, C., Bernhardt, P. V., Wu, G., Yeh, S. R., et al. (2010). The single-domain globin from the pathogenic bacterium *Campylobacter jejuni*: Novel D-helix conformation, proximal hydrogen bonding that influences ligand binding, and peroxidase-like redox properties. *The Journal of Biological Chemistry, 285,* 12747–12754.

Shepherd, M., Bernhardt, P. V., & Poole, R. K. (2011). Globin-mediated nitric oxide detoxification in the foodborne pathogenic bacterium *Campylobacter jejuni* proceeds via a dioxygenase or denitrosylase mechanism. *Nitric Oxide, 25,* 229–233.

Sies, H. (1991). *Oxidative stress II. Oxidants and antioxidants.* London: Academic Press.

Simonsen, J., Teunis, P., van Pelt, W., van Duynhoven, Y., Krogfelt, K. A., Sadkowska-Todys, M., et al. (2011). Usefulness of seroconversion rates for comparing infection pressures between countries. *Epidemiology and Infection, 139,* 636–643.

Singh, R. J., Hogg, N., Joseph, J., & Kalyanaraman, B. (1996). Mechanism of nitric oxide release from S-nitrosothiols. *The Journal of Biological Chemistry, 271,* 18596–18603.

Skirrow, M. B. (2006). John McFadyean and the centenary of the first isolation of *Campylobacter* species. *Clinical Infectious Diseases, 43,* 1213–1217.

Smith, M. A., Finel, M., Korolik, V., & Mendz, G. L. (2000). Characteristics of the aerobic respiratory chains of the microaerophiles *Campylobacter jejuni* and *Helicobacter pylori.* *Archives of Microbiology, 174,* 1–10.

Smith, A., Hill, S., & Anthony, C. (1990). The purification, characterization and role of the *d*-type cytochrome oxidase of *Klebsiella pneumoniae* during nitrogen fixation. *Journal of General Microbiology, 136,* 171–180.

Smith, H. K., Shepherd, M., Monk, C., Green, J., & Poole, R. K. (2011). The NO-responsive hemoglobins of *Campylobacter jejuni*: Concerted responses of two globins to NO and evidence in vitro for globin regulation by the transcription factor NssR. *Nitric Oxide, 25,* 234–241.

Spiro, T. G., & Wasbotten, I. H. (2005). CO as a vibrational probe of heme protein active sites. *Journal of Inorganic Biochemistry, 99,* 34–44.

Stevanin, T. M., Ioannidis, N., Mills, C. E., Kim, S. O., Hughes, M. N., & Poole, R. K. (2000). Flavohemoglobin Hmp affords inducible protection for *Escherichia coli* respiration, catalyzed by cytochromes *bo'* or *bd*, from nitric oxide. *The Journal of Biological Chemistry, 275,* 35868–35875.

Stuehr, D. J. (1999). Mammalian nitric oxide synthases. *Biochimica et Biophysica Acta, 1411,* 217–230.

Suschek, C. V., Schewe, T., Sies, H., & Kroncke, K. D. (2006). Nitrite, a naturally occurring precursor of nitric oxide that acts like a 'prodrug'. *Biological Chemistry, 387,* 499–506.

Svensson, L., Poljakovic, M., Save, S., Gilberthorpe, N., Schon, T., Strid, S., et al. (2010). Role of flavohemoglobin in combating nitrosative stress in uropathogenic Escherichia coli—Implications for urinary tract infection. *Microbial Pathogenesis, 49,* 59–66.

Tam, C. C., O'Brien, S. J., Petersen, I., Islam, A., Hayward, A., & Rodrigues, L. C. (2007). Guillain-Barré syndrome and preceding infection with *Campylobacter*, influenza and Epstein-Barr virus in the general practice research database. *PLoS One, 2,* e344.

Tarricone, C., Galizzi, A., Coda, A., Ascenzi, P., & Bolognesi, M. (1997). Unusual structure of the oxygen-binding site in the dimeric bacterial hemoglobin from *Vitreoscilla* sp. *Structure, 5,* 497–507.

Thomas, M. T., Shepherd, M., Poole, R. K., van Vliet, A. H., Kelly, D. J., & Pearson, B. M. (2010). Two respiratory enzyme systems in *Campylobacter jejuni* NCTC 11168 contribute to growth on L-lactate. *Environmental Microbiology, 13,* 48–61.

Tsai, P. S., Kallio, P. T., & Bailey, J. E. (1995). Fnr, a global transcriptional regulator of *Escherichia coli*, activates the *vitreoscilla* hemoglobin (VHb) promoter and intracellular

VHb expression increases cytochrome *d* promoter activity. *Biotechnology Progress, 11*, 288–293.

Tsaneva, I. R., & Weiss, B. (1990). *soxR*, a locus governing a superoxide response regulon in *Escherichia coli* K-12. *Journal of Bacteriology, 172*, 4197–4205.

van den Berg, J. M., van Koppen, E., Ahlin, A., Belohradsky, B. H., Bernatowska, E., Corbeel, L., et al. (2009). Chronic granulomatous disease: The European experience. *PLoS One, 4*, e5234.

van Wonderen, J. H., Burlat, B., Richardson, D. J., Cheesman, M. R., & Butt, J. N. (2008). The nitric oxide reductase activity of cytochrome *c* nitrite reductase from *Escherichia coli*. *The Journal of Biological Chemistry, 283*, 9587–9594.

Véron, M., & Chateline, R. (1973). Taxonomic Study of the genus *Campylobacter* Sebald and Véron and designation of the neotype strain for the type species, *Campylobacter fetus* (Smith and Taylor) Sebald and Véron. *International Journal of Systematic Bacteriology, 23*, 122–134.

Vinogradov, S. N., Hoogewijs, D., Bailly, X., Arredondo-Peter, R., Guertin, M., Gough, J., et al. (2005). Three globin lineages belonging to two structural classes in genomes from the three kingdoms of life. *Proceedings of the National Academy of Sciences of the United States of America, 102*, 11385–11389.

Vinogradov, S. N., Tinajero-Trejo, M., Poole, R. K., & Hoogewijs, D. (2013). Bacterial and Archaeal globins—A revised perspective. *Biochimica et Biophysica Acta, 1834*, 1789–1800.

Volkert, M. R., Loewen, P. C., Switala, J., Crowley, D., & Conley, M. (1994). The delta (*argF-lacZ*)205(U169) deletion greatly enhances resistance to hydrogen peroxide in stationary-phase *Escherichia coli*. *Journal of Bacteriology, 176*, 1297–1302.

Wainwright, L. M., Elvers, K. T., Park, S. F., & Poole, R. K. (2005). A truncated haemoglobin implicated in oxygen metabolism by the microaerophilic food-borne pathogen *Campylobacter jejuni*. *Microbiology, 151*, 4079–4091.

Wainwright, L. M., Wang, Y., Park, S. F., Yeh, S. R., & Poole, R. K. (2006). Purification and spectroscopic characterization of Ctb, a group III truncated hemoglobin implicated in oxygen metabolism in the food-borne pathogen *Campylobacter jejuni*. *Biochemistry, 45*, 6003–6011.

Wang, P. G., Xian, M., Tang, X., Wu, X., Wen, Z., Cai, T., et al. (2002). Nitric oxide donors: Chemical activities and biological applications. *Chemical Reviews, 102*, 1091–1134.

Wassenaar, T. M., & Blaser, M. J. (1999). Pathophysiology of *Campylobacter jejuni* infections of humans. *Microbes and Infection, 1*, 1023–1033.

Webster, D. A., & Hackett, D. P. (1966). The purification and properties of cytochrome *o* from *Vitreoscilla*. *The Journal of Biological Chemistry, 241*, 3308–3315.

Weerakoon, D. R., Borden, N. J., Goodson, C. M., Grimes, J., & Olson, J. W. (2009). The role of respiratory donor enzymes in *Campylobacter jejuni* host colonization and physiology. *Microbial Pathogenesis, 47*, 8–15.

Weingarten, R. A., Taveirne, M. E., & Olson, J. W. (2009). The dual-functioning fumarate reductase is the sole succinate:quinone reductase in *Campylobacter jejuni* and is required for full host colonization. *Journal of Bacteriology, 191*, 5293–5300.

Weiss, B. (2006). Evidence for mutagenesis by nitric oxide during nitrate metabolism in *Escherichia coli*. *Journal of Bacteriology, 188*, 829–833.

Winkelstein, J. A., Marino, M. C., Johnston, R. B., Jr., Boyle, J., Curnutte, J., Gallin, J. I., et al. (2000). Chronic granulomatous disease. Report on a national registry of 368 patients. *Medicine (Baltimore), 79*, 155–169.

Wittenberg, J. B., Bolognesi, M., Wittenberg, B. A., & Guertin, M. (2002). Truncated hemoglobins: A new family of hemoglobins widely distributed in bacteria, unicellular eukaryotes, and plants. *The Journal of Biological Chemistry, 277*, 871–874.

Wosten, M. M., Boeve, M., Koot, M. G., van Nuenen, A. C., & van der Zeijst, B. A. (1998). Identification of *Campylobacter jejuni* promoter sequences. *Journal of Bacteriology*, *180*, 594–599.

Yamada, H., Makino, R., & Yamazaki, I. (1975). Effects of 2,4-substituents of deuteropheme upon redox potentials of horseradish peroxidases. *Archives of Biochemistry and Biophysics*, *169*, 344–353.

Yang, J., Webster, D. A., & Stark, B. C. (2005). ArcA works with Fnr as a positive regulator of *Vitreoscilla* (bacterial) hemoglobin gene expression in *Escherichia coli*. *Microbiology Research*, *160*, 405–415.

Zheng, J., Meng, J., Zhao, S., Singh, R., & Song, W. (2008). *Campylobacter*-induced interleukin-8 secretion in polarized human intestinal epithelial cells requires *Campylobacter*-secreted cytolethal distending toxin- and Toll-like receptor-mediated activation of NF-kappaB. *Infection and Immunity*, *76*, 4498–4508.

Zheng, M., Wang, X., Templeton, L. J., Smulski, D. R., LaRossa, R. A., & Storz, G. (2001). DNA microarray-mediated transcriptional profiling of the *Escherichia coli* response to hydrogen peroxide. *Journal of Bacteriology*, *183*, 4562–4570.

CHAPTER FIVE

Haemoglobins of Mycobacteria: Structural Features and Biological Functions

Kelly S. Davidge[*,1], Kanak L. Dikshit[†,1]

[*]School of Life Sciences, Centre for Biomolecular Sciences, University of Nottingham, University Park, Nottingham, United Kingdom
[†]CSIR-Institute of Microbial Technology, Chandigarh, India
[1]Corresponding authors: e-mail address: kelly.davidge@nottingham.ac.uk; kanak@imtech.res.in

Contents

1. Introduction: Discovery of Haemoglobins in Mycobacteria   148
2. Co-occurrence of Multiple Hbs in Mycobacteria   149
3. trHbs of Mycobacteria: Small Hbs with Novel Characteristics   152
   3.1 Group I trHbs: trHbN   153
   3.2 Group II trHbs: trHbO   163
   3.3 Type III trHbs: trHbP   169
   3.4 Genetic regulation of the trHbs   169
4. Conventional and Novel FlavoHbs of Mycobacteria   173
   4.1 Structural features   174
   4.2 Functional properties   180
   4.3 Genomic organisation and genetic regulation   180
5. Biological Functions of trHbs and FlavoHbs   182
   5.1 Nitric oxide scavenging   182
   5.2 Oxygen metabolism   186
   5.3 Redox signalling and stress management   188
   5.4 Other functions   188
6. Concluding Remarks   189
References   189

Abstract

The genus *Mycobacterium* is comprised of Gram-positive bacteria occupying a wide range of natural habitats and includes species that range from severe intracellular pathogens to economically useful and harmless microbes. The recent upsurge in the availability of microbial genome data has shown that genes encoding haemoglobin-like proteins are ubiquitous among Mycobacteria and that multiple haemoglobins (Hbs) of different classes may be present in pathogenic and non-pathogenic species. The occurrence of truncated haemoglobins (trHbs) and flavohaemoglobins (flavoHbs) showing distinct haem active site structures and ligand-binding properties suggests

Advances in Microbial Physiology, Volume 63
ISSN 0065-2911
http://dx.doi.org/10.1016/B978-0-12-407693-8.00005-4

that these Hbs may be playing diverse functions in the cellular metabolism of Mycobacteria. TrHbs and flavoHbs from some of the severe human pathogens such as *Mycobacterium tuberculosis* and *Mycobacterium leprae* have been studied recently and their roles in effective detoxification of reactive nitrogen and oxygen species, electron cycling, modulation of redox state of the cell and facilitation of aerobic respiration have been proposed. This multiplicity in the function of Hbs may aid these pathogens to cope with various environmental stresses and survive during their intracellular regime. This chapter provides recent updates on genomic, structural and functional aspects of Mycobacterial Hbs to address their role in Mycobacteria.

ABBREVIATIONS

flavoHb flavohaemoglobin
Mb *Mycobacterium bovis*
Ms *Mycobacterium smegmatis*
Mtb *Mycobacterium tuberculosis*
MtbFHb *Mycobacterium tuberculosis* flavohaemoglobin
NO nitric oxide
NOD nitric oxide dioxygenation
trHb truncated haemoglobin

1. INTRODUCTION: DISCOVERY OF HAEMOGLOBINS IN MYCOBACTERIA

The genus *Mycobacteria* consists of a group of acid-fast, Gram–positive bacteria, some of which are severe human pathogens, while others are either opportunistic pathogens or environmental saprophytes (Grange, 1996). Members of this group have been categorised into three distinct classes, on the basis of their growth characteristics: the fast growers, the slow growers and those strains or species that have never been cultured in *in vitro* conditions. The O_2 availability in the natural niche of these Mycobacteria ranges from O_2-rich to near anaerobic, and although all species are considered to have an obligate requirement of O_2 for their life cycle (Berney & Cook, 2010; Park, Myers, & Marzella, 1992), paradoxically, the majority of them have the ability to grow and metabolise under hypoxia (Voskuil, Visconti, & Schoolnik, 2004). Some species, such as *Mycobacterium tuberculosis* and *Mycobacterium leprae*, are human pathogens that encounter severe hypoxia and environmental stresses within macrophages and avascular calcified granuloma where these bacilli primarily reside (Ehrt & Schnappinger, 2009; Rustad, Sherrid, Minch, & Sherman, 2009); they also get exposed to an

O_2-rich environment during lung infection (Balasubramanian, Wiegeshaus, Taylor, & Smith, 1994). The remarkable ability of these Mycobacteria to adapt their aerobic metabolism in response to the changing level of O_2 and environmental stresses depends on systems that can efficiently sense, procure and conserve O_2 in the cell. For example, haemoglobin (Hb)-like proteins are generally involved in binding and/or transporting O_2 for energy generation and management of environmental stresses against reactive oxygen and nitrogen species via the nitric oxide dioxygenation (NOD) reaction (Poole & Hughes, 2000; Weber & Vinogradov, 2001; Wittenberg & Wittenberg, 1990). Although Hbs in bacteria were discovered in 1986 (Wakabayashi, Matsubara, & Webster, 1986), no such O_2-binding molecule had been detected in any species of Mycobacteria before the unravelling of the genome sequence of *M. tuberculosis* in 1998 (Cole et al., 1998) which revealed two genomic loci, *glbN* and *glbO*, that could encode two distinct Hb-like proteins, later termed truncated haemoglobins (trHbs) (Wittenberg, Bolognesi, Wittenberg, & Guertin, 2002). More recently, the presence of two-domain haemoglobins (flavoHbs) with novel structural features has been identified in many Mycobacterial genomes (Gupta et al., 2012; Gupta, Pawaria, Lu, Yeh, & Dikshit, 2011). The diversity in the number and types of Hb-encoding genes in the genomes of Mycobacteria suggests that they have distinct cellular functions and reflect the difference in the functions of these O_2-binding proteins in individual Mycobacterial species. This chapter deals with recent genomic, biochemical, structural and functional information on Hb-like proteins and addresses the functional relevance of Hbs in cellular metabolism, stress management and pathogenicity of Mycobacteria.

2. CO-OCCURRENCE OF MULTIPLE Hbs IN MYCOBACTERIA

While trHbs were first identified in cyanobacteria with the characterisation of cyanoglobin in *Nostoc commune* (Thorsteinsson et al., 1999), the distinctive features and structural characteristics of this class of globins were only unravelled after the discovery of small Hbs in Mycobacteria (Cole et al., 1998; Couture et al., 1999) and the determination of their three-dimensional structures (Milani, Pesce, Ouellet, Guertin, & Bolognesi, 2003; Milani et al., 2001), along with other small Hbs from the ciliated protozoa *Paramecium caudatum* and the unicellular algae *Chlamydomonas eugametos* (Pesce et al., 2000). Since then, a number of plants, unicellular eukaryotes and bacterial species, including Mycobacteria, have been shown to contain trHbs (Vuletich & Lecomte, 2006), both through the analysis of sequenced genomes and experimental data; these include *Campylobacter jejuni* (Wainwright, Elvers, Park, & Poole, 2005), *Pseudoalteromonas haloplanktis* (Giordano et al., 2007), *Thermobifida fusca* (Droghetti et al., 2010) and other Mycobacterial species such as *M. leprae* (Visca et al., 2002). Phylogenetic analysis of Mycobacterial Hbs revealed that the trHb family branches into three main groups, designated as type I, II and III (named as HbN, HbO and HbP, respectively); the trHbO gene (*glbO*) has been predicted as the ancestral gene, with group I and group III genes appearing as a result of duplication and transfer events (Vuletich & Lecomte, 2006). All three groups of trHbs are represented in Mycobacterial species, and in some cases, two or three groups of trHbs coexist in the same organism, indicating that they have evolved to perform distinct cellular functions (Table 5.1). Detailed analysis of the co-occurrence of these three classes of trHbs in Mycobacteria indicates that all Mycobacterial genomes sequenced so far carry genes for the group II (trHbO), and the majority of them have group I (trHbN) trHbs, the exceptions being *M. leprae*, *Mycobacterium abscessus*, *Mycobacterium massiliense* and *Mycobacterium thermoresistibile* that lack the gene encoding for trHbN (Table 5.1).

In addition to these single-domain trHbs, the occurrence of more than one flavohaemoglobin (flavoHb)-encoding gene has been identified in many pathogenic and non-pathogenic Mycobacteria (Gupta et al., 2012, 2011). An interesting pattern in the occurrence of flavoHb-encoding genes in pathogenic and non-pathogenic Mycobacteria has been observed. The majority of opportunistic pathogens and fast-growing Mycobacteria display the presence of two flavoHb-encoding genes: one for conventional flavoHbs (Type I) similar to other microbial flavoHbs, and the other one for a new class of flavoHb (Type II), carrying unusual structural features in their haem and reductase domains (Gupta et al., 2012). Notably, the new class of flavoHb is present in the majority of Mycobacterial species and co-exists along with a conventional flavoHb in several Mycobacterial species. The occurrence of trHbs and flavoHbs in Mycobacterial genomes varies in different species and, to some extent, depends on their natural niche. Based on the available genome data, the distribution and co-existence of single-domain trHb and two-domain flavoHbs in various Mycobacterial species are presented in Table 5.1.

Table 5.1 Distribution of truncated haemoglobins and flavohaemoglobins among various species of Mycobacteria

Mycobacterium species	trHbN	trHbO	trHbP	FlavoHbs Type I (*E. coli* type)	Type II (*Mtb* type)	Status
Mycobacterium tuberculosis H37Rv	1	1	–	–	1	Virulent
Mycobacterium bovis	1	1	–	–	1	Virulent
Mycobacterium abscessus	–	1	–	–	1	Virulent
Mycobacterium marinum	1	1	1	–	1	Virulent
Mycobacterium leprae	–	1	–	–	–	Virulent
Mycobacterium avium	1	1	1	–	1	Virulent
Mycobacterium africanum	1	1	–	–	1	Virulent
Mycobacterium smegmatis	1	1	–	1	1	Avirulent
Mycobacterium liflandii	1	1	1	–	1	Virulent
Mycobacterium intracellulare	1	1	1	–	1	Virulent
Mycobacterium indicus pranii	1	1	1	–	1	Avirulent
Mycobacterium massiliense	–	1	–	–	–	Virulent
Mycobacterium sp.KMS	1	1	1	1	1	Avirulent
Mycobacterium sp.MCS	1	1	1	1	1	Avirulent
Mycobacterium sp.JLS	1	1	1	1	1	Avirulent
Mycobacterium gilvum	1	1	–	1	1	Avirulent
Mycobacterium vanbaalenii	1	1	–	1	1	Avirulent

Continued

Table 5.1 Distribution of truncated haemoglobins and flavohaemoglobins among various species of Mycobacteria—Cont'd

Mycobacterium species	trHbN	trHbO	trHbP	FlavoHbs Type I (*E. coli* type)	Type II (*Mtb* type)	Status
Mycobacterium hassiacum	1	1	–	–	1	Avirulent
Mycobacterium ulcerans	1	1	1	1	1	Virulent
Mycobacterium sp. MOTT36Y	1	1	1	–	1	Virulent
Mycobacterium sp. JDM 6-1	1	1	–	–	1	Virulent
Mycobacterium chubuense	1	1	–	1	1	Virulent
Mycobacterium colombiense	1	1	1	–	–	Virulent
Mycobacterium thermoresistibile	–	1	1	1	1	Virulent
Mycobacterium vaccae	1	1	–	1	–	Avirulent
Mycobacterium tusciae	1	1	1	–	1	Virulent
Mycobacterium phlei	1	1	–	1	1	Avirulent
Mycobacterium parascrofulaceum	1	1	1	–	1	Virulent
Mycobacterium fortuitum	1	1	1	1	1	Virulent

3. trHbs OF MYCOBACTERIA: SMALL Hbs WITH NOVEL CHARACTERISTICS

Unsurprisingly, considering their name, trHbs are smaller than typical Hbs, usually by around 20–40 residues. However, contrary to what the name suggests, they are not simply 'truncated' versions of conventional globins: their structures are based on a 2-on-2 α-helical fold that contributes to the stability of the protein. As Wittenberg et al. (2002) explain, trHbs show a universal conservation of the helix pairs B/E and G/H, deletion of almost the entire A-helix and CD–D region and a shortening of the F-helix. In addition, there are only a small number of conserved residues as shown by the alignments in Fig. 5.1; these include TyrB10 and PheE15 and will be discussed later.

TrHbs from the three groups have varying functions not restricted to O_2 carriage, and as we will see in this section, all three are represented in the Mycobacteria (Ascenzi, Bolognesi, Milani, Guertin, & Visca, 2007). Table 5.1 shows the Mycobacterial species which contain either one or more trHbs. While a few species, such as *Mycobacterium avium,* have genes encoding for all three trHbs others, such as *M. leprae*, have only one (Fabozzi, Ascenzi, Renzi, & Visca, 2006). The authors suggest that *M. leprae* has lost trHbN and can survive with only trHbO, which has concurrently acquired the ability to detoxify NO (Fabozzi et al., 2006), and that this probably reflects the reductive evolution of its genome (Cole et al., 2001). Some of the truncated globins are similar to each other, such as trHbN from *M. tuberculosis*, *M. avium* and *Mycobacterium smegmatis*, where high-sequence identity points towards similar functions in the cell (Pathania, Navani, Gardner, Gardner, & Dikshit, 2002). Interestingly, in the Mycobacteria, sequence identity between different trHbN proteins is less—between 21% and 32% (Couture et al., 1999)—than that between trHbO proteins, and most trHbs from other bacterial species share similarity with trHbO rather than trHbN, which is more related to globins from algae, protozoa and cyanobacteria (Pathania, Navani, Rajamohan, & Dikshit, 2002). Currently, there is no published data on the function of trHbP, and its existence has only been garnered from genome sequence data. Therefore, this section of the review will concentrate on the trHbs, N and O.

3.1. Group I trHbs: trHbN

3.1.1 Structural features and dynamics

TrHbs in the Mycobacteria have conserved structural features that maintain the globin fold. The first paper regarding *M. tuberculosis* trHbN (Mtb trHbN) was published in 1999. Alignment with other Hbs showed the conservation of the proximal HisF8 and the distal PheCD1 (Couture et al., 1999). Clues as to the function of important residues can be found from these alignments. For example, in vertebrate Hbs, the E7 position distal to the haem is always occupied by His or Gln, residues which contribute to the stabilisation of bound O_2 via hydrogen bonds; in Mtb trHbN, the residue at E7 is Leu, not capable of hydrogen bonds; therefore this position in Mtb trHbN cannot be involved in stabilising O_2 to the haem.

```
                                1   5    10   15    1   5    10    1   5    1
                                AAAAAAAAAAAAAAAA   BBBBBBBBBBBBBBBBBCCCCCCC   E
M.tuberculosisHbN        --MGLLSRLRKREPISIYDKIGGHEAIEVVVEDFYVRVLADDQLSAFFSG
M.bovisHbN               --MGLLSRLRKREPISIYDKIGGHEAIEVVVEDFYVRVLADDQLSAFFSG
M.ulceransHbN            --MGLLSRFRKRAPVSIYDKIGGYEAIEAVVEDFYVRVLADDQLGGFFTG
M.africanumHbN           --MGLLSRLRKREPISIYDKIGGHEAIEVVVEDFYVRVLADDQLSAFFSG
M.SUMu001HbN             -------MRKKREPISIYDKIGGHEAIEVVVEDFYVRVLADDQLSAFFSG
M.canettiiHbN            --MGLLSRLRKREPISIYDKIGGHEAIEVVVEDFYVRVLADDQLSAFFSG
M.aviumHbN               MVMKILARFRKAEPASIYDRIGGHEALEVVVEDFYVRVLADEQLSGFFTG
M.intracellulareHbN      --MKILARFRKPEPGTIYDRIGGHEAIEVAVDDFYVRVLADDELAGFFAG
M.MOTT36YHbN             --MKILARFRKPEPGTIYDRIGGHEAIEVAVDDFYVRVLADDELAGFFAG
M.colombienseHbN         --MKILARFRKSAPAPVYDRIGGHEAIEVVVEDFYVRVLADDELSGFFTG
M.parascrofulaceumHbN    --MKILARFRRPQPVAVYDRIGGREAIEVVVEDFYARVLADDQLCGFFTG
M.haemophilumHbN         --MRILSRFGRREPASIYDRIGGYEAIEVVVEDFYARVLADDQLSSFFAG
M.xenopiHbN              --MGMLPRFRTRKARSIYDRIGGYEAIEVVVEDFYSRVLADAQLAGFFAG
M.smegmatisHbN           ------------MTSIYEQIGGAEALEVVVEDFYRRVLADDELAGFFTG
M.sp.JLSHbN              ------------MTTIYEQIGGAEALEGVVEDFYGRVLADEQLSGFFTG
M.sp.KMSHbN              ------------MTTIYEQIGGAEALEGVVEDFYGRVLADEQLSGFFTG
M.sp.MCSHbN              ------------MTTIYEQIGGAEALEGVVEDFYGRVLADEQLSGFFTG
M.fortuitumHbN           ------------MTTIYDQIGGAEALETVVEDFYRRVLADDELAGFFTG
M.tusciaeHbN             ------------MPTIFDQIGGYEALEVVVADFYDRVLADSELAGFFTG
M.phleiHbN               ------------MTTIYDQIGGHEALEVVVDDFYRRVLHDPELAGFFTG
M.hassiacumHbN           ------------MTTIFEEIGGYEALEAVVEDFYRRVLDDPELNGFFTG
M.rhodesiaeHbN           ------------MPTVFAQVGGYEGLEVVVADFYDRVLTDAELAHFFTG
                                                        B10            CD1
```

```
                                5    10   15    1   5        1   5    10
                                EEEEEEEEEEEEEEEEEE   FFFFFFFF   GGGGGGGGGGG
M.tuberculosisHbN        TNMSRLKGKQVEFFAAALGGPEPYTGAPMKQVHQGRGITMHHFSLVAGHL
M.bovisHbN               TNMSRLKGKQVEFFAAALGGPEPYTGAPMKQVHQGRGITMHHFSLVAGHL
M.ulceransHbN            TNMNRLKGKQAEFFAAALGGPEPYTGAPMKQVHQGRGITMAHFSLVAGHL
M.africanumHbN           TNMSRLKGKQVEFFAAALGGPEPYTGAPMKQVHQGRGITMHHFSLVAGHL
M.SUMu001HbN             TNMSRLKGKQVEFFAAALGGPEPYTGAPMKQVHQGRGITMHHFSLVAGHL
M.canettiiHbN            TNMSRLKGKQVEFFAAALGGPEPYTGAPMKQVHQGRGITMHHFSLVAGHL
M.aviumHbN               TNMNRLKGKQVEFFAAALGGPHPYTGAPMKQVHQGRGITMHHFGLVAGHL
M.intracellulareHbN      TNMNRLKGKQVEFFAAALGGPEPYTGAPMKQVHQGRGITMHHFGLVAAHL
M.MOTT36YHbN             TNMNRLKGKQVEFFAAALGGPQPYTGAPMKQVHQGRGITMHHFGLVAAHL
M.colombienseHbN         TNMNRLKGKQVEFFAAALGGPEPYTGAPMKQVHQGRGITMHHFGLVAGHL
M.parascrofulaceumHbN    TNMNRLKGKQVEFFAAALGGPEPYTGAPMRQVHRGRGITMHHFNLVAGHL
M.haemophilumHbN         TNMSRLKGKQAEFIAAALGGPAPYTGASMKQVHQGRGITMHHFTLVAGHL
M.xenopiHbN              TNVNRLKGKQAEFLAAALGGPLPYTGVSMKQAHKGRGITMHHFNLVAGHL
M.smegmatisHbN           TNMSRLKGRQVEFFATALGGPDEYTGAPMRQVHQGRGITMHHFDLVAGHL
M.sp.JLSHbN              TNMARLKGRQVEFFAAALGGPEPYTGAPMRQVHQGRGITMHHFNLVAGHL
M.sp.KMSHbN              TNMARLKGRQVEFFAAALGGPEPYTGAPMRQVHQGRGITMHHFNLVAGHL
M.sp.MCSHbN              TNMARLKGRQVEFFAAALGGPEPYTGAPMRQVHQGRGITMHHFNLVAGHL
M.fortuitumHbN           TNMARLKGKQVEFFAAALGGPVPYSGAGMRQVHQGRGIGMHHFNLVAGHL
M.tusciaeHbN             TNMARLKGKQVEFFAAALGGPEPYSGAPMRQVHQGRGITMHHFALVANHL
M.phleiHbN               TNMSRLKGRQVEFFAAALGGPEPYTGAPMRQVHQGRGITMHHFNLVAGHL
M.hassiacumHbN           TNMARLKGKQVEFFAAALGGPDPYTGAPMRQVHQGRGITMHHFNLVAGHL
M.rhodesiaeHbN           TNMARLKGKQVEFFAAALGGPEPYTGAPMRQVHQGRGITMHHFTLVANHL
                         E7     E14                                F8
```

```
                                15    1   5    10   15    20
                                GGGGGGGG  HHHHHHHHHHHHHHHHHHHHHHHHHHHHH
M.tuberculosisHbN        ADALTAAGVPSETITEILGVIAPLAVDVTSGESTTAPV
M.bovisHbN               ADALTAAGVPSETITEILGVIAPLAVDVTSGESTTAPV
M.ulceransHbN            GDSLTAAGVPSETVTNILKLVAPLATDIASGETTTAGV
M.africanumHbN           ADALTAAGVPSETITEILGVIAPLAVDVTSGESTTAPV
M.SUMu001HbN             ADALTAAGVPSETITEILGVIAPLAVDVTSGESTTAPV
M.canettiiHbN            ADALTAAGVPSETITEILGVIAPLAVDVASGESTAAPV
M.aviumHbN               ADALTAAGVPSETVSEILGAIAPLAPEIATGEA-KATV
M.intracellulareHbN      ADALAAAGVPPETVTEILGAIAPLAPEIATGDA-KATV
M.MOTT36YHbN             ADALAAAGVPPETVTEILGAIAPLAPEIATGDA-KATV
M.colombienseHbN         ADALAAAGVPSETVTEILGAIAPLAPQIATGDAGKVRI
M.parascrofulaceumHbN    SDALAAAGVPSGTVTEIIGAIAPLAPEIATGEPNTATV
M.haemophilumHbN         SDALTTAGILAETVTEILGVIAGLAPDVVSGETTTAQV
M.xenopiHbN              SDSLTAAGVPDETVAEILAVVAPLASDIAS-DAEPARV
```

*M.smegmatis*HbN	GDALSAAGMPGATTSQIIAAIAPLAPEIATARTA----
*M.sp.JLS*HbN	TDALHAAGVPEALVGQIIGAVAPLADEIATARTA----
*M.sp.KMS*HbN	TDALHAAGVPEALVGQIIGAVAPLADEIATARTA----
*M.sp.MCS*HbN	TDALHAAGVPEALVGQIIGAVAPLADEIATARTA----
*M.fortuitum*HbN	SDSLAGAGVPEPIVGQIIAAIAPLADEIATARTA----
*M.tusciae*HbN	AASLSGAGVPAETVDKILAAIAPLSTDIATASVA----
*M.phlei*HbN	SDSLTAAGVPGEIVEQILAAVAPLSADIASS-VA----
*M.hassiacum*HbN	ADSLVAAGVGAATVEQILAAVAPLSAEIATASVA----
*M.rhodesiae*HbN	AASLSAAGVPGATVDQILGAIAPLSADIATA-VA----

Figure 5.1 Structure-based sequence alignment of group I (trHbN) haemoglobins of Mycobacteria. (For colour version of this figure, the reader is referred to the online version of this chapter.)

The crystal structure of oxygenated Mtb trHbN was solved in 2001; it contains a 2-over-2 α-helical fold which retains the main protein regions (B, E, G and H) for stabilisation of the trHb fold (Milani et al., 2001). As in other trHbs, these exist as two anti-parallel helix pairs (B/E and G/H) (Pesce et al., 2000). The protein is homodimeric and contains a short α-helix, termed the pre-A region, at the N-terminal end; deletion of the A-helix is also seen in a number of other trHbs, along with the removal of the traditional F-helix (Pesce et al., 2000). Features that contribute to the stabilisation of the globin fold include the C-helix that supports PheCD1, and the E-helix that supports LeuE7 and PheE14. The haem iron coordinating residue, a proximal unstrained HisF8 (Couture et al., 1999), is supported by the pre-F region, a 10-residue extension on the haem proximal side that is followed by the F-helix (Milani et al., 2001); TyrB10 is involved in ligand stability via hydrogen bonds (Couture et al., 1999). Analysis of trHb amino acid sequences from *P. caudatum* and *C. eugametos* by Pesce et al. (2000) showed that important structural features on the haem distal site include TyrB10, GlnE7, PheE14 and the residue located at E11. The trHbs from these species do not, however, feature the pre-A region, nor are they homodimeric (Milani et al., 2001).

The X-ray structure provides clues as to how small gaseous ligands such as NO and O_2 can gain access to the buried haem group. Initial work showed a unique tunnel system with two main branches leading from the haem group to distinct areas (one between the AB and GH hinge regions and the other between the G and H helices) on the surface of the protein (Milani et al., 2001). Inside the distal pocket, the two tunnels intersect at the haem-bound O_2 molecule near the two important residues TyrB10 and GlnE11; it is this intersection that suggests the tunnels function as a hydrophobic pathway through the protein for O_2 and maybe NO, and perhaps for escape of nitrate. The long tunnel is 20 Å long and has the biggest volume, with the short tunnel being only 13 Å long with the surface entry occurring between the G and H helices and defined by PheG5, AlaG9, LeuH8 and AlaH12 (Daigle, Guertin, & Lagüe, 2009). This tunnel system may function as a way to create high concentrations of ligands in a local environment; its hydrophobicity also slows the access of water to the haem pocket (Dantsker et al., 2004). As Bidon-Chanal et al. comment, ligand diffusion appears to be a limiting factor in the reaction of trHbN and certain aspects of the protein appear to have evolved to ensure that gaseous ligands can access the haem cavity. For example, the TyrB10–GlnE11 pair undergoes a conformation change upon binding of O_2 to the haem, accessed via the short tunnel, which causes the main tunnel to open via movement of the PheE15 gate (Bidon-Chanal, Marti, Estrin, & Luque, 2007).

Molecular dynamics studies have shown that the long tunnel provides the lowest energy pathway for NO migration to the haem (Lama et al., 2009). The proposed mechanism—that the main tunnel opens only after O_2 has bound to the haem—suggests that the protein may have evolved this control to ensure that NO can enter the protein only after O_2 has bound to the haem, resulting in the NOD reaction and the production of nitrate (see Section 5.1). The coordination state of the haem is the ultimate factor controlling the positions of key residues which allow or block ligand binding: when the protein is in the deoxy form, molecular dynamics show that the long tunnel is completely closed (Bidon-Chanal et al., 2006). A paper from 2011 proposes some additional conclusions about the tunnel system in trHbN, suggesting the presence of two extra tunnels, termed EH and BE, that lead from the surface of the protein to the haem distal pocket that are not seen in the crystal structure (Daigle et al., 2009). The EH tunnel is located between the E and H helices and the entrance is defined by the residues PheE14, AlaE18, ValH10 and LeuH14. It is 15 Å long and access by ligands is controlled by PheE15 (Crespo et al., 2005). In the open state, this tunnel can merge with the long tunnel and the authors suggest that ligands gain access to the haem via a two-stage process, first moving from the substrate into the protein itself, and then across the PheE15 barrier into the haem distal pocket (Daigle et al., 2009). The BE tunnel, between B and E helices, has its entrance defined by TyrB10, LeuB14 and MetE4. It is 10.5 Å long and has two conformations determined by the positioning of TyrB10, and movements only appear to occur when simulations were done with deoxy trHbN (Daigle et al., 2009). The paper is in variance with another molecular dynamics study on the tunnel system: results here show that in the open state, PheE15 is parallel to the long tunnel axis and in the closed state is orthogonal (Daigle et al., 2009); however, Bidon-Chanal et al. propose that the open state occurs when the phenyl ring of PheE15 is parallel to the axis of the tunnel and the closed state when it is closer to the haem (Bidon-Chanal et al., 2006). This disagreement could be due to the pre-A region, which appears to behave differently in the two simulations and will be discussed later.

More recent work, using solution NMR and molecular dynamics to further probe the structure of trHbN, has added to our understanding. These authors propose that the F and G helices provide flexibility to the protein, with the B and E helices being more rigid (Savard et al., 2011). The molecular dynamics study identified that helices B, G and H and residue TyrB10 have slow motions, and that residues in helices G and H, through which the short tunnel passes, are orientated away from the solvent. This could perhaps slow ligands passing through the tunnel or the movement of this part of the protein, which is potentially vital for the release of the relatively bulky nitrate once the reaction is complete (Savard et al., 2011). Their data suggest that this mixture of flexibility and rigidity is required for the formation of the four hydrophobic tunnels, allowing ligand migration that links the haem centre to the solvent. This agrees with the conclusions reached in Daigle et al., which showed that protein rigidity is required to generate and maintain the tunnels with the flexibility of the protein able to reshape the tunnels to allow passage of various ligands (Daigle et al., 2009). A previous study also commented on the complexity of the flexibility of trHbN (Crespo et al., 2005).

The control of ligand binding, and ligand access to the haem, has been studied extensively. One important residue conserved among Mycobacterial trHbs (Fig. 5.1) is TyrB10, which is located buried in the haem distal pocket and forms hydrogen bonds with GlnE11, thereby creating a polar environment in the haem pocket (Mukai, Ouellet, Ouellet, Guertin, & Yeh, 2004). TyrB10 is unique to the Mycobacterial Hbs, and Resonance Raman spectral analysis of Mtb trHbN mutants (where TyrB10 was mutated to Leu or Phe) showed that a change in this residue gives the globin similar properties to other haem-containing proteins, suggesting that TyrB10 plays an important role in ligand binding by Mycobacterial trHbN (Yeh, Couture, Ouellet, Guertin, & Rousseau, 2000). Data also showed that the binding properties between bound O_2 and TyrB10 in the haem pocket are unique. The haem group of Mtb trHbN can form H-bonds with many ligands (Yeh et al., 2000) including O_2 (Crespo et al., 2005), and NO (Mukai et al., 2004) and TyrB10 has been shown to stabilise bound O_2 at the distal site

(Couture et al., 1999). Complementing a *Mycobacterium bovis glbN* mutant with *M. bovis* trHbN (Mb trHbN) containing the amino acid substitution TyrB10Phe did not restore NO detoxifying activity, which points towards the important role of TyrB10 in the function of trHbN (Ouellet et al., 2002). Studies using Resonance Raman spectroscopy, optical spectroscopy and X-ray crystallography have shown that TyrB10, along with GlnE11, controls the binding of water molecules to ferric Mtb trHbN, plus its ionisation state (Ouellet, Milani, Couture, Bolognesi, & Guertin, 2006). These properties are important as they ensure that the sixth coordination position is kept empty in deoxy trHbN, in order to allow the binding of O_2 to the haem. Laser flash-photolysis analysis of Mtb trHbN TyrB10Phe and TyrB10Leu mutants showed that substitution of the Tyr at this position increases the k_{on} value for O_2 by approximately 10-fold, indicating that TyrB10 is a barrier to O_2 binding to the haem (Ouellet et al., 2008). The k_{on} for CO was also increased in these mutants. Using kinetics and molecular dynamics, the authors showed that ligand binding and geminate re-binding are discouraged by the displacement of a non-coordinated distal site water molecule, stabilised mostly by TyrB10. When TyrB10 is changed to Phe, geminate re-binding of both O_2 and CO is increased; along with additional data which showed increased NO binding rates to the ferric haem of a TyrB10Leu/GlnE11Val double mutant, these suggest that TyrB10 plays a major role in access of ligands to the haem iron (Ouellet et al., 2008).

Molecular dynamics studies of the TyrB10Phe mutant showed a decrease in binding energy compared with the wild type (Crespo et al., 2005). Two studies in 2009 found extensive hydrogen bonding between haem-bound O_2, TyrB10 and GlnE11 (Daigle et al., 2009; Mishra & Meuwly, 2009); they are dynamic in nature with the bonds being broken and reformed many times during the simulation. These bonds may contribute to the ligand chemistry by supporting the different protein conformations required in the NOD reaction (Mishra & Meuwly, 2009). Indeed, there is hydrogen bond formation between TyrB10 and GlnE11; GlnE11 may therefore interact with NO and allow it to be positioned correctly with respect to O_2, leading to the chemistry required for the detoxification to nitrate (Crespo et al., 2005). A molecular dynamics study of Mtb trHbN in explicit water investigating TyrB10 and GlnE11 showed that these two amino acids and the hydrogen bonds they form exist in two different configurations, controlled by O_2 coordination to the haem (Daigle et al., 2009). These changes in conformation, especially of GlnE11, control the positioning of PheE15, and therefore whether the large tunnel is open or closed; in addition, it was discovered that in deoxy trHbN, PheE15 can only be in the closed position, preventing NO access to the haem cavity. Another molecular dynamics study, previously discussed, proposed that TyrB10 has slow motions, which may play a role in how this residue modulates trHbN function (Savard et al., 2011). With respect to NO binding to trHbN, quantum mechanics/molecular mechanics calculations showed that NO interacts with TyrB10 and GlnE11, and that NO oxidation is a favourable process in the trHbN haem environment (Crespo et al., 2005). The same calculations with the TyrB10Phe mutant had similar results to wild type, which suggest that this residue is not important in the chemical reaction.

Another important residue for the function of trHbN is PheE15, thought to be involved in controlling ligand migration. PheE15 is located in the main tunnel, and in the crystal structure was present in two different conformations (Milani et al., 2001), leading to a suggested function as a 'gate' to the haem (Bidon-Chanal et al., 2007) controlling ligand access to the haem pocket; this function is supported by a 2006 paper which implicates both TyrB10 and GlnE11 in the ability of PheE15 to influence the protein tunnel dynamics (Ouellet et al., 2006). PheE15 exists in two conformations: molecular dynamics simulations showed that when in the closed conformation, the phenyl ring blocks the tunnel, whereas in the open conformation, there is no blockage and the ligand can move freely into the tunnel (Mishra & Meuwly, 2009). The authors investigated when such a change in conformation was likely to occur and found a greater than 70% probability that PheE15 is in the open conformation when the protein is in the oxy form. They also concluded that there is a large energy barrier preventing the switch from open to closed state (greater than 3 kcal mol^{-1}), and this change is more rare than the switch from closed to open (around 1.2 kcal mol^{-1}) (Mishra & Meuwly, 2009). Daigle et al. (2009) state that the orientation of PheE15 defines the open or closed state of the protein, as determined by both the redox state of the haem and whether O_2 is bound, whereas Bidon-Chanal et al. (2007) concluded that this state is controlled solely by O_2 binding.

To further study this gate residue, site-directed mutants were made where PheE15 was replaced by Ala, Ile, Tyr or Trp. None of these mutants showed any differences in the binding of O_2 or CO; however, all mutants showed significant decreases in NO detoxification activity even in the conservative PheE15Tyr mutation, suggesting that specifically Phe at this position is important for the function of the protein (Oliveira et al., 2012). This reduction in activity was mirrored in the resistance of the mutant proteins to oxidation when exposed to NO. The mutants were analysed further by molecular dynamics studies. In the PheE15Tyr mutant, the protein consistently favoured the closed state, even when the simulation was started in the open conformation. In addition, the hydrogen bond network formed between TyrB10, GlnE11 and the bound ligand was not disrupted by any of the mutants and there was no change in the dynamic behaviour of trHbN, suggesting that PheE15 does not play a role in these interactions. In contrast, it appears that the reduction in NO detoxification activity for the PheE15 mutants was caused by interruptions in the long tunnel, forcing NO to gain access by one of the other three routes (Oliveira et al., 2012).

The crystal structure of Mtb trHbN showed the presence of the pre-A motif, which consists of 12 residues at the N-terminus (Milani et al., 2001). Residues 6–10 are polar and form a highly polar sequence motif that could be possible for the breakdown of the α-helical structure in this region, as the pre-A region was shown to protrude out from the protein structure (Milani et al., 2001). The region is present only in pathogenic Mycobacteria (e.g. in addition to trHbN of *M. tuberculosis*, it is found in trHbN of *M. bovis* and *M. avium* but not *M. smegmatis*) and is not found in other organisms that also contain trHbs (Lama et al., 2009). The same authors showed that deletion of this region from Mtb trHbN leads to a reduction in NO scavenging activity when expressed in *Escherichia coli hmp$^-$* and addition to *M. smegmatis* trHbN improves NOD activity, suggesting that it may have a role in efficient NO detoxification; however, its deletion from Mtb trHbN did not alter the overall protein structure, as studied using circular dichroism spectroscopy. Molecular dynamics simulations showed that the pre-A region is highly flexible and that deleting pre-A from the protein alters the ability of the PheE15 residue to act as a gate, trapping it in the 'closed' state and blocking access to the tunnel for NO (Lama, Pawaria, & Dikshit, 2006). Indeed, it was argued that this region of the protein provides trHbN with a large fraction of its structural flexibility (Crespo et al., 2005). However, a study from 2011, which generated a Δpre-A mutant and investigated its structure and dynamics using NMR spectroscopy, disagrees with this conclusion. This group showed that deletion of the pre-A region did not alter the structural properties of Mtb trHbN when compared with wild-type protein, and they did not discover any interaction of the pre-A region with other parts of the protein; in addition, changes in the electronic environment were limited to only neighbouring residues Asp17 and His22 (Savard et al., 2011). They also found that the region does not display any secondary structure and is disordered; the authors attribute the disagreement with previous data to crystal contracts in the crystal structure, but do not deny its potential importance in the function of Mtb (and other) trHbN (Savard et al., 2011).

In addition to Mtb trHbN, the trHb from *M. smegmatis* (Ms trHbN) has also been studied. As already mentioned, Ms trHbN is smaller than Mtb trHbN due to the deletion of the pre-A region; adding the pre-A region from Mtb trHbN onto Ms trHbN improves NO detoxification activity when over-expressed in *E. coli* (8.9 nmol NO haem^{-1} s^{-1} compared to 3.8 nmol NO haem^{-1} s^{-1}) and *M. smegmatis* (11.8 nmol NO haem^{-1} s^{-1} compared to 4.9 nmol NO haem^{-1} s^{-1}) (Lama et al., 2009). Bidon-Chanal and co-authors speculate that the difference between the abilities of wild-type trHbN from *M. tuberculosis* and *M. smegmatis* to detoxify NO may be due to the pre-A region, with the proviso that until the function of the pre-A region is determined, this is speculation (Bidon-Chanal et al., 2007). However, Ms trHbN retains all of the already mentioned important structural features for the function of trHbN and for the preservation of the globin fold (Lama et al., 2006), lending further proof that the pre-A region is important for the function of trHbN proteins that are found in pathogenic Mycobacteria species.

3.1.2 Biochemical characteristics

An early paper expressed Mtb trHbN in *E. coli* and then purified the protein (Couture et al., 1999). Analysis by SDS-PAGE determined that the protein is 14.4 kDa in size. It purified as a dimer, shown by gel filtration analysis, and was an iron-protoporphyrin IX protein, as judged by maxima at 525 and 556 nm and a minimum at 539 nm in the oxidised minus reduced pyridine-haemochromogen spectra, with the haem non-covalently bound to the protein (Couture et al., 1999). Optical spectra showed that Mtb trHbN can bind a number of ligands, including O_2, CO and NO. In the ferric state, the haem is six-coordinate high-spin with a water molecule acting as the sixth ligand, and in the ligand-free ferrous form, the haem is in the five-coordinate high-spin state (Couture et al., 1999). An optical spectrum of the oxy-ferrous form shows a maximum in the Soret region of 416 nm, with peaks in the α/β region at 545 and 581 nm. These conclusions were confirmed by a Resonance Raman study, which also showed that the ferric form changes from high- to low-spin state upon increase in pH (Yeh et al., 2000).

TrHbN has been shown to act in an NO detoxification reaction, turning harmful NO into nitrate via reaction with O_2 (Pathania, Navani, Gardner, et al., 2002). During this reaction, the protein itself is oxidised

from ferrous to ferric (Ouellet et al., 2002). When NO was added to purified, fully oxygenated Mtb trHbN, spectral analysis clearly showed oxidation of the protein, with the disappearance of peaks in the α/β region
(Lama et al., 2009; Ouellet et al., 2002). In addition, when NO was added
to fully oxygenated trHbN, the wild-type protein was oxidised as deduced
by a flattening of the α/β region, whereas a mutant with the pre-A region
removed showed no change in this region, suggesting that the protein cannot either bind NO or proceed with the reaction without the pre-A region
intact (Lama et al., 2009). Resonance Raman spectroscopy predicted that
the ligand on-rate for trHbN would be fast due to the unstrained His residue acting as the proximal haem ligand; consequently, the iron atom can
be brought into a fully planar conformation upon ligand binding, thus
avoiding the constraint that is present in other globins (Couture et al.,
1999). Indeed, the same authors present experimental data which show that
trHbN has a high affinity for O_2, with the binding process being cooperative (Couture et al., 1999). The authors speculate that this property could
be useful for the protein to function in the low O_2 environments that
Mycobacteria operate in, for example, in the hypoxic granuloma environment in the lung, as it will be able to capture O_2 for the reaction with NO
even when O_2 concentrations are relatively low. When the kinetics of CO
re-binding to trHbN after photodissociation were investigated, it was
shown that CO re-binds in a single exponential phase, assigned to solvent
phase recombination (Ouellet et al., 2008).

Analysis of the formation of nitrate by Mb trHbN in the presence of NO
and O_2 showed a stoichiometric formation of nitrate when protein was present in excess of NO, and stopped-flow spectroscopy detected the production of the aquomet form of Mb trHbN without detection of any
intermediates (Ouellet et al., 2002). Similar to Mtb trHbN, Ms trHbN
was purified in the oxy-ferrous form when expressed in E. coli, albeit smaller
at 12.8 kDa as predicted by the amino acid sequence, particularly due to the
lack of pre-A region (Lama et al., 2006). In addition, antibodies raised against
Mtb trHbN can also react with Ms trHbN, reiterating their similarities. As is
the case with M. leprae, where reductive evolution has led to the loss of all
trHbs except trHbO (Wittenberg et al., 2002), perhaps the loss of the pre-A
region is a result of the non-pathogenic lifestyle of M. smegmatis compared
with M. tuberculosis (Lama et al., 2006). Its optical spectra are also very similar
to that of Mtb trHbN, with peaks in the Soret region at 416 nm and in the
α/β region at 543 and 580 nm. Other values obtained with the deoxy ferrous and CO-bound forms also conformed to known values for the trHbs.

As expected from this reported data, the protein is six-coordinated and the haem is in the high-spin form when in the ferric state. However, compared with Mtb trHbN, Ms trHbN shows a faster rate of autoxidation (Lama et al., 2006).

The NOD reaction of trHbN has also been studied using molecular dynamics. Briefly, the reaction is as follows: when the protein is in the oxy form, NO binds which leads to a peroxynitrite intermediate. This is rearranged to the nitrato complex, with nitrate finally dissociating, leaving trHbN in the ferric form (Mishra & Meuwly, 2010). This reaction is facilitated by the side chains TyrB10 and GlnE11, which act to orientate the ligands in the correct position via the dynamic hydrogen bonding network, especially in the third step where GlnE11 stops the ligand from moving too much and allowing the reaction to continue with all the ligands in the correct place. The second step disagrees with the accepted mechanism which predicts a dissociated state; molecular dynamics simulations on this scale suggest that this is energetically unfavourable and the new mechanism, involving simply a rearrangement, is more likely (Mishra & Meuwly, 2010). During the final step, the haem changes to a six-coordinate species from five-coordinated. This study therefore shows that the protein environment plays a role in the NOD reaction, in agreement with a study which showed that upon NO binding, the protein undergoes a large-scale conformational change that extends to the E and B helices (Mukai et al., 2004). However, both of these are in contrast with another study (Crespo et al., 2005) which suggests that the protein environment is not important.

3.2. Group II trHbs: trHbO

3.2.1 Structural features and dynamics

Less is known about the structure and dynamics of trHbO. As shown in Fig. 5.2, as for trHbN, there is extensive structural conservation of the important residues required for the formation of the 2-over-2 globin fold; these include PheB9 and TyrB10, PheE14 and HisF8 but not CD1, which is often Phe in other globins but here it is Tyr (Mukai, Savard, Ouellet, Guertin, & Yeh, 2002). Unlike trHbN, trHbO does not have a pre-A region but does have an EF-loop region which is highly conserved within the group and contains a highly charged, polar sequence motif (Pathania, Navani, Rajamohan, et al., 2002), potentially important to the function of trHbO, which is probably different to that of trHbN (see Section 5). Resonance Raman spectroscopy showed that the proximal ligand is His and analysis

```
                                           1          5    10  1    5    10  1
                                           AA         AAAAAAAA BBBBBBBBBBBBBBBBCC
M.tuberculosisHbO          --------------MP-----KSFYDAVGGAKTFDAIVSRFYAQVAEDE
M.canettiiHbO              --------------MP-----KSFYDAVGGAKTFDAIVSRFYAQVAEDE
M.colombienseHbO           ------------MDQVA-----ESFYDAVGGAETFHAIVARFYAQVPEDE
M.intracellulareHbO        -----------MDEVG-----QSFYDAVGGAETFHAIVSRFYAQVPEDE
M.aviumHbO                 ------------MDQV-----SFYDAVGGAETFQAIVSRFYAQVPEDE
M.parascrofulaceumHbO      ---------MEVMDSEP-----QSFYDAVGGAETFHAIVSRFYAQVAEDE
M.lepraeHbO                -----------MDQVQ-----QSFYDAIGGAETFKAIVSRFYAQVPEDE
M.marinumHbO               --------------MT-----EPFYDAVGGAKTFETIVSRFYELVAEDE
M.ulceransHbO              --------------MT-----EPFYDAVGGAKTFETIVSRFYELVAEDE
M.fortuitumHbO             ---------MSDVTQVQ-----RSFYDEVGGHETFAAIVSRFYQLVREDE
M.smegmatisHbO             -----------MTQVQ-----RSFYDEVGGHDTFHAIVSRFYQLVREDE
M.hassiacumHbO             -----------MTQPQ-----RSFYDEVGGHETFRRIVSRFYELVREDE
M.phleiHbO                 --------MGSVTQPQ-----GSFYDEVGGHETFRRIVARFYELVAEDE
M.thermoresistibileHbO     ----------------M-----QSFYDEVGGHETFRAIVSRFYELVRSDE
M.gilvumHbO                -------METMDAHSKQ-----QSFYDAVGGHATFHTLVARFYELVRDDE
M.vanbaaleniiHbO           -------MEPMDAQSKQ-----QSFYDAVGGHDTFRTIVSRFYQLVRDDE
M.vaccaeHbO                ---------MEAQNKQ-----QSFYDAVGGQDTFRTLVSRFYELVREDE
M.tusciaeHbO               -----------MTQPQ-----RSFYDEVGGHDTFHAIVSRFYQLVREDE
M.sp.MCSHbO                -----------MTQPT-----RSFYDEVGGHETFRAIVARFYELVREDE
M.rhodesiaeHbO             MGRAGADVHNGGVTETQANGDGQTFYDAVGGAATFFRKIVSRFYELVREDE
M.massilienseHbO           --------MENVAELSPAPTPDNFYDAVGGAATFHAIVARFYKLVADDE
M.abscessusHbO             -----------MAELSPAPTPDNFYDAVGGAATFHAIVARFYKLVADDE
                                                                            B10
```

```
                            5    1    5    10 15              1    5
                            CCCCC EEEEEEEEEEEEEEEEEE            FFFFFFF
M.tuberculosisHbO          VLRRVYPEDDLAGAEERLRMFLEQYWGGPRTYSEQRGHPRLRMRHAPFRI
M.canettiiHbO              VLRRVYPEDDLAGAEERLRMFLEQYWGGPRTYSEQRGHPRLRMRHAPFRI
M.colombienseHbO           ILRELYPLDDLEGAEERLRMFLEQYWGGPRTYSDRRGHPRLRMRHVPFRI
M.intracellulareHbO        ILSELYPLDDLEGAEERLRMFLEQYWGGPRTYSDRRGHPRLRMRHVPFRI
M.aviumHbO                 ILRELYPLDDLEGAEERLRMFLEQYWGGPRTYSDRRGHPRLRMRHVPFRI
M.parascrofulaceumHbO      ILRQLYPEDDLAGAEERLRMFLEQYWGGPRTYSDRRGHPRLRMRHVPFRI
M.lepraeHbO                ILRELYPADDLAGAEERLRMFLEQYWGGPRTYSSQRGHPRLRMRHAPFRI
M.marinumHbO               ILRPLYPEEDLSGAEERLRMFLEQYWGGPRTYSEQRGHPRLRMRHAPFRI
M.ulceransHbO              ILRPLYPEEDLSGAEERLRMFLEQYWGGPRTYSEQRGHPRLRMRHMPFRI
M.fortuitumHbO             ILLPLYPEDDIDGAEERLRMFLEQYWGGPRTYSDQRGHPRLRMRHAPFRI
M.smegmatisHbO             ILHPLYPEDDFEGAEERLRMFLEQYWGGPRTYSDQRGHPRLRMRHAPFRI
M.hassiacumHbO             ILRPLYPEEDLGPAEERLRMFLEQYWGGPRTYSELRGHPRLRMRHAPFRI
M.phleiHbO                 ILRPLYPEEDLGPAEERLRMFLEQYWGGPRTYSEQRGHPRLRMRHMPFRI
M.thermoresistibileHbO     ILRPLYPEDDLLDGAEERLRMFLEQYWGGPRTYSEQRGHPRLRMRHAPFRI
M.gilvumHbO                ILRPLYPEDDLVGAEERLRMFLEQYWGGPRTYSDQRGHPRLRMRHAPFRI
M.vanbaaleniiHbO           ILRPLYPEDDLDGAEERLRMFLEQYWGGPRTYSDQRGHPRLRMRHAPFRI
M.vaccaeHbO                ILRPLYPEDDLDGAEERLRMFLEQYWGGPRTYSDQRGHPRLRMRHAPFRI
M.tusciaeHbO               ILRPLYPEDDLDGAEVRLRMFLEQYWGGPRTYSEQRGHPRLRMRHAPFVI
M.sp.MCSHbO                ILRPLYPDDELEAAEVRLRMFLEQYWGGPRTYSDQRGHPRLRMRHAPFRI
M.rhodesiaeHbO             VLLPLYPPDELDDAEDRLRMFLEQYWGGPRTYSDQRGHPRLRMRHAPFRI
M.massilienseHbO           VLRPLYPEEDLTGAEDRLRMFLEQYWGGPRTYSDQRGHPRLRMRHAPFRI
M.abscessusHbO             VLRPLYPEEDLTGAEDRLRMFLEQYWGGPRTYSDQRGHPRLRMRHAPFRI
                                CD1                                        F8
```

```
                            GGGGGGGGGGGGGGGGGGGG    HHHHHHHHHHHHHHHHHHHHHHHH
M.tuberculosisHbO          SLIERDAWLRCMHTAVASIDSETLDDEHRRELLDYLEMAAHSLVNSPF
M.canettiiHbO              SLIERDAWLRCMQTAVASIDSETLDDEHRRELLDYLEMAAHSLVNSPF
M.colombienseHbO           TPLARDAWLRCMHTAVASIDSQTLDDEHRRELLNYLEMAADSLVNSPF
M.intracellulareHbO        TPLARDAWLRCMHTAVASIDSATLDDEHRRELLSYLEMAADSLVNSPF
M.aviumHbO                 TPLARDAWLRCMHTAVASIDSKTLDDEHRRELLDYLEMAAHSLVNSPL
M.parascrofulaceumHbO      TPIERDAWLRCMHTAVASIDSETLDDEHRRQLLDYLEMAAHSLVNSPF
M.lepraeHbO                TAIERDAWLRCMHTAVASIDSHTLDNEHRRELLDYLEMAAHSLVNSAS
M.marinumHbO               TPIERDAWLRCMHTAVAAVDSQTLDDDHRRELLDYLEMAAHSLVNSPF
M.ulceransHbO              TPIERDAWLRCMHTAVAAVDPQTLDDDHRRELLDYLEMAAHSLVNSPF
M.fortuitumHbO             GYLERDAWLRCMHTAVASIDSQTLDDPHRRALLDYLEMAADSMVNSAF
M.smegmatisHbO             GFLERDAWLRCMHTAVAEIDSQTLDDRRALLDYLQMAADSMVNSAF
M.hassiacumHbO             GYLERDAWLRCMHTAVASVDSDTLDDEHRKVLLDYLEMAAHSLVNSPF
M.phleiHbO                 GFLERDAWLRCMHTAVASIDAQTLDDAHSRVLLDYLEMAADSMVNSAF
```

```
M.thermoresistibileHbO    GYLERDAWLRCMHTAVASIDSQTLDDAHREQLLDYLEMAADSLVNSPF
M.gilvumHbO               GYLERDAWLRCMHTAVAEIDSATLDDAHRRELITYLEMAAESMVNSPF
M.vanbaaleniiHbO          GYLERDAWLRCMHTAVAEIDSRTLDDRHRQELIAYLEMAADSMVNSPF
M.vaccaeHbO               GYLERDAWLRCMHTAVAEIDTETLDDRHRRALLDYLEMAAESMVNSPF
M.tusciaeHbO              GYIERDAWLRCMHTAVAEIDSQTLDDEHRAELLAYLEMAAQSMVNSPF
M.sp.MCSHbO               GYIERDAWLRCMHTAVAEIDAVTLDDEHRRELLAYLEMAAQSMVNSPF
M.rhodesiaeHbO            GFIERDAWLRCMHTAVASIDSATLDDEHRRELLAYLDMAAQSMVNSAF
M.massilienseHbO          GPIERDAWLRCMRTAVDSIDRNTLDDDHRAQLWSYLEMAAQSMVNSPF
M.abscessusHbO            GPIERDAWLRCMRTAVDSIDRNTLDDDHRAQLWSYLEMAAQSMVNSPF
```

Figure 5.2 Structure-based sequence alignment of group II (trHbO) haemoglobins of Mycobacteria. (For colour version of this figure, the reader is referred to the online version of this chapter.)

of the wild type showed the presence of an unusual O–O stretching mode, suggesting an atypical distal haem pocket, which is rigid and polar and significantly different to that in trHbN (Mukai et al., 2002). In TyrCD1Phe, the O–O stretching mode disappeared, suggesting that this residue is important in the unique structure of the haem pocket. In wild-type protein, when CO is bound to the haem, TyrCD1 appears to be forming hydrogen bonds with the haem-bound CO and the authors postulate that a hydrogen bond is also formed between TyrCD1 and TyrB10, contributing to the rigidity of this region of the protein (Mukai et al., 2002). The crystal structure of cyano-met Mtb trHbO showed that the protein exists as a compact dodecamer with a covalent bond linking TyrB10 to TyrCD1 in the haem distal cavity (Milani, Savard, et al., 2003). A study using UV-enhanced Resonance Raman spectroscopy confirmed that TrpG8 is involved in stabilising the ligand upon binding due to a conformational change upon ligation of the ligand, in concert with TyrB10 and TyrCD1, residues to which TrpG8 forms hydrogen bonds (Ouellet et al., 2003). TyrCD1 is also thought to be involved in the modulation of O_2 binding to the haem of trHbO (Ouellet et al., 2003). The protein fold is conserved, with some modifications, including a pre-F region which is pushed away from the haem and the protein core due to a number of insertions and might be reflected in a shift of the haem group, at some points greater than 7 Å, in relation to its position in trHbN. In addition, the haem group is located closer to the surface than in other trHbs, which could reflect the functional role of trHbO (Milani, Savard, et al., 2003). Another paper showed, using Resonance Raman spectroscopy, that CO bound to haem was stabilised by both TyrCD1 and TrpG8 residues, via hydrogen bonds; in addition, the TrpG8 residue also anchors TyrCD1 and LeuE11, stabilising the ligand bound to the haem via hydrogen bonds (Guallar, Lu, Borrelli, Egawa, & Yeh, 2009). Other authors suggest that these hydrogen bonds may represent a considerable barrier to ligand movement towards and away from the haem and may result in the high affinity for O_2 that the protein displays (Ouellet et al., 2003). Upon escape from the haem pocket, the ligand has three docking sites to choose from upon dissociation, where it can then either re-bind to the haem or escape through the protein into solution (Guallar et al., 2009). The unique hydrogen bond network displayed in trHbO in comparison with trHbN shows that the distal haem pocket in trHbO is crowded and rigid and quite distinct from the haem region in trHbN (Ouellet et al., 2003), adding further support for the difference in functions.

Additional information is available about the importance of TrpG8 in the stability and structure of trHbO. Protein stability was investigated and it was shown that when TrpG8 was mutated to Phe, the protein was less stable than wild-type protein (Ouellet, Milani, et al., 2007). Analysis of wild-type trHbO and both TyrCD1Phe and TrpG8Phe mutants by Resonance Raman spectroscopy showed that bound CO potentially interacts with TyrCD1 and TrpG8 through hydrogen bonds; however, no polar interactions between the distal residues and the bound CO were detected, further characterising the haem pocket in trHbO (Ouellet, Milani, et al., 2007). Studies were also made of the O_2-bound ferrous complex of wild-type and mutant proteins; data suggest that TrpG8 interacts with the proximal O_2 atom bound to the haem and is involved in stabilisation of bound O_2 (Ouellet, Milani, et al., 2007).

The tunnel system in trHbO is more restricted than in trHbN due to the orientation of the G and H helices and six amino acid changes; in trHbO, the short tunnel is inhabited by TrpG8 and the long tunnel is split into two smaller cavities which inhabit the same space as in trHbN (Milani, Savard, et al., 2003). Entrance into the tunnel by ligands appears to be controlled by the AlaE7 gate (Boechi et al., 2008; Milani, Savard, et al., 2003). Molecular dynamics investigations of the entrance of O_2 into deoxy Mtb trHbO found there to be two access tunnels for O_2 from the solvent into the haem: a long tunnel (16 Å long) was also identified in the crystal structure (Milani, Savard, et al., 2003) along with a short tunnel (around 10 Å long), thought to be blocked in the crystal structure (Boechi et al., 2008). The long tunnel connects the haem with the solvent between the B, E and G helices, and ligand movement through the tunnel appears to be controlled by the distance between PheB2 and LeuH3—when these are close together, the ligand cannot pass between them and is forced to choose another path (Boechi et al., 2008). Following from the suggestion that TrpG8 is involved in blocking the short tunnel and the distal haem pocket (Milani, Savard, et al., 2003), *in silico* mutants (TrpG8Phe and TrpG8Ala) were analysed; results showed that when TrpG8 was replaced by a smaller amino acid, the long tunnel was unblocked and was able to connect the haem pocket to the solvent (Boechi et al., 2008). As trHbN undergoes changes in conformation following O_2 binding to the haem, allowing NO passage into the active site, simulations were performed to determine whether trHbO also undergoes such changes. In Mtb trHbO, after O_2 has bound to the haem, TrpG8 is displaced—moving away from LeuE11 and forming a hydrogen bond with O_2—which leads to the opening of the long tunnel; this leads to a decrease in the free energy barrier for ligand diffusion towards the haem (Boechi et al., 2008). The authors suggest that the first ligand (i.e. O_2) gains access to the deoxy haem via the short tunnel, which opens the long tunnel and allows the second ligand (NO) into the haem pocket via either the long or the short tunnels.

3.2.2 Biochemical characteristics

The trHbO from *M. tuberculosis* has also been studied in great detail by a number of different groups. Its expression in *E. coli* lends the cells a reddish brown colour (Pathania, Navani, Rajamohan, et al., 2002), as does trHbN (Lama et al., 2006), giving researchers an easy visual test for the over-expression of these proteins in heterologous hosts. TrHbO usually purifies as a monomer (Ouellet et al., 2003) and was identified as an iron-protoporphyrin IX (Mukai et al., 2002); analysis of cells over-expressing Mtb trHbO showed a 14.5-kDa protein which corresponds to the predicted size of trHbO (Pathania, Navani, Rajamohan, et al., 2002). Optical absorption spectra showed that Mtb trHbO has a Soret peak at 413 nm with peaks in the α/β region at 544 and 581 nm, suggesting that, again similarly to trHbN, trHbO is expressed and purified in the oxy-ferrous form. Upon reduction using dithionite, the peak in the Soret region shifts to 422 nm and the two distinct peaks in the α/β region merge to give a single, broad peak at 554 nm; this behaviour is consistent with other globins (Pathania, Navani, Rajamohan, et al., 2002). The function of trHbO is thought to be an involvement in O_2 transport or delivery, proposed due to its association with the cell membrane during purification from over-expression in *E. coli* (Pathania, Navani, Rajamohan, et al., 2002). However, trHbO has a very high affinity for O_2 compared with other globins (Mukai et al., 2002). The protein also displays low O_2 dissociation rate constants (Ouellet et al., 2003) and extremely fast ligand re-binding after photoexcitation (Jasaitis et al., 2012), suggesting that O_2 is tightly bound to the haem and also that it will re-bind with extreme efficiency; this is perhaps evidence against a role in O_2 delivery—it is hard to envisage a protein playing a role in O_2 delivery if it is unlikely to dissociate and Ouellet et al. (2003) instead suggest a role as a redox sensor.

After photodissociation of CO from CO-bound wild-type Mtb trHbO, CO re-binds to the haem as expected; a high proportion of this can be attributed to geminate re-binding (Guallar et al., 2009). In fact, around 95% of CO removed from the haem by photoexcitation re-binds (Jasaitis et al., 2012). It is thought that the distal pocket residues are responsible for this fast geminate re-binding of CO, and for the slow ligand on- and off-rates (Guallar et al., 2009). A molecular dynamics study of the CO complex showed that interactions between water molecules play an important role in this high rate of re-binding, in addition to the rotational freedom employed by the protein (Jasaitis et al., 2012). Again, the hydrogen bond network is important in this aspect of the biochemistry of trHbO: CO is orientated so that it points towards the haem. Interestingly, only 1% of photoexcited O_2 re-binds, and less than 1% of NO (Jasaitis et al., 2012). In concert with this fast re-binding, it follows that trHbO displays very little exchange of ligands with the environment (Jasaitis et al., 2012).

Perhaps it is proper to discuss the biochemistry of Ml trHbO separately, as it may perform multiple functions and therefore show slightly different functional characteristics to other trHbO proteins. Oxy-ferrous Ml trHbO binds NO and is itself oxidised in the production of stoichiometric amounts of nitrate (Ascenzi, Bocedi, et al., 2006; Fabozzi et al., 2006) as is the case for trHbN. Analysis of the transient species formed in this reaction suggests the production of a peroxynitrite intermediate, Fe(II)OONO (Ascenzi, Bocedi, et al., 2006). The k_{on} value for NO oxidation $(2.1 \times 0^6 \, M^{-1} \, s^{-1})$ is lower than for trHbN, perhaps reflecting the different function of trHbO, and is probably due to the differences in haem pocket arrangement and tunnel availability. Due to this discovery, the peroxynitrite scavenging ability of Ml trHbO was investigated. Upon exposure of ferrous–NO-bound Ml trHbO with peroxynitrite, the protein immediately forms the ferric–NO-bound form before switching to the ferric form with assumed release of NO (Ascenzi, Milani, & Visca, 2006). After addition of peroxynitrite to oxyferrous Ml trHbO, the oxy-ferryl state was formed; the authors concluded that both oxy-ferrous and ferrous–NO-bound trHbO are able to scavenge peroxynitrite (Ascenzi, Milani, et al., 2006). Investigations into the denitrosylation and O_2-mediated oxidation of ferrous–NO-bound trHbO presented a possible reaction mechanism: NO binds to ferrous trHbO but is then displaced by O_2; the NO may, however, stay trapped in a cavity close to the haem, as suggested by the structural features already discussed. This oxy-ferrous trHbO then reacts with the displaced NO to give the above-mentioned ferric peroxynitrite-bound intermediate, before the ferric species alone is formed (Ascenzi, Bolognesi, & Visca, 2007). The k_{off} for NO was found to be 1.3×10^{-4} s^{-1} at pH 7, and this is thought to be the rate-limiting step in the proposed reaction mechanism (Ascenzi, Bolognesi, et al., 2007).

3.3. Type III trHbs: trHbP

So far, no member of the group III trHb (trHbP) from Mycobacteria has been isolated and characterised. The occurrence of trHbP is limited to only a few species of Mycobacteria where it is present along with its other molecular relatives, trHbN and trHbO, and may therefore have distinct cellular function(s). Inspection of the available Mycobacterial genome data indicated the presence of trHbP in 8 out of 22 Mycobacterial species (Table 5.1). The identity of trHbP in Mycobacteria was checked on the basis of their homology with other type III trHbs (Fig. 5.3), in particular, the presence of specific structural features, such as the absence of the Gly-Gly motif present in group I and group II trHbs, conservation of residues at key structural positions (B10, E7, F8, etc.) and similarities in the residues building the proximal and distal haem pockets. Sequence alignment of Mycobacterial trHbP from various Mycobacterial species is presented in Fig. 5.3. At present, only the X-ray structure of trHbP of *C. jejuni* has been unravelled (Nardini et al., 2006). Contrary to what has been observed in group I and II trHbs, the functional residues of classical globins, for example, PheCD1 and E7His, are invariant among this group and no distinct protein matrix tunnels are formed. It has been proposed that E7His might operate as a gating residue for ligand entry to the haem distal site, very similar to classical Hbs. All these functional residues are conserved in Mycobacterial trHbP. In the absence of any information on the biochemical and structural properties of Mycobacterial group III trHbs, the specific *in vivo* function of trHbP in the cellular metabolism of Mycobacteria remains an open issue.

3.4. Genetic regulation of the trHbs

TrHbs were first identified using a bioinformatics approach, where they were found in many genomes of the Mycobacteria (Wittenberg et al., 2002). The gene encoding for trHbN is designated *glbN* (Rv1542c) and for trHbO, *glbO* (Rv2470). Bioinformatic probing of the regions up- and down-stream of both *glbN* and *glbO* genes did not provide any clues towards

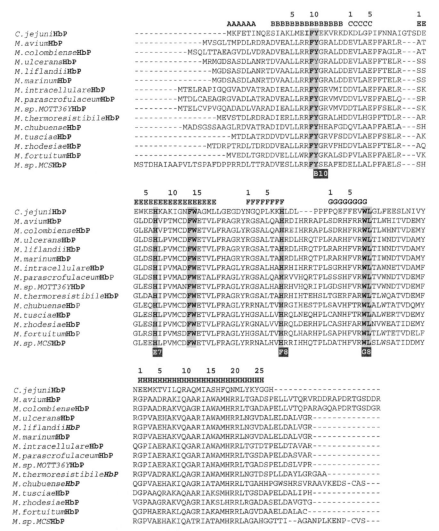

Figure 5.3 Structure-based sequence alignment of group III (trHbP) haemoglobins of Mycobacteria. (For colour version of this figure, the reader is referred to the online version of this chapter.)

the regulation of either gene; there are no recognition sequences involved in the response to stress or O_2 dependence (Pawaria, Lama, Raje, & Dikshit, 2008). Further study into the genetic regulation of the trHbs showed that the organisation of the *glbN* gene is conserved in *M. tuberculosis* and *M. bovis* (Couture et al., 1999; Ouellet et al., 2002), and in other pathogenic Mycobacteria such as *M. avium* (Pawaria et al., 2008; see Fig. 5.4). However,

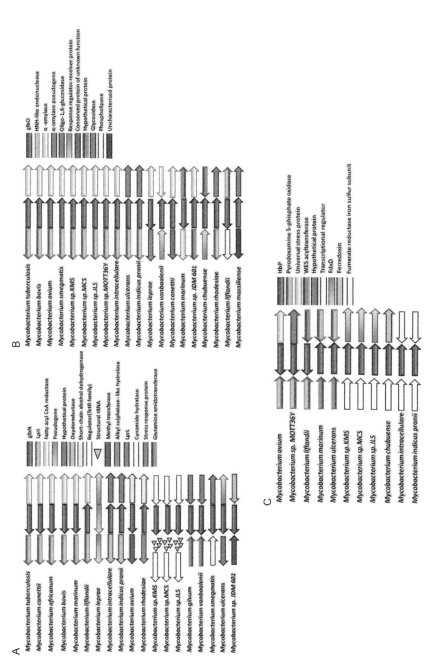

Figure 5.4 Genomic organisation of Mycobacterial truncated haemoglobins: (A) group I, (B) group II, and (C) group III. (For colour version of this figure, the reader is referred to the online version of this chapter.)

The following labels appear with the figure:

A
- glbN
- Lpd
- Fatty acyl CoA reductase
- Pseudogene
- Hypothetical protein
- Oxydoreductase
- Short-chain alcohol dehydrogenase
- Regulator(TetR family)
- Structural tRNA
- Methyl transferase
- Alkyl sulphatase-like hydrolase
- Lpri
- Cynamide hydratase
- Stress response protein
- Glutamine amidotransferase

Organisms (A):
- Mycobacterium tuberculosis
- Mycobacterium conettii
- Mycobacterium africanum
- Mycobacterium bovis
- Mycobacterium marinum
- Mycobacterium tliflandii
- Mycobacterium leprae
- Mycobacterium intracellulare
- Mycobacterium indicus pranii
- Mycobacterium avium
- Mycobacterium rhodesiae
- Mycobacterium sp.KMS
- Mycobacterium sp.MCS
- Mycobacterium sp.JLS
- Mycobacterium glivum
- Mycobacterium vanbaalenii
- Mycobacterium smegmatis
- Mycobacterium ulcerans
- Mycobacterium sp. JDM 601

B
- glbO
- HNH-like endonuclease
- α-amylase
- α-amylase pseudogene
- Oligo-1,6-glucosidase
- Response regulator receiver protein
- Conserved protein of unknown function
- Hypothetical protein
- Glycosidase
- Phospholipase
- Uncharacterised protein

Organisms (B):
- Mycobacterium tuberculosis
- Mycobacterium bovis
- Mycobacterium ovium
- Mycobacterium smegmatis
- Mycobacterium sp.KMS
- Mycobacterium sp.MCS
- Mycobacterium sp.JLS
- Mycobacterium sp.MOTT36Y
- Mycobacterium intracellulare
- Mycobacterium indicus pranii
- Mycobacterium ulcerans
- Mycobacterium leprae
- Mycobacterium vanbaalenii
- Mycobacterium conettii
- Mycobacterium marinum
- Mycobacterium sp. JDM 601
- Mycobacterium chubuense
- Mycobacterium rhodesiae
- Mycobacterium tlifandii
- Mycobacterium massillense

C
- HbP
- Pyrodoxamine 5-phisphate oxidase
- Universal stress protein
- WES acyltransferase
- Hypothetical protein
- Transcriptional regulator
- FdxD
- Ferredoxin
- Fumerate reductase iron sulfur subunit

Organisms (C):
- Mycobacterium avium
- Mycobacterium sp. MOTT36Y
- Mycobacterium tlifandii
- Mycobacterium marinum
- Mycobacterium ulcerans
- Mycobacterium sp.KMS
- Mycobacterium sp.MCS
- Mycobacterium sp.JLS
- Mycobacterium chubuense
- Mycobacterium intracellulare
- Mycobacterium indicus pranii

the genetic organisation is different in the non-pathogenic bacteria; for example, in *M. tuberculosis*, a short-chain alcohol dehydrogenase is transcribed in the opposite direction to *glbN*, whereas in *M. smegmatis*, the gene in this position is *lprL* (Pawaria et al., 2008). In *M. leprae*, this region of the genome is occupied by a pseudogene (Pawaria et al., 2008). In both *M. tuberculosis* and *M. bovis*, *lprI* is positioned directly adjacent to *glbN* (Pawaria et al., 2008); it is co-transcribed in *M. bovis* but has not been implicated in trHbN activity. Western Blot analysis was used to prove when Mb trHbN is expressed in *M. bovis* BCG growing in aerobic cultures; expression is low both at the beginning of growth and in exponential phase, peaking when cells are in stationary phase (Couture et al., 1999). This suggests that in *M. bovis*, and probably in other pathogenic species of Mycobacteria, trHbN is expressed when the bacteria is entering a period of growth where, for example, nutrients are scarce and O_2 levels are low, corresponding to the environment encountered in low O_2 granulomas. Less is known about the genetic regulation of *glbO*, the gene encoding for trHbO. The region around the *glbO* gene is conserved across pathogenic and non-pathogenic Mycobacteria (Pawaria et al., 2008).

In order to probe the genetic regulation of both *glbN* and *glbO* from *M. tuberculosis*, potential promoter regions containing the -10 and -35 sites and Shine-Dalgarno sequences were cloned in front of a promoter-less *gfp* gene and expressed in *M. smegmatis*. In agreement with data obtained in *M. bovis* (Ouellet et al., 2002), *glbN* was up-regulated in late exponential and stationary phases of growth; its expression was also increased in these phases when *M. smegmatis* was exposed to nitrite and SNP (Pawaria et al., 2008). In contrast, *glbO* appeared to be more consistently expressed over growth phases, and showed increased expression in response to nitrite in all growth phases but SNP only in stationary phase, probably reflecting the fact that trHbO is not thought to be involved in the detoxification of NO (Pawaria et al., 2008). The same experiments were performed in *M. tuberculosis* H37Ra and the authors comment that both *glbN* and *glbO* promoters behaved in the same way as in *M. smegmatis* (Pawaria et al., 2008). This strain of *M. tuberculosis* H37Ra was used to infect PMA-activated THP-1 cells, which mimic human alveolar macrophages, to determine the promoter activities *in vivo*. The results follow those obtained *in vitro*: FACS analysis showed fluorescence in cells infected with *glbN* and *glbO*, although this experiment was only done once (Pawaria et al., 2007). This suggests that both promoters are active during macrophage infection, although more work needs to be presented to further analyse this activity.

An important system for the expression of genes involved in the response to hypoxia and NO is Dos; it consists of the sensor kinases DosS and DosT (Roberts, Liao, Wisedchaisri, Hol, & Sherman, 2004) and the response regulator DosR (Kendall et al., 2004). These proteins are involved in the response by *M. tuberculosis* to reactive nitrogen intermediates (Ohno et al., 2003). Perhaps surprisingly, *glbN* was not up-regulated in response to exposure to 0.005–5 mM DETA/NO for 40 min, and *glbO* was not up-regulated in response to exposure to 0.05–50 mM H_2O_2 for 40 min (Voskuil, Bartek, Visconti, & Schoolnik, 2011). Neither *glbN* nor *glbO* has been implicated in the Dos regulon, suggesting that they are regulated by some other system; however, in a *dosR* mutant, *glbN* was down-regulated, perhaps suggesting that *glbN* is somewhat controlled by this system (Kendall et al., 2004).

In the case of Ml trHbO, a putative promoter was found and cloned into a shuttle vector in front of the *lacZ* gene, which was then transformed into both *M. smegmatis* and *E. coli*. Activity of the LacZ protein, measured using the β-galactosidase assay, was found in *M. smegmatis* during early and late exponential phase but not in *E. coli* (Fabozzi et al., 2006), suggesting the absence of some cellular machinery in *E. coli* required for promoter activity. The promoters for *M. smegmatis glbN* and *glbO* were cloned in the same way and showed β-galactosidase activity which was highest in late exponential phase and increased by NO donors SNP and SNAP; the *glbN* promoter was activated to a higher extent than the *glbO* promoter. Interestingly, the *M. leprae glbO* promoter also showed increased activation to SNP and SNAP, suggesting again that Ml trHbO may be involved in NO detoxification (Fabozzi et al., 2006). Similar experiments were done with promoters for *glbN*, *glbO* and *groEL*, a heat shock protein, cloned into *M. smegmatis*. Cells were subjected to different stresses and the up-regulation of each gene detected by RT-PCR: exposure to 10-mM sodium nitrite led to the biggest increase in *glbN* expression, with 30 mM leading to the biggest increase in *glbO* expression (Joseph, Madhavilatha, Kumar, & Mundayoor, 2012). In the absence of any stress, and agreeing with previous data, both promoters showed increased activity in late exponential phase.

4. CONVENTIONAL AND NOVEL FlavoHbs OF MYCOBACTERIA

Less is known about the flavoHbs in Mycobacteria. Computational and sequence analysis of available Mycobacterial genomes and protein data indicates that more than one flavoHb-encoding gene may be present in many Mycobacterial species (Table 5.1). An interesting pattern in the occurrence of flavoHbs can be observed in slow- and fast-growing Mycobacteria. Opportunistic pathogens and fast-growing Mycobacteria carry two flavoHb-encoding genes, one encoding flavoHb similar to conventional flavoHbs present in other microbes (Type I), and the second one encoding an unusual flavoHb, displaying unconventional globin and reductase domains (Type II). The occurrence of two distinct classes of flavoHbs in Mycobacteria suggests their distinct cellular functions. Notably, the new class of Type II flavoHb is present in the majority of Mycobacterial species and also co-exists with a conventional flavoHb (Type I) in some of the Mycobacteria. Unlike conventional and Type I flavoHbs, Type II flavoHbs are not widely distributed among microbial systems and are restricted to Mycobacterial and certain species of actinomycetes (Gupta et al., 2011).

4.1. Structural features

Sequence alignment of Type I and Type II flavoHbs of various Mycobacterial species are presented in Fig. 5.5, which indicates that structural features required for adopting a classical 3-over-3 globin fold and the signature sequences of typical microbial globin fold (e.g. TyrB10, PheCD1, GlnE7 and HisF8) are well preserved in both classes of Mycobacterial flavoHbs. Type I flavoHbs of Mycobacteria appear very similar to conventional flavoHbs present in other microbes, and display conservation of sequences including the hydrogen bonding network between HisF8, TyrG5 and GluH23 that is required for the architecture of the haem pocket in the proximal region (Bonamore & Boffi, 2008; Frey & Kallio, 2003; Mukai, Mills, Poole, & Yeh, 2001). Additionally, the FAD- and NADH-binding sites and critical residues involved in electron transfer (YS pair in the FAD-binding domain) are present at the topological positions required for the electron transfer from the FAD to the haem domain. Thus, Type I flavoHbs of Mycobacteria are structurally and functionally similar to conventional flavoHbs.

Type II flavoHbs of Mycobacteria display several differences from conventional flavoHbs in the functionally conserved region of their haem and FAD-binding sites (Fig. 5.5). The most prominent change is the lack of hydrogen bonding interactions between HisF8, TyrG5 and GluH23 within the proximal site of its globin domain that may change the redox properties of the haem. This indicates that Type II flavoHbs do not have the

A

```
M.smegmatis         --MTVTTPEP---ADVRVDLEPEHAEIVSATLPLIGANIDAITSEFYRRLFTNHPELLRN
M.fortuitum         --MTVTSPEP---SAVRDELEPKHAEIVAATLPLIGAHIDEITSEFYRRLFANHPELLRN
M.thermoresistibile --MTVTTPGPRPTAVAHAELEPQHAEIVAATLPLVGAHIDEITAEFYRRMFGAHPELLRN
M.vanbaalenii       --MTVTA-------APAELEPAHAEIVTATLPLIGAHIDEITKEFYRRMFGAHPELLRN
M.vaccae            --MTVT--------APAELEPAHAQIISATLPLIGAHIDEITTEFYRRMFGAHPELLRN
M.gilvum            MAPNVTA-------APAELEAAHAEMIAATLPLVGAHIDEITTEFYRRMFAAHPELLRN
M.chubuense         --MTVIA-------APAELEPAHAQIVSATLPLIGAHIDEITTEFYRGMFSAHPELLRN
M.phlei             -MTAIAA-------ASLELEPQHAEIVSATLPLIGAHIDEITTEFYRRMFTNHPELLRN
M.sp.JLS            --MTVTAY-----VAVGEELDPQHAEIITATLPLVGAHIDEITKVFYSRMFAARPELLRN
M.sp.MCS            --MTVTAN-----VAVGEELDPQHAEIITATLPLVGAHIDEITTVFYSRMFAARPELLRN
M.rhodesiae         --MTTTEA------AEVQELEPGHAEIIRATLPLVGAHVDEITQEFYRRMFGRHPELLRT
                                                                     B9B10

M.smegmatis         LFNRGNQAQGAQQRALAASIATFAKHLVDPDLPHPAALLSRIGHKHASLGVTAEQYPIVH
M.fortuitum         LFNRGNQAQGAQQRALAASIATFATHLVNPDLPHPSALLARIGHKHASLGITADQYPIVH
M.thermoresistibile LFNRGNQAQGAQQRALAASIATFATHLVDPELPHPAELLSRIAHKHASLGVTADQYPIVH
M.vanbaalenii       LFNRGNQAQGAQQRALAASIATFASHLVDPELPHPAELLSRIGHKHASLGVTADQYPVVH
M.vaccae            LFNRGNQAQGAQQRALAASIATFASHLIDPGLPHPAELLSRIGHKHASLGVTADQYPVVH
M.gilvum            LFNRGNQAQGAQQRALAASIATFATHLIDPDLPHPAELLSRIGHKHASLGVTADQYPIVH
M.chubuense         LFNRGNQAQGAQQRALAASIATFASHLVDPDLPHPAELLSRIGHKHASLGVTADQYPIVQ
M.phlei             LFNRGNQAQGAQQRALAASIATFATHLVDPDLPHPAELLSRIGHKHASLGVTADQYPIVH
M.sp.JLS            LFNRGNQAQGAQQRALAASIATYATHLVDPKLPHPAELLSRIGHKHASLGITADQYEIVH
M.sp.MCS            LFNRGNQAQGAQQRALAASIATYATHLVDPNLPHPAELLSRIGHKHASLGITADQYEIVY
M.rhodesiae         LFNRGNQAQGSQQRALAASIATFATHLVNPDLPHPSELLSRIGHKHASLGVTADQYPIVH
                    CD1CD2     E7                                F7F8

M.smegmatis         DNLFAAIVAVLGADTVTPEVAAAWDRVFWIMADTLIALERNLYDEAGVEPGDVYRRLRVV
M.fortuitum         DNLFAAIVEVLGADTVTPEVAEAWDRVYWIMADTLIALERDLYAGAGVEPGDVFRRLRVV
M.thermoresistibile EHLFAAIVAVLGADTVTAEVAAAWDRVYWIMADVLIDLERSLYAEAGVPAGDVYRRARVT
M.vanbaalenii       EHLFAAIVAVLGADTVTADVAAAWDRVYWIMADTLIALEHDLYRDAGVADGDVYRRTRVV
M.vaccae            EHLFAAIVEVLGADTVTEDVAEAWDRVYWIMADTLIGMERDLYRSAGVADGDVYRRARVV
M.gilvum            EHLFAAIVEVLGADTVTADVAAAWDRVYWIMADTLIALERDLYKAAGVTDGDVYRRATVV
M.chubuense         EHLFAAIVAVLGADTVTEEVAAAWDRVYWIMADTLIALERQLYRSAGVDDGDVYRRARVV
M.phlei             ENLFAAIVAVLGADTVTDDVAAAWDRVYWIMAQTLIDLEHALYEDAGVADGDVFRRARVV
M.sp.JLS            EHLFAAIVEVLGADTVTAEVAGAWDAVYWMMARTLIELERNLYAEAGVDDGDVYRRASVV
M.sp.MCS            EHLFAAIVEVLGADTVTAPVAEAWDAVYWMMARTLIELERHLYAEAGVDDGDVYRRARVV
M.rhodesiae         EHLFAAIVEVLGADTVTAEVAAAWDRVYWIMADTLIGLEHELYRAAGVDDGDVFRRLRVT
                                          H12              H23 H27
```

                                FAD Binding motif

```
M.smegmatis         GRVDDPSGAVLVTVRAAEP--VNFLPGQYVSVGVTLPDGARQLRQYSLVSVA-GATDLTF
M.fortuitum         SRVDDPSGAVLVTVRAAEP--VNFIAGQYVSVGVTLPDGARQLRQYSLVGAP-GSTDLTF
M.thermoresistibile ARVDDPSGAVLLTVRPLGERFAEFRPGQYISVGVTLPDGARQLRQYSLVNPP-GADELTF
M.vanbaalenii       SRVDDPSGAVLITVRPMGEPPFADFRPGQYVSVGVTLPDGARQLRQYSLVNVP-GDDGLTF
M.vaccae            SRVDDPSGAVLITVRPLGGENVDFLPGQYVSIGVTMPDGARQLRQYSLVNAA-GRDVLTF
```

Figure 5.5—Cont'd

```
M.gilvum                SRVDDPSGVVLITVRPEGRPFTGSIPAQYVSVGVTLPDGARQLRQYTLVNAP-GTADLTF
M.chubuense             SRVDDPSGAVLITVRPAGRPFAEFSPAQYVSVGVTMPDGARQLRQYSLVNAP-ESDELTF
M.phlei                 ARVDDPSGAVLLTVRSTGAAFPKFRPGQYVSVGVVMPDGARQLRQYSLINAP-GSEDLTF
M.sp.JLS                AREDDPSGAVLLTVRSTGRPFATFRPGQYVSVGVTLPDGARQLRQYSLINTP-ADGDLTF
M.sp.MCS                AREDDPSGAVLLTVRSTAQPFSGFRPGQYVSVGVTLPDGARQLRQYSLINTPGGNGDLTF
M.rhodesiae             AREDDPSGAVLLTVRGDGN--SNTGAAQYVSVGVTLPDGARQLRQYSLLNAP-GAGELTF
```

```
                                                      NADH Binding motif
M.smegmatis             AVKPVEATAAQPAGEVSNWIRDNVCVGDLLDVTLPFGDLHTDDLSTGPVMLISAGIGITP
M.fortuitum             AVKPVDAVADQPAGEVSSWIRANLCIGDLLDVTLPFGDLYTGGRPDGPVVLVSAGIGITP
M.thermoresistibile     AVRPVAATPDQPAGEVSTWIAANVCVGDILDITAPFGDLPTP-VAGEPLVLISAGIGITP
M.vanbaalenii           AVKPIAAGADAPAGEVSSWIAANVCVGDLLDVTVPFGDLPAP-DGSAPVVLISAGIGVTP
M.vaccae                AVKPVAAVGESPAGEVSSWIAANVCVGDILDVTVPFGDLPAP-DGSRPVVLISAGIGITP
M.gilvum                AVKPVASVAESPAGEVSSWISANLCVGDIVDVTVPFGDLPAP-DGAAPLVLISAGIGITP
M.chubuens              AVKPVAATAGQPAGEVSSWIAANVCVGDILDVTVPFGDLPAP-VAGQPLVLISAGIGVTP
M.phlei                 AVKPVAGTDTQPAGEVSTWIAANVCVGDLLDVTVPFGDLPAP-DG-RPVVLISAGIGITP
M.sp.JLS                AVR-------PLGEVSNWIRANVQAGDVLDVTVPFGDLPDPEAHHRPLVLVSAGIGITP
M.sp.MCS                AVR-------PLGEVSNWIRANVRAGDVLDVTVPFGDLPDPEQHHRPLVLVSAGIGITP
M.rhodesiae             AVKPVDAEGPNPAGEVSNWIRRCVRVGDLLDVTVPFGDLPKP-RSGTPVVLISAGIGITP
```

```
M.smegmatis             MIGILEHLAAERPDAHVRVLHADRSDTAHPLRERQRELVDALPEASLDIWYEDGVTAGMP
M.fortuitum             MVGILEFLAVEAPDTQLRVLHADRSDSGQPLRERQHELIEALPNASLDLWYEDGVTGGAP
M.thermoresistibile     MIGILEYLAAEEARATPVRVLHADRGDRSHPLRERQRELVAALPDATLDIWYEDGLTAGRP
M.vanbaalenii           MIGILEHLAASAPDTVVRVLHADRSDQTHPLRERQLELIAAMPDARLDIWYEDGLTAGTP
M.vaccae                MVGILEYLAASAPDTPVRVLHADRSDQTHPLRERQHELAAALSDARLDIWYEDGMTAGTP
M.gilvum                MIGILEYLAASAPEASVQVLHADRSDRVHPLRERQHELAEALPNARLDVWYEDGVTAGAP
M.chubuense             MVGILEYLDSQAPGTPVQVLHADRSDQTHPLRERQMELAAQLPHASLDVWYEDGLTAGSP
M.phlei                 MIGILEYLAAEKPQTAVRVLHADRGDATHPLRERQRELLAQLPNGTLDVWYEDGVTAGLP
M.sp.JLS                MVGILEHLAAVAPATSVTVWHADRSAVTHPLKERQRELVAALPDATLEVWYEE----ADA
M.sp.MCS                MIGILEHLASVAPATSVQVWHADRSAATHPLKDRQQELVAALPDATLELWYEE----ADA
M.rhodesiae             MVGLLEYLVAQGYQAPVQVLHADRSEDSHPLRKRHRELVDILPNATLQLWYAEGVEPDSA
```

```
                                      NADH Binding motif
M.smegmatis             GVHAGLLDLDGIELPVEPQIYLCGGAGFVDAVRGQLAARGVPAERVHCELFAPNDWLLD
M.fortuitum             GVHAGLLKLDGIELPDDAAYYLCGGAGFVEAVRTQLSDRGIAHERVHCELFAPNDWLLD
M.thermoresistibile     GVRPGLLDLDAVELPAGAAIYLCGGTGFVQAVRDQLMARGVPAERVHCELFAPNDWLV-
M.vanbaalenii           GAHAGLMNVGTVELPKDAQIYLCGNNGFVQAVRAQLLDRGVVAERVHCELFSPNDWLLG
M.vaccae                DSHAGLMDLAGIELPTDADVYLCGNNGFVQAVRAQLLDRGIAGDRLHCELFSPDDWLLG
M.gilvum                GAHAGLMNLSAIDLPADAHVYLCGNNGFVQAVRGQLTDRGFISERVHCELFSPNDWLLG
M.chubuense             GVHPGLMNLDGIEIAPEAHVYLCGGNGFVHAVRSQLTAKGVATERVHCELFSPNDWLLD
M.phlei                 GVHAGLLTLDGIEVPADAEIYLCGNNGFVQAVRGQLIDRGVPAARVHSELFSPNDWLLG
M.sp.JLS                DAREGLLDVSAVELPDDAEVYLCGGNGFVQAVRAQLQERGVPNERLHCELFSPNDWLLA
M.sp.MCS                HARPGLLNVADVELPEDAEVYLCGGNGFVQAVRAQLQERGVPNERLHCELFSPNDWLLA
M.rhodesiae             GTREGLMNLEGVEIADGAEIYLCGNAGFVQAVRAQLTGRGIAPGRVHCELFSPNDWLLD
```

Figure 5.5—Cont'd

B

```
E.coli             MQKKEDHMLDAQTIATVKAT--IPLLVETGPKLTAHFYDRMFTHNPELKEIFN--MSNQR
R.opacus           MISRLRASFSSIVSTEHGH-----------NRLTTTFYATLFAEHPEFRALFPAALDQQA
M.tuberculosis     MGLEDRDALRVLQNAFKL---------DDPELVRRFYAHWFALDASVRDLFPPDMGAQR
M.fortuitum        MGLDDRQALQTLQNAVDPK--------QGSEPLIRDFYTRWFATDLSARDLFPPDMDSQR
M.smegmatis        MGLEDRESLQVLRAAVDPA--------NDSDPLIRDFYTNWFAADLSVRDLFPPEMAEQR
M.vaccae           MGLEDRESMRILRDAFDFD--------RGSDRLIADFYLRWFSADLSARDLFPPDMHKQR
M.vanbaalenii      MGLEDRESMRILRDALDLD--------LGSEALIGDFHTRWFAADLSARDLFPPDLEKQR
M.chubuense        MGLEDREALRVVRDALDVD--------HGGETLIAEFYTRWFARDLSARDLFPPDMHKQR
M.gilvum           --------MRILREAFEPG--------PGSERLVGEFYTRWFAADLTARDLFPPDMAGQR
M.avium            MSLEDADALRVLRDAFAP---CDAGRPGDSGDLVHRFYTHWFALDPSVRDLFPPEMGAQR
M.parascrofulaceum MGLEDTDALRVLRDAFAP------DEPGTGNELVRRFYTHWFGLDVSVRDLFPPEMGAQR
M.marinum          MGLEDRDALRVLRDAFAPQ--PDLDHKTQSSELVRSFYTNWFSLDSSVRDLFPPEMSGQR
M.liflandii        MGLEDRDALRVLRDAFAPQ--PDLDHKTQSSELVRSFYTNWFSLDSSVRDLFPPEMSGQR
M.ulcerans         MGLEDRDALRVLRDAFAPQ--PDLDHKTQSSELVRSFYTNWFSLDSSVRDLFPPEMSGQR
M.rhodesiae        MGLEDRDALRVLQAAFEES----AAPESASDELLRRFYTRWFALDTSVRDHFPPELAPQR
M.abscessus        MGLEDVEALHDLVGGVELS---------GNKLVRSFYSRWFAIDPTVGDLFPADMSAQR
                                                    B9B10          CD1CD2 E7
```

```
E.coli             NGDQREALFNAIAAYASNIENLPALLPAVEKIAQKHTSFQIKPEQYNIVGEHLLATLDEM
R.opacus           HRLFQALQYVIDNLEDE--DRVLGFL---GQLGRDHRKYGVEARHYFASGHALLVAVQSS
M.tuberculosis     AAFGQALHWVYGELVAQRAEEPVAFLAQLG---RDHRKYGVLPTQYDTLRRALYTTLRDY
M.fortuitum        KVFASALTWLFGELIAQRAEEPVAFLAQLG---RDHRKYGVVQSHYDSMQEALYNTLRDH
M.smegmatis        RVFAHALTWLFGELIAQRAEDPITFLAQLG---RDHRKYGVTQQHYDSMQSALYGALKAR
M.vaccae           EAFSQAMCWLCDQLIAQRAEEPVAFLAQLG---RDHRKYGVTQAHYATLQPALLTTLRTA
M.vanbaalenii      QVFAQALCWLCDELIAQRAEEPVAFLAQLG---RDHRKYGVTQSHYATLQDALLTTLRKA
M.chubuense        RVFAEALCWLCDELIAQRAEEPVGFLAQLG---RDHRKYGVTQAHYESLKDALGATLRAR
M.gilvum           QVFAHAMRWLCDELIAQRAEEPVAFLAQLG---RDHRKYGVTPEHYAGLQDALLAAVRTT
M.avium            VAFGHALHWVYGELVARRAQEPVAFLAQLG---RDHRKYGVLPRHYQTLGRALLATLRSH
M.parascrofulaceum AAFAHALHYVYGELVAQRAQEPVAFLAQLG---RDHRKYGVLPRHYDTLRRALLSTLRTS
M.marinum          AAFTRALHWVYSELVAQRAEEPIAFLAQLG---RDHRKYGVQPTQYETLRRALQTTLRSH
M.liflandii        AAFTRALHWVYSELVAQRAEEPIAFLAQLG---RDHRKYGVQPTQYETLRRALQTTLRSH
M.ulcerans         AAFTRALHWVYSELVAQCAEEPIAFLAQLG---RDHRKYGVQPTQYETLRRALQTTLRSH
M.rhodesiae        AAFGQAMHWVFGEFVAQRAHEPVAFLAQLG---RDHRKYGVTQQHYDSMRQAWYSTMRSH
M.abscessus        EHFRQALQFVLWEMAAYRTEGLVNFLAQLG---RDHRKFGATDSQYDTMRQALLDTTREV
                                                    F7F8        G5 G8
```

```
E.coli             FSPGQEVLDAWGKAYGVLANVFINREAEIYNENASKAGGWEGTRDFRIVAKTPRSALITS
R.opacus           FT-----QAIWTPSLTAAWTELLDLLLHTMAD-AADNDDLPAAWGATVVDHERRLDDLAI
M.tuberculosis     LGHPS--RGAWTDAVDEAAGQSLNLIIGVMSG-AADADDAPAWWDGTVVEHIRVSRDLAV
M.fortuitum        LQ------DDWDDALAEATHDAVALVIGVMRG-AADAEDSPAYCDGTVIEHHRVTRDVSV
M.smegmatis        LA------DRWTDRLAAATRDAVALFIGVMRG-AADAEESPAYCDGTVVETHRLTRDVSV
M.vaccae           LA------DQWDRRLAEAVDDVVQLVVGVMSG-AADAETIPPFCDGTVLEHLRPTRDVSV
M.vanbaalenii      LV------EQWDRRLAEAVDDVVQLVVGVMSG-AADAETSPPFCDGTVLEHLRPARDISV
M.chubuense        LA------GIWEPRLAEVVDDVVTLVVGVMSG-AADAETSPPFCDGTVLEHLRPTRDVSV
M.gilvum           LT------DRWDARLEEAATDVVTLAVGVMSG-AGEAETAPPFCDGTVLEHLRPTRDVSV
```

Figure 5.5—Cont'd

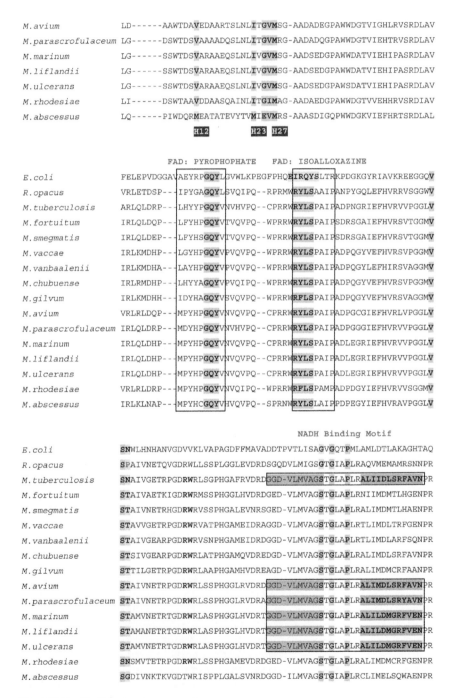

Figure 5.5—Cont'd

```
E.coli              VNWFHAAENGDVHAFADEVKELGQSLPRFTAH---------TWYRQPSEADRAKGQFDSE
R.opacus            VHMFVGGKYPCDLYDIETLWHLSLSNPWLTVVPVSEEDEDP-WWHTIPAPEAPPGLHQRL
M.tuberculosis      VHLFFGARYACELYDLPTLWQIAAHNPWLSVSPVSEYNGDPAWAADYPDVSAPRGLHVRQ
M.fortuitum         VHLFFGGRYPCELYDLKTLWHIASTNPWLSVTPVSEYSTDPPWAGQYPDVQPPRGLHVRQ
M.smegmatis         VHLFFGGRFPCDLYDLKTLWTIASTNPWLSVTPVSEYSTDPPWARDYPDPTPPRGLHVRQ
M.vaccae            VHLFFGAKYPCELYDLRTLWEIASVNPWLSVTPVSELNFDPPWAADYPDVTPPRGLHVRQ
M.vanbaalenii       VHLFFGAKYPCELYDLRTLWEIASTNPWLSVTPVSELNFDPPWAADYPDVTPPRGLHVRQ
M.chubuense         VHLFFGAKHPCELYDLRTLWEIASTNPWLSVTPVSELNFDPPWASRYPDVTPPRGLHVRQ
M.gilvum            VHLFFGAKYPCELYDLPTLWEVASMNPWLSVTPVSEFDFDPPWAAEYPDNRPPRGLHVRQ
M.avium             VHLFFGARYRCELYDLPTLWQIASHNPWLSVSPVSEYRADPPWAADYPDVTPPRGLHVHQ
M.parascrofulaceum  VHLFFGARYRCELYDLPTLWQVASHNPWLSVSPVSEYGADPAWAADYPDCTPPRGLHVRQ
M.marinum           VHLFFGARYFCELYDLRTLWQVAAHNPWLSVSPVAEYKLDPSWATDYPDVSPPRGLHVRQ
M.liflandii         VHLFFGARYFCELYDLRTLWQVAAHNPWLSVSPVAEYKLDPSWATDYPDVSPPRGLHVRQ
M.ulcerans          VHLFFGARYFCELYDLRTLWHVAAHNPWLSVSPVAEYKLDPSWATDYPDVSPPRGLHVRQ
M.rhodesiae         VHVFFGARYPCELYDLRTLWEIAASSPWLSVSPVSEYAGDPPWAADYPDVQPPRGLHVRQ
M.abscessus         VHLFYGARYPQELYDLWTLWHIASTNPWLSVTPVTEYPRNPDWAAEYHDPTPPRGLHVRQ
```

```
                                     NADPH:ADENINE                    Glu 394
E.coli              G--LMDLSKLEGAFSDPTMQFYLCGPVGFMQFTAKQLVDLGVKQENIHYECFGPHKVL
R.opacus            QGQLGRVVARFGSWAD--RQIQISGGSPAMIKTTVYALQGGGTPPELIRHDPLI-----
M.tuberculosis      TGRLPDVVSRYGGWGD--RQILICGGPAMVRATKAALIAKGAPPERIQHDPLSR----
M.fortuitum         TGTLAEVVTRYGNWGD--RQILICGGPDMVTATKAALVERGAPAERIQHDPLTR----
M.smegmatis         TGTLADVVTRYGNWGD--RQILICGGPQMVEATKAALIAKGAPPERIQHDPLTAR---
M.vaccae            TGLLPEVVTRYGNWGD--RQILLCGRPEMVRATKAALIAKGAPADRIQHDPLAS----
M.vanbaalenii       LGLLPEVVTRYGNWSD--RQILLSGRPAMVQATKAALITKGAPAERIQHDPLAN----
M.chubuense         TGTLPEVVTRYGSWSD--RQVLLCGGPAMVAATKAALIAKGTPPERIHHDPPAS----
M.gilvum            TGRLDEVVTRYGSWGD--RQILVCGRPEMVAATRSALIAKGAPAERIQHDPLGN----
M.avium             TGRLPEVVTKYGGWGD--RQILICGGPRMVAATKAALIAKGAAAQRIQHDPLSR----
M.parascrofulaceum  TGRLPDVVTKYGGWGD--RQILICGGPAMVRATKAALIAKGAPPERIQHDPLWR----
M.marinum           TGRLAEVVTNYGNWGD--RQILICGGPAMVRTTKAALIAKGAPRERIQHDPLP-----
M.liflandii         TGRLAEVVTNYGNWGD--RQILICGGPAMVRTTKAALIAKGAPRERIQHDPLP-----
M.ulcerans          TGRLAEVVTNYGNWGD--RQILICGGPAMVRTTKAALIAKGAPRERIQHDPLP-----
M.rhodesiae         TGRLPDVVTKYGAWGD--RQILICGGPAMVRATYDALVAKGAPPERIQHDPLV-----
M.abscessus         TGLLSEVVTAYGGWGD--RQILIGGSASMIQATKEALVSRGADASRIQHDPL------
```

Figure 5.5—Cont'd Structure based sequence alignment of Type I (A) and Type II (B) flavohemoglobins of mycobacteria. Conserved regions and co-factor binding sites are highlighted and shown in boxes. (For colour version of this figure, the reader is referred to the online version of this chapter.)

peroxidase-like imidazolate character present in a typical flavoHb (Gupta et al., 2012). Thus, the architecture and the redox properties of Mycobacterial Type II flavoHbs might be distinct from other microbial flavoHbs. The reductase domain of Type II flavoHbs of Mycobacteria is also modified and devoid of critical residues needed for the binding of co-factors FAD and NADH. For example, the RxYS motif, the hall mark of FAD binding, and the GxGxxP motif for NADH binding are mutated in Type II flavoHbs. Additionally, an RKY/F sequence motif, known as a high-affinity lipid-binding site (Hunte, 2005) appears conserved within the proximal site of haem in Type II flavoHbs, suggesting the possibility of its strong association with membrane lipids. Interestingly, a co-factor binding site, resembling the FAD-binding motif of respiratory lactate dehydrogenase, has been found in flavoHb of *M. tuberculosis* (Gupta et al., 2012). Interestingly, this FAD-binding motif is conserved in Type II flavoHbs of pathogenic Mycobacteria, whereas in many flavoHbs of non-pathogenic Mycobacteria and certain actinomycetes, these FAD-binding motifs are not fully conserved. These unusual features suggest that the structural and functional properties of co-factor-binding domains of type II flavoHbs are different from other microbial flavoHbs.

4.2. Functional properties

Not much investigation has been done on the flavoHbs of Mycobacteria and only primary information on the flavoHb of *M. tuberculosis* is available (Gupta et al., 2012, 2011). Close similarity of Type I flavoHbs of Mycobacteria with other microbial flavoHbs indicates that they may be structurally and functionally similar. Type II flavoHb of *M. tuberculosis* (MtbFHb) is a monomeric protein of 43 kDa that has been found closely associated with the cell membrane (Gupta et al., 2012). At variance with conventional flavoHbs that display the pentacoordinated high-spin form, MtbFHb unusually displays the hexa-coordinated form in both ferric and ferrous states, indicating that it may not be able to carry out the NOD reaction similar to other flavoHbs. Consistent with this, experimental studies confirmed that MtbFHb lacks this activity. Instead, it was found to confer resistance against oxidative stress through a novel electron transfer mechanism via consumption of D-lactate that may accumulate during membrane lipid peroxidation (Gupta et al., 2011). In the absence of information on other Mycobacterial flavoHbs, the physiological function of this new class of flavoHbs cannot be generalised. Further studies are required to understand the precise role of MtbFHb in electron transfer and the oxidative stress response of Mycobacteria.

4.3. Genomic organisation and genetic regulation

Genomic organisation of Type I and Type II flavoHbs has been obtained from the available Mycobacteria genome data. The flanking genes of Type

I flavoHbs of seven Mycobacterial species are presented in Fig. 5.6, showing close similarity in the genes present in the flanking sites of flavoHb-encoding genes. The flavoHb gene has a transcriptional regulator at the upstream region in five out of the seven strains of Mycobacteria examined that may possibly control its genetic regulation. It is likely that similar regulatory controls exist for Type I flavoHbs, and that they perform similar functions in these Mycobacterial species.

Until now, only one Type II flavoHb, from *M. tuberculosis*, has been characterised. It carries a gene encoding *clpB* upstream and a gene for a regulatory protein down-stream. The divergent *clpB* gene is present upstream of

Figure 5.6 Genomic organisation of Mycobacterial flavohaemoglobins: (A) Type I and (B) Type II. (For colour version of this figure, the reader is referred to the online version of this chapter.)

the majority of Type II flavoHbs of pathogenic bacteria. The flavoHb in the non-pathogenic species has an aldose epimerase gene down-stream and a gene encoding a hypothetical protein (proposed to be an integral membrane protein) upstream of it. Thus, a clear-cut distinction is present in the genomic organisation of flavoHb-encoding genes of pathogenic and non-pathogenic Mycobacteria.

Genetic regulation studies on the MtbFHb-encoding gene (Rv0385) indicated that its promoter activity increases gradually up to 24 h post-infection, showing a three-fold increase in promoter activity, which was reduced to basal level after 48 h (Pawaria, Lama and Dikshit, unpublished). This suggests a stage-specific and high level up-regulation of Rv0385 during intracellular infection of *M. tuberculosis*.

5. BIOLOGICAL FUNCTIONS OF trHbs AND FlavoHbs

This is not the place for a detailed review of the host response to infection by Mycobacteria; for more details, see Liu and Modlin (2008). However, in order to understand the biological functions of trHbN and trHbO, we must have an appreciation of the pathogenicity of, in particular, *M. tuberculosis*. A 1999 report for WHO documented that in 1997, there were just under 8 million new cases of TB worldwide, in addition to 16.2 million existing cases of TB; 1.87 million people died of TB in the same year (Dye, Scheele, Dolin, Pathania, & Raviglione, 1999), emphasising how important it is to study mechanisms of disease. Upon infection of the host, *M. tuberculosis* can either cause TB or enter into a dormancy mode inside macrophage phagosomes, where they will encounter hypoxic conditions and probably NO from the NOS enzymes. It is thought that trHbN and trHbO are important proteins for *M. tuberculosis* when inside the macrophage; their biological functions will be discussed here.

5.1. Nitric oxide scavenging

Studies into the biological functions of trHbN have been done in *M. tuberculosis*, *M. bovis* and *M. smegmatis*. The first evidence for function came in *M. bovis*. The *glbN* gene was inactivated via gene insertion of a kanamycin cassette; no difference in growth in liquid or in solid media was observed compared with wild-type cells, suggesting that Mb trHbN is not essential for growth under normal laboratory conditions (Ouellet et al., 2002). As other globins, notably flavoHb from *E. coli*, are implicated in NO detoxification (Stevanin et al., 2000), NO metabolism was investigated by comparing NO consumption of wild type, ΔHbN and the complemented strains. Under aerobic conditions, after addition of NO-saturated solution to wild-type cells, the NO signal generated showed that NO is rapidly removed from solution; the case in the complemented cells is more extreme, showing that trHbN can efficiently remove NO from solution (Ouellet et al., 2002). However, in ΔHbN cells, NO removal appears to be the same as buffer alone when exposed to 1 μM NO, indicating that there is no NO detoxification activity in these cells; however, when these cells are exposed to a lower concentration of NO (0.5 μM), there is some NO detoxification activity, indicating that there is another system in *M. bovis* that can detoxify low concentrations of NO, which could be trHbO. Interestingly, and agreeing with data discussed in Section 3.1.1, a ΔHbN mutant complemented with trHbN with TyrB10 replaced with Phe showed no improvement in NO removal over the ΔHbN cells (Ouellet et al., 2002), which is further proof that the TyrB10 residue is essential for the function of trHbN (Ouellet et al., 2006). The same group showed the stoichiometric production of nitrate by purified Mb trHbN, showing that the reaction between NO and O_2 leads to this product (Ouellet et al., 2002).

When Mtb (Pathania, Navani, Gardner, et al., 2002) or Ms trHbN was expressed in an *E. coli hmp⁻* strain, NO uptake was increased; Mtb trHbN appears more efficient at NO detoxification than Ms trHbN, as determined by NO uptake assays using an NO electrode (Lama et al., 2006). Under normal laboratory conditions, that is, without any NO stress, growth was neither improved nor impaired by the expression of the two globins under aerobic conditions. The survival of these *E. coli* strains carrying the two trHbN was determined after exposure to SNP, an NO donor. As expected, expression of Mtb trHbN confers protection against SNP compared to wild-type cells, and the protection afforded by Ms trHbN is almost negligible (Lama et al., 2006). When the same two globins were over-expressed in *M. smegmatis*, there was an increase in the NO detoxification ability of the cells, and again this ability was more improved in cells expressing Mtb trHbN (18.6 nmol NO haem^{-1} s^{-1}) than Ms trHbN (5.1 nmol NO haem^{-1} s^{-1}) (Lama et al., 2006). Wild-type *M. smegmatis* cells showed low NO uptake (0.09 nmol NO haem^{-1} s^{-1}), perhaps again a reflection of the lack of pre-A region and the fact that non-pathogenic Mycobacteria might not require efficient NO removal activities. When the pre-A region was removed from Mtb trHbN, NO consumption activity of the purified protein was lower than wild type; expression of wild-type trHbN in

M. smegmatis showed NO uptake activity of 27.82 nmol NO haem^{-1} s^{-1}, but when the protein with the pre-A region deleted was subjected to the same tests, rates of only 3.43 nmol NO haem^{-1} s^{-1} were observed (Lama et al., 2009). Conversely, adding the pre-A region onto the *M. smegmatis* trHbN increased its NO detoxification abilities: expression of wild-type trHbN on a plasmid gave rates of 4.9 nmol NO haem^{-1} s^{-1}, with the modified protein showing an NO consumption of 11.8 nmol NO haem^{-1} s^{-1} (Lama et al., 2009). These results confirm the authors' idea that the pre-A region is a requirement for efficient NO detoxification.

As NO is a respiratory inhibitor, the ability of trHbN to protect respiration of cells exposed to NO has been investigated. In *M. bovis*, respiration was initially inhibited by 1 and 2 μM NO and then recovered; however, when *glbN* was knocked out, cells lost the ability to detoxify NO which showed that the trHbN protein is responsible for the NO uptake activity in wild-type cells (Ouellet et al., 2002). Furthermore, complementation of the mutant and therefore probable over-expression of trHbN leads to cells that are more resistant to respiratory inhibition by NO than wild-type cells. Under high O_2 concentrations, the *glbN* mutant recovered somewhat and there was less inhibition of respiration, suggesting that ambient O_2 concentrations are an important factor in protection against NO toxicity (Ouellet et al., 2002).When Mtb trHbN was expressed in both *E. coli hmp$^-$* and *M. smegmatis*, O_2 uptake was protected from inhibition by NO; for example, in *E. coli hmp$^-$* when 5 μM NO was added to cells, O_2 uptake was 0.2 μmol min^{-1} 10^{-8} cells^{-1}, whereas expression of trHbN increased this to 6.8 μmol min^{-1} 10^{-8} cells^{-1}; control values without addition of NO were 7.3 and 7.5 μmol min^{-1}10^{-8} cells^{-1}, respectively (Pathania, Navani, Gardner, et al., 2002). Although the authors do not specify where on the O_2 concentration curves the rates were calculated from, these data clearly show that the expression of Mt trHbN can protect respiration from NO inhibition.

Expression of trHbN in *M. tuberculosis* and *M. bovis* is higher in late exponential and early stationary phases than in early exponential phase (Pathania, Navani, Gardner, et al., 2002), and *glbN* shows an early response to nitrosative stress (Joseph et al., 2012), reflecting the point at which these species may require NO detoxification activities during infection. Haem staining did not show any detectable amount of Hb-like protein in *M. smegmatis*, although we know it possesses a trHbN homologue. The presence of Mtb trHbN in an *E. coli hmp$^-$* mutant protected against NO toxicity under aerobic conditions. Importantly, this protection was sustained over repetitive NO additions, which suggests that the redox state of the protein can be regenerated after the reaction to recycle the protein into the ferrous state (Pathania, Navani, Gardner, et al., 2002).

Salmonella enterica serovar Typhimurium (hereafter *S. typhimurium*) also has NO detoxification abilities in the form of the Hmp protein (Stevanin, Poole, Demoncheaux, & Read, 2002); when an *S. typhimurium hmp⁻* strain was transformed with a vector expressing *glbN* from *M. tuberculosis*, no growth advantage was conferred under normal laboratory conditions (Pawaria et al., 2007). The NO uptake activity of *S. typhimurium hmp⁻* at ambient O_2 levels (0.02 nmol haem^{-1} s^{-1}) was significantly improved upon expression of Mtb trHbN (19.7 nmol haem^{-1} s^{-1}) and Mtb trHbO (18.4 nmol haem^{-1} s^{-1}); however, under low O_2 levels, trHbN continues to efficiently remove NO, whereas trHbO does not (Pawaria et al., 2007). These data suggest that the function of trHbN is NO removal even under low O_2 levels, which perhaps reflect the hypoxic conditions in which it will be required by the cell to function, and while trHbO can remove NO from solution, it cannot do so when O_2 levels are low. Growth curves were performed in the presence of acidified nitrite, which should mimic nitrosative stress within the cell. When exposed to 30 mM acidified nitrite at pH 7, the growth of *S. typhimurium* was inhibited, showing a long lag phase before recovering after about 20 h, but when expressing trHbN, the lag phase was shortened and cells recovered from the stress after around 7.5 h, suggesting that trHbN is aiding in the response to acidified nitrite. Under the same conditions, trHbO provides no protection (Pawaria et al., 2007). When *S. typhimurium* cells carrying the globins were used to infect activated macrophages, expression of *glbN* but not *glbO* improved survival after 30, 60 and 90 min (Pawaria et al., 2007), again implicating trHbN but not trHbO in NO detoxification, as the activated macrophages were shown to contain elevated levels of NO.

As already suggested, the reductive evolution of the *M. leprae* genome has led to the loss of trHbN and the expression of only trHbO. It is conceivable that in this organism, trHbO has evolved to detoxify NO, replacing the lost detoxification activity provided by trHbN. Indeed, purified oxy-ferrous Ml trHbO can detoxify NO and produce nitrate (Fabozzi et al., 2006). Expression of *glbO* from *M. leprae* in an *E. coli hmp⁻* strain protected cells from the toxicity of SNP and SNAP, but only when induced by IPTG, suggesting that the levels of trHbO inside the cell must be high in order for NO detoxification to occur. There are two reasons that this could happen: as trHbO is being expressed in *E. coli*, it is unlikely that any partner proteins would be present and perhaps trHbO requires a partner, for example a reductase, for efficient function. Second, perhaps trHbO is not as efficient as trHbN at removing NO and so has to be present at high concentrations to be effective. As has already been discussed, homologues of trHbN are less likely to be found in the genomes of other bacteria than trHbO; Pathania, Navani, Gardner, et al. (2002) suggest that this is because NO scavenging is a more niche function than the proposed function of trHbO, possibly O_2 scavenging or delivery (see Section 5.2).

5.2. Oxygen metabolism

Due to its unusual O_2 binding features, trHbO has been implicated in O_2 metabolism. When Mtb trHbO was expressed in a *S. typhimurium hmp⁻* mutant, O_2 uptake was slightly higher than in wild-type cells (Pawaria et al., 2007). Uptake of O_2 was increased when Mtb trHbO was expressed in *E. coli* (control: 4.4 μmol min^{-1} 10^{10} cells^{-1}, plus trHbO: 5.6 μmol min^{-1} 10^{10} cells^{-1}) and *M. smegmatis* (control: 2.6 μmol min^{-1} 10^{10} cells^{-1}, plus: trHbO 3.5 μmol min^{-1} 10^{10} cells^{-1}); this was also the case with *E. coli* membrane vesicles expressing trHbO (Pathania, Navani, Rajamohan, et al., 2002). The O_2 uptake of mutants of *E. coli* which expressed only cytochrome *bo'* (*cyd⁻*) or only cytochrome *bd*-type (*cyo⁻*) was also determined. In *E. coli cyo⁻*, O_2 uptake rates were increased from 76 μm min^{-1} mg protein^{-1} in control cells to 88 μm min^{-1} mg protein^{-1} in those cells expressing trHbO; in the *E. coli cyd⁻* strain, rates went from 111 μm min^{-1} mg protein^{-1} to 192 μm min^{-1} mg protein^{-1}, suggesting that trHbO is able to sustain and even improve respiration in these single mutants (Pathania, Navani, Rajamohan, et al., 2002). Another study found that Mtb trHbO was only able to improve growth in the stationary phase when expressed in wild-type *E. coli* cells, and that this improvement was not seen in a *cyo⁻* mutant (Liu, He, & Chang, 2004). These results both point towards an interaction with cytochrome *bo'* rather than *bd*, and this conclusion was confirmed when it was shown that purified trHbO can interact, albeit weakly, with purified *E. coli* CyoB; CyoB is however not required for the initial interaction as trHbO can bind to artificial phospholipid membranes as well (Liu et al., 2004). During purification of trHbO from *E. coli* cells, large amounts of the protein were found to be located with the membrane fraction and were only removed after treatment with 0.2 M potassium thiocyanate which interrupts hydrophobic interactions, suggesting that the function of trHbO is involved with the membrane; cells expressing trHbO also grew faster and reached a higher final OD than control cells (Pathania, Navani, Rajamohan, et al., 2002), suggesting a contribution to growth of the cells, perhaps by more efficient aerobic respiration. As previously discussed, trHbO is expressed throughout growth in *M. bovis* as detected by Western Blotting (Mukai et al., 2002; Pathania, Navani, Rajamohan, et al., 2002), which adds credence to the assumption that trHbO can aid respiration. Liu et al. (2004) additionally comment that trHbO may act as an O_2 collector or reservoir to support respiration in hypoxic conditions.

However, if trHbO is to be implicated in aerobic metabolism, we must be able to link this function back to the biochemistry of the protein. The combination rate for Mtb trHbO and O_2 is $<1.0 \, \mu M^{-1} \, s^{-1}$ and the dissociation rate is $<0.006 \, s^{-1}$; coupled with the very high affinity for O_2 (Ouellet et al., 2003), these data suggest that once trHbO has trapped O_2 in the haem pocket, it will stay there which is contrary to a role in O_2 capture and delivery. These authors suggest that trHbO may play a role in sensing the redox state of the cell. However, as Liu et al. (2004) found, trHbO dissociates from a dimer into a monomer during its interaction with membrane lipids, and this change may alter the affinity of the globin for O_2. A definitive answer remains to be found.

As has already been discussed, flavoHbs are traditionally thought of as NO detoxification proteins, as shown in *E. coli* (Stevanin et al., 2000) and *S. typhimurium* (Stevanin et al., 2002). However, the recently discovered flavoHb of *M. tuberculosis* appears to have a different function. When expressed in *E. coli*, cells expressing MtbFHb showed NO consumption of $5.32 \, nmol \, NO \, haem^{-1} \, s^{-1}$, in comparison with wild-type values of $1.14 \, nmol \, NO \, haem^{-1} \, s^{-1}$, a difference that was not deemed significant (Gupta et al., 2011). The same cells showed slightly improved O_2 uptake ($7.1 \, \mu mol \, min^{-1} \, 10^{-9} \, cells^{-1}$ compared with $5.3 \, \mu mol \, min^{-1} \, 10^{-9} \, cells^{-1}$ in wild type). These data suggest that MtbFHb has a different function altogether. A second paper from the same group implicated MtbFHb in the oxidation of D-lactate into pyruvate; this was first proposed due to the similarities between the FAD-binding domain of MtbFHb and D-lactate dehydrogenase (Gupta et al., 2012). Indeed, during oxidative stress, as delivered by sub-lethal concentrations of hydrogen peroxide, *E. coli* and *M. smegmatis* cells carrying MtbFHb showed improved survival; further investigations showed that there were reduced levels of lipid peroxidation and the authors suggested that MtbFHb is protecting the cell membrane from lipid peroxidation during oxidative stress. Additionally, the promoter of the MtbFHb gene showed increased activity in the presence of hydrogen peroxide and in macrophages during infection, adding extra proof to the hypothesis that MtbFHb is part of the *M. tuberculosis* oxidative stress defence (Gupta et al., 2012).

5.3. Redox signalling and stress management

Haem proteins, including Hbs, have been found to play an important role in sensing gaseous O_2 and other ligands. The interaction between the haem-Fe-bound ligand and haem distal amino acid residues and/or haem pocket geometry play a vital role in carrying out distinct redox reactions and signal triggering. TrHbO of *M. tuberculosis* has been found to have some properties that identify a possible function as a redox sensor (Ouellet, Ranguelova, et al., 2007). It has been shown that trHbO could be prone to redox reactions due to the presence of oxidisable residues in the haem pocket, for example, TyrB10, TrpG8 and the unusual Tyr CD1 (Ouellet, Ranguelova, et al., 2007). Optical spectra suggest that Mtb trHbO can react with hydrogen peroxide and form a novel intermediate, containing ferric haem iron and two oxidising equivalents, that is similar to the product of the reaction of peroxidases and other Hbs with high concentrations of H_2O_2, indicating that trHbO may have peroxidase activity. Similar oxidation/reduction function has also been suggested for trHbO of *M. leprae* (Ascenzi, De Marinis, Coletta, & Visca, 2008; Ascenzi, Milani, et al., 2006) and may be true for other group II trHbs of Mycobacteria.

5.4. Other functions

As trHbO appears to have some properties that identify a possible function as a redox sensor, the reaction of purified protein with H_2O_2 was investigated. Optical spectra suggest that Mtb trHbO can react with H_2O_2 and that it forms Compound III, a common product of the reaction of peroxidases and other Hbs with high concentrations of H_2O_2 (Ouellet, Ranguelova, et al., 2007). Peroxidase activity was also shown. Additional evidence showed that cross-linked dimers were produced after trHbO was exposed to only 0.25 molar equivalents, common in haemoproteins that are exposed to H_2O_2 due to tyrosine–tyrosine cross-links formed after certain amino acid residues have become radicals due to the transfer of electrons during the reaction. Indeed, Mtb trHbO has many tyrosine residues which are available for potential conversion to tyrosol radicals, and that could be responsible for this dimerisation (Ouellet, Ranguelova, et al., 2007). A function as a peroxidase has also been suggested for Ml trHbO. Both oxy-ferrous and ferrous-

NO-bound Ml trHbO can catalyse peroxynitrite scavenging (Ascenzi, Milani, et al., 2006). The time course for peroxide-mediated oxidation of Ml trHbO is a monophasic process, and oxy-ferryl Ml trHbO can facilitate NO oxidation (Ascenzi et al., 2008). It has also been implicated in peroxynitrite detoxification (Ascenzi, De Marinis, Visca, Ciaccio, & Coletta, 2009), perhaps reflecting the multi-functional nature of trHbO in *M. leprae*.

At present, only three Mycobacterial trHbs (Mtb trHbN, trHbO and *M. leprae* trHbO) have been characterised and cellular functions have been proposed on the basis of their functionality. No study has been conducted on the group III trHb of Mycobacteria. It is likely that trHbP has functions that are very distinct from trHbN and trHbO. Also, the function of flavoHbs in the physiology of Mycobacteria has not been fully explored, except in the case of flavoHb of *M. tuberculosis* that represents a new class among the flavoHb family (Gupta et al., 2012), having a hexacoordinated haem domain and unusual reductase domain that carries an FAD-binding motif similar to D-lactate dehydrogenase (Gupta et al., 2012). It has been proposed that MtbFHb utilises D-lactate to carry out electron cycling and may be involved in protecting the cellular membrane from oxidative damage by preventing the accumulation of D-lactate in the cell and its interactions with free iron that can generate superoxide ions. MtbFHb homologues are ubiquitous among Mycobacteria and appear restricted to actinomycetes. Further studies are required to understand the physiological role of this new class of flavoHbs in Mycobacterial cellular metabolism.

6. CONCLUDING REMARKS

In conclusion, although the role of trHbN in the response to NO in *M. bovis* and *M. tuberculosis* and the involvement of trHbO in O_2 chemistry are well established, the functions of the other Hbs in Mycobacteria are still to be fully characterised. In particular, there have not been any studies done on the group III trHbP, and there are only two published papers on the Type II flavoHb from *M. tuberculosis*. Further studies on the structure, biochemistry and functions of both truncated and flavohaemoglobins are on-going and should provide further insight into the reasons why multiple haemoglobins exist in many species of Mycobacteria.

REFERENCES

Ascenzi, P., Bocedi, A., Bolognesi, M., Fabozzi, G., Milani, M., & Visca, P. (2006). Nitric oxide scavenging by *Mycobacterium leprae* GlbO involves the formation of the ferric heme-bound peroxynitrite intermediate. *Biochemical and Biophysical Research Communications*, *339*, 450–456.

Ascenzi, P., Bolognesi, M., Milani, M., Guertin, M., & Visca, P. (2007). Mycobacterial truncated hemoglobins: From genes to functions. *Gene*, *398*, 42–51.

Ascenzi, P., Bolognesi, M., & Visca, P. (2007). NO dissociation represents the rate limiting step for O_2-mediated oxidation of ferrous nitrosylated *Mycobacterium leprae* truncated hemoglobin O. *Biochemical and Biophysical Research Communications*, *357*, 809–814.

Ascenzi, P., De Marinis, E., Coletta, M., & Visca, P. (2008). H_2O_2 and NO scavenging by *Mycobacterium leprae* truncated hemoglobin O. *Biochemical and Biophysical Research Communications*, *373*, 197–201.

Ascenzi, P., De Marinis, E., Visca, P., Ciaccio, C., & Coletta, M. (2009). Peroxynitrite detoxification by ferryl *Mycobacterium leprae* truncated hemoglobin O. *Biochemical and Biophysical Research Communications*, *380*, 392–396.

Ascenzi, P., Milani, M., & Visca, P. (2006). Peroxynitrite scavenging by ferrous truncated hemoglobin GlbO from *Mycobacterium leprae*. *Biochemical and Biophysical Research Communications*, *351*, 528–533.

Balasubramanian, V., Wiegeshaus, E. H., Taylor, B. T., & Smith, D. W. (1994). Pathogenesis of tuberculosis: Pathway to apical localization. *Tubercle and Lung Disease*, *75*, 168–178.

Berney, M., & Cook, G. M. (2010). Unique flexibility in energy metabolism allows mycobacteria to combat starvation and hypoxia. *PLoS One*, *5*, e8614.

Bidon-Chanal, A., Martí, M. A., Crespo, A., Milani, M., Orozco, M., Bolognesi, M., et al. (2006). Ligand-induced dynamical regulation of NO conversion in *Mycobacterium tuberculosis* truncated hemoglobin-N. *Proteins: Structure, Function, and Bioinformatics*, *64*, 457–464.

Bidon-Chanal, A., Marti, M. A., Estrin, D. A., & Luque, F. J. (2007). Dynamical regulation of ligand migration by a gate-opening molecular switch in truncated hemoglobin-N from *Mycobacterium tuberculosis*. *Journal of the American Chemical Society*, *129*, 6782–6788.

Boechi, L., Martí, M. A., Milani, M., Bolognesi, M., Luque, F. J., & Estrin, D. A. (2008). Structural determinants of ligand migration in *Mycobacterium tuberculosis* truncated hemoglobin O. *Proteins: Structure, Function, and Bioinformatics*, *73*, 372–379.

Bonamore, A., & Boffi, A. (2008). Flavohemoglobin: Structure and reactivity. *IUBMB Life*, *60*, 19–28.

Cole, S. T., Brosch, R., Parkhill, J., Garnier, T., Churcher, C., Harris, D., et al. (1998). Deciphering the biology of *Mycobacterium tuberculosis* from the complete genome sequence. *Nature*, *393*, 537–544.

Cole, S. T., Eiglmeier, K., Parkhill, J., James, K. D., Thomson, N. R., Wheeler, P. R., et al. (2001). Massive gene decay in the leprosy bacillus. *Nature*, *409*, 1007–1011.

Couture, M., Yeh, S.-R., Wittenberg, B. A., Wittenberg, J. B., Ouellet, Y., Rousseau, D. L., et al. (1999). A cooperative oxygen-binding hemoglobin from *Mycobacterium tuberculosis*. *Proceedings of the National Academy of Sciences of the United States of America*, *96*, 11223–11228.

Crespo, A., Martí, M. A., Kalko, S. G., Morreale, A., Orozco, M., Gelpi, J. L., et al. (2005). Theoretical study of the truncated hemoglobin HbN: Exploring the molecular basis of the NO detoxification mechanism. *Journal of the American Chemical Society*, *127*, 4433–4444.

Daigle, R., Guertin, M., & Lagüe, P. (2009). Structural characterization of the tunnels of *Mycobacterium tuberculosis* truncated hemoglobin N from molecular dynamics simulations. *Proteins: Structure, Function, and Bioinformatics*, *75*, 735–747.

Dantsker, D., Samuni, U., Ouellet, Y., Wittenberg, B. A., Wittenberg, J. B., Milani, M., et al. (2004). Viscosity-dependent relaxation significantly modulates the kinetics of CO recombination in the truncated hemoglobin TrHbN from *Mycobacterium tuberculosis*. *The Journal of Biological Chemistry*, *279*, 38844–38853.

Droghetti, E., Nicoletti, F. P., Bonamore, A., Boechi, L., Arroyo Mañez, P., Estrin, D. A., et al. (2010). Heme pocket structural properties of a bacterial truncated hemoglobin from *Thermobifida fusca*. *Biochemistry, 49*, 10394–10402.

Dye, C., Scheele, S., Dolin, P., Pathania, V., & Raviglione, R. C. (1999). Global burden of tuberculosis—Estimated incidence, prevalence, and mortality by country. *JAMA: The Journal of the American Medical Association, 282*, 677–686.

Ehrt, S., & Schnappinger, D. (2009). Mycobacterial survival strategies in the phagosome: Defence against host stresses. *Cellular Microbiology, 11*, 1170–1178.

Fabozzi, G., Ascenzi, P., Renzi, S. D., & Visca, P. (2006). Truncated hemoglobin GlbO from *Mycobacterium leprae* alleviates nitric oxide toxicity. *Microbial Pathogenesis, 40*, 211–220.

Frey, A. D., & Kallio, P. T. (2003). Bacterial hemoglobins and flavohemoglobins: Versatile proteins and their impact on microbiology and biotechnology. *FEMS Microbiology Reviews, 27*, 525–545.

Giordano, D., Parrilli, E., Dettaï, A., Russo, R., Barbiero, G., Marino, G., et al. (2007). The truncated hemoglobins in the Antarctic psychrophilic bacterium *Pseudoalteromonas haloplanktis* TAC125. *Gene, 398*, 69–77.

Grange, J. M. (1996). The biology of the genus Mycobacterium. *Journal of Applied Microbiology, 81*, 1S–9S.

Guallar, V., Lu, C., Borrelli, K., Egawa, T., & Yeh, S.-R. (2009). Ligand migration in the truncated hemoglobin-II from *Mycobacterium tuberculosis*: The role of G8 tryptophan. *The Journal of Biological Chemistry, 284*, 3106–3116.

Gupta, S., Pawaria, S., Lu, C., Hade, M. D., Singh, C., Yeh, S.-R., et al. (2012). An unconventional hexacoordinated flavohemoglobin from *Mycobacterium tuberculosis*. *The Journal of Biological Chemistry, 287*, 16435–16446.

Gupta, S., Pawaria, S., Lu, C., Yeh, S.-R., & Dikshit, K. L. (2011). Novel flavohemoglobins of mycobacteria. *IUBMB Life, 63*, 337–345.

Hunte, C. (2005). Specific protein-lipid interactions in membrane proteins. *Biochemical Society Transactions, 33*, 938–942.

Jasaitis, A., Ouellet, H., Lambry, J. C., Martin, J. L., Friedman, J. M., Guertin, M., et al. (2012). Ultrafast heme-ligand recombination in truncated hemoglobin HbO from *Mycobacterium tuberculosis*: A ligand cage. *Chemical Physics, 396*, 10–16.

Joseph, S. V., Madhavilatha, G. K., Kumar, R. A., & Mundayoor, S. (2012). Comparative analysis of mycobacterial truncated hemoglobin promoters and the groEL2 promoter in free-living and intracellular mycobacteria. *Applied and Environmental Microbiology, 78*, 6499–6506.

Kendall, S. L., Movahedzadeh, F., Rison, S. C. G., Wernisch, L., Parish, T., Duncan, K., et al. (2004). The *Mycobacterium tuberculosis dosRS* two-component system is induced by multiple stresses. *Tuberculosis, 84*, 247–255.

Lama, A., Pawaria, S., Bidon-Chanal, A., Anand, A., Gelpi, J. L., Arya, S., et al. (2009). Role of Pre-A motif in nitric oxide scavenging by truncated hemoglobin, HbN, of *Mycobacterium tuberculosis*. *The Journal of Biological Chemistry, 284*, 14457–14468.

Lama, A., Pawaria, S., & Dikshit, K. L. (2006). Oxygen binding and NO scavenging properties of truncated hemoglobin, HbN, of *Mycobacterium smegmatis*. *FEBS Letters, 580*, 4031–4041.

Liu, C., He, Y., & Chang, Z. (2004). Truncated hemoglobin O of *Mycobacterium tuberculosis*: The oligomeric state change and the interaction with membrane components. *Biochemical and Biophysical Research Communications, 316*, 1163–1172.

Liu, P. T., & Modlin, R. L. (2008). Human macrophage host defense against *Mycobacterium tuberculosis*. *Current Opinion in Immunology, 20*, 371–376.

Milani, M., Pesce, A., Ouellet, Y., Ascenzi, P., Guertin, M., & Bolognesi, M. (2001). *Mycobacterium tuberculosis* hemoglobin N displays a protein tunnel suited for O_2 diffusion to the heme. *The EMBO Journal, 20*, 3902–3909.

Milani, M., Pesce, A., Ouellet, H., Guertin, M., & Bolognesi, M. (2003). Truncated hemoglobins and nitric oxide action. *IUBMB Life, 55*, 623–627.

Milani, M., Savard, P.-Y., Ouellet, H., Ascenzi, P., Guertin, M., & Bolognesi, M. (2003). A TyrCD1/TrpG8 hydrogen bond network and a TyrB10–TyrCD1 covalent link shape the heme distal site of *Mycobacterium tuberculosis* hemoglobin O. *Proceedings of the National Academy of Sciences, 100*, 5766–5771.

Mishra, S., & Meuwly, M. (2009). Nitric oxide dynamics in truncated hemoglobin: Docking sites, migration pathways, and vibrational spectroscopy from molecular dynamics simulations. *Biophysical Journal, 96*, 2105–2118.

Mishra, S., & Meuwly, M. (2010). Atomistic simulation of NO dioxygenation in Group I truncated hemoglobin. *Journal of the American Chemical Society, 132*, 2968–2982.

Mukai, M., Mills, C. E., Poole, R. K., & Yeh, S.-R. (2001). Flavohemoglobin, a globin with a peroxidase-like catalytic site. *The Journal of Biological Chemistry, 276*, 7272–7277.

Mukai, M., Ouellet, Y., Ouellet, H., Guertin, M., & Yeh, S.-R. (2004). NO binding induced conformational changes in a truncated hemoglobin from *Mycobacterium tuberculosis*. *Biochemistry, 43*, 2764–2770.

Mukai, M., Savard, P.-Y., Ouellet, H., Guertin, M., & Yeh, S.-R. (2002). Unique ligand–protein interactions in a new truncated hemoglobin from *Mycobacterium tuberculosis*. *Biochemistry, 41*, 3897–3905.

Nardini, M., Pesce, A., Labarre, M., Richard, C., Bolli, A., Ascenzi, P., et al. (2006). Structural determinants in the Group III truncated hemoglobin from *Campylobacter jejuni*. *The Journal of Biological Chemistry, 281*, 37803–37812.

Ohno, H., Zhu, G., Mohan, V. P., Chu, D., Kohno, S., Jacobs, W. R., Jr., et al. (2003). The effects of reactive nitrogen intermediates on gene expression in *Mycobacterium tuberculosis*. *Cellular Microbiology, 5*, 637–648.

Oliveira, A., Singh, S., Bidon-Chanal, A., Forti, F., Martí, M. A., Boechi, L., et al. (2012). Role of PheE15 gate in ligand entry and nitric oxide detoxification function of *Mycobacterium tuberculosis* truncated hemoglobin N. *PLoS One, 7*, e49291.

Ouellet, Y. H., Daigle, R., Lague, P., Dantsker, D., Milani, M., Bolognesi, M., et al. (2008). Ligand binding to truncated hemoglobin N from *Mycobacterium tuberculosis* is strongly modulated by the interplay between the distal heme pocket residues and internal water. *The Journal of Biological Chemistry, 283*, 27270–27278.

Ouellet, H., Juszczak, L., Dantsker, D., Samuni, U., Ouellet, Y. H., Savard, P.-Y., et al. (2003). Reactions of *Mycobacterium tuberculosis* truncated hemoglobin O with ligands reveal a novel ligand-inclusive hydrogen bond network. *Biochemistry, 42*, 5764–5774.

Ouellet, Y., Milani, M., Couture, M., Bolognesi, M., & Guertin, M. (2006). Ligand interactions in the distal heme pocket of *Mycobacterium tuberculosis* truncated hemoglobin: Roles of TyrB10 and GlnE11 residues. *Biochemistry, 45*, 8770–8781.

Ouellet, H., Milani, M., LaBarre, M., Bolognesi, M., Couture, M., & Guertin, M. (2007). The roles of Tyr(CD1) and Trp(G8) in *Mycobacterium tuberculosis* truncated hemoglobin O in ligand binding and on the heme distal site architecture. *Biochemistry, 46*, 11440–11450.

Ouellet, H., Ouellet, Y., Richard, C., Labarre, M., Wittenberg, B., Wittenberg, J., et al. (2002). Truncated hemoglobin HbN protects *Mycobacterium bovis* from nitric oxide. *Proceedings of the National Academy of Sciences of the United States of America, 99*, 5902–5907.

Ouellet, H., Ranguelova, K., LaBarre, M., Wittenberg, J. B., Wittenberg, B. A., Magliozzo, R. S., et al. (2007). Reaction of *Mycobacterium tuberculosis* truncated hemoglobin O with hydrogen peroxide: Evidence for peroxidatic activity and formation of protein-based radicals. *The Journal of Biological Chemistry, 282*, 7491–7503.

Park, M. K., Myers, R. A. M., & Marzella, L. (1992). Oxygen tensions and infections: Modulation of microbial growth, activity of antimicrobial agents, and immunologic responses. *Clinical Infectious Diseases, 14*, 720–740.

Pathania, R., Navani, N. K., Gardner, A. M., Gardner, P. R., & Dikshit, K. L. (2002). Nitric oxide scavenging and detoxification by the *Mycobacterium tuberculosis* haemoglobin, HbN in *Escherichia coli*. *Molecular Microbiology, 45*, 1303–1314.

Pathania, R., Navani, N. K., Rajamohan, G., & Dikshit, K. L. (2002). *Mycobacterium tuberculosis* hemoglobin HbO associates with membranes and stimulates cellular respiration of recombinant *Escherichia coli*. *The Journal of Biological Chemistry, 277*, 15293–15302.

Pawaria, S., Lama, A., Raje, M., & Dikshit, K. L. (2008). Responses of *Mycobacterium tuberculosis* hemoglobin promoters to in vitro and in vivo growth conditions. *Applied and Environmental Microbiology, 74*, 3512–3522.

Pawaria, S., Rajamohan, G., Gambhir, V., Lama, A., Varshney, G. C., & Dikshit, K. L. (2007). Intracellular growth and survival of *Salmonella enterica* serovar Typhimurium carrying truncated hemoglobins of *Mycobacterium tuberculosis*. *Microbial Pathogenesis, 42*, 119–128.

Pesce, A., Couture, M., Dewilde, S., Guertin, M., Yamauchi, K., Ascenzi, P., et al. (2000). A novel two-over-two α-helical sandwich fold is characteristic of the truncated hemoglobin family. *The EMBO Journal, 19*, 2424–2434.

Poole, R. K., & Hughes, M. N. (2000). New functions for the ancient globin family: Bacterial responses to nitric oxide and nitrosative stress. *Molecular Microbiology, 36*, 775–783.

Roberts, D. M., Liao, R. P., Wisedchaisri, G., Hol, W. G. J., & Sherman, D. R. (2004). Two sensor kinases contribute to the hypoxic response of *Mycobacterium tuberculosis*. *The Journal of Biological Chemistry, 279*, 23082–23087.

Rustad, T. R., Sherrid, A. M., Minch, K. J., & Sherman, D. R. (2009). Hypoxia: A window into *Mycobacterium tuberculosis* latency. *Cellular Microbiology, 11*, 1151–1159.

Savard, P.-Y., Daigle, R., Morin, S., Sebilo, A., Meindre, F., Lagüe, P., et al. (2011). Structure and dynamics of *Mycobacterium tuberculosis* truncated hemoglobin N: Insights from NMR spectroscopy and molecular dynamics simulations. *Biochemistry, 50*, 11121–11130.

Stevanin, T. M., Ioannidis, N., Mills, C. E., Kim, S. O., Hughes, M. N., & Poole, R. K. (2000). Flavohemoglobin Hmp affords inducible protection for *Escherichia coli* respiration, catalyzed by cytochromes *bo'* or *bd*, from nitric oxide. *The Journal of Biological Chemistry, 275*, 35868–35875.

Stevanin, T. M., Poole, R. K., Demoncheaux, E. A. G., & Read, R. C. (2002). Flavohemoglobin Hmp protects *Salmonella enterica* serovar typhimurium from nitric oxide-related killing by human macrophages. *Infection and Immunity, 70*, 4399–4405.

Thorsteinsson, M. V., Bevan, D. R., Potts, M., Dou, Y., Eich, R. F., Hargrove, M. S., et al. (1999). A cyanobacterial hemoglobin with unusual ligand binding kinetics and stability properties. *Biochemistry, 38*, 2117–2126.

Visca, P., Fabozzi, G., Petrucca, A., Ciaccio, C., Coletta, M., De Sanctis, G., et al. (2002). The truncated hemoglobin from *Mycobacterium leprae*. *Biochemical and Biophysical Research Communications, 294*, 1064–1070.

Voskuil, M. I., Bartek, I., Visconti, K., & Schoolnik, G. K. (2011). The response of *Mycobacterium tuberculosis* to reactive oxygen and nitrogen species. *Frontiers in Microbiology, 2*, 1–12.

Voskuil, M. I., Visconti, K. C., & Schoolnik, G. K. (2004). *Mycobacterium tuberculosis* gene expression during adaptation to stationary phase and low-oxygen dormancy. *Tuberculosis, 84*, 218–227.

Vuletich, D., & Lecomte, J. J. (2006). A phylogenetic and structural analysis of truncated hemoglobins. *Journal of Molecular Evolution, 62*, 196–210.

Wainwright, L. M., Elvers, K. T., Park, S. F., & Poole, R. K. (2005). A truncated haemoglobin implicated in oxygen metabolism by the microaerophilic food-borne pathogen *Campylobacter jejuni*. *Microbiology, 151*, 4079–4091.

Wakabayashi, S., Matsubara, H., & Webster, D. A. (1986). Primary sequence of a dimeric bacterial haemoglobin from Vitreoscilla. *Nature*, *322*, 481–483.

Weber, R. E., & Vinogradov, S. N. (2001). Nonvertebrate hemoglobins: Functions and molecular adaptations. *Physiological Reviews*, *81*, 569–628.

Wittenberg, J. B., Bolognesi, M., Wittenberg, B. A., & Guertin, M. (2002). Truncated hemoglobins: A new family of hemoglobins widely distributed in bacteria, unicellular eukaryotes, and plants. *The Journal of Biological Chemistry*, *277*, 871–874.

Wittenberg, J. B., & Wittenberg, B. A. (1990). Mechanisms of cytoplasmic hemoglobin and myoglobin function. *Annual Review of Biophysics and Biophysical Chemistry*, *19*, 217–241.

Yeh, S.-R., Couture, M., Ouellet, Y., Guertin, M., & Rousseau, D. L. (2000). A cooperative oxygen binding hemoglobin from *Mycobacterium tuberculosis*. Stabilization of heme ligands by a distal tyrosine residue. *The Journal of Biological Chemistry*, *275*, 1679–1684.

CHAPTER SIX

The Globins of Cyanobacteria and Algae

Eric A. Johnson, Juliette T.J. Lecomte[1]

T.C. Jenkins Department of Biophysics, Johns Hopkins University, Baltimore, Maryland, USA
[1]Corresponding author: e-mail address: lecomte_jtj@jhu.edu

Contents

1.	Overview	196
2.	Historical Perspective	199
3.	Phylogeny	204
	3.1 Tying photosynthesis to cyanobacterial and algal globins	204
	3.2 Evolution and phylogeny	206
4.	Physiological Characterization	217
	4.1 *Nostoc commune*	217
	4.2 *Nostoc punctiforme*	219
	4.3 *Synechococcus* sp. strain PCC 7002	220
	4.4 *Chlamydomonas eugametos*	222
	4.5 *Chlamydomonas reinhardtii*	223
5.	*In Vitro* Characterization	226
	5.1 Structural properties	226
	5.2 Ligand-binding properties	244
	5.3 Auto-oxidation, redox potential and electron transfer	249
	5.4 Chemical reactivity and possible function	251
6.	Needs and Opportunities	254
	6.1 Physiological studies	255
	6.2 Biophysical studies	260
	6.3 Concluding remarks	262
	References	262

Abstract

Approximately, 20 years ago, a haemoglobin gene was identified within the genome of the cyanobacterium *Nostoc commune*. Haemoglobins have now been confirmed in multiple species of photosynthetic microbes beyond *N. commune*, and the diversity of these proteins has recently come under increased scrutiny. This chapter summarizes the state of knowledge concerning the phylogeny, physiology and chemistry of globins in cyanobacteria and green algae. Sequence information is by far the best developed and the most rapidly expanding aspect of the field. Structural and ligand-binding properties have been described for just a few proteins. Physiological data are available for

Advances in Microbial Physiology, Volume 63
ISSN 0065-2911
http://dx.doi.org/10.1016/B978-0-12-407693-8.00006-6

even fewer. Although activities such as nitric oxide dioxygenation and oxygen scavenging are strong candidates for cellular function, dedicated studies will be required to complete the story on this intriguing and ancient group of proteins.

ABBREVIATIONS

5c five-coordinate, pentacoordinate
6803 sp. strain PCC 6803
6c six-coordinate, hexacoordinate
7002 sp. strain PCC 7002
CM-H$_2$DCFDA 5- (and 6-) chloromethyl-2′,7′-dichlorodihydrofluorescein diacetate, acetyl ester
CT charge transfer
CtrHb haem domain of *Chlamydomonas eugametos* LI637 haemoglobin
DCMU (3,4-dichlorophenyl)-1,1-dimethylurea
EPR electron paramagnetic resonance
FHb flavohaemoglobin
GCS globin-coupled sensor
Hb haemoglobin
HIF hormogonia-inducing factor
IFT intraflagellar transport
LI light induced
M myoglobin-like
Mb myoglobin
NMR nuclear magnetic resonance
NOD nitric oxide dioxygenase
nsHb non-symbiotic haemoglobin
PDB protein data bank
Pgb protoglobin
PTM post-translational modification
RNS reactive nitrogen species
ROS reactive oxygen species
RR resonance Raman
S sensor
SDgb single-domain globin
sHb symbiotic haemoglobin
SSDgb sensor single-domain globin related to the GCS globin domain
T truncated
UV–vis ultraviolet–visible

1. OVERVIEW

The presence of globins in oxygen-evolving photosynthetic unicellular organisms may come as a surprise to the unprepared reader. Why would globins be useful in such cells, not only permeable to dioxygen but generating it under the action of light? The question derives from the expectation that a globin's sole purpose is the transport or storage of O_2 for aerobic respiration. We now know this view of the superfamily to be incomplete. All globins, even vertebrate haemoglobins and myoglobins, appear to have multiple functions in addition to the reversible binding of dioxygen. In the microbial world, these auxiliary roles as well as others conveyed by the reactive haem cofactor dominate the functional properties (Vinogradov & Moens, 2008). This chapter summarizes the state of knowledge concerning the globins of cyanobacteria and green algae and outlines the many opportunities that they offer for a deeper understanding of haem proteins and the evolution of the haemoglobin superfamily.

Cyanobacterial and algal globin research is only a couple of decades old. In contrast to vertebrate haemoglobins and myoglobins, which are immediately noticeable by their colour and obtainable in high quantities by direct extraction from the organism, cyanobacterial and algal globins occur at catalytic levels in cells replete with photosynthetic pigments. It is through their genes that they first become apparent. Thus, their discovery had to wait for the development of modern molecular biology tools and efficient genome sequencing techniques. With a few detailed biochemical studies of Cyanobacteria and Chlorophyta exemplars and a large number of sequences, we have nevertheless caught a glimpse of the remarkable properties exhibited by this group of proteins.

It is necessary at this stage to specify what we call 'haemoglobin'. The term is generally reserved for protein sequences that belong to the haemoglobin superfamily and contain the 'proximal histidine'. The proximal histidine serves as a haem axial ligand (Fig. 6.1) and presumably indicates that the protein's functional state contains that particular cofactor. Sequences related to algal globins but lacking the proximal histidine exist, though little is known about these proteins, and their competency as haem binders is unclear. They are only mentioned briefly in this chapter. We do not review haem proteins related to *Escherichia coli* Dos, *Acetobacter xylinum* phosphodiesterase and *Bradyrhizobium japonicum* FixL, all of which have a proximal histidine and are capable of reversible dioxygen binding for sensing purposes (Tomita, Gonzalez, Chang, Ikeda-Saito, & Gilles-Gonzalez, 2002) but do not belong to the haemoglobin superfamily. We also exclude the relatives of the stress signalling protein RsbR, which does not bind haem (Murray, Delumeau, & Lewis, 2005) and is a globin in fold only.

In Section 2, we present a historical overview of the field. Progress in the study of globins from unicellular photosynthetic organisms is entwined with broad advances in the structural biology and phylogeny of the haemoglobin superfamily. After a review of these advances, Sections 3

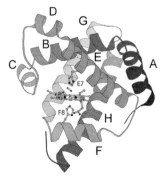

Figure 6.1 Ribbon diagram of sperm whale myoglobin in the oxy state (PDB ID 1MBO, Phillips, 1980). The haem, proximal histidine (F8), distal histidine (E7), and dioxygen molecule are shown with sticks. Helices are labelled from A to H and form the classical 3/3 orthogonal sandwich. This figure and Figs. 6.2, 6.9, and 6.12 were prepared with Molscript (Kraulis, 1991). (For colour version of this figure, the reader is referred to the online version of this chapter.)

and 4 focus on phylogenetic and physiological studies of the cyanobacterial globins (Cyanobacteria phylum) and the green algae (a class of the Chlorophyta division). The majority of globins detected in cyanobacteria and green algae belong to the 'truncated' lineage of the haemoglobin superfamily. This has provided insightful sequence analyses revealing evolutionary relationships between cyanobacterial and Archaeplastida globins (Vinogradov, Fernandez, Hoogewijs, & Arredondo-Peter, 2011). Through modern biochemical investigations, however, the accumulation of sequence information has considerably outpaced the laboratory work of physiological and molecular characterization. To date, these studies remain comparatively few.

In vitro data are available for a few Cyanobacteria and Chlorophyta globins. We discuss the features of purified proteins in Section 5, addressing first their structural properties. Numerous experimental techniques can provide fragmentary descriptions of medium size haem proteins. Optical absorbance spectroscopy is one of the methods that respond to the state of the iron. Practically all globin studies begin with the collection of optical data, which we compiled for the most important protein forms. Resonance Raman (RR), magnetic circular dichroism and electron paramagnetic resonance (EPR) are effective methods for ligand identification and spin-state determination. For full description at atomic resolution, traditional X-ray crystallography offers frozen views of proteins. As an alternative for complete structure determination, nuclear magnetic resonance (NMR) spectroscopy has been applied to several globins in solution. NMR spectroscopy, however, is most powerful in the characterization of local features, in particular the haem group and its direct environment (La Mar, Satterlee, & de Ropp, 2000). It is also well suited for monitoring the perturbations caused by a change in conditions, such as pH or temperature, and for describing the dynamic properties of the structure on a wide range of timescales. Cyanobacteria and Chlorophyta globins have benefitted from most of these experimental approaches.

For any globin, the kinetics of ligand binding is of paramount functional importance. Naturally, the focus is on oxygen, but other small ligands must be considered as well, either because they are physiological or because they serve as alternative chemical probes and provide valuable physical insight. Ligand-binding data are available for a subset of wild-type and variant proteins and are next inspected in light of the structural data. In the same section, we organize the limited and scattered information published on the reactivity of the haem group, for example, reduction potential and auto-oxidation properties. These parameters can be compared with those of other globins to distinguish trends and guide functional hypotheses. The data, though limited, suggest many possible directions for deepening the characterization of the globins from Cyanobacteria and Chlorophyta. In the final section, we emphasize the most pressing questions and offer comments on the future of this field of research.

2. HISTORICAL PERSPECTIVE

The year 1992 marked a turning point in the study of the haemoglobin superfamily. Potts, Angeloni, Ebel, and Bassam (1992), who were then investigating nitrogen metabolism in the cyanobacterium *Nostoc commune*, discovered that a globin gene (*glbN*) was present in the *nif*UHD gene cluster of the nitrogen fixation operon. The amino acid sequence of 'cyanoglobin', as it became known, was found to be related to that of protozoan globins (Iwaasa, Takagi, & Shikama, 1989, 1990). Protozoan globins had been detected much earlier by optical measurements on cell suspensions of *Paramecium caudatum* (Sato & Tamiya, 1937) and *Tetrahymena pyriformis* (Keilin & Ryley, 1953). By 1992, a few sequences of protozoan globins had been determined (Iwaasa et al., 1989, 1990), and it was recognized that they were shorter than the vertebrate proteins and distantly related to them.

The relationship of *N. commune* GlbN, closer to the protozoan globins than vertebrate globins or a bacterial globin such as that from *Vitreoscilla* (Wakabayashi, Matsubara, & Webster, 1986), was immediately intriguing. The seminal contribution attracted attention to the presence of myoglobin-like proteins in organisms that were not, up to then, known to use this particular kind of haem protein.

Two years later, Guertin and co-workers published the observation of light-induced (LI) globins in the unicellular green alga *Chlamydomonas eugametos* (Couture, Chamberland, St-Pierre, Lafontaine, & Guertin, 1994). Like the *N. commune* and protozoan globins, the haem domains of the *C. eugametos* Hbs were shorter than their vertebrate counterparts. For this reason, these odd proteins were collectively called 'truncated' globins. Several years later, the same group (Couture et al., 2000) and ours (Scott & Lecomte, 2000) reported the preliminary characterization of the globin from the cyanobacterium *Synechocystis* sp. PCC 6803. In contrast to *N. commune glbN*, the position of *Synechocystis* sp. PCC 6803 *glbN* within the genome (Kaneko et al., 1996) held no functional hint. Soon thereafter, a *glbN* gene was also found in the cyanobacterium *Synechococcus* sp. PCC 7002, and the gene product was prepared by recombinant means (Scott et al., 2002). Again, gene location did not suggest a functional role. The two strains of *Synechocystis* sp. and *Synechococcus* sp. added new members to the growing truncated globin family, but little insight into the possible roles of the globins in cyanobacteria.

Truncated globins are present not only in cyanobacteria and green algae but also in many bacteria (Vinogradov, Tinajero-Trejo, Poole, & Hoogewijs, 2013), fungi (Hoogewijs, Dewilde, Vierstraete, Moens, & Vinogradov, 2012) and plants (Vinogradov, Fernandez, et al., 2011). As early as 2002, it became possible to initiate a phylogenetic analysis of the family. Thus, Wittenberg, Bolognesi, Wittenberg, & Guertin, 2002 determined that truncated haemoglobins (TrHbs) could be separated into three distinct groups, referred to as N (Group I), O (Group II) and P (Group III). Later phylogenetic analyses (Vinogradov et al., 2007; Vuletich & Lecomte, 2006) based on an increased number of sequences confirmed this interpretation of the genomic record. To this date, all the truncated globins of cyanobacteria belong to Group I. The majority of truncated globins in green algae are also Group I proteins; however, a few Group II globins are found in these organisms (Group II proteins dominate in diatoms and higher plants, Vinogradov, Fernandez, et al., 2011).

A second turning point was met in 2000 when Bolognesi and co-workers published the structures of *P. caudatum* haemoglobin and the haem domain of *C. eugametos* LI637 haemoglobin, which is depicted in Fig. 6.2 (Pesce et al., 2000). These first two structures of TrHbs anchored subsequent studies of the entire family. With a three-dimensional framework, progress could be made in the interpretation of the ligand-binding data being collected in several laboratories on *N. commune*, *C. eugametos*, and *Synechocystis* sp. PCC 6803 globins. These systematic studies combined with the structural information allowed for the design of targeted, hypothesis-driven experiments aimed at delineating the functional roles of the globins and exploring the physical properties conveyed by the new fold.

In parallel with the characterization of globins in unicellular photosynthetic organisms, rapid advances were taking place in the field of haemoglobin at large. New sequences were deposited in databases on what seemed to be a daily basis, and previously unsuspected members of the superfamily emerged in all kingdoms of life. Globin E (Kugelstadt, Haberkamp, Hankeln, & Burmester, 2004), globin X (Fuchs, Burmester, & Hankeln, 2006), globin Y (Fuchs et al., 2006), cytoglobin (Burmester, Ebner,

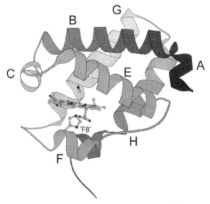

Figure 6.2 Ribbon diagram of ferric *C. eugametos* LI637 haemoglobin (haem domain only, CtrHb) in the cyanomet state (PDB ID 1DLY, Pesce et al., 2000). The haem, proximal histidine ('F8'), and cyanide ligand are shown with sticks. Helices are labelled from A to H. Compared to Fig. 6.1, the A helix is shorter, the D helix is missing, the EF loop is extended, and the F helix is a reduced to one turn. The topology is a 2/2 orthogonal sandwich. (For colour version of this figure, the reader is referred to the online version of this chapter.)

Weich, & Hankeln, 2002; Trent & Hargrove, 2002) and neuroglobin (Burmester, Weich, Reinhardt, & Hankeln, 2000) made their appearance in the vertebrate set; multiple sequences and structures for Group II and Group III truncated globins became available (as reviewed in Nardini, Pesce, Milani, & Bolognesi, 2007); and globin-coupled sensors (GCSs) and protoglobins (Pgbs) were discovered (as reviewed in Freitas, Saito, Hou, & Alam, 2005). All these developments provided context and material for a broader understanding of the haemoglobin superfamily. The vast increase in the number of sequenced genomes allowed Vinogradov and colleagues to develop an exhaustive model of haemoglobin evolution (Vinogradov et al., 2005) and iterate on this model ever since (Hoogewijs et al., 2012; Vinogradov & Moens, 2008; Vinogradov, Fernandez, et al., 2011; Vinogradov et al., 2007, 2013). They concluded that the haemoglobin superfamily is composed of three lineages. The best known includes the flavohaemoglobins (FHbs), bacterial proteins that contain an N-terminal haem-binding domain and a flavin-binding domain, and the single-domain globins (SDgbs), which are related to the haem domain of the FHbs. Myoglobin belongs to this lineage. A second lineage comprises the GCS proteins. These are also chimeric and contain an N-terminal canonical globin domain. The single-domain Pgbs (Hou et al., 2001) and sensor single-domain globins related to the GCS globin domain (SSDgbs) fit in this family. Finally, the truncated lineage is divided into three groups as mentioned earlier.

In light of sequence alignments and the structural information, the name two–over–two (or 2/2) globin surfaced as an alternative name for the truncated globins. This is an apt reference to the helical topology and the fact that truncated globin domains differ considerably in length and the position of insertions and deletions. The full-size globins exhibit a three-over-three (or 3/3) fold and are often referred to as 3/3 globins for contrast with the 2/2 proteins. As pointed out in a recent publication (Vinogradov et al., 2013), the field of haemoglobin research suffers from haphazard and evolving nomenclature. In this chapter, we follow the new recommendations proposed in that work. The three lineages are therefore referred to as the M family (i.e. myoglobin-like, including FHb and SDgb), the S family (i.e. sensor, including GCS, Pgb, SSDgb), and the T family (i.e. truncated, classified as TrHb1, TrHb2 or TrHb3 for Group I, II or III, respectively). TrHb1s, SDgbs and SSDgbs are found in Cyanobacteria. TrHb1s outnumber proteins from the M and S lineages, but as we will see, the presence of proteins from these two 3/3 lineages raises interesting phylogeny questions. In green algae, SDgbs (M lineage) are found but no protein from the S lineage. These eukaryotes contain TrHb1s, which outnumber TrHb2s, unlike in higher plants where rare TrHb1s are present (Vazquez-Limon, Hoogewijs, Vinogradov, & Arredondo-Peter, 2012). The distribution of globins, discussed in greater detail in Section 3.2, continues to provide models of evolution and the transition between anaerobic and aerobic life.

Meanwhile, a novel structural aspect of the Hb superfamily was slowly emerging. In the canonical vertebrate myoglobin, the proximal histidine is the only proteinaceous ligand (Fig. 6.1). The deoxy state, that is, the ferrous protein without added exogenous ligand, has a 'distal' coordination site either emptied or occupied by a loosely bound water molecule readily displaced by dioxygen. Coordination of a residue on the distal side interferes with dioxygen binding and is a non-functional, occasionally pathological, situation. In many of the recently identified haemoglobins, however, endogenous hexacoordination, that is, the ligation of the iron by two protein residues, is observed. Examples of these so-called hexacoordinate globins include cytoglobin (Trent & Hargrove, 2002), neuroglobin (Fuchs et al., 2004), *Drosophila melanogaster* intracellular haemoglobin (Hankeln et al., 2002), *Oryza sativa* non-symbiotic haemoglobin (nsHb; Arredondo-Peter et al., 1997), *Hordeum* sp. (Duff, Wittenberg, & Hill, 1997) and *Solanum lycopersicon* nsHb (Ioanitescu et al., 2005), to name just a few.

Synechocystis sp. PCC 6803 GlbN (hereafter *Synechocystis* 6803 GlbN; Scott & Lecomte, 2000), *Synechococcus* sp. PCC 7002 GlbN (hereafter *Synechococcus* 7002 GlbN; Scott et al., 2002) and *C. eugametos* LI637 Hb (Couture & Guertin, 1996) are also hexacoordinate globins. *Synechocystis* 6803 GlbN and *Synechococcus* 7002 GlbN use two histidines as axial ligands, as in cytochrome b_5, whereas *C. eugametos* LI637 Hb uses a tyrosine as the distal ligand. Thus, on the basis of these three instances, the classical myoglobin coordination scheme does not seem to hold well in the world of cyanobacterial and algal proteins. The presence of a labile protein ligand on the distal side has several important consequences. The work of Hargrove and colleagues, among others, in this specialized area of globin research, is shedding light on the control of iron coordination by the globin fold (Kakar, Hoffman, Storz, Fabian, & Hargrove, 2010).

Haem post-translational modification (PTM) is non-existent in the globin superfamily, except in *Synechocystis* 6803 and *Synechococcus* 7002 GlbNs, where irreversible attachment to the protein matrix via a histidine linkage occurs under certain conditions (Vu, Jones, & Lecomte, 2002; Vu, Vuletich, Kuriakose, Falzone, & Lecomte, 2004). This feature, discovered in 2002, is likely to represent a response to specific metabolic needs. Endogenous hexacoordination and covalent attachment illustrate original ways in which haem reactivity can be co-opted to condition globin function. Since it was first observed, the unusual chemistry supported by T globins has been the subject of several studies and has fostered new functional hypotheses.

Few laboratories are actively engaged in the study of algal and cyanobacterial globins. Although information becomes occasionally available as a by-product of other investigations such as global gene expression studies, much dedicated work will be necessary to mine the rich chemical and functional landscape these proteins have to offer.

3. PHYLOGENY

3.1. Tying photosynthesis to cyanobacterial and algal globins

The cyanobacteria and algae considered here are first and foremost photosynthetic organisms. During photosynthesis, they are actively extracting electrons from water and generating molecular oxygen within cellular compartments undergoing rapid metabolism. Given the probable role of certain cyanobacterial globins in mitigation of reactive oxygen and nitrogen molecules (reactive oxygen species (ROS)/reactive nitrogen species (RNS); Hill et al., 1996; Scott et al., 2010; Smagghe, Trent, & Hargrove, 2008), recognition of the photosynthesis occurring in these cells seems essential for a finer definition of relevant activities. Within this context, it is also important to stress that although cyanobacteria and algae are both photosynthetic, they possess distinctive physical characteristics and do not utilize identical machineries.

Cyanobacteria are photosynthetic bacteria commonly referred to as blue-green algae. Although the term 'algae' can be, and is, applied to a wide variety of aquatic organisms, in this discussion we will use the term cyanobacteria to highlight the prokaryotic lineage of these organisms. Likewise, we will define 'algae' as selectively referring to the green algae (as discussed below). As prokaryotes, cyanobacteria lack specialized compartments such as the chloroplasts of eukaryotes, but do have dedicated membranes for photosynthesis called thylakoids (Beck, Knoop, Axmann, & Steuer, 2012; Vothknecht & Westhoff, 2001). Cyanobacteria are found in virtually every terrestrial and aquatic ecosystem on the planet and are extraordinarily prolific; by some estimates they account for over 20% of the Earth's annual oxygen production (Pisciotta, Zou, & Baskakov, 2010). The 'blue-green' moniker for cyanobacteria comes from the pigments found in specialized structures called phycobilisomes that are mostly used to harvest sunlight and funnel the resulting energy into chlorophyll-based reaction centres for conversion into chemical energy (Bryant & Frigaard, 2006; Neilson & Durnford, 2010). Cyanobacteria can also fix atmospheric nitrogen (Herrero, Muro-Pastor, & Flores, 2001), though this is not a trait common to all such organisms. Cyanobacteria are hypothetically linked to eukaryotic algae and land plants through an ancient endosymbiotic pairing of a non-photosynthetic eukaryote with a cyanobacterium that then gave rise to the plastids (including the chloroplast; Neilson & Durnford, 2010; Reyes-Prieto, Weber, & Bhattacharya, 2007).

In this chapter the term 'algae' will refer solely to green algae, because it is so far the only type of alga for which any physiological information exists concerning globins. The green algae are part of the supergroup Archaeplastida, composed of land plants, red algae, green algae and a class of organisms known as the glaucophytes (Yoon, Hackett, Ciniglia, Pinto, & Bhattacharya, 2004). The Archaeplastidae include all the species we classically consider plants (sometimes, the supergroup is referred to a *Plantae sensu lato* or 'plants in the broad sense'). The green algae are a class of unicellular eukaryotes belonging to the phylum Chlorophyta, and are the most closely related algae to land plants (Keeling, 2004; Rodriguez-Ezpeleta et al., 2005). Green algae depend solely on chlorophyll for light capture. Instead of the phycobilisomes of cyanobacteria, green algae have large chlorophyll-filled antenna complexes (light-harvesting complexes) attached to their photosystems. Green algae have dedicated chloroplasts with their photosynthetic thylakoid membranes structured into highly organized stacks called granum (Vothknecht & Westhoff, 2001).

As mentioned earlier, green algae acquired their chloroplasts through an ancient endosymbiotic event. As such these cells have undergone several rounds of horizontal gene transfer, where genetic information has been incorporated into the nuclear genome during the evolution of both mitochondria and the chloroplasts (Richards & van der Giezen, 2006; Rodriguez-Ezpeleta et al., 2005). In addition, there is evidence for more than one endosymbiotic event involving the evolution of the chloroplast, or at least multiple horizontal gene transfer events during its evolution (Vinogradov, Fernandez, et al., 2011). This makes the heritage of algal globins complicated by an intertwining polyphyletic history. We note that this heritage is surpassed in complexity by other aquatic photosynthetic microbes such as cryptophytes and chlorarachniophytes. These organisms are endosymbionts arising from the assimilation of photosynthetic eukaryotes by non-photosynthetic eukaryotes. Their phylogeny is particularly difficult to unravel (Curtis et al., 2012) and will not be considered here.

As is apparent from this section, oxygenic photosynthetic organisms are a diverse lot. What is common in all of these species is the fact that highly specialized structures are actively extracting electrons from water and creating molecular oxygen—all in close proximity to a host of competing metabolic pathways within the cell. Given that globins have been shown (at least in some cases) to help mitigate the production of—or damage from—reactive molecules within a cell (Hill et al., 1996; Scott et al., 2010) and that some are chloroplastic and induced by light (Couture et al., 1994; Couture & Guertin, 1996), it is quite possible that globins have important and unique roles within the regulation and efficiency of photosynthesis. Involvement in other metabolic processes, such as assimilation of nitrogen from nitrogen oxide sources, is also emerging and indicates related functions.

3.2. Evolution and phylogeny

Since Vinogradov and co-workers advanced their classification of globins into three distinct families (Vinogradov et al., 2005), hundreds of additional globin genes have been sequenced and their protein products categorized accordingly (Vinogradov et al., 2013). As mentioned earlier, the FHbs together with the single-domain globins (SDgbs) form the M (myoglobin-like) family. The S family is comprised of proteins known as GCSs but has recently grown to include Pgb and SSDgb. The third family is the T family, made up of proteins known as the truncated globins. Whereas both the M and S families of globins contain a canonical 3/3 protein structure, the truncated globins possess the smaller 2/2 structure (Wittenberg et al., 2002). Globins as an evolutionary group are quite ancient, having emerged at the very beginning of life on this planet, more than 3 billion years ago (Vazquez-Limon et al., 2012). This places their appearance well before the development of multicellular organisms and even unicellular eukaryotes. Because of the linkage of all the three families of globins, it is thought that both the canonical 3/3 fold and the truncated 2/2 fold bear a single ancient common ancestor (Vinogradov et al., 2007) although these two types of domains separated from each other very early in evolution (Vinogradov et al., 2005). The fact that globins evolved long before the advent of aerobic respiration—before the oxygenation of our atmosphere—indicates far more diverse roles for these proteins than the oxygen transport or storage that is typically associated with myoglobin and haemoglobin (Vinogradov & Moens, 2008). Although stable and reversible oxygen binding is likely to be at the root of many of these functions (Shikama & Matsuoka, 2004), it is also possible that activities unrelated to dioxygen stabilization were, or still are, represented in the superfamily.

The currently available genomic data for cyanobacteria and algae have identified globins that show strong homology to both the M family with its canonical 3/3 fold and the T family identified by its unique 2/2 topology (Vinogradov, Fernandez, et al., 2011). Until recently, cyanobacteria and algae were not known to contain genes for globins from the S family. This has now changed with the discovery of SSDgb in three separate cyanobacteria (Vinogradov et al., 2013). However, not all cyanobacteria have globins; in fact only about half of the cyanobacterial genomes that have been sequenced to this date contain one or more globin genes (Vinogradov et al., 2013). In this section, we will discuss the common phylogeny of proteins predicted to be globins or to contain globin domains.

A succinct list of globins from cyanobacteria and algae compiled from the UniProt Knowledgebase suffices to assess the state of the field. In cyanobacteria, 52 species with 75 predicted proteins possessing either canonical or 2/2 predicted structures are listed. The green algae of Chlorophyta are represented by eight species within the databases accounting for a total of 33 known or predicted globins. We have no doubt that by the time of publication of this chapter these numbers will be outdated owing to the ever-increasing volume of genomic databases. With this reservation, Fig. 6.3 shows a simple phylogenetic tree constructed from the simultaneous alignments of these ~100 proteins.

3.2.1 The M lineage

Within land plants, the canonical Hbs are often associated with nitrogen fixing metabolism (Vazquez-Limon et al., 2012). Nitrogen fixation in land plants takes place through a symbiotic relationship between the plants and certain bacteria. Because legumes were among the first plants recognized to have this symbiotic relationship, these canonical haemoglobins are referred to as leghaemoglobin (Virtanen & Laine, 1946) or as symbiotic canonical haemoglobins (sHb; Vinogradov, Fernandez, et al., 2011). They are classified as oxygen transport proteins. Land plants also harbour nsHbs

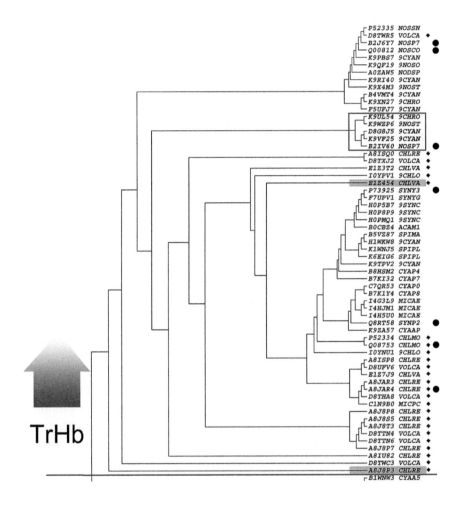

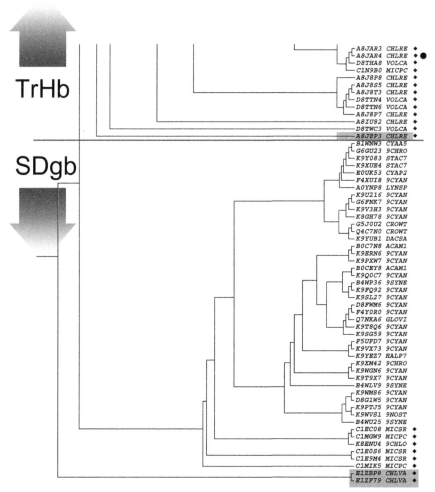

Figure 6.3 Phylogenetic tree of available predicted and confirmed globins from cyanobacteria and algae. Compiled from InterPro database (www.ebi.ac.uk/interpro) from all proteins listed within the Globin (IPR000971) and Globin, truncated bacteria-like (IPR001486) families and categorized as either a cyanobacterium or a chlorophyte. Resulting sequences were aligned using ClustalW (www.genome.jp/tools/clustalw) and rendered with the same program. Proteins are listed by UniProt identification code. Chlorophyte proteins are denoted by a black diamond; black circles highlight those proteins for which some physiologic information is known. Shaded grey boxes denote TrHb2 type proteins. Outlined box shows group of TrHb1-2 proteins.

(non–oxygen transport proteins), the phylogeny of which has been extensively studied (Garrocho-Villegas, Gopalasubramaniam, & Arredondo-Peter, 2007; Hoy, Robinson, et al., 2007; Vinogradov, Fernandez, et al., 2011). sHbs are derived from one of two classes of nsHbs. Both the sHb and nsHb are part of the M family of haemoglobins (Vazquez-Limon et al., 2012).

Based upon previous analysis, the M family globins found in cyano-bacteria and algae are typically represented by the SDgb class of globins, and these globins do not seem to either exclude or imply the presence of globins from the T lineage (Vinogradov et al., 2013). Very few M globins have been identified in cyanobacteria and green algae. With one exception among Chlorophyta, these proteins are all SDgbs rather than the chimeric FHbs (Vinogradov, Fernandez, et al., 2011).

In the green algae, the T globins are more prevalent than the M (and S) globins (Vinogradov, Fernandez, et al., 2011). The M globins align well with each other and relative to the cyanobacterial proteins as seen in Fig. 6.4. They also display strong similarity to the nsHbs found in land plants (not shown). This suggests a common heritage for these globins and gives strength to the theory that these 3/3 globins arose within Archaeplastida via a single horizontal gene transfer event (Vinogradov, Fernandez, et al., 2011). This is mentioned because it is at odds with the results seen when algae TrHbs are fit into the same tree, as discussed below. In connection with the alignment of the M globins shown in Fig. 6.4, it is interesting to note that a structural homologue of the cyanobacterial 3/3 globins is Hell's Gate glo-bin I, a haemoglobin from *Methylacidiphilum infernorum* (Verrucomicrobia group). Hell's Gate globin I has been related to vertebrate neuroglobin (Teh et al., 2011) and illustrates a key aspect of the evolution of M globins (Vinogradov, Hoogewijs, & Arredondo-Peter, 2011).

The phylogenetic tree drawn in Fig. 6.3 is clearly split into two different branches. Based upon published sequence analysis of some of the proteins found in the lower branch of the tree, these proteins should all contain a canonical 3/3 fold and belong to the M family of globins (none of the recently described SSDgbs were used to generate the figure). A small set of six proteins from chlorophytes group together at the bottom of the tree, showing the distinct evolutionary divergence of the eukaryotes from the cyanobacteria. However, these are still within the grouping for the

```
                     AAAAAAAAAAAAAAAA-BBBBBBBBBBBBBBBBBB-CCCCCC--E-EEEEEEEEEEEEEEEEEE------F
                       ::  *: :        ** *:  *   **      **   * *     *     :
B4WP36_9SYNE/1-135   MALDVELLEQSFELVKPKADDFVASFYNNLFTDYPDAKPLFEHTNMAAQQQMLKGALVMVVDNLRRPEVL
K9Q0C7_9CYAN/9-143   TGLQVELLESSFEQVKPKANEFASSFYENLFTDYPAAKPLFENTDVKEQSKKLLASLVFVVENLKNPEAL
K9VX73_9CYAN/1-135   MSLNVELLEQSFEKIKPHADEFAASFYENLFQLYPEVQPLFANTEMAKQQKKLLNALVLVVENLRSPEAL
F4Y0R0_9CYAN/1-135   MSLQVELLEQSFDKVKPRATEFVASFYENLFTDYPAAQPLFDTTDMVAQKKKLLASLVLVVENLRKPDTL
K9SL27_9CYAN/1-135   MSLPVEILQESFNKVKPYAGELGDRFYENLFTMYPEAKPLFAHTAMGKQKQMLVGSLVMTIDNLTNPEVL
K9SG59_9CYAN/1-135   MSLKVKLLEDSFDRIKPKARAFSASFYHNLFEMYPVAKPLFANTDMIAQREKLIKSLVLVTSNLRSPDVL
D8FWM6_9CYAN/1-135   MSLNIELLEESFNRIKPNAPEFATSFYDNLFADYPHVKPLFANANMAEQKKKLIASLVLVIQNLRKPDAL
K9XM42_9CHRO/1-135   MALQVEVLEQSFERVKPYANEFAASFYNNLLTDYPQLQPLFAKTDMDQQHQKLIMSLILVVSNLRNPELL
K9T8Q6_9CYAN/1-135   MSLKIDLLEQSFESVKQREAEFTTQFYANLFSDYPIVKPLFANTHMEEQGKKLFASLVLVVDALRKPEVL
Q7NKA6_GLOVI/8-142   MALQVKLLEQSFEGVKPNAHAFAASFYDNLFSDFPQTQALFAHSDMQAQQQKLLASLVLVVENLRQPQVL
K9T9X7_9CYAN/1-135   MPLNADVLEKSFNLVEPRANEFAASFYETLFTDSPEAKPLFANTDMEKQQQKLIMSLVYVVTNLRYPEEL
B0C7N8_ACAM1/1-135   MALQTELLTNSFDLLRENEAEFTQVFYGTLFTDYPQVKPLFSNTHMDEQAKKLFASLLLVVNNLTKPDAL
K9ERN6_9CYAN/22-156  MALNTNLLKTSFTLLKEDQSAFSDLFYSTLFSDYPQVKPLFAHTNMDEQPKKLFASLVLVVENLVKPDVL
K9PXW7_9CYAN/1-135   MSLNTELLETSFALLRDHKTEFTQHFYHHLFADYPQVKPLFKETQMDKQAAKLFASLVLVVDNLKKTDTL
K9WGN6_9CYAN/2-134   PLSDIKVLEVSFALIQPQATEFASKFYKNLFTDYPQLQPLFAYTHIEVQEKKLITALVLVINNLRKLTYL
B4WLV9_9SYNE/1-133   --MDVALLEKSFEQISPRAIEFSASFYQNLFHHHPELKPLFAETSQTIQEKKLIFSLAAIIENLRNPDIL
D8G1W5_9CYAN/380-513 VPSQVELVQSSFEKVKPIADKAAELFYQRLFELSPSLRPLFKG-EMKEQERKLMATLALAVEGLRRPDRI
B4WU25_9SYNE/369-503 AVRQVELVQRSFAKVEPISEQVGKLFYEHLFETRPDFKPLFSSTDMETQQRKLMMTLATAVEGLRHPEEI
B3DUZ7_METI4/3-133   DQKELIKESWKRIEPNKNEIGLLFYANLFKEEPTVSVLFQNP-ISSQSRKLMQVLGILVQGIDNLEGL
                     1.......10........20........30.......40.......50........60.......70


                     FFFFFFFFFFFFFFFF----GGGGGGGGGGGGGGGGGGGG---HHHHHHHHHHHHHHHHHHHHHHHH----
                       :  :*  *    *       *   :*  ::  ::      :.    **   :    : *
B4WP36_9SYNE/1-135   SKSLKGLGARHIKYGALPEHYPLVGNSLIKTLEQYAGPAWNSKLESAWAGAYSAITELMLEGASY
K9Q0C7_9CYAN/9-143   TDALKGLGARHVKYGALPEHYPLVGNTLLKTFEQFLGDAWTEPVKGAWVNAYGVITEVMLDGADY
K9VX73_9CYAN/1-135   EPVLKVLGERHIGYGAIANSYPAVGEALITTFEQYLQQDWTTEVKQAWIDAYGAITALMLKGAGV
F4Y0R0_9CYAN/1-135   SSALKGLGARHVQYGALPEHYPLVGNSLLKTFEQYLGADWTPEVKQAWVDAYGAITTIMLDGADY
K9SL27_9CYAN/1-135   TSELKGLGARHVKYGALPAHYPLVGNALLATLEQYLKADWTPEVKEAWVAAYGAITAIMLDGADY
K9SG59_9CYAN/1-135   TETLAGLGSRHVKYGALPEHYPLVGNALLATFEEYLGTAWTEEVKGAWVAAYAAIITELMLEGADY
D8FWM6_9CYAN/1-135   TGALKGMGARHVQYGTLPEHYPLVGASILKTFESYLGPDWTPEVKQAWVDAYGAIANLMLEGADY
K9XM42_9CHRO/1-135   KITLQNLGARHVSYGTLQQHYPMVGAALLKTFESYLGKDWTPEVKQAWADAYGVLAEMMLEGAQS
K9T8Q6_9CYAN/1-135   ENALKGLGTRHVQYGVLPQHYPMVGGALLKTFEALLGSDWTPELKQAWIDAYGSVTQLMLEGADY
Q7NKA6_GLOVI/8-142   STALQDLGNRHAGYGIVPEHYPMVGTSLLKTFETYLGDAWTPEVKQAWVDAYGAITGLMLTGAES
K9T9X7_9CYAN/1-135   TKVLREMGEKHATYGAKAEHYPIVGAALLKTLEAYLGADWTPEVKQAWTDAYEEISYLMLEGAKR
B0C7N8_ACAM1/1-135   SSALKGLGTRHVKYGVLPEHYPLVGSTLLKSMAATLKDQWTPDIEAAWTDAYGAITEIMLEGTDY
K9ERN6_9CYAN/22-156  TAALQGLGTRHIKYGVLPEHYPMVGGTLLKSMETILQDDWTPEISAAWTEAYAAITEIMLEGADY
K9PXW7_9CYAN/1-135   THALQGLGTRHVKYGVLPEHYPMVGRTLLKAMAIALDEQWTTEFSEAWAEAYATITEIMLDGLEY
K9WGN6_9CYAN/2-134   KNILKDLGTRHVRYGTIQEHYPMVGGTLLKTLESFLGKEWTPEVKRAWTHGYKAIANLMQEEH--
B4WLV9_9SYNE/1-133   QPALKSLGARHAEVGTIKSHYPLVGQALIETFAEYLAADWTEQLATAWWEAYDVIASTMIEGADN
D8G1W5_9CYAN/380-513 ILAVQDLGRRHAGYGVKAEYYDIVGEALLWTLGQGLGVEFTIPVRKAWEEAYTFLSEIMKEAAAE
B4WU25_9SYNE/369-503 ISKVQALGRSHQGYGVKAEDYEAIGATLLWTLKEKLGDDFTPEVKRAWEVAFQFLSKIMINAAAQ
B3DUZ7_METI4/3-133   IPTLQDLGRRHKQYGVVDSHYPLVGDCLLKSIQEYLGQGFTEEAKAAWTKVYGIAAQVMTAE---
                     .......80........90.......100.......110.......120.......130....
```

Figure 6.4 Alignment of 18 cyanobacterial M globins using *Synechococcus* sp. PCC 7335 B4WP36 as the query. The top line indicates secondary structure as found in *M. infernorum* Hb I (PDB ID 3S1I, Teh et al., 2011), the closest homologue with available three-dimensional structure. *M. infernorum* is a bacterium of the Verrucomicrobia group. Residues in bold are at positions B10, E7 and F8, as numbered by structural homology to the canonical 3/3 fold. The proteins and organisms are: B4WP36_9SYNE *Synechococcus* sp. PCC 7335; K9Q0C7_9CYAN *Leptolyngbya* sp. PCC 7376; K9VX73_9CYAN *Crinalium epipsammum* PCC 9333; F4Y0R0_9CYAN *Moorea producens* 3L; K9SL27_9CYAN *Pseudanabaena* sp. PCC 7367; K9SG59_9CYAN *Pseudanabaena* sp. PCC 7367; D8FWM6_9CYAN *Oscillatoria* sp. PCC 6506; K9XM42_9CHRO *Gloeocapsa* sp. PCC 7428; K9T8Q6_9CYAN *Pleurocapsa* sp. PCC 7327; Q7NKA6_GLOVI *Gloeobacter violaceus* strain PCC 7421; K9T9X7_9CYAN *Pleurocapsa* sp. PCC 7327; B0C7N8_ACAM1 *Acaryochloris marina* strain MBIC 11017; K9ERN6_9CYAN *Leptolyngbya* sp. PCC 7375 (preliminary); K9PXW7_9CYAN *Leptolyngbya* sp. PCC 7376; K9WGN6_9CYAN *Microcoleus* sp. PCC 7113; B4WLV9_9SYNE *Synechococcus* sp. PCC 7335; D8G1W5_9CYAN *Oscillatoria* sp. PCC 6506; B4WU25_9SYNE *Synechococcus* sp. PCC 7335; B3DUZ7_METI4 *Methylacidiphilum infernorum* (isolate V4). This set was filtered at 75% redundancy to limit the number of sequences shown.

M family proteins and serve to demonstrate the distinct phylogenetic division seen within individual globin families.

3.2.2 The T lineage

The class of globin with the 2/2 structural fold is referred to as the TrHbs as they are on average 20–40 amino acids shorter than the myoglobin-like haemoglobins (Wittenberg et al., 2002; Wu, Wainwright, & Poole, 2003). These deletions are not uniform truncations from either end of the peptide but rather are distributed throughout the primary structure. In addition, they vary widely across the T family. The result is distinct structural differences between 2/2 proteins and 3/3 proteins, and among 2/2 proteins as a set. Essential structural elements of the 2/2 fold are recapitulated in the 3/3 fold. These constant parts hold important clues for the construction of functional haem pockets and the stabilization of bound oxygen, whereas the variable features allow for the exploration of a broader functional space (Lecomte, Vuletich, & Lesk, 2005).

The TrHbs are the only globins found in all three of the Bacteria, Archaea, and Eukaryota families. Archaea do not appear to have any M family globins, and Eukaryotes are missing S family globins (except for SSDgbs in fungi, Vinogradov et al., 2007, 2013). This has been used to suggest that the truncated forms of globins may be more ancient than the 3/3 canonical forms (Vuletich & Lecomte, 2006), but the alternative view has since gained much support from additional sequences and in-depth phylogenetic analyses (Vinogradov et al., 2007).

Regardless of the nature of the common globin ancestor, all phylogenetic analyses agree that the TrHbs branch into three distinct groups designated as TrHb1 (with gene names often ending with an N), TrHb2 (with gene names often ending in an O) and TrHb3 (with gene names often ending in a P; Vuletich & Lecomte, 2006; Wittenberg et al., 2002; Wu et al., 2003). In Bacteria, the TrHb2 is the dominant group from the T family, occurring roughly twice as frequently as either TrHb1 or TrHb3. This is in contrast to Cyanobacteria, where to date only TrHb1s have been identified. It should be noted that these groupings are based upon phylogeny, and we do not as yet understand fully what different functional characteristics each of these globins possess. It is still interesting to note that Cyanobacteria TrHbs are so divergent from the rest of the bacterial world. In algae both TrHb1 and TrHb2 have been found; however, TrHb1 are by far the dominant group (Vinogradov, Fernandez, et al., 2011). Figure 6.5 presents a

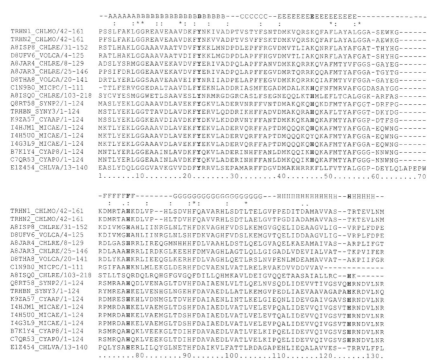

Figure 6.5 Alignment of 18 green algae (GA) and cyanobacterial (CYA) TrHb1s using *Chlamydomonas eugametos* LI637 Hb as the query. The top line indicates secondary structure as found in the query protein (PDB ID 1DLY, Pesce et al., 2000). Residues in bold are at positions B10, E10, F8 and H16, as numbered by structural homology to the canonical 3/3 fold. The proteins and organisms are: TRHN1_CHLMO *Chlamydomonas moewusii* (GA); TRHN2_CHLMO *Chlamydomonas moewusii* (GA); A8ISP8_CHLRE *Chlamydomonas reinhardtii* (GA); D8UFV6_VOLCA *Volvox carteri* (GA); A8JAR4_CHLRE *C. reinhardtii* (GA); A8JAR3_CHLRE *C. reinhardtii* (GA); D8THA8_VOLCA *V. carteri* (GA); C1N9B0_MICPC *Micromonas pusilla* strain CCMP1545 (GA); A8ISQ0_CHLRE *C. reinhardtii* (GA); Q8RT58_SYNP2 *Synechococcus* sp. strain PCC 7002 (CYA); TRHBN_SYNY3 *Synechocystis* sp. strain PCC 6803 (CYA); K9ZA57_CYAAP *Cyanobacterium aponinum* strain PCC 10605 (CYA); I4HJM1_MICAE *Microcystis aeruginosa* PCC 9808 (CYA); I4H5U0_MICAE *M. aeruginosa* PCC 9807; I4G3L9_MICAE *M. aeruginosa* PCC 9443 (CYA); B7K1Y4_CYAP8 *Cyanothece* sp. strain PCC 8801 (CYA); C7QR53_CYAP0 *Cyanothece* sp. strain PCC 8802 (CYA); E1Z454_CHLVA *Chlorella variabilis* (GA). Note the absence of the proximal histidine in *Chlamydomonas reinhardtii* A8ISQ0.

sampling of cyanobacterial and algal TrHb1s. Figure 6.6 contains representative TrHb2s.

Phylogenetically, the TrHb1 group further segregates into two subgroups or classes (Vuletich & Lecomte, 2006). To follow the new naming

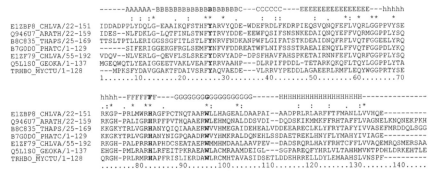

Figure 6.6 Alignment of TrHb2s using *Chlorella variabilis* E1ZBP8 as the query. The top line indicates secondary structure as found in *Mycobacterium tuberculosis* trHbO (PDB ID 1NGK, Milani et al., 2003). Residues in bold are at positions B10, F8 and G8, as numbered by structural homology to the canonical 3/3 fold. The proteins and organisms are: E1ZBP8_CHLVA *C. variabilis* (GA); Q946U7_ARATH *Arabidopsis thaliana* (Embryophyta); B8C835_THAPS *Thalassiosira pseudonana* (Stramenopiles); B7G0D0_PHATC *Phaeodactylum tricornutum* strain CCAP 1055/1 (Stramenopiles); E1ZF79_CHLVA *C. variabilis* (GA); Q5L1S0_GEOKA *Geobacillus kaustophilus* strain HTA426 (Firmicutes); TRHBO_MYCTU *Mycobacterium tuberculosis* (Actinobacteria).

recommendations, we refer to these as TrHb1-1 and TrHb1-2 when the distinction between the two classes is necessary. Figure 6.7 compares the sequences of selected TrHb1-1s and TrHb1-2s. Analysis of the separation of the two subgroups of TrHb1 has suggested that they are not the result of evolutionary divergence but are the result of two separate instances of horizontal gene transfer (Vinogradov, Fernandez, et al., 2011). These two gene transfers are suggested to be the result of two separate endosymbiotic events. This conclusion is in contrast with the idea that the rise of the Archaeplastida group came from a single endosymbiotic event, making algae and land plants monophyletic. Based upon the existing sequence information, the endosymbiotic events that gave rise to the algae could have occurred not only with cyanobacteria (as is the current hypothesis) but may also have happened following the incorporation of an ancestor of gamma-proteobacteria. In the same study, it was noted that TrHb2 proteins (not to be confused with TrHb1-2) do not fit into either the monophyletic or the (cyanobacterial/gamma-proteobacteria) polyphyletic model (Vinogradov, Fernandez, et al., 2011). Rather it appears that TrHb2 genes are the result of an additional independent instance of horizontal gene transfer that occurred via a bacterial ancestor of Chloroflexi, Firmicutes or epsilon-proteobacteria (Vinogradov, Hoogewijs, et al., 2011). Consequently, the origin of the primary endosymbiotic event at the beginning of Archaeplastida has been put

```
                --AAAAAABBBBBBBBBBBBBBBBBBBBBB---CCCCCC--------EEEEEEEBEEEEEEEEE-hhhhh-FF
                     *   :      :    *  :.          .   .         ..   :  **
B2IV60_NOSP7/56-179  GSSLYKRLGGYNAIAAVIDDSAKNIFADPLIGKYFIGLSTN---SKQRLRELLIAQFCQAAGGPCIYTGR
D8G8J5_9CYAN/58-178  GISLYNRLGGYNAIASVIDRAANYIFNDPLIGKYFIGLSTN---SKLRLRQLLVDQFCQAAGGPCVYTGR
K9VF25_9CYAN/1-115   --------MGYNAIAAVIDDSATFIFADPVIGKYFIGLSTN---SQQRLRQLLVDQFCQAAGGPCVYTGR
K9WZP6_9NOST/1-120   -MTLYERLGGYDAISAVADNLLPRLQADPQLGRFWQYRGD---DGLKREKQLLVDFLCASSGGPMYYTGR
K9UL54_9CHRO/7-133   AKSLYERLGGVYSIATVVDDFIDRIMVDPRLNSNPLVDEAHHRVSPAGFKYLVTEMSCWAIGGPQQYSGR
B4VMT4_9CYAN/6-123   NLTLYEKLGGQPVVAQIVDDFYQRVLADDTVSHFFANT------DMEKQRRHQTAFVSHALGGPNQYTGR
K9XN27_9CHRO/3-120   ATTLYEKLGGEQAIKQVVDDFYTRVLADDTVNSFFAHT------DMEKQRRHQTAFISFALGSPTPYTGR
B2J6Y7_NOSP7/1-118   MSTLYDNIGGQPAIEQVVDELHKRIATDSLLSPIFAGT------DMAKQRNHLVAFLGQIFEGPKQYGGR
Q8RT58_SYNP2/1-124   MASLYEKLGGAAAVDLAVEKFYGKVLADERVNRFFVNT------DMAKQKQHQKDFMTYAFGGTDRFPGR
                     1.......10........20........30........40........50........60........70

                FFFFFFFFFFFFFFF-GGGGGGGGGGGGGGGGGGGG--HHHHHHHHHHHHHHHHHHHHHHHH-----
                     *  :*   .   .: :: ::            .  :   : . .:   : :.
B2IV60_NOSP7/56-179  TMKLSHSGI--GRGLTNGEFYAFVNDIALALDKNGVKQPEKNQVLGFANSL--RDQIVEKP--
D8G8J5_9CYAN/58-178  TMNLSHSGI----GLTNDEFNAFANGISQSLDDNRVNQTEKLEVLTFVNSL--RSKIVEG---
K9VF25_9CYAN/1-115   TMKLSHSGM--DGGLTNNEFNAFANDVSQALDKNRVNPPEKAEVLAFVNSL--RSKIVEK---
K9WZP6_9NOST/1-120   NMMTSHQGM----KISENDWSSFLAHLNATLEFFQLPQNERDEVIAFIQTT--KLDIVES---
K9UL54_9CHRO/7-133   SMYDAHAHL----KITREEWAAFMEDLDATFDKFNVPQPERAEFIAIIEST--KPDIVLPTAS
B4VMT4_9CYAN/6-123   SMEKAHAGL----DLQPEHFDAIAKHLGESLDEYGLTQEEINSVLERISTL--KEAVLYK---
K9XN27_9CHRO/3-120   SMEKAHAGL----NLQPEHFDAIVKHLSEALEVHHVPPAEINKILDRITTL--KEAVLYK---
B2J6Y7_NOSP7/1-118   PMDKTHAGL----NLQQPHFDAIAKHLGEAMAVRGVSAENTKAALDRVTNM--KGAILNK---
Q8RT58_SYNP2/1-124   SMRAAHQDLVENAGLTDVHFDAIAENLVLTLQELNVSQDLIDEVVTIVGSVQHRNDVLNR---
                     ........80........90........100.......110.......120.......130...
```

Figure 6.7 Alignment of five cyanobacterial TrHb1-2s (top) and four TrHb1-1s (bottom) using *Nostoc punctiforme* TrHb1-2 as the query. The top line indicates secondary structure as found in *Synechococcus* sp. strain PCC 7002 GlbN (PDB ID 4IOV). Residues in bold are at positions B10, E10, F8 and H16, as numbered by structural homology to the canonical 3/3 fold. The proteins and organisms are: B2IV60_NOSP7 *Nostoc punctiforme* strain ATCC 29133/PCC 73102; D8G8J5_9CYAN *Oscillatoria* sp. PCC 6506; K9VF25_9CYAN *Oscillatoria nigro-viridis* PCC 7112; K9WZP6_9NOST *Cylindrospermum stagnale* PCC 7417; K9UL54_9CHRO *Chamaesiphon minutus* PCC 6605; B4VMT4_9CYAN *Coleofasciculus chthonoplastes* PCC 7420; K9XN27_9CHRO *Gloeocapsa* sp. PCC 7428; B2J6Y7_NOSP7 *N. punctiforme* strain ATCC 29133/PCC 73102; Q8RT58_SYNP2 *Synechococcus* sp. strain PCC 7002.

into doubt according to the phylogenetic analysis of these photosynthetic-truncated globins. It is remarkable that the intron structure of the eukaryotic globin genes has proved practically useless for the purpose of understanding these events (Vinogradov, Fernandez, et al., 2011). As more genetic information becomes available and a more diverse group of genomic sequences is incorporated into phylogenetic analyses, it will be fascinating to learn if globins and their evolution can be used to shed light on the development of eukaryotic photosynthesis.

Thus within the primary endosymbionts, there is a single lineage for the 3/3 Hbs, arguing that from a canonical Hb perspective, the Archaeplastida are in fact monophyletic. Since this is at odds with the heterogeneity seen in the phylogenetic data for the TrHbs, it is now proposed that the origin of the canonical Hbs is from a horizontal gene transfer resulting from the original mitochondrial endosymbiotic event, probably involving a precursor of a present-day proteobacteria (Vinogradov, Fernandez, et al., 2011). The overall picture therefore becomes one where, in the case of photosynthetic eukaryotes, the canonical 3/3 Hbs were acquired first as the result of mitochondrial assimilation, and TrHbs were acquired later as the result of the chloroplast assimilation.

All of this can be seen to play out within the phylogenetic tree shown in Fig. 6.3. The bottom of the tree groups the M family globins, and the top portion of the tree groups proteins from the T family. The exception to this is a small grouping of two proteins (seen at the very bottom of the tree diagram). These two proteins (Uniprot E1ZBP8 and E1ZF79, both from the genome of *Chlorella* sp. NC64A) clearly segregate from both portions of the phylogenetic tree. Previous analysis of these predicted proteins by Vinogradov, Fernandez, et al. (2011) has determined that they are related to TrHb2-type globins. Two other eukaryotic hypothetical proteins are predicted to be TrHb2 globins (Uniprot identifier E1Z454 from *Chlorella* sp. NC64A, and A8J8P3 from *C. reinhardtii*), but in Fig. 6.3, both of these proteins cluster within the TrHb1 proteins, as noted in the phylogenetic analysis by Vinogradov, Fernandez, et al. (2011).

The remainder of the proteins within the top portion of the tree are all TrHb1 globins. Interspersed within this section are the GlbN proteins from *N. commune* (Uniprot ID Q00812), *Nostoc punctiforme* (Uniprot ID B2J6Y7), *Synechococcus* sp. PCC 7002 (Uniprot ID Q8RT58), and *Synechocystis* sp. PCC 6803 (Uniprot ID P73925), and the LI637 protein from *C. eugametos* (Uniprot ID Q08753). The eukaryotic TrHb1s cluster together into two smaller groups (eukaryote proteins identified by small black diamond) with the exception of one protein from *Volvox carteri* (Uniprot ID D8TWR5) that clusters with a group of TrHb1 proteins from three different species of the Cyanobacteria *Nostoc*. The significance of this cluster is unknown, but given the simplistic approach used to generate this figure, further analysis would be required in order to speculate. A small cluster also forms around the TrHb1 protein from *N. punctiforme* (Uniprot ID B2IV60), known to belong to the subgroup TrHb1-2. The four proteins (K9UL54, K9WZP6, D8G8J5, K9VF25) that cluster with the TrHb1-2 from *N. punctiforme* were subsequently aligned and found to possess attributes of TrHb1-2 (Fig. 6.7).

The globins of cyanobacteria and green algae fall mostly within the grouping of TrHb1-1 and SDgb type globins, with a few TrHb1-2, TrHb2 and recently added SSDgb examples. In Section 4, we examine what physiological information exists for these globins. Although we are constantly increasing the genomic information available for expanding our phylogenetic classification of these proteins and making great strides in that domain, the physiologic research conducted on cyanobacterial and algal globins is limited. Nevertheless, it is a key part in our current understanding of the function of these enzymes, and once coupled to additional phylogenetic findings, it is expected to lead us to truly dramatic discoveries.

4. PHYSIOLOGICAL CHARACTERIZATION

The current state within the field of globin biology is decidedly one-sided. In the previous section, we discussed the recent explosion of genomic data that stretch across all kingdoms of life, giving an unparalleled view of evolution and phylogeny at a level of detail unimaginable only a decade ago. However, when we look at experimental data, structural analysis and physiological analysis, we see a relative handful of examples. Without detracting from the importance of these results, we must emphasize that a deep and vast field of information about the function and physiological value of globins remains virtually untapped. In a recent publication reviewing the status of bacterial globin research, the authors bemoaned the fact that though thousands of genomic sequences are currently available, the sum of the 'gold standard' of globin research—functional determination by mutagenesis and phenotypic analysis—is contained in only a few experiments (Vinogradov et al., 2013). In this section, we review past and current work conducted on cyanobacteria and algae at the physiological, proteomic and/or transcriptomic level. We believe that these are the bedrock experiments on which the future understanding of globin biology will be built.

4.1. *Nostoc commune*

The first published discovery of a globin in cyanobacteria also contained the first physiological hint at one of the roles of these proteins. The 1992 *Science* paper by Potts et al. (1992) described the *glbN* gene discovered as an open reading frame (ORF) within the *nifUHD* gene cluster of *N. commune* UTEX 584. The location of the ORF within a gene cluster dedicated to nitrogen fixation (Angeloni & Potts, 1994) and the deduced amino acid similarity to globins previously identified in ciliated protozoa *P. caudatum* (Iwaasa et al., 1989) and *T. pyriformis* (Iwaasa et al., 1990) led to the supposition that the gene product for the *glbN* gene is involved in nitrogen metabolism.

The resulting protein, which we call here GlbN according to the corresponding gene, was found through immunoblotting to be expressed only during prolonged anaerobic growth under conditions of nitrogen starvation. The relative expression of GlbN shows marked correlation to the expression of both NifH and PetH proteins. NifH is a dinitrogen reductase and PetH is a ferredoxin NADP$^+$ oxidoreductase, both integral components of the nitrogen fixation pathway. The concomitant accumulation of these proteins with GlbN further strengthens the involvement of GlbN in nitrogen metabolism (Hill et al., 1996).

Analysis of the context of the *glbN* gene revealed that the sequence located approximately 100 basepairs upstream of the start of the gene is similar to the NtcA-binding domain found in *glnA* promoter of the fresh water cyanobacterium *Synechococcus elongatus* (Hill et al., 1996). The NtcA domain acts as a global transcriptional activator for the regulation of the genes for nitrogen metabolism in cyanobacteria (Flores & Herrero, 1994). This suggests that *glbN* is regulated by nitrogen metabolism in a manner similar to nitrate reductase. Mobility shift analysis of the upstream sequence using recombinant NtcA confirmed that this sequence binds the regulatory protein (Hill et al., 1996).

Given these results, the authors proposed a function for the globin of *N. commune*. GlbN is predicted to serve as an oxygen scavenger, perhaps acting as a delivery molecule for a terminal cytochrome oxidase complex, to aid in ATP production during the anaerobic conditions that exist during nitrogen fixation (Hill et al., 1996). Although this proposed function still requires experimental validation, it does fit with the evidence thus far accumulated for this protein.

Polyclonal antibodies raised against a recombinant form of GlbN were used to stain ultrathin-sectioned cells of *N. commune* UTEX 584. With these gold-labelled antibodies, the GlbN protein was localized to the peripheral membrane of the cells via transmission electron microscopy (Hill et al., 1996). The presence of GlbN on the peripheral membrane is seen in both heterocysts and vegetative cells. Subsequent fractionations of cellular compartments demonstrate that GlbN is a soluble protein with all cellular GlbN found in aqueous fraction of the cell and no protein found within the membrane fractions.

The polyclonal antibodies were also used to scan for cross-reactivity in a series of cyanobacterial strains. No other species contained a cross-reacting protein of equivalent molecular weight to the GlbN protein found in *N. commune* UTEX 584; however, several species, including the diazotroph nonheterocyst-forming filamentous *Trichodesmium thiebautii*, did show the presence of a slightly larger (approximately 18 vs. 12 kDa) protein (Hill et al., 1996). This cross-reacting protein appears to be constitutively expressed even under oxic conditions, whereas the GlbN of *N. commune* UTEX 584 is only induced during anaerobic growth. Although it is possible that these cross-reacting proteins are also globins, no further investigation has been undertaken. A query of the *Trichodesmium erythraeum* genome with the *N. commune* sequence retrieves no significant hit, which leaves the question unresolved.

4.2. *Nostoc punctiforme*

The wealth of genomic information available with the advent of large-scale sequencing has also resulted in the ability to use a global analysis of gene expression within the whole cell. These gene expression profiles can give transcriptional information about the regulation of globins in cyanobacteria. Although this does not necessarily highlight function, it does show when the cell chooses to up-regulate, or down-regulate, the expression of a gene. This in turn gives valuable information about conditions under which the protein's activity is useful. Following the completion of the genome of *N. punctiforme* strain ATCC 29133 (PCC 73102; Meeks et al., 2001), the genomic sequences were used to construct DNA microarrays. These arrays can detect more than 6500 specific genes, with two of these genes predicted to be globins (Campbell, Summers, Christman, Martin, & Meeks, 2007). Gene NpR0416 codes for a TrHb1-1 globin (Uniprot ID B2J6Y7) analogous to the *glbN* gene product found in *N. commune* UTEX 584, and gene NpR1005 for a TrHb1-2 globin (Uniprot ID B2IV60; Vuletich & Lecomte, 2006). Based upon clustering with orthologous expression groups within the microarray, NpR0416 is classified as part of the adaptive metabolism of the cell, linked to metabolic protection, while NpR1005 is a core protein involved in nitrogen metabolism (Campbell et al., 2007).

In these experiments, RNA is isolated from cells under three different growth conditions and then compared with RNA isolated from cells grown in replete medium. Under the first condition, ammonia is withheld from the medium to favour diazotrophic growth and induce nitrogen fixation. The second growth condition favours akinete (a spore-like cell type transiently formed during nutrient stress) formation by extended incubation in the dark (Argueta & Summers, 2005). The third condition favours a switch to a third cell type, termed hormogonia. This type is formed when the vegetative filaments differentiate into small filaments capable of gliding and buoyant motility (Campbell & Meeks, 1989). The three cell types each alter gene regulation and give insight into the relative dependence of the cells on the two globins. The TrHb1-1 globin (NpR0416) shows up-regulation during nitrogen depletion, while the TrHb1-2 globin (NpR1005) shows down-regulation after 3 days in akinete differentiation medium and after 24 h within hormogonia-inducing media (Campbell et al., 2007; Christman, Campbell, & Meeks, 2011). The down-regulation of NpR1005 was further verified by subsequent experimentation on hormogonia induction using both a hormogonia-inducing factor (HIF) as well as nitrogen stress induction (Campbell, Christman, & Meeks, 2008). The up-regulation of NpR0416 follows the same pattern as the nitrogen fixing (*nif*) genes, which can be correlated to the fact that the *glbN* gene, coding for the NpR0416 protein, is found within the *nif* gene cluster (Christman et al., 2011).

A recent gene expression study in *N. punctiforme* (Soule, Gao, Stout, & Garcia-Pichel, 2013) assessed stress from exposure to ultraviolet light and found that exposure to UVA radiation causes an up-regulation of the NpR1005 TrHb1-2 protein. Ultraviolet light is known to cause significant damage to the cells and, although not investigated in this study, up-regulation of the TrHb may in fact be linked to minimizing this radiation damage.

4.3. *Synechococcus* sp. strain PCC 7002

The cyanobacterium *Synechococcus* sp. strain PCC 7002 contains a TrHb1-1 protein referred to as GlbN and coded for by the *glbN* gene (Scott et al., 2002). The GlbN protein has been produced heterologously in *E. coli* and the recombinant GlbN protein studied *in vitro* (See Section 5). However, *Synechococcus* sp. PCC 7002 is also the first of the cyanobacterial species in which physiological evidence exists for the function of the GlbN protein *in vivo* (Scott et al., 2010). As a first step, the *glbN* gene was deleted from the *Synechococcus* strain by introducing an antibiotic resistance gene (the *aadA* gene for the aminoglycoside adenyl transferase protein), which confers resistance to the antibiotic spectinomycin (Goldschmidt-Clermont, 1991), into the genome in place of the native *glbN* gene. In brief, the antibiotic resistance gene was part of an engineered recombinant plasmid, situated between two fragments of genomic DNA from *Synechococcus* sp. PCC 7002. The two fragments were located directly upstream and downstream of the native *glbN* gene. When the recombinant DNA was delivered into the *Synechococcus* sp. PCC 7002 strain, homologous recombination between the plasmid and the host genome inserted the antibiotic resistance gene in place of the *glbN* gene, thereby removing the native gene from the cell's genome and creating the novel *ΔglbN* strain. Subsequent PCR analysis verified that the $\Delta glbN$ strain is a homologous deletion mutant. This strain allows side-by-side comparison with the wild-type strain to determine the physiological effects resulting from the complete loss of the $glbN$ gene.

Under standard growth conditions, no obvious differences exist between the $\Delta glbN$ and wild-type strains when grown in replete medium. However, when medium is limited, either in carbon dioxide or iron, the growth rate of $\Delta glbN$ noticeably slows relative to wild type. Transcriptional analysis suggests the $glbN$ gene expression is constitutive, so altered growth rates would indicate that the loss of $glbN$ inhibits growth under these nutrient limitations. Low-temperature chlorophyll fluorescence measurements of both strains demonstrate that the $\Delta glbN$ strain alters its chlorophyll composition when starved of these nutrients, limiting the amount of photosystem I present in this strain. Both the fluorescence emission change and altered growth rates suggest that the $\Delta glbN$ strain experiences more stress under nutrient-limiting conditions than does the wild-type strain (Scott et al., 2010). Further testing using spermine NONOate (which decomposes to release nitric oxide) demonstrates that the stress in $\Delta glbN$ is consistent with nitrosative damage to the cell. Quantitation of this damage was estimated using 5- (and 6-) chloromethyl-$2',7'$- dichlorodihydrofluorescein diacetate, acetyl ester (CM-H_2DCFDA) to measure ROS/RNS content in CO_2/Fe-limited $\Delta glbN$. CM-H_2DCFDA experiments find the $\Delta glbN$ strain to have a reactive molecule content over sixfold higher than in the control cells. Thus, with reservations associated with the interpretation of data obtained with fluorescent probes (Wardman, 2007), it appears that the GlbN protein confers protection from oxidative damage to the *Synechococcus* sp. PCC 7002 cells.

Both the wild-type strain and the $\Delta glbN$ strain can grow in media supplemented with sodium nitrate; however, as the concentration of nitrate increases to about 90 mM, the growth rate of the $\Delta glbN$ strain slows significantly (Scott et al., 2010). The $\Delta glbN$ strain cannot grow at media concentrations of 240 mM nitrate, and the ROS/RNS content of the $\Delta glbN$ strain is noticeably higher than the wild type with the highest ROS/RNS content being seen when the $\Delta glbN$ strain is placed under 240 mM nitrate and the media is made micro-oxic by sparging with nitrogen and CO_2. These results are strengthened by the transformation of the $\Delta glbN$ strain with a plasmid containing the functional $glbN$ gene, with the resulting rescued strain regaining its ability to grow under the nutrient stress lethal to the $\Delta glbN$ strain. The processing of nitrate appears to require the GlbN protein for effective nitrogen metabolism, with excessive nitrosative stress resulting from the deletion of the *glbN* gene, which is consistent with the predicted use of TrHb1 by *Synechococcus* 7002.

4.4. *Chlamydomonas eugametos*

Among algal Hbs, the most thoroughly studied is a TrHb found in the chlorophyte *C. eugametos*. This globin was initially identified through the screening of genes induced during the dark-to-light transition of the dark-adapted algae (Gagné & Guertin, 1992). Using cDNA libraries created from polyA RNA in cells blocked with cyclohexamide (to limit detection to genes in the primary response to light), four genes from *C. eugametos* were positively identified with light-stimulated expression. Of the four genes isolated from this initial screen, two, *LI637* and *LI410*, were subsequently found to have products homologous to globins from the ciliates *P. caudatum* (Iwaasa et al., 1989), *T. pyriformis* (Iwaasa et al., 1990) and the cyanobacterium *N. commune* (Potts et al., 1992). The *LI637* and *LI410* genes code for TrHb1-1 type proteins (Couture et al., 1994). Of these two genes, *LI637* is more extensively studied, as *LI410* has been found to have a much lower LI expression than *LI637* and may have split-expression characteristics due to gene duplication (Couture et al., 1994).

Genetic analysis of *LI637* shows a greater than twofold increase in transcriptional activity within the first 2 h after the transition from dark-to-light of dark-adapted cells (Couture et al., 1994). It is also noted that the expression of *LI637* is blocked by the addition of 3-(3,4-dichlorophenyl)-1,1-dimethylurea (DCMU), a compound that inhibits photosynthetic electron transport (Gagné & Guertin, 1992). The demonstration of a nuclear gene being controlled by the inhibitor of chloroplast metabolism suggests the involvement of photosynthetic metabolites, with the LI637 protein potentially playing a role in this second messenger pathway. The gene product of *LI637* does appear to have a chloroplast targeting sequence, consisting of approximately 44 amino acids with biochemical characteristics of transit peptides for chloroplast precursors (Tardif et al., 2012). The full-length protein consists of 167 amino acids and has a predicted molecular weight of 17.9 kDa. Polyclonal antibodies raised against the recombinant LI637 protein do recognize a single protein of the approximately correct molecular weight within whole-cell extracts of *C. eugametos*. The antibodies recognize native antigens that, through immunolabelled electron micrographs, are localized within the chloroplast of *C. eugametos*, reacting with the thylakoid membranes and pyrenoid (Couture et al., 1994). These findings have further strengthened the idea that LI637 is involved in regulation of photosynthesis—or in mitigation of photosynthetic stressors—as the pyrenoid consists largely of ribulose 1,5-biphosphate carboxylase/oxygenase (Rubisco), the enzyme responsible for the critical step in carbon fixation and photosynthetic efficiency (Meyer et al., 2012). However, neither direct characterization of the native protein nor identification of its metabolic purpose has been undertaken.

Biochemical characterization and subsequent crystallization of the recombinant haem domain of the LI637 protein (CtrHb) resolved the first structure of a TrHb and demonstrated the existence of a distinctive 2/2 fold with an associated haem group buried in the core of the protein (Pesce et al., 2000). CtrHb is also unique in that it displays a high O_2 binding affinity when compared with other TrHbs (such as that from *P. caudatum*) even though it shares, on the distal site, the tyrosine and glutamine residues interacting with exogenous ligands. Mutational studies of distal residues coordinated to the haem group within CtrHb suggest that the modulation of substrate binding by the protein is central to its activity (Das et al., 1999). Among possible functions, a protective role against dioxygen damage has emerged as a plausible hypothesis.

4.5. *Chlamydomonas reinhardtii*

Both *C. eugametos* and a distantly related species within the same genus, *Chlamydomonas reinhardtii*, have been used as model organisms for the study of eukaryotic green algae. But *C. eugametos* is an obligate photoautotroph, whereas *C. reinhardtii* has an added benefit of heterotrophic or mixotrophic growth on acetate-supplemented medium (Harris, 2001). In *C. reinhardtii*, this has allowed the isolation of many photosynthetic mutants and given rise to the dominant use of *C. reinhardtii* over *C. eugametos* in algal genetics (Grossman et al., 2003). The nuclear, chloroplastic and mitochondrial genomes of *C. reinhardtii* have now been completely sequenced (Grossman et al., 2010). In contrast, only a few genes have been isolated from *C. eugametos* and only partial sequencing of its genome has been completed. Although initial investigation of TrHbs in algae utilized *C. eugametos*, further genetic characterization of TrHb is focusing on homologous genetic sequences in *C. reinhardtii*.

Within the nuclear genome of *C. reinhardtii*, a family of 10 putative globin genes has been identified (Vinogradov, Fernandez, et al., 2011). All 10 are categorized as TrHb proteins. Within the annotations of the

C. reinhardtii genome, four of these putative genes have been annotated as truncated globin-like proteins and given the designation 'THB'; THB1 (ORF 81856), THB2 (ORF 196750), THB3 (ORF 145701) and THB4 (ORF 145700), which have been given Uniprot identifiers A8JAR4, A8JAR3, A8ISQ0 and A8ISP8, respectively. Within the nuclear genome, genes *THB1* and *THB2* are located together on chromosome 14 while *THB3* and *THB4* are located together on chromosome 4. Both THB1 and THB2 show high similarity to CtrHb (81% and 79%, respectively); in contrast, THB3 and THB4 similarities are lower (64% and 73%, respectively). One of the putative proteins from these genes (THB4) has a potential transit peptide that suggests chloroplast targeting (Tardif et al., 2012), while THB1 does not appear to have any potential transit peptide (E. Johnson, personal observation).

In addition to having a low similarity to known TrHbs, THB3 is missing some conserved residues found in TrHb1 proteins, notably the proximal histidine (Fig. 6.5). This histidine is thought to be essential for haem binding; however, a handful of globin sequences (approximately 1–5%) do not possess this residue even though the remainder of the sequence aligns very well with known globin proteins (Vinogradov et al., 2013). Whether these proteins are capable of haem binding, whether they associate with a different cofactor, or what function they might have, if any, remains to be determined.

The four THB genes mentioned above are for predicted proteins, but the product of one of these genes, THB1, has been tentatively located within the cell. In a study by Lechtreck et al. (2009), THB1 was found to accumulate in the flagella of *C. reinhardtii* strains impaired in intraflagellar transport (IFT). These strains are defective in a protein complex termed the BBSome, which is linked to a rare human disorder known as Bardet–Biedl syndrome (Tobin & Beales, 2007; Zaghloul & Katsanis, 2009). This syndrome has itself been linked to cilia defects, and BBSomes are believed to be involved in IFT trafficking (Beales, 2005; Rosenbaum & Witman, 2002).

Specifically, the strain used in the Lechtreck study is defective in the *BBS4* gene and exhibits a non-phototactic phenotype, meaning that the *Chlamydomonas* cells cannot perform the native alteration of its directional motility following changes in light direction or intensity. The genetic transformation of the *bbs4* strain with a native *BBS4* gene restored the strong phototaxis characteristics of the parental strain. Further screening reveals the same phenotype from additional strains, as deletions within genes *BBS1* and *BBS7* (also linked to BBSome complex) resulted in loss of phototaxis. Analysis of these strains showed that the loss of BBSome function does not result in any general defect in transport of IFT or flagellar membrane proteins; however, the *bbs4* strain does show accumulation of four specific proteins within the flagella. Three of the proteins appear to be membrane-bound and were identified as putative signalling proteins. The fourth protein accumulates within the flagella is not membrane-bound and was identified as the gene product of the *THB1* gene using mass spectrometry. The failure of *bbs4* to clear the THB1 protein results in its accumulation in a process apparently leading to progressive loss of phototaxis. This study, as well as a subsequent investigation (Lechtreck et al., 2013), suggests that the THB1 protein is a component of a signalling pathway within the flagella and it reinforces, along with the DCMU-dependent expression of *LI637* (Gagné & Guertin, 1992), a novel role for the TrHb proteins in green algae.

An additional advantage in using *C. reinhardtii* is the genome-wide gene expression analysis available for this species (Grossman et al., 2003). In a screen of the effects of anoxia on gene regulation, the gene for a predicted TrHb (*THB8*) showed over 1000-fold increase in expression when oxygen was removed from the system (Hemschemeier et al., 2013). These researchers also created a novel strain of *C. reinhardtii* where the transcript level of the *THB8* gene was substantially reduced through artificial microRNA silencing. The resulting strain showed greatly impaired growth under low oxygen conditions and increased sensitivity to nitric oxide. The work demonstrates the great value that a model organism such as *C. reinhardtii* plays in the functional studies of haemoglobins in algae.

Several additional *C. reinhardtii* reports, although not directly relating to the topic of this chapter, provide supplemental information including data on THBs in response to stress conditions. In one study (Gonzalez-Ballester et al., 2010), the transcript levels of a mutant of *C. reinhardtii* impaired in sulphur accumulation are compared to those of wild-type cells under both sulphur-replete and sulphur-deprived conditions. The transcript level of *THB1* increases by several fold when sulphur is depleted from the medium. The authors note that extreme sulphur deprivation caused cell stress due to an increase in reactive oxygen formation within the chloroplast. A second study (Fischer et al., 2012) investigated a mutant of *C. reinhardtii* with singlet oxygen resistance. The mutant in this chapter (*sor1*) causes the up-regulation of genes related to detoxification using the glutathione pathway. The THB proteins were not identified in this pathway, but the genome-wide analysis does reveal that the expression of *THB1* increases significantly when wild-type cells are exposed to chemicals that create reactive oxygen in the algal medium (Fischer et al., 2012). A dedicated analysis of TrHb proteins in *Chlamydomonas* is clearly needed, and it is fascinating that at least some TrHb proteins appear to be regulated in response to environmental conditions in and around the cell.

5. *IN VITRO* CHARACTERIZATION

5.1. Structural properties

5.1.1 Nostoc commune GlbN

Electronic absorbance spectra provide a low-resolution assessment of haemoglobin properties. The position, intensity and width of bands due to the haem group are influenced by the oxidation state of the iron and its axial ligands. The strongest absorption on the low energy side of 300 nm, generally appearing in the 300- to 500-nm range, is called the Soret band. Lower intensity bands occur in the visible region (in the 500- to 700-nm range). Weak lower energy features appearing in the 600- to 700-nm range correspond to charge transfer (CT) transitions observed in high-spin species. In contrast, aromatic residues (tryptophan, tyrosine and phenylalanine) absorb in the near-UV (275–290 nm) and their contribution can be roughly estimated based on amino acid composition. The ratio of the Soret intensity at its maximum ($A_{\text{Soret max}}$) to that of the maximum aromatic intensity ($A_{\text{aro max}}$) is often used to evaluate the degree of haem incorporation in globins. Thus, optical spectra are well suited for rapid preparative and comparative purposes.

For reference, it is useful to review the characteristics of sperm whale Mb. The physiological state contains ferrous iron. In the absence of exogenous ligand (i.e. O_2 or CO), the ferrous state is a five-coordinate (5c) paramagnetic species (four unpaired electrons, $S=2$) referred to as the deoxy state. Deoxymyoglobin can bind dioxygen (yielding a diamagnetic complex, $S=0$), carbon monoxide ($S=0$) and nitric oxide ($S=1/2$). In the ferric state, also referred to as 'met', myoglobin contains either a hydroxide ion or a water molecule as the sixth ligand to the iron. The aquomet complex is 'high-spin' (five unpaired electrons, $S=5/2$), whereas the hydroxymet complex exists as an $S=1/2$, $S=5/2$ spin equilibrium. The ionization of the water ligand occurs with a pK_a of ~9, but the value depends strongly on the protein (Antonini & Brunori, 1971). Cyanide is a strong-field ligand to ferric iron, which leads to a low-spin ($S=1/2$) complex. Cyanomet globins are generally stable and regularly used as an isoelectronic version of the carbonmonoxy complex. Liganded azide and fluoride yield a mixed spin and a high-spin complex, respectively. NO can also bind to the ferric protein. The optical signatures of the various species are summarized in Table 6.1. In addition to these states, haemoglobin and certain variants of haemoglobin and myoglobin can undergo aberrant ligation of the haem iron by a protein side chain on the distal side (Culbertson & Olson, 2010; Rachmilewitz & Harari, 1972; Rachmilewitz & White, 1973; Sugawara et al., 2003). These

Table 6.1 Optical properties of sperm whale myoglobin

	λ_{max} (nm) (ε, mM^{-1} cm^{-1})				Conditions
Ferrous state		Soret	Visible		
	δ	γ	β	α	
Mb (deoxy)		434 (115)	556 (11.8)		pH 7, 20 °C
MbO$_2$ (oxy)	348 (26)	418 (128)	543 (13.6)	581 (14.6)	pH 7, 20 °C
MbCO (carbonmonoxy)	345 (26.9)	422 (201)	540 (14)	576 (12.1)	pH 7, 4 °C
MbNO (nitrosyl)		420 (133)	548 (11.9)	580 (10.7)	pH 7, 4 °C
Ferric state		Soret	Visible		
	δ	γ			
Mb H$_2$O (aquomet)[a]		409 (157)	505 (9.5)	635 (3.6)	pH 6, 25 °C
Mb (OH$^-$) (hydroxymet)	358 (33.5)	414 (97.2)	542 (9.48)	582 (9.10)	pH 11.5, 25 °C
Mb (CN$^-$) (cyanomet)	359 (29.4)	424 (113)	541 (10.4)		pH 7, 25 °C
Mb (N$_3^-$) (azidomet)		422 (116)	542 (10.3)	574 (8)	pH 7, 4 °C
Mb (F$^-$) (fluorimet)		406 (133)	489 (8.8)	606 (8.5)	pH 7, 4 °C
Mb NO		420 (149)	532 (11.5)	573 (12.5)	pH 7, 4 °C

[a]The pK_a of the aquomet–hydroxymet transition is ~9.
Taken from Antonini and Brunori (1971) and Dawson, Kadkhodayan, Zhuang, and Sono (1992).

endogenous, hexacoordinate species are referred to as haemichrome (ferric iron) and haemochrome (ferrous iron).

The physical characterization of recombinant *N. commune* cyanoglobin (hereafter *N. commune* GlbN) began with optical spectroscopy (Potts et al., 1992; Thorsteinsson, Bevan, Ebel, Weber, & Potts, 1996). The original procedure for preparation of the protein resulted in purified material containing a ferric haem incapable of cyanide binding at neutral pH. In fact, the ferric spectrum seemed to agree with endogenous hexacoordination. At pH 5.5, the spectrum develops a CT band consistent with the coordination of a water molecule, and the protein is then able to associate with cyanide. A subsequent study (Thorsteinsson et al., 1999), however, demonstrates that the properties of the material depend on the purification protocol. If the protein is obtained in the oxy state, the iron can be oxidized to the ferric state, and the protein is then able to bind other ligands at neutral pH. This illustrates a drawback of the recombinant approach in that it is possible to generate irreversibly misfolded globin species, a potential problem for the many globins that cannot be isolated directly from the source organism.

N. commune GlbN prepared in the 1999 study conforms to expectations of ligand binding and spectral appearance. In the ferric state at neutral pH, the UV–visible (UV–vis) spectrum is consistent with a mixture of low-spin and high-spin species. Sperm whale myoglobin has a similar behaviour, with the low-spin species favoured at high pH and containing a hydroxide ligand on the distal side, and the high-spin species favoured at low pH and containing a bound water molecule. The oxy complex of *N. commune* GlbN, on the other hand, is distinct from that of oxy myoglobin; specifically, the α/β ratio of intensities is significantly lower than 1.0. Summary optical data for this and other proteins are listed in Table 6.2.

Reduced *N. commune* GlbN binds dioxygen reversibly with affinity comparable to that of mammalian myoglobin (Thorsteinsson et al., 1996). The Hill coefficient is close to 1, in agreement with a monomeric protein or non-cooperative binding by a multimeric assembly. What appeared to be an unusual oxy spectrum at the time, that is, one exhibiting a low α/β intensity ratio, has since been observed in other TrHb1s and is a hallmark of this group of globins.

The far-UV circular dichroism spectrum of ferric *N. commune* GlbN is indicative of helical secondary structure ($\sim$60%; Thorsteinsson et al., 1996) as expected for a globin. This is confirmed by NMR studies carried out in the La Mar laboratory on the cyanomet complex (Yeh, Thorsteinsson, Bevan, Potts, & La Mar, 2000). NMR data, greatly aided by the

Table 6.2 Optical properties of cyanobacterial and green algal haemoglobins

Ferrous state	λ_{max} (nm) (ε, mM^{-1} cm^{-1}) Soret		λ_{max} (nm) Visible		Comment	References
	δ	γ	β	α		
N. commune (deoxy)		422				Potts et al. (1992)
C. eugametos (deoxy)		426 (108)	529 (sh)[a]	557 (15)		Couture & Guertin (1996)
		428 (80.6)	556 (10.9)		pH 6.0	Couture et al. (1999)
		423 (100)	529 (sh)	556 (15.5)	pH 7.5	Couture et al. (1999)
		424 (170.5)	528 (13.5)	557 (29.6)	pH 9.5	Couture et al. (1999)
Synechocystis (deoxy)		424	526	555	pH 5.5,[b]	Couture et al. (2000)
		426 (162)	528 (13)	560 (24)	pH 7.4,[b]	Scott & Lecomte (2000)
Synechococcus (deoxy)		425 (132)	528 (11)	558 (21)	pH 7.5,[b]	Scott et al. (2002)
N. commune (oxy)					–[c]	Thorsteinsson et al. (1996)
C. eugametos (oxy)		412 (102)	545 (13)	581 (10)	–[d]	Couture & Guertin (1996)
Synechocystis (oxy)		409	550	585	–[e]	Couture et al. (2000)
Synechococcus (oxy)		408	550	585	pH 7.1,[b]	
N. commune (carbonmonoxy)		419				Potts et al. (1992)
C. eugametos (carbonmonoxy)		420 (142)	542 (11)		–[d]	Couture & Guertin (1996)

Continued

Table 6.2 Optical properties of cyanobacterial and green algal haemoglobins—cont'd

	λ_{max} (nm) (ε, mM^{-1} cm^{-1})				Comment	References
Ferrous state	Soret		Visible			
	δ	γ	β	α		
Synechocystis (carbonmonoxy)		416	508 (sh)	540	pH 7.4,[e]	Couture et al. (2000)
		420 (184)	544 (14)	568 (sh)	pH 7.4,[b,e]	Scott & Lecomte (2000)
Synechococcus (carbonmonoxy)	397 (sh)	418 (165)	543 (13)		pH 7.5,[b,e]	Scott et al. (2002)

	λ_{max} (nm) (ε, mM^{-1} cm^{-1})					Comment	References
Ferric state	Soret		Visible		CT		
	δ	γ					
N. commune (met)		408 (123)	536 (10.3)	570 (sh)	632 (3.28)	pH 7	Thorsteinsson et al. (1999)
C. eugametos (met)		406 (130)	529 (6)		624 (2.6)	pH 5,[d,e]	Couture & Guertin (1996)
		410 (110)	535 (8.5)			pH 7.5,[d,e]	Couture & Guertin (1996)
Synechocystis (met)	362 (sh)	409.5 (100)	544 (11)	578 (sh)		pH 7.4,[f]	Couture et al. (2000) & Scott & Lecomte (2000)
Synechococcus (met)	365 (sh)	411 (96)	544 (10)	568 (sh)		pH 7.5,[g]	Scott et al. (2002)
N. commune (cyanomet)							Thorsteinsson et al. (1996)
C. eugametos (cyanomet)		416 (97.6)	547 (11)		−[e]		Couture & Guertin (1996)
Synechocystis (cyanomet)	360	416	546		−[g]		
Synechococcus (cyanomet)	363	416	553		−[g]		
	359	415	550		−[b]		
C. eugametos (azidomet)		413 (105)	543 (9)	580 (sh)	−[d]		Couture & Guertin (1996)

[a]Shoulder.
[b]Given the method of preparation, the spectrum is for the post-translationally modified protein (covalent attachment of the haem group to His117, Vu et al., 2002).
[c]Spectrum presented in the given reference.
[d]Data for the H19 construct, with protein product starting at Ser44.
[e]The pK_a of the transition is 6.3.
[f]Extent of post-translational modification unknown.
[g]No haem post-translational modification.

paramagnetism of the complex, characterize the haem environment and determine unambiguously the orientation of the cofactor in its pocket.

Because of the pseudo C_{2v} symmetry of the haem group (Fig. 6.8) and the relatively small size difference between the porphyrin substituents at positions 1 (methyl) and 4 (vinyl), and 2 (vinyl) and 3 (methyl), the haem can sit in its protein cavity in two orientations differing approximately by a 180° rotation about the $\alpha-\gamma$ axis. The energetic difference between the two isomers depends on the protein and can be relatively small. As a result, many b haem proteins exhibit 'haem orientational disorder' and exist as mixtures (La Mar et al., 2000). The two isomers have practically identical absorption spectra, but they have distinct NMR signatures, readily identified as such in paramagnetic states. At equilibrium, cyanomet $N.$ commune GlbN solutions contain two haem orientational isomers in the ratio 4:1 (Yeh et al., 2000). When gauged by the relative disposition of secondary structure elements, for example, the C-(D)-E corner, the major isomer of cyanomet $N.$ commune GlbN corresponds to the minor isomer of sperm whale Mb.

Other structural features are extracted from the analysis of NMR spectra. In the cyanomet state, the chemical shift of the haem methyl substituents can be used to determine the orientation of the proximal histidine with respect to the haem group (Bertini, Luchinat, Parigi, & Walker, 1999). The data

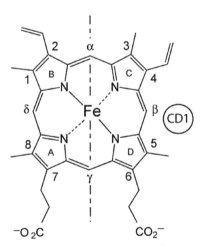

Figure 6.8 The structure of b haem (iron protoporphyrin IX, protohaem) and the nomenclature used in the text. The dashed vertical line indicates the pseudo C_{2v} axis of symmetry. An axial histidine with imidazole plane oriented along this axis or in the perpendicular direction offers the least steric clash with the porphyrin ring. The circled CD1 indicates the position of a conserved Phe in sperm whale myoglobin and serves to orient the haem in its pocket.

suggest that the angle formed by the projection of the imidazole ring onto the porphyrin plane is approximately $60°$ measured from the $N_A - N_C$ axis of reference, pointing towards the $8\text{-}CH_3$ and the 4-vinyl group. This orientation avoids major steric conflicts between the pyrrole nitrogens and the $C\delta 2H$ and $C\varepsilon 1H$ units of the imidazole ring, presumably allowing greater freedom of motion to the proximal histidine and the iron (Samuni, Ouellet, Guertin, Friedman, & Yeh, 2004). The NMR study was performed before the X-ray structure of a truncated globin became available. In their conclusions, the authors point out that '... the folding topology of the compact globin adheres well to the conventional globin fold'. Although this is not strictly the case (Figs. 6.1 and 6.2), an overlay of the three-dimensional structures of CtrHb and sperm whale myoglobin illustrates that the cages formed around the haem group by the two proteins are strikingly similar. As the first cyanobacterial TrHb1 to be characterized in some detail, *N. commune* GlbN illustrates both the pitfalls of *in vitro* studies and the power of UV–vis, CD and NMR spectroscopy as structural tools. No additional structural information is available on *N. commune* GlbN at this time.

5.1.2 C. eugametos LI637 Hb

The optical properties of CtrHb ('H19' construct starting at Ser44), like those of *N. commune* GlbN, immediately indicated a behaviour distinct from that of myoglobin (Couture & Guertin, 1996). The purified protein in the ferric state has a pH-dependent spectrum showing evidence for a coordinated water at low pH and endogenous coordination at high pH. This sixth ligand can be displaced by cyanide. The reduced state spectrum resembles that of *N. commune* GlbN, also supporting a mixture of low- and high-spin species. Ferrous CtrHb is able to bind dioxygen reversibly and exhibits a depressed α/β intensity ratio.

In a follow-up study, Couture et al. (1999) explored further the behaviour of the ferrous state. Three distinct protein forms are observed as the pH is varied: a four-coordinate species at acidic pH, a 5c species with high–spin character at neutral pH and a low-spin six-coordinate (6c) species at high pH. This complex response and that of *N. commune* GlbN support the extraordinary plasticity of the distal haem pocket. An additional study resorting to a combination of site-directed mutagenesis, RR and EPR added to UV–vis absorbance spectroscopy identified the distal ligand to the ferric iron as Tyr B10. Ionization of the ligating phenol to phenolate for coordination is presumably aided by a lysine at position E10 (Das et al., 1999). Replacement of Tyr B10 with a leucine yields an EPR spectrum resembling that of the basic form of cytochrome c, which has a histidine and a lysine for axial ligands (Brautigan et al., 1977). The likely candidate is Lys E10 in a demonstration that the orientation of the E helix can be adjusted under the driving force of ligation. The exact configuration of the haem pocket in the reduced state is not known.

The chemical properties of haem proteins are exquisitely dependent on structural details. The power of prediction—regarding how an amino acid replacement, a change in pH, or some apparently benign perturbation, will affect reactivity—is therefore remarkably limited. Despite the educated expectations based on extensive sequence information, knowledge of canonical globin behaviour and spectral comparison, the T globins presented uncharted and mysterious territory at the time of the first *N. commune* and *C. eugametos* studies. Three-dimensional structural information was greatly needed to rationalize spectroscopic data. The first structures of T globins, from *P. caudatum* and *C. eugametos* (Pesce et al., 2000), were therefore much welcomed by the globin research community.

The three-dimensional structure of CtrHb revealed how the 'truncated' fold differs from the canonical myoglobin fold (Fig. 6.2). The latter is described as a 3/3 orthogonal sandwich, placing three of eight α helices (A, E, F in the Perutz nomenclature) against another three (B, G, H). In the truncated globin domain, the A helix is cut short, reduced from over four turns to a single one, and loss of secondary structure occurs in the F helix region. Thus, the sandwich is best described with B–E/G–H pairing, and therefore the 2/2 denomination. There are other distinctive features, specifically two glycines separating the A and B helices, the absence of the D helix and the resulting tight connection between the C helix and the E helix, and the unstructured long loop between the E and F helices (Nardini et al., 2007; Vuletich & Lecomte, 2006). The protein, however, is large enough to provide the haem group with a sheltered environment limiting solvent access to the distal side.

For stability reasons, the protein form of CtrHb that was chosen for crystallization was the cyanomet complex. Haem orientational isomerism is observed, the proportions of the two forms being roughly 1:1 (Pesce et al., 2000). Compared with the *N. commune* GlbN results and assuming that the equilibrium was reached in both cases, this indicates a less discriminating haem pocket in the algal protein. It is tempting to attribute the difference to the immediate neighbours of the haem group. A survey of b haem proteins and their variants, however, indicates that remote effects can be responsible for the imbalance (La Mar et al., 2000). At this stage, the significance of the varying ratio is unclear and is likely to remain so until a definite function can be attributed to the proteins, and the properties of the two isomers are determined. The fact that the *b* haem is in a dynamic equilibrium in the holoprotein goes against the rigid view provided by solid-state structures. From the *in vitro* experimental standpoint, the coexistence of two forms should not be overlooked because it can cause heterogeneous and complicated responses such as additional kinetic phases in ligand-binding studies.

An important feature of the cyanomet structure of CtrHb is the network of interactions established between the cyanide ligand and distal residues. Figure 6.9 illustrates the most important of these specific contacts, which involve Tyr B10, Gln E7 and Gln E11 in a rigid network. Tyr B10 has the central role as it forms a hydrogen bond with bound cyanide. TrHb1s contain a nearly conserved Tyr B10 and a glutamine at either E7 or E11, or at both positions. The network presented by CtrHb is therefore expected to occur in many other TrHb1s.

Interestingly, the structure of cyanomet CtrHb did not shed much light on the optical properties mentioned earlier. In the absence of cyanide ligand and at low pH, water appears to be the sixth ligand to the ferric iron (Couture & Guertin, 1996). The structure of *P. caudatum* Hb, solved in

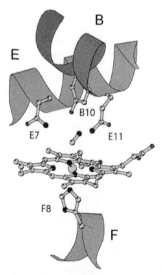

Figure 6.9 The distal hydrogen bond network in CtrHb as formed in the cyanomet complex (PDB ID 1DLY, Pesce et al., 2000) involves Tyr B10, Gln E7 and Gln E11. This network is expected to form in the majority of TrHb1s. (For colour version of this figure, the reader is referred to the online version of this chapter.)

the aquomet state (Pesce et al., 2000), serves as an appropriate model for this species. It features Tyr B10 in the distal pocket, hydrogen bonded to the axial water. The Tyr-Lys arrangement anticipated in CtrHb at high pH (Das et al., 1999) requires the elimination of the water molecule and the collapse of Tyr B10 Oη onto the haem iron. Furthermore, the ligation of Lys E10 to the iron in the Leu B10 variant requires a substantial rearrangement of the structure. Is such conformational change feasible? We will return to this question in the discussion of *Synechococcus* and *Synechocystis* GlbN.

Cyanomet CtrHb revealed an unexpected property of the T globins. The structure contains large internal apolar tunnels that lead from the surface of the protein to the distal side of the iron (Pesce et al., 2000). These branched tunnels are also observed in the related protein from *Mycobacterium tuberculosis* and are thought to control ligand access with the aid of gating residues (Milani et al., 2001; Milani, Pesce, et al., 2004). The long tunnel, shown in Fig. 6.10, has a volume of 400 Å^3 (Milani et al., 2005) and is well suited for the purpose of channelling diatomic molecules to the active site. To probe the nature of the cavities, the X-ray structure was solved under xenon pressure. Four atoms are located in the long tunnel, at locations that appear conserved among several TrHb1s (Milani, Pesce, et al., 2004).

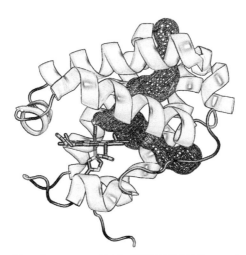

Figure 6.10 The cavities of cyanomet CtrHb (PDB ID 1DLY, Pesce et al., 2000) calculated with the program CAVER (Petrek et al., 2006). Tunnels provide access to the distal side of the haem. The structure is oriented as in Fig. 6.2. This figure was prepared with PyMOL (DeLano, 2002). (For colour version of this figure, the reader is referred to the online version of this chapter.)

As one last property of the TrHb1 folds, it is worth mentioning that the haem is distorted from the planar geometry. In general, several types of haem distortions can be observed (e.g. saddling, doming and ruffling; Shelnutt et al., 1998), and it appears that each type of distortion is associated with a specific group of proteins (Jentzen, Ma, & Shelnutt, 1998). For example, M globins tend to harbour domed haems, whereas c cytochromes contain ruffled haems. The haem in CtrHb is ruffled, a characteristic that may condition its chemical properties.

5.1.3 Synechocystis *sp. strain PCC 6803 GlbN*

Synechocystis sp. PCC 6803 is a non-nitrogen fixing mesophilic cyanobacterium that contains one *glbN* gene. The gene product has optical spectra similar to those of *N. commune* and *C. eugametos* globins (Table 6.2). The met form of the protein remains low spin over a broad range of pHs, and the spectra were interpreted early on as due to the coordination of His E10 to the iron (Couture et al., 2000). The coordination scheme survives reduction of the iron so that a bis-histidine complex is present regardless of the protein redox state. Nevertheless, the distal histidine can be displaced by an exogenous ligand and GlbN binds cyanide in the ferric state as well as dioxygen and carbon dioxide in the ferrous state (Couture et al., 2000; Scott & Lecomte, 2000). Once more, the α/β intensity observed in the oxy state is unlike that in sperm whale myoglobin.

The structure of the haemichrome was solved by NMR methods (Falzone, Vu, Scott, & Lecomte, 2002). The average of the structural ensemble resembles the structure of CtrHb except for perturbations caused by the coordination of His46 (E10) to the iron. Haem isomerism is highly in favour of one haem orientation (95:5) corresponding to that found in myoglobin and to the major isomer of CtrHb. The similarity led to the expectation that exogenous ligand binding would induce a large conformational change transforming a tight haemichrome into an open structure (Falzone et al., 2002).

Recombinant *Synechocystis* 6803 GlbN exhibits one additional distinctive property, apparent in various preparations for NMR analysis. The overexpressed apoprotein can be produced in high yield in *E. coli* cells. The resulting inclusion bodies are readily solubilized in urea and refolded during a chromatographic step. Addition of haemin chloride to solutions of apoprotein generates the ferric holoprotein (Scott & Lecomte, 2000). This mode of preparation affords consistently a mixture of the two haem orientational isomers, identified by their distinct NMR signatures. Reduction to the ferrous state, followed by re-oxidation to the ferric state, however, leads to a different spectrum, indicating that a modification of the holoprotein takes place to completion. A combination of NMR spectroscopy and mass spectrometry identified the modified GlbN (referred to as GlbN-A) as a covalent adduct involving His117, a histidine in the C-terminal portion of the H helix (H16; Vu et al., 2002): iron reduction caused the addition of this residue onto the haem 2-vinyl (Fig. 6.11). The reaction results in a pure protein as the small population of alternative *b* haem isomer eventually becomes depleted by mass action and reaction. Holoprotein can also be obtained by purification from soluble *E. coli* cell extracts. Once oxidized to the ferric state, the protein is revealed to be a mixture of both unmodified and modified protein. The presence of GlbN with covalently attached haem in the cytoplasm of *E. coli* suggests that the post-translational modification occurs *in vivo*. Once more, however, it illustrates the risks in utilizing recombinant material and the exquisite versatility of the haem group in globins.

The NMR structure of the protein containing a *b* haem can be compared with the X-ray structure of the protein in the ferric bis-histidine state with modified haem (Hoy et al., 2004). The unreacted His117 is not held in a rigid position near the vinyl group (Falzone et al., 2002) but instead appears

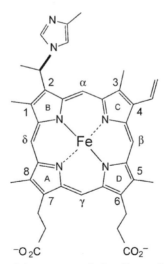

Figure 6.11 The structure of the post-translationally modified haem in *Synechocystis* 6803 GlbN and *Synechococcus* 7002 GlbN (Vu et al., 2002). A histidine from the H helix (position H16) adds to the 2-vinyl to form a covalent bond. The modification is irreversible.

to adopt multiple orientations. In addition, differences unrelated to the PTM are observable in the A helix and the EF loop, which sample a large conformational space in solution and are restrained in the solid state.

What happens to the bis-histidine structure once an exogenous ligand binds? Hargrove and colleagues pursued the X-ray structures of the cyanomet and azidomet complexes (Trent et al., 2004). The result shows clearly the resemblance of the bound state to the cyanomet CtrHb structure. The conformational change (Fig. 6.12) is of unusual amplitude for a globin as the tertiary structure of these proteins is typically not affected to an appreciable extent by ligand binding. This raises the possibility of a signalling function coupled to the presence of exogenous ligand. But it also suggests that rotation of the E helix and B helix can allow for coordination of the residue at position E10, as surmised for the Tyr B10 Lys variant of CtrHb. Once again, the capacity of the haem pocket for deformation in TrHbs is confirmed.

X-ray structures were obtained for the H117A variant in the reduced and oxidized states (Hoy, Smagghe, Halder, & Hargrove, 2007). The replacement or change in redox state does not appear to have an effect on the conformation of the protein. Removal of the bulky histidine group, however, opens a cavity that is filled by water molecules. Overall, the TrHb1 from the cyanobacterium resembles closely that from the green alga, despite the low level of sequence identity (Table 6.3). Also of note is the opening of tunnels and cavities in *Synechocystis* 6803 GlbN only upon exogenous ligand binding.

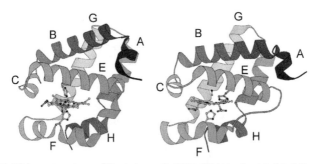

Figure 6.12 Ribbon structure of *Synechocystis* 6803 GlbN in the bis-histidine state (PDB ID 1RTX, Hoy, Kundu, Trent, Ramaswamy, & Hargrove, 2004) and the cyanomet state (PDB ID 1S69, Trent, Kundu, Hoy, & Hargrove, 2004). The conformational change upon exogenous ligand binding affects the B and the E helix. His E10 does not interact with the bound ligand. Note the resemblance of the cyanomet structures of CtrHb (Fig. 6.2) and *Synechocystis* 6803 GlbN. (For colour version of this figure, the reader is referred to the online version of this chapter.)

Table 6.3 Sequence identity of the TrHb1s considered in this chapter

	N. com. GlbN	N. pun. TrHb1-1	N. pun. TrHb1-2	S. 7002 GlbN	S. 6803 GlbN	CtrHb	THB1
N. com. GlbN	1						
N. pun. TrHb1-1	96% 113/118	1					
N. pun. TrHb1-2	29% 35/122	31% 38/122	1				
S. 7002 GlbN	30% 36/119	31% 37/119	25% 30/119	1			
S. 6803 GlbN	40% 40/100	41% 41/100	31% 33/106	60% 74/124	1		
CtrHb	33% 39/118	33% 39/118	27% 22/120	47% 57/122	45% 55/122	1	
THB1	37% 44/118	36% 43/118	27% 34/124	46% 52/114	49% 53/109	48% 55/114	1

N. com. GlbN, *Nostoc commune* UTEX 584 GlbN (Q00812); N. pun. TrHb1-1, *Nostoc punctiforme* ATTC 29133 TrHb1-1 (B2J6Y7); N. pun. TrHb1-2, *Nostoc punctiforme* strain ATTC 29133TrHb1-2 (B2IV60); S. 7002 GlbN, *Synechococcus* sp. strain PCC 7002 GlbN (Q8RT58); S. 6803 GlbN, *Synechocystis* sp. strain PCC 6803 GlbN (P73925); CtrHb, *Chlamydomonas eugametos* LI637 Hb (Q98753), haem domain; THB1, *Chlamydomonas reinhardtii* THB1 (A8JAR4).

5.1.4 Synechococcus *sp. strain PCC 7002 GlbN*

The level of identity between *Synechococcus* 7002 GlbN and *Synechocystis* 6803 GlbN is modest (60%, Table 6.3) but the main features of bis-histidine coordination in the absence of exogenous ligand and the alkylation of His117 by the haem group are conserved. Optical data are similar to those of the *Synechocystis* protein (Scott et al., 2002; Table 6.2). The three-dimensional structure of the posttranslationally modified protein has been solved by both NMR spectroscopy and X-ray crystallography. The differences noted in the case of *Synechocystis* 6803 GlbN, specifically the greater flexibility of the EF loop and the variable position of the A helix relative to the rest of the protein, are also observed in this TrHb1.

A study of backbone dynamics assessed by ^{15}N NMR relaxation provides a complementary view of GlbN in solution (Pond, Majumdar, & Lecomte, 2012). This method is well suited for comparative purposes, and six different states were inspected (Table 6.4). In all cases, the structure is rigid on the picosecond to nanosecond timescale, though additional motions are detected in the CE turn, the EF loop, and the kink in the H helix. Motions on the microsecond to millisecond timescale appear to depend on the distal ligand and the oxidation state of the iron. The ferrous states, in particular, are endowed with higher flexibility.

5.1.5 Summary of structural information

There are now almost a dozen cyanobacterial and green algal globin structures in the Protein Data Bank (PDB), but they are from only three different organisms and all are TrHb1s (Table 6.5). They offer a narrow view of the globins found in unicellular photosynthetic organisms. Nevertheless, along with related proteins from different sources, they illustrate essential features that can be kept in mind when inspecting other globins.

5.1.5.1 Two-over-two fold

In addition to the three TrHb1 structures mentioned earlier, the PDB holds coordinates for several T globins. Variations in loops and helix length do occur, generally leaving the 2/2 topology intact. Thus, it is reasonable to expect that the available structures are adequate templates for the modelling of the three-dimensional properties of other cyanobacterial and algal T globins. This, however, can only serve as a rough approach because of the structural plasticity mentioned earlier.

5.1.5.2 Endogenous hexacoordination

Sequence alignment of cyanobacterial TrHb1s related to *N. commune* GlbN reveals that the histidine at position E10 is conserved in many instances (Fig. 6.13). In view of the behaviour of *Synechococcus* 7002 GlbN (30% identity with *N. commune* GlbN) and *Synechocystis* 6803 GlbN (40% identity with *N. commune* GlbN), it can be proposed that the spurious haemichrome

Table 6.4 Forms of *Synechococcus* sp. PCC 7002 GlbN inspected by ^{15}N NMR relaxation[a]

Cross-linked	Spin state	Non-cross-linked	Spin state
Fe(III)GlbN–A–His	1/2	Fe(III)GlbN–R–His	1/2
Fe(III)GlbN–A–CN	1/2	Fe(III)GlbN–R–CN	1/2
Fe(II)GlbN–A–His	0	–	
Fe(II)GlbN–A–CO	0	–	

[a]The post-translationally modified protein is referred to as 'cross-linked' or GlbN-A. The unmodified protein is referred to as GlbN-R. (Pond et al., 2012).

Table 6.5 Three-dimensional structures of cyanobacterial and algal globins

Source	Protein	PDB ID	Method	Resolution (Å)	Characteristics	References
Chlamydomonas eugametos	CtrHb	1DLY	X-ray	1.8	Cyanomet complex	Pesce et al. (2000)
C. eugametos	CtrHb	1UVX	X-ray	2.45	Cyanomet complex, Xe pressure	Milani, Ouellet, et al. (2004)
Synechocystis sp. PCC 6803	GlbN	1MWB	NMR	–	Ferric bis–histidine	Falzone et al. (2002)
Synechocystis sp. PCC 6803	GlbN	1RTX	X-ray	1.8	Ferric bis–histidine, haem PTM[a]	Hoy et al. (2004)
Synechocystis sp. PCC 6803	GlbN	1S69	X-ray	1.68	Cyanomet complex, haem PTM	Trent et al. (2004)
Synechocystis sp. PCC 6803	GlbN	1S6A	X-ray	1.69	Azidomet complex, haem PTM	Trent et al. (2004)
Synechocystis sp. PCC 6803	GlbN	2HZ1	X-ray	1.80	Ferrous bis–histidine complex haem PTM	Hoy, Smagghe, et al. (2007)
Synechocystis sp. PCC 6803	GlbN	2HZ2	X-ray	2.00	H117A variant, ferric bis–histidine complex	Hoy, Smagghe, et al. (2007)
Synechocystis sp. PCC 6803	GlbN	2HZ3	X-ray	1.90	H117A variant, ferrous bis–histidine complex	Hoy, Smagghe, et al. (2007)
Synechococcus sp. PCC 7002	GlbN	2KSC	NMR	–	Ferric bis–histidine, haem PTM	Scott et al. (2010)
Synechococcus sp. PCC 7002	GlbN	4I0V	X-ray	1.45	Ferric bis–histidine, haem PTM	–[b]
Synechococcus sp. PCC 7002	GlbN	4L2M	X-ray	2.25	Cyanomet complex, haem PTM	–[b]

[a]The haem post-translational modification is the covalent attachment of the haem group to His117 (Vu et al., 2002)
[b]Structures solved by B. B. Wenke and J. L. Schlessman in the Lecomte laboratory.

obtained in the original preparation of *N. commune* GlbN (Thorsteinsson et al., 1996) corresponds to the coordination of His E10 on the distal side. However, this also indicates that the degree of endogenous coordination cannot be anticipated from the primary structure. In the absence of exogenous ligand, it is not obvious whether modelling based on the open conformation of CtrHb or the closed conformation of *Synechocystis* 6803 GlbN (or any intermediate state) should be selected. Sequence retrieval and alignment using CtrHb as the query show that lysine is a common residue at position E10 and that tyrosine is a conserved residue at B10. In this group of proteins as well, some degree of endogenous hexacoordination may be expected.

5.1.5.3 Tunnels and cavities

With a similar degree of reservation, it is reasonable to assume that TrHb1s display access tunnels when they are in the conformation containing an exogenous ligand. The volume of the tunnels and cavities and the identity of gating residues allowing or blocking access to the haem distal side may vary. For example, Phe E15 is a gating residue in *M. tuberculosis* trHbN (a TrHb1; Milani, Pesce, et al., 2004), but it is not conserved in the cyanobacterial and algal set. Whether gating also occurs in other structures is not known.

5.1.5.4 Distal hydrogen bond network

In the TrHb1s that contain hydrogen-bonding residues at positions B10, E7 and E11, it is likely that the hydrogen bond network observed in cyanomet CtrHb (PDB ID 1DLY, Pesce et al., 2000), cyanomet *Synechocystis* 6803 GlbN (PDB ID 1S69, Trent et al., 2004) or cyanomet *Synechococcus* 7002 GlbN (PDB ID 4L2M) is formed. The conservation of these residues in TrHb1-1s supports this expectation. The strength of the network and its exact geometry, however, are difficult to predict. The geometry of the TrHb1-2s bound state is unknown since these proteins have a hydrophobic residue in place of Tyr or His B10.

5.1.5.5 Covalent haem attachment

The same words of caution hold for the formation of the histidine–haem bond described for *Synechococcus* 7002 and *Synechocystis* 6803 GlbNs. A histidine is found at the identical position in several cyanobacterial TrHb1s (Fig. 6.13), and the modification follows a simple mechanism (Preimesberger, Wenke, Gilevicius, Pond, & Lecomte, 2013; see

```
          --AAAAAAABBBBBBBBBBBBBBBBBBBB--CCCCCCC--EEEEEEEEEEEEEEEEE-hhhhh-FFFFFFFF
           :*:::::**     :    *: ::     :: *  :.  **   ** :*  :*  *.:  :. .  : * * :*
Q8RT58_SYNP2/1-124    MASLYEKLGGAAAVDLAVEKFYGKVLADERVNRFFVNTDMAKQKQHQKDFMTYAFGGTDRFPGRSMRAAH
I4HJM1_MICAE/1-124    MKTLYEKLGGAAAVDLAVEKFYEKVLADERVQRFFAPTDMQQQKQHQKAFMTYAFGGAEQWNGRPMRDAH
I4H5U0_MICAE/1-124    MKTLYEKLGGAAAVDLAVEKFYEKVLADERVQRFFAPTDMQQQKQYQKAFMTYAFGGAEQWNGRPMRDAH
K9ZA57_CYAAP/1-124    MSSLYEKLGGKEAVDIAVDKFYDKVLKDDRVNYFFANTDMKKQKAHQKAFMTYAFGGPDTFSGRDMRESH
C7QR53_CYAP0/1-124    MNTLYERLGGEAAINLAVDKFYQKVLADERINHFFANLDMKQQIKHQKAFMTYAFGGGNNWNGRSMRQAH
B8HSM2_CYAP4/1-124    MSTLFEKLGGSAAIQLAVDKFYERVLQDDRVKHFFADVDMDKQKDHQRAFLTYAFGGAPQYDGRTMRKAH
TRHBN_SYNY3/1-124     MSTLYEKLGGTTAVDLAVDKFYERVLQDDRIKHFFADVDMAKQRAHQKAFLTYAFGGTDKYDGRYMREAH
B0CBZ4_ACAM1/1-124    MPTLFDKLGGSAAVDLAVDKFYERVLQDERINHFFANTDMVRQRAHQKAFLTYAFGGTDKYDGRHMRAAH
L8N569_9CYAN/1-124    MTTLFEKLGGAAAVDLAVDRFYERVLQDDRIKHFFADVDMKKQRSHQKAFLTYAFGGTDKYDGQLMRQAH
B7KI32_CYAP7/1-124    MSSLFEQLGGQEAVDLTVDKFYERVLKDERVKHFFDDVDMVKQRQHQKQFLTYAFGGSSKYSGKAMRQSH
H1WKW8_9CYAN/1-124    MATLFEKLGGKDAVDLAVDKFYERVLNDDRIKHFFVNTDMKKQRSHQKAFLTYAFGGSDKYDGRYMREAH
K9TPV2_9CYAN/1-124    MATLFEKLGGKAGVETAVDKFYQRVLNDDRIKHFFEGVDMVKQRAHQRAFLTYAFGGTDRYDGLQMREAH
K6EIG6_SPIPL/1-124    MTTLFDKLGGKDAVYLVVDKFYERVLQDNRIKHFFANIDMKKQLSHQKAFLTYAFGGTDKYDGRYMREAH
L8LUN7_9CHRO/1-124    MTTLFDKLGGADAVDLAVDKFYERVLQDDRIKHFFEHIDMVKQRSHQKAFLTYAFGGTDTYNGRYMREAH
B4VMT4_9CYAN/6-123    NLTLYEKLGGQPVVAQIVDDFYQRVLADDTVSHFFANTDMEKQRRHQTAFVSHALGGPNQYTGRSMEKAH
F5UFJ7_9CYAN/2-119    PTTLYEKMGGESAIKEMVDDFYRDVLADEIVSHFFDHTDMEKQRRHQTAFISYALGGSQQYSGRSMEKAH
K9XN27_9CHRO/3-120    ATTLYEKLGGEQAIKQVVDDFYTRVLADDTVNSFFAHTDMEKQRRHQTAFISFALGGSPTPYTGRSMEKAH
K9PBS7_9CYAN/1-117    -MSLYDKLGGKPAIEKVVDELHKRILADGSLKPFFAKTDMAKQRNHQVAFFTQIFEGPKEYKGRAMDKTH
K9RI40_9CYAN/1-118    MSTLYEKLGGQPTIEKVVDDFHNRIMADSTVSGFFANTDMEKQRDHQIGFFSLILGGPKDYKGRSMDKTH
K9QF19_9NOSO/2-119    SATLYEKLGGQPTIEKIVDDLHKRIVADNTLKPFFANTDMAKQRAHQIAFFSLIFEGPKQYTGRPMDKTH
                      1.......10........20........30........40........50........60........70
```

```
          FFFFFFFF-GGGGGGGGGGGGGGGGGGGGG--HHHHHHHHHHHHHHHHHHHHHHHH--
               .*  ****: :.*  ::           : :*
Q8RT58_SYNP2/1-124    QDLVENAGLTDVHFDAIAENLVLTLQELNVSQDLIDEVVTIVGSVQHRNDVLNR
I4HJM1_MICAE/1-124    KELVAEMGLTDSHFDAIAEDLVATLVELEVPQALIDEVVQIVGSVTHRNDVLNR
I4H5U0_MICAE/1-124    KELVAEMGLTDSHFDAIAEDLVATLVELEVTQALIDEVVQIVGSVTHRNDVLNR
K9ZA57_CYAAP/1-124    KHLVDNMGLTDVHFDAIAENLINTLKELGIEQNLIDEVGAIVGAVSHRNDVLNR
C7QR53_CYAP0/1-124    QKLVEEKGLTDIHFDAVAEDLVATLVELKIPQELIDEVVQIVGSVSHRNDVLNR
B8HSM2_CYAP4/1-124    QKLVAEQGLNGNHFDAIAEDLVLTLQELGISQDLIDQVVAIAGAPEHRSDVLNQ
TRHBN_SYNY3/1-124     KELVENHGLNGEHFDAVAEDLLATLKEMGVPEDLIAEVAAVAGAPAHKRDVLNQ
B0CBZ4_ACAM1/1-124    NELVEKQGLKGEHFDAVAENLIATLKDMGVSEELMAEVAAVAAAPQHKKDVLNQ
L8N569_9CYAN/1-124    KELVEKQGLSGDEHFDAVAEDLILTLREMGVSDELIDEVAAVAAAPQRKKDVLNG
B7KI32_CYAP7/1-124    KQLVQEKGLSDEHFDAIVEDLVETLKELEVSENLIEQVKSIAGDIHHRNDILNR
H1WKW8_9CYAN/1-124    KALVEEQGLSSEHFDAVAEDLMETLKEMGVPDDLLAEVAAVAAAPQHKKDVLNQ
K9TPV2_9CYAN/1-124    KELVEERGLKSEHFDAVAENLLETLREMGVSEDLIAQVATVAAAPQHKKDVLNQ
K6EIG6_SPIPL/1-124    QALVQEQGLNSEHFDAVAEDLIETLKEMGVADDLLAEVAAIVAAPQHQKDVLNQ
L8LUN7_9CHRO/1-124    KELVEKQGLNSEHFDAVAENLIETLKEMGVPAELIAEVAAVAAAPEHKKDLLNQ
B4VMT4_9CYAN/6-123    AG----LDLQPEHFDAIAKHLGESLDEYGLTQEEINSVLER--ISTLKEAVLYK
F5UFJ7_9CYAN/2-119    AG----LNLQPEHFDAIVKHLTDAIIAHGASQEDLDKALAK--IATLKDAVLYK
K9XN27_9CHRO/3-120    AG----LNLQPEHFDAIVKHLSEALEVHHVPPAEINKILDR--ITTLKEAVLYK
K9PBS7_9CYAN/1-117    TG----MSLQQPHFDAILKHLSESMAAAGVSADNTKAALAG--VTALKGSILNK
K9RI40_9CYAN/1-118    TG----MGLQQPHFDAISKHLSSAMTSSGVSADDAKAAMAE--VEKLKPAILNK
K9QF19_9NOSO/2-119    TG----MSLQEQHFDAIAKHLTGAMSVAGVSADDTKAALDR--VASLKGAILNK
                      ........80........90........100........110........120...
```

Figure 6.13 Alignment of 20 cyanobacterial globins using *Synechococcus* sp. strain PCC 7002 as the query. The top line indicates secondary structure as found in the query protein (PDB ID 4I0V). Residues in bold are at positions B10, E10, F8 and H16, as numbered by structural homology to the canonical 3/3 fold. The proteins and organisms are: Q8RT58_SYNP2 *Synechococcus* sp. strain PCC 7002; I4HJM1_MICAE *Microcystis aeruginosa* PCC 9808; I4H5U0_MICAE *M. aeruginosa* PCC 9807; K9ZA57_CYAAP *Cyanobacterium aponinum* strain PCC 10605; C7QR53_CYAP0 *Cyanothece* sp. strain PCC 8802; B8HSM2_CYAP4 *Cyanothece* sp. strain PCC 7425/ATCC 29141; TRHBN_SYNY3 *Synechocystis* sp. strain PCC 6803; B0CBZ4_ACAM1*Acaryochloris* marina strain MBIC 11017; L8N569_9CYAN *Pseudanabaena biceps* PCC 7429; B7KI32_CYAP7 *Cyanothece* sp. strain PCC 7424; H1WKW8_9CYAN *Arthrospira* sp. PCC 8005; K9TPV2_9CYAN *Oscillatoria acuminata* PCC 6304; K6EIG6_SPIPL *Arthrospira platensis* str. Paraca; L8LUN7_9CHRO *Gloeocapsa* sp. PCC 73106; B4VMT4_9CYAN *Coleofasciculus chthonoplastes* PCC 7420; F5UFJ7_9CYAN *Microcoleus vaginatus* FGP-2; K9XN27_9CHRO *Gloeocapsa* sp. PCC 7428; K9PBS7_9CYAN *Calothrix* sp. PCC 7507; K9RI40_9CYAN *Rivularia* sp. PCC 7116; K9QF19_9NOSO *Nostoc* sp. PCC 7107.

Section 5.4.2). However, only experimental data can ascertain whether or not haem attachment occurs in these proteins.

5.2. Ligand-binding properties

5.2.1 Theoretical and practical aspects of ligand-binding studies

The overwhelming majority of haemoglobins bind dioxygen and several other exogenous ligands. (It is interesting that two exceptions have been recently noted among the many globins of *Caenorhabditis elegans*: Glb-6 is a bis-histidine haemoglobin that does not bind any ligand (Yoon et al., 2010), whereas Glb-23, also a bis-histidine haemoglobin, binds CO but not O_2 (Kiger et al., 2011).) Because globin function is intimately linked to rate constants for association and dissociation of ligands, kinetic studies have been systematically performed on a large number of purified globins having wild-type or altered sequences. The aggregate data provide a highly informative view of structure–activity relationship in the superfamily, and they help formulate or eliminate hypotheses for proteins without obvious function.

Two methods were developed many years ago for measuring rate constants in canonical haemoglobins and myoglobins. The first involves mixing solutions of protein and ligand in a rapid-mixing stopped-flow apparatus and monitoring the changes in optical spectra as the ligand associates with the protein (Olson, 1981). Because of dead time and sensitivity consideration, the rapid-mixing approach is suitable for relatively slow ($k < 10\ \mu M^{-1}\ s^{-1}$) association events. Rapid-mixing stopped-flow experiments can also determine dissociation rate constant by ligand displacement (i.e. O_2 in oxy Mb displaced by CO). Flash photolysis on the other hand consists in applying a short pulse of intense light to cause practically instantaneous dissociation of a distal ligand (Gibson, Olson, McKinnie, & Rohlfs, 1986). This is the method of choice for determining faster rate constants. Rapid mixing applies best to ferric states as these tend to exhibit slower kinetics, whereas flash photolysis is reserved for ferrous states, which tend to have photolabile ligands. The two methods are complementary and have been used to obtain rate and binding constants. Myoglobin and numerous variants thereof have been extensively studied with these methods and we now have a detailed energetic landscape for this protein (Olson & Phillips, 1996; Olson, Soman, & Phillips, 2007; Scott, Gibson, & Olson, 2001; Tsai, Martin, Berka, & Olson, 2012).

Endogenous hexacoordination has multiple physico-chemical consequences, the most obvious one being the requirement to displace the distal

$$Hb(6c) \underset{k_{+X}}{\overset{k_{-X}}{\rightleftarrows}} Hb(5c)$$

$$Hb(5c) + L \underset{k_{-L}}{\overset{k_{+L}}{\rightleftarrows}} Hb\text{-}L$$

Figure 6.14 The simplest reaction scheme for exogenous ligand binding to a haemoglobin exhibiting endogenous hexacoordination. The axial ligands to the iron in the starting material, Hb(6c), are the proximal histidine and X (e.g. Tyr B10 in CtrHb or His E10 in *Synechococcus* 7002 GlbN). The pentacoordinate state with free distal site is represented with Hb(5c). L represents an exogenous ligand (e.g. O_2). The rate constants are k_{-X} (s^{-1}) for the dissociation of X, k_{+X} (s^{-1}) for its association, k_{+L} ($M^{-1}s^{-1}$) for the binding of L, and k_{-L} (s^{-1}) for its dissociation. Equilibrium constants are $K_X = k_{+X}/k_{-X}$ for the association of X and $K_L = k_{+L}/k_{-L}$ for the association of L.

histidine in order to bind an exogenous ligand. This additional step in the binding process renders the evaluation of affinities and rate constants particularly complex. The simplest overall reaction is represented with two equilibria as in Fig. 6.14.

Depending on the relative values of the rate constants and the concentration of ligand L, certain simplifications can be made to the equation representing the observed changes in concentrations with time, and different kinetic regimes can be distinguished. Two common assumptions are that the dissociation of L is very slow (except of course when flashed off), and that L is in excess so that its concentration remains practically constant throughout the course of the experiment (Hargrove, 2000). Under those conditions, the second step is considered irreversible ($k_{-L} \sim 0$) and the product $k_{+L} \times [L] = k'_{+L}$ is a pseudo first-order rate constant. Solving the set of differential equations yields three eigenvalues, λ_1, λ_2 and λ_3. The first two are given by

$$\lambda_{1,2} = \frac{1}{2}\left(k_{-X} + k_{+X} + k'_{+L} \pm \sqrt{(k_{-X} + k_{+X} + k'_{+L})^2 - 4k_{-X}k'_{+L}} \right)$$

and the third, λ_3, is 0 by conservation of mass (Beard & Qian, 2008). For example, if k_{+X} and $k_{-X} \gg k'_{+L}$, the kinetics present a slow phase corresponding to the product of k'_{+L} by the rapid equilibrium fraction of Hb(5c). Additional complexity is introduced when the pseudo first-order approximation does not hold, and all scenarios are possible, depending on the protein and the chosen conditions for the experiment. Within the pseudo first-order regime and if the solution conditions (i.e. temperature, pH, etc.) are kept constant, the same eigenvalues should hold for all concentration conditions so that global fitting can resolve the individual constants. Furthermore, the kinetic information can be combined to yield thermodynamic parameters, that is, binding constants or free energy of binding.

In the rapid-mixing experiment, the protein sample is at equilibrium prior to encounter with the ligand L. Thus, the initial concentrations of Hb(6c) and Hb(5c) are dictated by $K_X = k_{+X}/k_{-X}$. Two versions of the flash photolysis experiment can be implemented. In the first, there is no added exogenous ligand. X is flashed off, leaving Hb(5c) behind. The kinetics of X rebinding is then obtained (Vos et al., 2008). In the second, HbL is prepared and L is flashed off, again leaving Hb(5c) behind and allowing for measurements of L rebinding. This has the advantage of simpler initial conditions for the determination of k_{+L}. Nevertheless, the kinetic behaviour of the haemoglobins considered in this chapter has not been easy to interpret. This is in part because the mechanism can be considerably more complicated than captured by the two-equation model and because the concentration regimes necessary for complete description may be difficult to access experimentally. Higher complexity is also expected of proteins that are not monomeric, although so far, the studied algal and cyanobacterial globins have no quaternary structure interfering with the measurements.

The function of a globin in the cell depends on the flux of ligands, itself dependent on all the rate constants pertaining to the system and the intracellular concentrations. For realistic modelling, rate-determining constants are required under the conditions that best represent cellular metabolic states. The organisms considered here vary widely in their physiology and the conditions they encounter. In fact, cyanobacteria form an extremely complex phylum (Bryant & Frigaard, 2006). The two cyanobacteria *Synechocystis* 6803 and *Synechococcus* 7002 differ in the enzymes participating in nitrogen metabolism, which compromises extension of physiological data from one globin to another. Even with well-determined *in vitro* rate constants, we are far from the goal of understanding the role of globins in the cell.

5.2.2 Binding of endogenous and exogenous ligands
Table 6.6 lists the binding and equilibrium constants for *N. commune* GlbN, CtrHb and *Synechocystis* 6803 GlbN along with selected reference proteins. The parameters were obtained with rapid-mixing and flash photolysis methods. The first observation is that dioxygen affinity spans a wide range of values (see also Table 6.7). As observed for other globins, this is due to

Table 6.6 Kinetic parameters of gaseous ligand binding to ferrous cyanobacterial, algal and selected globins

Protein	k'_{O_2} ($\mu M^{-1} s^{-1}$)	k_{O_2} (s^{-1})	K_{O_2} (μM^{-1})	Conditions	References
Sperm whale Mb	15	18	0.83	pH 7, 20 °C	Thorsteinsson et al. (1999)
Rice Hb1	68	0.038	1800	pH 7	Arredondo-Peter et al. (1997)
Soybean Lba	130	5.6	23	pH 7, 20 °C	Hargrove et al. (1997)
Nostoc commune UTEX 584 GlbN	390	79	4.8	pH 7, 20 °C	Thorsteinsson et al. (1999)
Synechocystis sp. PCC 6803 GlbN	240	0.014	~100[a]	pH 7, 20 °C	Hvitved, Trent, Premer, and Hargrove (2001) and Kundu, Premer, Hoy, Trent, and Hargrove (2003)
Chlamydomonas eugametos CtrHb	(192, 145)[b]	0.0141		pH 9.5, 20 °C	Couture et al. (1999)

Protein	k'_{CO} ($\mu M^{-1} s^{-1}$)	k_{CO} (s^{-1})	K_{CO} (μM^{-1})	Conditions	References
Sperm whale Mb	0.53	0.019	27	pH 7, 20 °C	Thorsteinsson et al. (1999)
Rice Hb1	7.2	0.001	7200	pH 7	Arredondo-Peter et al. (1997)
Soybean Lba	17	0.0078	2200	pH 7, 20 °C	Hargrove et al. (1997)
Nostoc commune UTEX 584 GlbN	41	0.010	4100	pH 7, 20 °C	Thorsteinsson et al. (1999)
Synechocystis sp. PCC 6803 GlbN	90	–	14,000[a]	pH 7, 20 °C	Hvitved et al. (2001), Kundu et al. (2003), and Smagghe, Sarath, Ross, Hilbert, and Hargrove (2006)
Chlamydomonas eugametos CtrHb	(165, 113)[b]	0.0022		pH 9.5, 20 °C	Couture et al. (1999)

Continued

Table 6.6 Kinetic parameters of gaseous ligand binding to ferrous cyanobacterial, algal and selected globins—cont'd

Protein	k'_{NO} $(\mu M^{-1}\,s^{-1})$	k_{NO} (s^{-1})	K_{NO} (μM^{-1})	Conditions	References
Sperm whale Mb	22	0.000098	220,000	pH 7, 20 °C	Thorsteinsson et al. (1999)
Soybean Lb*a*	170	0.00002	9,000,000	pH 7, 20 °C	Hargrove et al. (1997)
Nostoc commune UTEX 584 GlbN	600	0.00022	2,700,000	pH 7, 20 °C	Thorsteinsson et al. (1999)
Chlamydomonas eugametos CtrHb	(169, 101)[b]			pH 9.5, 20 °C	Couture et al. (1999)

[a]Equilibrium affinity measurement by competition (Kundu et al., 2003).
[b]Numbers are listed as $(k_{-X}, k_{+X}/k_{+L})$ in s^{-1} and μM. See de Sanctis et al. (2004) for an additional compilation.

Table 6.7 Oxygen affinity and partition coefficient

Protein	p_{50} (mm Hg)	CO/O_2 partition	References
Sperm whale myoglobin	0.5	25	Antonini and Brunori (1971) and Olson and Phillips (1997)
Nostoc commune GlbN	0.55	850	Thorsteinsson et al. (1996)
Synechocystis sp. PCC 6803 GlbN[a]	0.01	140	Kakar et al. (2010)
Chlamydomonas eugametos LI637 Hb, haem domain	0.03	5.0	Couture et al. (1999) and de Sanctis et al. (2004)

[a]The protein likely carried the haem post-translational modification (covalent attachment of the haem group to His117, Vu et al., 2002).

large variations in the dissociation rate constants. *N. commune* GlbN has a relatively low affinity, comparable to that of sperm whale myoglobin, whereas CtrHb and *Synechocystis* 6803 GlbN bind tightly. Interestingly, *N. commune* GlbN differs in the identity of the residues at position B10 (a histidine) and E11 (a leucine) compared with the other two TrHb1s. As mentioned earlier, these residues participate in the hydrogen bond network on the distal side of the haem (Fig. 6.9). This may account for the low dioxygen affinity.

Carbon monoxide (CO) and nitric oxide (NO) are excellent ligands for the haem ferrous iron. In fact, CO has a significantly higher affinity for the free haem than O_2. The discrimination factor (or $M = K_{CO}/K_{O_2}$) is greater than 20,000 in a non-polar solvent (Traylor et al., 1981), a value so large that if it were maintained by a globin, a function as reversible dioxygen binder would be impaired by irreversible CO poisoning. Discrimination among the three gaseous ligands, which have approximately the same size and are apolar, is therefore an essential property conveyed by the haem environment. In sperm whale myoglobin, the M for CO value drops to 25. The 800-fold reduction is attributed to hydrogen bonding by the distal histidine. The polarity of the distal histidine and other H-bonding residues in the distal pocket is also at work for the discrimination of NO (Olson & Phillips, 1997; Spiro & Soldatova, 2012; Tsai, Berka, Martin, & Olson, 2012).

Unfortunately, there is little information on the capacity of cyanobacterial and algal globins to discriminate among ligands. The numbers presented in Table 6.7 should be taken with caution as they are affected by the difficulty in measuring ligand affinities for these proteins. At face value, no specific pattern emerges, although it is interesting to note that *N. commune* GlbN does not discriminate as efficiently as the other two proteins, a property likely to be linked to the composition of the distal pocket mentioned earlier.

5.3. Auto-oxidation, redox potential and electron transfer
5.3.1 Auto-oxidation

A ferrous globin, as it combines with dioxygen, is susceptible to oxidation. The products are the ferric protein, unable to bind O_2, and the superoxide anion (Wever, Oudega, & Van Gelder, 1973). For the globins that perform transport and storage, the rate at which this process occurs is a physiologically important chemical characteristic. The reaction, known as auto-oxidation (George & Stratmann, 1954), has been the subject of numerous mechanistic studies (Brantley, Smerdon, Wilkinson, Singleton, & Olson, 1993; Shikama, 2006). In general, the rate constant for auto-oxidation, k_{ox}, depends strongly on pH and dioxygen concentration. The molecular explanation advanced for myoglobin is as follows. At low oxygen concentration, if the hydrogen bond between the bound O_2 molecule and the distal histidine breaks, dioxygen can dissociate from the iron. It leaves behind deoxymyoglobin, which contains a loosely associated water molecule in the distal pocket. Iron oxidation then occurs with free dioxygen in a bimolecular fashion resulting in aquomet myoglobin. At a higher concentration of oxygen, the dominant mechanism involves the protonation of bound dioxygen and dissociation of HO_2, leaving behind a ferric ion (Brantley et al., 1993), which then binds a water molecule. For reference, the auto-oxidation of sperm whale myoglobin occurs with a rate constant, k_{ox}, of 0.055 h^{-1} at 37 °C and neutral pH (Springer, Sligar, Olson, & Philips, 1994). This corresponds to a half-life of ~12 h. Details of the mechanism vary according to, among other features, the nature of the residues in the distal pocket and the ease of water access.

N. *commune* GlbN has a relatively high rate of auto-oxidation (half-life shorter than 12 h, Thorsteinsson et al., 1996). CtrHb remains stable in the oxy state for longer periods (half-life of 7 days at pH 8, Couture & Guertin, 1996). Replacement of Tyr B10 accelerates the rate of dioxygen dissociation and also enhances the rate of auto-oxidation. This is in line with the observation in other globins that weaker oxygen affinity is accompanied with faster auto-oxidation (Shibata et al., 2012). Auto-oxidation rates for *Synechococcus* and *Synechocystis* GlbNs are not available, but it should be noted that endogenous hexacoordination is expected to enhance the rate (Weiland, Kundu, Trent, Hoy, & Hargrove, 2004). The direct significance of auto-oxidation rates for globins that do not transport or store oxygen and may not exhibit a stringent necessity to remain in the ferrous state is not clear.

5.3.2 Electron transfer and redox potential

Canonical globins, which utilize the proximal histidine as the only protein ligand, have slow electron transfer rates due at least in part to the necessary change in exogenous ligand accompanying the switch between ferric and ferrous states. To return to the sperm whale example, the ferric state is stable as an aquomet complex,whereas the ferrous state has no distal ligand. This raises the re-organization energy associated with a change in redox state. Endogenous hexacoordination, however, can reduce the energetic cost. This is apparently the case in the bis-histidyl GlbNs (Preimesberger, Pond, Majumdar, & Lecomte, 2012). NMR measurements have determined the rate of electron self-exchange for these proteins to be slower than for cytochrome *c* but still sufficient to consider electron transfer as a possible component of function. Thus, it is conceivable that in a catalytic mechanism involving electron transfer, ferric and ferrous bis-histidine states are intermediates that can undergo facile reduction or oxidation.

The standard reduction potential of the cyanobacterial and algal proteins discussed here has not been exhaustively studied. Values of −195 mV (reduction midpoint potentials vs. the standard hydrogen electrode; Halder, Trent, & Hargrove, 2007) and −150 mV (Lecomte, Scott, Vu, & Falzone, 2001) have been reported for *Synechocystis* 6803 GlbN, although the latter measurement was made before awareness of the post-translational modification and may be inaccurate. Regardless, a thermodynamic preference for the ferric state is apparent in comparison with myoglobin. Additional oxidation–reduction measurements in *Synechocystis* 6803 GlbN have demonstrated that bis-histidine coordination is tighter in the ferric state than in the ferrous state (Halder et al., 2007). Weaker haem affinity in the ferrous state is also apparent in *Synechococcus* 6803 GlbN.

5.4. Chemical reactivity and possible function

Section 4 exposed how little is known about cyanobacterial and algal globins from focused investigations *in vivo*. Because of this lack of information, there have been attempts to circumscribe the function of cyanobacterial and algal globins using *in vitro* information. The approach suffers from the fact that haem proteins have a great capacity for performing chemistry, so that an activity that is detected in the test tube may not be effective *in vivo*. The cellular context, including localization, determines which of the various possibilities are physiologically relevant.

5.4.1 Endogenous hexacoordination

The coordination of the distal histidine has consequences for ligand binding and for electron transfer. Furthermore, blocking the distal site can result in diminished reactivity. Thus, bis-histidine *Synechocystis* 6803 and *Synechococcus* 7002 GlbNs react slowly with hydrogen peroxide, whereas variants substituting a non-ligating residue for the distal histidine tend to be more susceptible to haem damage (Nothnagel et al., 2011). Endogenous hexacoordination may be essential for proteins that function under high levels of oxidative stress as encountered during oxygenic photosynthesis.

5.4.2 Haem post-translational modification

A hallmark of *Synechococcus* 7002 and *Synechocystis* 6803 GlbNs is their ability to attach the haem covalently and irreversibly to the protein matrix. The mechanism of this post-translational modification has interest not only from a chemical point of view, but also because the modification conveys exceptional properties to the protein and therefore raises questions of molecular evolution. Unlike cytochrome *c*, which requires a dedicated machinery for maturation in the cell and complete formation of the correct thioether linkages (Kranz, Richard-Fogal, Taylor, & Frawley, 2009), GlbN undergoes the modification spontaneously in the ferrous state, without the aid of chaperones or enzymes. The reaction is adequately described as an electrophilic addition (Nothnagel et al., 2011). Interestingly, because electron transfer occurs readily in the bis-histidine state, mixtures of ferrous and ferric proteins convert fully to the post-translationally modified form via a linear redox chain reaction (Preimesberger et al., 2012). Thus far, the only known requirement for reaction is the ability of the reactive histidine and the haem to adopt a relative orientation such that sp^3 geometry is achieved at the vinyl site (Preimesberger et al., 2013). Sequence inspection (Fig. 6.13) suggests strongly that several other cyanobacterial GlbNs are capable of the same chemistry because it is likely that the topology of *Synechococcus* 7002 and *Synechocystis* 6803 GlbN is compatible with the listed primary structures.

The post-translational modification of the haem occurs in *Synechococcus* 7002 cells (Scott et al., 2010). A possible benefit of the covalent attachment is the retention of the haem in the haem pocket (Hoy, Smagghe, et al., 2007), although it is not clear that in the cell, haem loss is a problem that needs remedy. Also, as demonstrated by many TrHb1s, the 2/2 fold can be decorated by side chains that ensure high haem-binding constants. Low haem affinity, however, may be particularly problematic for photosynthetic organisms because *b* haem is used for the synthesis of linear tetrapyrrole pigments. Attaching the haem to the globin would prevent haem dioxygenases to act on this particular source of starting material. It is also possible that the attached haem is less prone to oxidative damage, another advantage in organisms that must tolerate high light and high O_2 conditions.

5.4.3 NO dioxygenation

Nitric oxide and other RNS are known to elicit a variety of responses in microbes (Bowman, McLean, Poole, & Fukuto, 2011). These responses involve a number of haem proteins, often including globins. *In vitro*, practically all globins have the ability to perform some degree of NO dioxygenation, a reaction that converts NO to nitrate as shown in the following equation

$$O_2 + NO + 1e^- \rightarrow NO_3^- \tag{6.1}$$

This is the main function of FHbs (Gardner, 2005), a large set of bacterial proteins that continue to inform on globin chemistry, electron transfer and flavin-containing haem enzymes (Mowat, Gazur, Campbell, & Chapman, 2010). NO dioxygenation appears to be an ancestral activity that was eroded over the years in the globins responsible for reversible dioxygen binding.

Among the properties that are consistent with efficient NO di-oxygenation are high O_2 affinity and high superoxo character to the bound oxygen, ability to limit the amount of bound NO competing with O_2 for the ferrous iron, high auto-oxidation rates and internal electron transfer rates appropriate for efficient catalysis (Gardner, 2012; Mowat et al., 2010). The set of favourable characteristics can be traced with some success to specific structural features, including a distal tyrosine interacting with the bound oxygen (Gardner, Martin, Gardner, Dou, & Olson, 2000; Mukai, Mills, Poole, & Yeh, 2001). It must be noted, however, that the nitric oxide dioxygenase (NOD) function requires the participation of a reductase or a reducing agent in order to return the iron to the ferrous state after dioxygenation.

NO dioxygenation naturally appears among the functions proposed for cyanobacterial and algal globins because these organisms have to cope with NO as a product of nitrate and nitrite metabolism (Mallick, Rai, Mohn, & Soeder, 1999; Sakihama, Nakamura, & Yamasaki, 2002) or other metabolic processes. The identification of a reductase remains a difficult step in solid-ifying the hypothesis of NOD function (Smagghe et al., 2008). In one case, that of the assimilatory nitrate reductase of the raphidophyte *Heterosigma akashiwo*, a TrHb1 domain is present between a cytochrome b_5 domain and a FAD-binding domain (Stewart & Coyne, 2011). NO dioxygenation is facilitated by the proximity of the latter domain. For globins that are not chimeric, the problem remains. It is possible that ferredoxin reductases play a role (Gardner, 2012), along with the pool of ferredoxin (Scott et al., 2010).

In an effort to explore the functional possibilities of *Synechocystis* 6803 GlbN, Hargrove and co-workers examined the response to NO challenge of an engineered *E. coli* strain deprived of its FHb and harbouring the *glbN* gene. GlbN rescues the wild-type phenotype as an indication that the protein is capable of protecting its heterologous host. A lag in the response also sug-gests that a reductase is needed to be produced for detoxification to occur; however, the nature of this reductase is not known. Although this experiment demonstrates the participation of GlbN in protection, it unfortunately does not inform on its role in the *Synechocystis* cell (Smagghe et al., 2008).

5.4.4 Other reactions
In addition to the NO dioxygenase activity, *Synechocystis* 6803 GlbN is capa-ble of nitrite reduction under anaerobic conditions (Sturms, DiSpirito, & Hargrove, 2011).

$$NO_2^- + 2H^+ + 1e^- \rightarrow NO + H_2O \qquad (6.2)$$

This activity has been observed in other globins (Gladwin, Grubina, & Doyle, 2009), although it often proceeds at too slow a rate to be considered of physiological relevance. The reaction uses nitrite and one (two) proton(s) to generate NO and a hydroxyl ion (water molecule). The ferrous protein provides the necessary electron and ends up in the ferric state. Thus, for turnover to occur, the protein must be restored to the ferrous state, and the issue of reduction is present for this particular activity as well as the NOD activity. *In vitro*, the rates of nitrite consumption by *Synechocystis* 6803 GlbN (as measured by the disappearance of ferrous protein) are faster than those for cytoglobin (Li, Hemann, Abdelghany, El-Mahdy, & Zweier, 2012) and neuroglobin (Petersen, Dewilde, & Fago, 2008). This reaction, however, does not seem to be useful for generating free NO (Sturms, DiSpirito, et al., 2011; Tiso, Tejero, Kenney, Frizzell, & Gladwin, 2012).

Another reaction in the realm of nitrogen oxides is hydroxylamine reduction.

$$NH_2OH + 3H^+ + 2e^- \rightarrow NH_4^+ + H_2O \qquad (6.3)$$

Hydroxylamine is an intermediate species in the reduction of nitrite to the ammonium ion. When presented with hydroxylamine under anaerobic conditions, *Synechocystis* 6803 GlbN releases an ammonium ion (Sturms, DiSpirito, Fulton, & Hargrove, 2011). Here as well, the *in vitro* reaction is carried out faster by GlbN than other globins. The two-electron process requires re-reduction of the protein. Although participation in anaerobic respiration has been proposed, it is unclear how an energy-yielding process could take place in the cytoplasm, and the nature of the electron donor is not known.

Many proteins are also capable of peroxidase activity. Because of the involvement of *Synechococcus* 7002 GlbN in protecting the cell from ROS/RNS, its activity towards hydrogen peroxide was tested. Modest activity was observed, which was not consistent with a physiological role (Scott et al., 2010).

6. NEEDS AND OPPORTUNITIES

Since the pioneering 1992 publication first identified a globin in the cyanobacterium *N. commune* (Potts et al., 1992), the dominant questions in the field have been: 'How prevalent are these proteins?', 'How do they differ from the globins we associate with mammalian oxygen transport?', and 'What function do they provide for the cells in which they are expressed?' To restate the issues in plain terms, 'How common are these globins, what do they look like, and what do they do?' The most progress has been made in addressing the first point, or the commonality of globins in these microbes. The advent of genomic sequencing since 1992 has provided substantial amounts of phylogenetic information on globins from photosynthetic microbes, with additional sequences becoming available at an accelerated pace. From this, researchers have identified the ancient origins of globins, traced their evolution and even begun to illuminate the evolution of eukaryotic photosynthetic cells through the phylogenetic linkage among cyanobacterial globins and their counterparts in algae and higher plants. These investigations will no doubt continue, and with an ever-more complete genetic picture of both cyanobacteria and algae, our understanding of their phylogenetic heritage will grow. But we must begin to balance this wealth of genomic knowledge with substantive information about the actual proteins.

6.1. Physiological studies

A physiological understanding of globin function within photosynthetic organisms such as cyanobacteria and green algae presents a true opportunity. The functional characterization of the GlbN protein in *Synechococcus* sp. PCC 7002 (Scott et al., 2010) demonstrates that such studies can help define the role of globins within these systems. Below is a 'wish list' of studies involving both cyanobacterial and algal species. All of these studies are based upon currently available techniques but will nonetheless be challenging. As discussed in Section 5, the functions of globins are most likely tied to dioxygen manipulation and reactive molecule processing. As discussed in Section 3, photosynthetic organisms are ripe with oxygen evolution, reactive molecule formation and electron transfer. They are also masters at adapting to diverse environments and conditions. Section 4 has shown tantalizing clues that genes for globins are regulated in response to these molecular events within photosynthetic cells. The challenge, therefore, is left for us to design and implement experiments that can effectively link the function of globins to the dynamic physiology occurring within actively photosynthesizing organisms.

6.1.1 Purification of globins from the source organisms

Specificity for the haem cofactor is a relatively unexplored aspect of globins in photosynthetic organisms. Sperm whale apomyoglobin binds chlorophyll-like molecules (Davis & Pearlstein, 1979; Markovic, Proll, Bubenzer, & Scheer, 2007; Pröll, Wilhelm, Robert, & Scheer, 2006; Wright & Boxer, 1981), but of course, myoglobin never encounters these pigments. Depending on their localization, cyanobacterial and algal proteins may need to discriminate among haem and its derivatives used in photosynthesis. The derivatives include linear tetrapyrroles that are used by phycocyanins. In this regard, it is fitting to recall that globins and phycocyanins are related (Pastore & Lesk, 1990) and the issue of cofactor recognition, in itself, is a topic worthy of exploration (Aloy, Ceulemans, Stark, & Russell, 2003). We have inspected the association of Zn pheophorbide *a* methyl ester (as a soluble mimic of chlorophyll) with GlbN and have not found strong interactions (Landfried, 2010). The possibility exists that in some cases, the globins do bind a cofactor different from the haem group, and we suggest that this should be kept in mind, in particular, when physiological information on recombinant proteins and divergent or suspicious sequences are studied in a non-photosynthetic host.

Since virtually all of the structural data discussed in Section 5 are based upon recombinant protein prepared in a non-photosynthetic host (*E. coli*), a control experiment may be as conceptually simple as verifying the nature of the prosthetic group contained within the native form of these proteins. In several cases, unique and highly selective antibodies already exist for these proteins, and it may be possible to isolate purified proteins from whole-cell extracts of actively growing cultures. Although the expression levels of globins is in almost all cases low, especially when compared with dominant protein complexes such as the cell's photosystems, purification of small amounts of material may be achievable. This is particularly so, given recent transcriptomic results for several species, as we now know growth conditions and media additives that stimulate the expression of several globin genes.

Instances such as those illustrated by *Synechococcus* 7002 and *Synechocystis* 6803 GlbNs are favourable because the haem group becomes covalently attached to the protein matrix. A simple procedure relying on the peroxidase activity of the haem group in the unfolded protein and coupled with chemiluminescence detection (Dorward, 1993) would provide a rapid assessment. In other cases, with small amounts of purified native globins, the most productive analyses would involve delicate measurements such as chromatography or mass spectrometry or both. These techniques can help us understand the presence of cofactors within the purified proteins and potentially identify any post-translational modifications that have occurred.

Such experiments may actually be fairly uneventful. The most probable result would be finding haem bound to protein that is consistent with the genetic data and the recombinant structural results. But the simple verification of the presence of these holoproteins within actively growing cultures will definitively link the structural results to the physiology of the native cells. This necessary bridge between the *in vivo* state of native cells and the *in vitro* structural and mechanistic characterization of globins will lead to a comprehensive understanding of the function of these proteins within the cell.

6.1.2 Genetic manipulation and phenotypic analysis

One of the exciting aspects of cyanobacteria and green algae is the genetic tractability offered by both of these systems. Within cyanobacteria, several species, such as *Synechococcus* sp. PCC 7002 and *Synechocystis* sp. PCC 6803, are capable of genetic transformation. In green algae, as discussed in Section 4, *C. reinhardtii* has an extensive molecular toolkit available for genetic manipulation (Harris, 2001). This genetic malleability will be valuable as a method for determining the necessity of globin genes by the generation of 'knock-out' mutants, such as the *ΔglbN* mutant created by for the experiments involving *Synechococcus* sp. PCC 7002 (Scott et al., 2010).

An important result from this initial study using *Synechococcus* sp. PCC 7002 was the identification of a distinctive phenotype following the deletion of the globin from the wild-type strain. Using variations to the conditions under which this phenotype is displayed allows the activity of GlbN to be investigated. This can now be built upon, using the well-characterized structure of GlbN to make predictions as to how alterations to the amino acid sequence of the peptide would affect the function of the protein. The rescue of the *ΔglbN* strain with the native *glbN* gene also demonstrates the ability to add back genetic information to the cell. By transforming cyanobacteria with a plasmid containing an altered form of the *glbN* gene, it will be possible to connect functional and physical properties.

The availability of *C. reinhardtii* is especially interesting because it is an easily accessible eukaryotic model system. This has been shown by the recent creation of a strain with reduced expression of *THB8* (Hemschemeier et al., 2013). The genome of *C. reinhardtii* offers additional candidate genes for investigation, as discussed in Section 4, with an initial review of transcriptomic data highlighting the expression of *THB1* and *THB4* for further study. Both of these genes are candidates for knock-out or knock-down, well-established procedures in *C. reinhardtii*. The challenge will not necessarily be to generate novel strains of *C. reinhardtii* in which TrHb genes have been deleted but rather to identify a definitive phenotype that results from the deletion of these genes. From studies such as that of *Synechococcus* sp. PCC 7002, we have seen that under normal growth conditions very little difference is detected between the deletion strain and the wild-type strain (Scott et al., 2010). Transcriptomic data are helpful in identifying conditions where globin gene expression changes, revealing when cell survivability may be enhanced by globin production. Using variations in environmental conditions, the possibility for an identifiable phenotype can be investigated. The generation of such novel strains from *C. reinhardtii* would allow the linkage of globins to individual metabolic pathways.

6.1.3 Linkage to pathways
6.1.3.1 Nitrogen metabolism
The *Synechococcus* sp. PCC 7002 study (Scott et al., 2010) suggests strongly the involvement of globins in the metabolism of nitrogen-containing compounds within cyanobacteria, but many questions remain as to the true function of such proteins. One possible role for GlbN is in mitigation of peroxynitrite, a reactive molecule present as a by-product in nitrogen metabolism and implicated in several disease pathways (Alvarez & Radi, 2003). Interestingly, peroxynitrite, which results from the combination of nitric oxide with the superoxide anion, is thought to participate in signalling in plants through tyrosine nitration (Vandelle & Delledonne, 2011). Whether peroxynitrite chemistry is relevant to globins in photosynthetic microbes is a possibility that has not been explored.

Within eukaryotic algae, a recent paper has begun to make inroads into understanding potential globin roles in nitrogen metabolism (Hemschemeier et al., 2013). As discussed in Section 4.5, this study looked at the effects of reduced protein expression (through RNA knockdown) on a TrHb1 (THB8) from the alga *C. reinhardtii*. The result was a dramatic decrease in cell survivability during conditions of hypoxia, and this survival was linked to the prevalence of nitric oxide. It still needs to be determined exactly how THB8 confers this survivability, and what interactions it may have with nitric oxide within the cell, but it does signal a new and growing interest in these proteins and their interactions with reactive nitrogen molecules.

Further study of globin function within nitrogen metabolism of cyanobacteria has many advantages as this metabolism is quite diverse in these organisms. With globins present in species both with and without nitrogen fixation pathways and both with and without nitric oxide reductases, there is the opportunity to look at similar proteins in different contexts. This can definitely be tied to the wealth of available structural and ligand-binding information. By studying the nitrogen metabolism pathways in cyano-bacteria, we have a clear route to deciphering the relationship between structure and metabolic function. It is probable that these metabolic func-tions will be analogous to the function of globins in other organisms, giving insight into other pathways such as disease resistance in bacteria.

6.1.3.2 Photosynthesis

In photosynthetic organisms, extreme environmental conditions such as high light intensities can be detrimental to the photosynthetic process (Krieger-Liszkay, 2005), reducing the photosynthetic efficiency and giving rise to deleterious by-products (e.g. superoxide radicals, singlet oxygen) and conditions for the formation of RNS. The localization of several globins from chlorophytes to the chloroplast, either by predicted chloroplast transit peptide or by confirmed immuno-staining of the organelle, suggests that these proteins are involved in mitigating the production of these molecules as a way of preserving photosynthetic efficiency under harsh conditions. The *C. reinhardtii* protein THB4 occurs in the most genetically tractable species, and is closely related to the LI637 protein from *C. eugametos* (Fig. 6.3) for which some physiological information is available. This protein may be the most promising target for understanding how these globins fit into photosynthetic metabolism.

It will also be interesting to see how the eukaryotic globins in chloro-plasts relate to the cyanobacterial globins in terms of function. Despite the large evolutionary distance between the two species, the structure of ligand-bound TrHb1 (GlbN) from *Synechocystis* sp. PCC 6803 (Hoy et al., 2004) is very similar to the structure of the TrHb1 (CtrHb) from *C. eugametos* (Pesce et al., 2000). The question becomes whether we can draw similarities between the functions of these proteins as well. In this sense, eukaryotic globins operating within the chloroplast are exposed to an environment similar to that of their counterparts in cyanobacteria. The eukaryotic globin sequences in Fig. 6.3 were analyzed by the predictive software PredAlgo (Tardif et al., 2012). Five of the eukaryotic globins do, in fact, have predicted chloroplast targeting sequences. Of these, three are TrHb1s (from *Chlamydomonas* species), one is a TrHb2 (from *Chlorella vulgaris*), and one is predicted to be an SDgb (M globin; this protein is in *Micromonas* sp. RCC299). If globins in the chloroplast are used to mitigate similar reactive molecules, then why is there such a phylogenetically diverse group of globins targeted to the chloroplast among different species of chlorophytes?

6.1.3.3 Signal transduction

The observation that the THB1 protein accumulates within the flagella of *C. reinhardtii* (Lechtreck et al., 2009) offers an exciting opportunity to investigate the possible role of globins in signal transduction. While *C. reinhardtii* has long been a model organism for photosynthetic research, it also has a rich history as a model organism for developmental biology, especially in the study of flagella and cilia (Harris, 2001). The flagella of *C. reinhardtii* are coated with sensory molecules, and maintenance of these molecules constitutes a major role for much of the trafficking that occurs in the flagella (Rosenbaum & Witman, 2002). Thus, *C. reinhardtii* is a perfect host for investigation into the role of globins in signal transduction. Much of the groundwork may already have been completed with the creation of the *bbs4* strain of *C. reinhardtii* (Lechtreck et al., 2009). In the discussion of their results, the authors postulate that the observed accumulation of THB1 may in fact disrupt normal signalling within the flagella, noting that flagella are rich in redox proteins such as flavoproteins and identifying two potential redox proteins, AGG2 and AGG3, which have already been confirmed in the regulation of phototaxis (Iomini, Li, Mo, Dutcher, & Piperno, 2006). Light-dependent redox changes are also known to alter the frequency and beat duration of flagella by altering the disulphide-based interactions of the outer dynein arms (Wakabayashi & King, 2006), meaning that THB1 could easily play a key role in the modulation of phototaxis and that the accumulation of THB1 may cause loss of phototaxis. Even more enticing is the authors' comment '. . .similar disruption of ciliary signalling could be responsible for the diverse and complex phenotypes observed when BBS proteins are defective in other organisms' (Lechtreck et al., 2009). In a recent study that follows on this work, the same group has identified an additional protein involved in this pathway, phospholipase D (Lechtreck et al., 2013). Although it has not been shown that phospholipase D interacts with THB1, both proteins accumulate within the cilia of the *bbs4* strain (as discussed in Section 4.5). The phospholipase D protein is directly involved in signal transduction and lends credibility to the conclusion that THB1 is involved in signalling pathways. The study of the possible roles for globins in signal transduction could indeed be a rich field of research.

6.2. Biophysical studies

6.2.1 Electron transfer

The recognition that certain globins exhibit endogenous hexacoordination and are unable to bind exogenous ligand suggests a role in electron transfer.

Electron transfer may also be a feature of enzymatic reactions of hexacoordinate globins that can, like *Synechococcus* 7002 GlbN, bind ligands. Additional studies of electron transfer rates will be necessary to define fully the role that the globins play in the cell.

6.2.2 Structures and dynamic properties

The large conformational change that occurs upon ligand binding in endogenous hexacoordinate TrHb1s constitutes a structural aspect worthy of consideration. The two extreme structures, illustrated by *Synechocystis* 6803 and *Synechococcus* 7002 GlbNs in the bis-histidine and cyanomet states, are stabilized either by the coordination of the distal histidine or by the formation of the distal hydrogen bond network. Yet, the replacement of the distal histidine (Couture et al., 2000; Hvitved et al., 2001) or substitution of zinc for iron (Lecomte, Vu, & Falzone, 2005) leads to species with ambiguous structural properties. The determinants of flexibility are of interest in connection with an enzymatic function requiring the timing of substrate access to the haem, release of the product, and interactions with a reductase. Added to NMR investigations, molecular dynamic studies as performed with other globins (see, e.g. Nadra, Marti, Pesce, Bolognesi, & Estrin, 2008; Savard et al., 2011) will shed some light on protein motions as they relate to enzymatic activity, explore the role of solvent in conditioning the properties of the haem, and provide atomic level explanation for the consequences of specific amino acid replacements. It is also possible that large conformational changes modulate interactions with other proteins by altering the surface properties of the molecule. A search for specific partners once additional physiological information restricts the functional possibilities may be enlightening, particularly for the proteins that may require a dedicated reductase for enzymatic turnover.

It should be stressed that there are large gaps in the structural descriptions. Not only are few structures available (Table 6.5), but they are all of the T family and the same group within that family (TrHb1s). The structure of the proteins from the M lineage would provide a basis for further *in vitro* experiments. Also interesting would be a description of the globins that lack the proximal histidine. Does haem bind to these proteins? It is known that it can bind to a distal histidine and there is no *a priori* reason that chemistry cannot occur on the 'dark side' of the structure (Ascenzi, Leboffe, & Polticelli, 2013). Thus, although the proteins without proximal histidine have typically not been included in any extensive study, if there is any evidence that they are functional, they would be rich in structural, chemical and evolutionary clues.

In view of possible signalling roles, further covalent modifications, as observed in mammalian proteins (Ascenzi et al., 2013), should be considered as well. Tyrosine nitration, phosphorylation, carbamation, glycosylation, etc., may all play a role in globins.

6.3. Concluding remarks

The diversity of globins throughout all kingdoms of life strongly suggests that these proteins perform fundamentally necessary roles within cells, yet especially in photosynthetic microbes, we are only beginning to define what those roles might be. The great need in this field is to formulate an understanding of the native function of globins in photosynthetic microbes. Growing reservoirs of genetic information have provided us with a great opportunity by revealing multiple globin genes within model photosynthetic organisms. These organisms, such as *Synechococcus* sp. PCC 7002 and *C. reinhardtii*, have been extensively used for physiological characterization, and many analytical techniques are already established for the *in vivo* characterization of proteins. By identifying orthologous proteins in model organisms and enhancing our structural information about their characteristics, we can finally begin to address the root questions asked throughout this chapter. From phylogenetic studies, we know that these proteins are important. From structural and recombinant studies, we understand what potential activity they exhibit. We must now focus on what functions they perform and how they enhance the survival of both cyanobacteria and algae throughout the diverse environments in which they exist.

REFERENCES

Aloy, P., Ceulemans, H., Stark, A., & Russell, R. B. (2003). The relationship between sequence and interaction divergence in proteins. *Journal of Molecular Biology, 332,* 989–998.

Alvarez, B., & Radi, R. (2003). Peroxynitrite reactivity with amino acids and proteins. *Amino Acids, 25,* 295–311.

Angeloni, S. V., & Potts, M. (1994). Analysis of the sequences within and flanking the cyanoglobin-encoding gene, *glbN,* of the cyanobacterium *Nostoc commune* UTEX 584. *Gene, 146,* 133–134.

Antonini, E., & Brunori, M. (1971). *Hemoglobin and myoglobin in their reactions with ligands.* Amsterdam: North-Holland.

Argueta, C., & Summers, M. L. (2005). Characterization of a model system for the study of *Nostoc punctiforme* akinetes. *Archives of Microbiology, 183,* 338–346.

Arredondo-Peter, R., Hargrove, M. S., Sarath, G., Moran, J. F., Lohrman, J., Olson, J. S., et al. (1997). Rice hemoglobins. Gene cloning, analysis, and O_2-binding kinetics of a recombinant protein synthesized in *Escherichia coli*. *Plant Physiology*, *115*, 1259–1266.

Ascenzi, P., Leboffe, L., & Polticelli, F. (2013). Reactivity of the human hemoglobin "Dark side" *IUBMB Life*, *65*, 121–126.

Beales, P. L. (2005). Lifting the lid on Pandora's box: The Bardet-Biedl syndrome. *Current Opinion in Genetics & Development*, *15*, 315–323.

Beard, D. A., & Qian, H. (2008). *Chemical biophysics. Quantitative analysis of cellular systems.* New York: Cambridge University Press.

Beck, C., Knoop, H., Axmann, I. M., & Steuer, R. (2012). The diversity of cyanobacterial metabolism: Genome analysis of multiple phototrophic microorganisms. *BMC Genomics*, *13*, 56.

Bertini, I., Luchinat, C., Parigi, G., & Walker, F. A. (1999). Heme methyl ^{1}H chemical shifts as structural parameters in some low-spin ferriheme proteins. *Journal of Biological Inorganic Chemistry*, *4*, 515–519.

Bowman, L. A. H., McLean, S., Poole, R. K., & Fukuto, J. M. (2011). The diversity of microbial responses to nitric oxide and agents of nitrosative stress: Close cousins but not identical twins. *Advances in Microbial Physiology*, *59*, 135–219.

Brantley, R. E., Jr., Smerdon, S. J., Wilkinson, A. J., Singleton, E. W., & Olson, J. S. (1993). The mechanism of autooxidation of myoglobin. *The Journal of Biological Chemistry*, *268*, 6995–7010.

Brautigan, D. L., Feinberg, B. A., Hoffman, B. M., Margoliash, E., Peisach, J., & Blumberg, W. E. (1977). Multiple low spin forms of the cytochrome *c* ferrihemochrome. EPR spectra of various eukaryotic and prokaryotic cytochromes *c*. *The Journal of Biological Chemistry*, *252*, 574–582.

Bryant, D. A., & Frigaard, N. U. (2006). Prokaryotic photosynthesis and phototrophy illuminated. *Trends in Microbiology*, *14*, 488–496.

Burmester, T., Ebner, B., Weich, B., & Hankeln, T. (2002). Cytoglobin: A novel globin type ubiquitously expressed in vertebrate tissues. *Molecular Biology and Evolution*, *19*, 416–421.

Burmester, T., Weich, B., Reinhardt, S., & Hankeln, T. (2000). A vertebrate globin expressed in the brain. *Nature*, *407*, 520–523.

Campbell, E. L., Christman, H., & Meeks, J. C. (2008). DNA microarray comparisons of plant factor- and nitrogen deprivation-induced hormogonia reveal decision-making transcriptional regulation patterns in *Nostoc punctiforme*. *Journal of Bacteriology*, *190*, 7382–7391.

Campbell, E. L., & Meeks, J. C. (1989). Characteristics of hormogonia formation by symbiotic *Nostoc* spp. in response to the presence of *Anthoceros punctatus* or its extracellular products. *Applied and Environmental Microbiology*, *55*, 125–131.

Campbell, E. L., Summers, M. L., Christman, H., Martin, M. E., & Meeks, J. C. (2007). Global gene expression patterns of *Nostoc punctiforme* in steady-state dinitrogen-grown heterocyst-containing cultures and at single time points during the differentiation of akinetes and hormogonia. *Journal of Bacteriology*, *189*, 5247–5256.

Christman, H. D., Campbell, E. L., & Meeks, J. C. (2011). Global transcription profiles of the nitrogen stress response resulting in heterocyst or hormogonium development in *Nostoc punctiforme*. *Journal of Bacteriology*, *193*, 6874–6886.

Couture, M., Chamberland, H., St-Pierre, B., Lafontaine, J., & Guertin, M. (1994). Nuclear genes encoding chloroplast hemoglobins in the unicellular green alga *Chlamydomonas eugametos*. *Molecular and General Genetics*, *243*, 185–197.

Couture, M., Das, T. K., Lee, H. C., Peisach, J., Rousseau, D. L., Wittenberg, B. A., et al. (1999). *Chlamydomonas* chloroplast ferrous hemoglobin. Heme pocket structure and reactions with ligands. *The Journal of Biological Chemistry*, *274*, 6898–6910.

Couture, M., Das, T. K., Savard, P. Y., Ouellet, Y., Wittenberg, J. B., Wittenberg, B. A., et al. (2000). Structural investigations of the hemoglobin of the cyanobacterium

Synechocystis PCC 6803 reveal a unique distal heme pocket. *European Journal of Biochemistry, 267,* 4770–4780.

Couture, M., & Guertin, M. (1996). Purification and spectroscopic characterization of a recombinant chloroplastic hemoglobin from the green unicellular alga *Chlamydomonas eugametos. European Journal of Biochemistry, 242,* 779–787.

Culbertson, D. S., & Olson, J. S. (2010). Role of heme in the unfolding and assembly of myoglobin. *Biochemistry, 49,* 6052–6063.

Curtis, B. A., Tanifuji, G., Burki, F., Gruber, A., Irimia, M., Maruyama, S., et al. (2012). Algal genomes reveal evolutionary mosaicism and the fate of nucleomorphs. *Nature, 492,* 59–65.

Das, T. K., Couture, M., Lee, H. C., Peisach, J., Rousseau, D. L., Wittenberg, B. A., et al. (1999). Identification of the ligands to the ferric heme of *Chlamydomonas* chloroplast hemoglobin: Evidence for ligation of tyrosine-63 (B10) to the heme. *Biochemistry, 38,* 15360–15368.

Davis, R. C., & Pearlstein, R. M. (1979). Chlorophyllin-apomyoglobin complexes. *Nature, 280,* 413–415.

Dawson, J. H., Kadkhodayan, S., Zhuang, C., & Sono, M. (1992). On the use of iron octa-alkylporphyrins as models for protoporphyrin IX-containing heme systems in studies employing magnetic circular dichroism spectroscopy. *Journal of Inorganic Biochemistry, 45,* 179–192.

DeLano, W. L. (2002). *The PyMOL molecular graphics system.* San Carlos, CA: DeLano Scientific.

de Sanctis, D., Pesce, A., Nardini, M., Bolognesi, M., Bocedi, A., & Ascenzi, P. (2004). Structure-function relationships in the growing hexa-coordinate hemoglobin sub-family. *IUBMB Life, 56,* 643–651.

Dorward, D. W. (1993). Detection and quantitation of heme-containing proteins by chemi-luminescence. *Analytical Biochemistry, 209,* 219–223.

Duff, S. M., Wittenberg, J. B., & Hill, R. D. (1997). Expression, purification, and properties of recombinant barley (*Hordeum* sp.) hemoglobin. Optical spectra and reactions with gaseous ligands. *The Journal of Biological Chemistry, 272,* 16746–16752.

Falzone, C. J., Vu, B. C., Scott, N. L., & Lecomte, J. T. J. (2002). The solution structure of the recombinant hemoglobin from the cyanobacterium *Synechocystis* sp. PCC 6803 in its hemichrome state. *Journal of Molecular Biology, 324,* 1015–1029.

Fischer, B. B., Ledford, H. K., Wakao, S., Huang, S. G., Casero, D., Pellegrini, M., et al. (2012). *SINGLET OXYGEN RESISTANT 1* links reactive electrophile signaling to singlet oxygen acclimation in *Chlamydomonas reinhardtii. Proceedings of the National Academy of Sciences of the United States of America, 109,* E1302–E1311.

Flores, E., & Herrero, A. (1994). Assimilatory nitrogen metabolism and its regulation. In D. A. Bryant (Ed.), *The molecular biology of cyanobacteria* (pp. 487–517). Dordrecht, The Netherlands: Kluwer Academic Publishers.

Freitas, T. A., Saito, J. A., Hou, S., & Alam, M. (2005). Globin-coupled sensors, protoglobins, and the last universal common ancestor. *Journal of Inorganic Biochemistry, 99,* 23–33.

Fuchs, C., Burmester, T., & Hankeln, T. (2006). The amphibian globin gene repertoire as revealed by the *Xenopus* genome. *Cytogenetic and Genome Research, 112,* 296–306.

Fuchs, C., Heib, V., Kiger, L., Haberkamp, M., Roesner, A., Schmidt, M., et al. (2004). Zebrafish reveals different and conserved features of vertebrate neuroglobin gene structure, expression pattern, and ligand binding. *The Journal of Biological Chemistry, 279,* 24116–24122.

Gagné, G., & Guertin, M. (1992). The early genetic response to light in the green unicellular alga *Chlamydomonas eugametos* grown under light dark cycles involves genes that represent direct responses to light and photosynthesis. *Plant Molecular Biology, 18,* 429–445.

Gardner, P. R. (2005). Nitric oxide dioxygenase function and mechanism of flavohemoglobin, hemoglobin, myoglobin and their associated reductases. *Journal of Inorganic Biochemistry*, *99*, 247–266.

Gardner, P. R. (2012). Hemoglobin, a nitric-oxide dioxygenase. *Scientifica*, *2012*, http://dx.doi.org/10.6064/2012/683729.

Gardner, A. M., Martin, L. A., Gardner, P. R., Dou, Y., & Olson, J. S. (2000). Steady-state and transient kinetics of *Escherichia coli* nitric-oxide dioxygenase (flavohemoglobin). The B10 tyrosine hydroxyl is essential for dioxygen binding and catalysis. *The Journal of Biological Chemistry*, *275*, 12581–12589.

Garrocho-Villegas, V., Gopalasubramaniam, S. K., & Arredondo-Peter, R. (2007). Plant hemoglobins: What we know six decades after their discovery. *Gene*, *398*, 78–85.

George, P., & Stratmann, C. J. (1954). The oxidation of myoglobin to metmyoglobin by oxygen. III. Kinetic studies in the presence of carbon monoxide, and at different hydrogen-ion concentrations with considerations regarding the stability of oxymyoglobin. *Biochemistry Journal*, *57*, 568–573.

Gibson, Q. H., Olson, J. S., McKinnie, R. E., & Rohlfs, R. J. (1986). A kinetic description of ligand binding to sperm whale myoglobin. *The Journal of Biological Chemistry*, *261*, 10228–10239.

Gladwin, M. T., Grubina, R., & Doyle, M. P. (2009). The new chemical biology of nitrite reactions with hemoglobin: R-state catalysis, oxidative denitrosylation, and nitrite reductase/anhydrase. *Accounts of Chemical Research*, *42*, 157–167.

Goldschmidt-Clermont, M. (1991). Transgenic expression of aminoglycoside adenine transferase in the chloroplast: A selectable marker of site-directed transformation of chlamydomonas. *Nucleic Acids Research*, *19*, 4083–4089.

Gonzalez-Ballester, D., Casero, D., Cokus, S., Pellegrini, M., Merchant, S. S., & Grossman, A. R. (2010). RNA-Seq analysis of sulfur-deprived *Chlamydomonas* cells reveals aspects of acclimation critical for cell survival. *The Plant Cell*, *22*, 2058–2084.

Grossman, A. R., Harris, E. E., Hauser, C., Lefebvre, P. A., Martinez, D., Rokhsar, D., et al. (2003). *Chlamydomonas reinhardtii* at the crossroads of genomics. *Eukaryotic Cell*, *2*, 1137–1150.

Grossman, A. R., Karpowicz, S. J., Heinnickel, M., Dewez, D., Hamel, B., Dent, R., et al. (2010). Phylogenomic analysis of the *Chlamydomonas* genome unmasks proteins potentially involved in photosynthetic function and regulation. *Photosynthesis Research*, *106*, 3–17.

Halder, P., Trent, J. T., III., & Hargrove, M. S. (2007). Influence of the protein matrix on intramolecular histidine ligation in ferric and ferrous hexacoordinate hemoglobins. *Proteins*, *66*, 172–182.

Hankeln, T., Jaenicke, V., Kiger, L., Dewilde, S., Ungerechts, G., Schmidt, M., et al. (2002). Characterization of *Drosophila* hemoglobin. Evidence for hemoglobin-mediated respiration in insects. *The Journal of Biological Chemistry*, *277*, 29012–29017.

Hargrove, M. S. (2000). A flash photolysis method to characterize hexacoordinate hemoglobin kinetics. *Biophysical Journal*, *79*, 2733–2738.

Hargrove, M. S., Barry, J. K., Brucker, E. A., Berry, M. B., Phillips, G. N., Jr., Olson, J. S., et al. (1997). Characterization of recombinant soybean leghemoglobin a and apolar distal histidine mutants. *Journal of Molecular Biology*, *266*, 1032–1042.

Harris, E. H. (2001). *Chlamydomonas* as a model organism. *Annual Review of Plant Physiology and Plant Molecular Biology*, *52*, 363–406.

Hemschemeier, A., Duner, M., Casero, D., Merchant, S. S., Winkler, M., & Happe, T. (2013). Hypoxic survival requires a 2-on-2 hemoglobin in a process involving nitric oxide. *Proceedings of the National Academy of Sciences of the United States of America*, *110*, 10854–10859.

Herrero, A., Muro-Pastor, A. M., & Flores, E. (2001). Nitrogen control in cyanobacteria. *Journal of Bacteriology*, *183*, 411–425.

Hill, D. R., Belbin, T. J., Thorsteinsson, M. V., Bassam, D., Brass, S., Ernst, A., et al. (1996). GlbN (cyanoglobin) is a peripheral membrane protein that is restricted to certain *Nostoc* spp. *Journal of Bacteriology, 178,* 6587–6598.

Hoogewijs, D., Dewilde, S., Vierstraete, A., Moens, L., & Vinogradov, S. N. (2012). A phylogenetic analysis of the globins in fungi. *PLoS One, 7,* e31856.

Hou, S., Belisle, C., Lam, S., Piatibratov, M., Sivozhelezov, V., Takami, H., et al. (2001). A globin-coupled oxygen sensor from the facultatively alkaliphilic *Bacillus halodurans* C-125. *Extremophiles, 5,* 351–354.

Hoy, J. A., Kundu, S., Trent, J. T., III., Ramaswamy, S., & Hargrove, M. S. (2004). The crystal structure of *Synechocystis* hemoglobin with a covalent heme linkage. *The Journal of Biological Chemistry, 279,* 16535–16542.

Hoy, J. A., Robinson, H., Trent, J. T., III., Kakar, S., Smagghe, B. J., & Hargrove, M. S. (2007). Plant hemoglobins: A molecular fossil record for the evolution of oxygen transport. *Journal of Molecular Biology, 371,* 168–179.

Hoy, J. A., Smagghe, B. J., Halder, P., & Hargrove, M. S. (2007). Covalent heme attachment in *Synechocystis* hemoglobin is required to prevent ferrous heme dissociation. *Protein Science, 16,* 250–260.

Hvitved, A. N., Trent, J. T., III., Premer, S. A., & Hargrove, M. S. (2001). Ligand binding and hexacoordination in *Synechocystis* hemoglobin. *The Journal of Biological Chemistry, 276,* 34714–34721.

Ioanitescu, A. I., Dewilde, S., Kiger, L., Marden, M. C., Moens, L., & Van Doorslaer, S. (2005). Characterization of nonsymbiotic tomato hemoglobin. *Biophysical Journal, 89,* 2628–2639.

Iomini, C., Li, L., Mo, W., Dutcher, S. K., & Piperno, G. (2006). Two flagellar genes, AGG2 and AGG3, mediate orientation to light in *Chlamydomonas*. *Current Biology, 16,* 1147–1153.

Iwaasa, H., Takagi, T., & Shikama, K. (1989). Protozoan myoglobin from *Paramecium caudatum*. Its unusual amino acid sequence. *Journal of Molecular Biology, 208,* 355–358.

Iwaasa, H., Takagi, T., & Shikama, K. (1990). Protozoan hemoglobin from *Tetrahymena pyriformis*. Isolation, characterization, and amino acid sequence. *The Journal of Biological Chemistry, 265,* 8603–8609.

Jentzen, W., Ma, J. G., & Shelnutt, J. A. (1998). Conservation of the conformation of the porphyrin macrocycle in hemoproteins. *Biophysical Journal, 74,* 753–763.

Kakar, S., Hoffman, F. G., Storz, J. F., Fabian, M., & Hargrove, M. S. (2010). Structure and reactivity of hexacoordinate hemoglobins. *Biophysical Journal, 152,* 1–14.

Kaneko, T., Sato, S., Kotani, H., Tanaka, A., Asamizu, E., Nakamura, Y., et al. (1996). Sequence analysis of the genome of the unicellular cyanobacterium *Synechocystis* sp. strain PCC6803. II. Sequence determination of the entire genome and assignment of potential protein-coding regions. *DNA Research, 3,* 109–136.

Keeling, P. J. (2004). Diversity and evolutionary history of plastids and their hosts. *American Journal of Botany, 91,* 1481–1493.

Keilin, D., & Ryley, J. F. (1953). Haemoglobin in protozoa. *Nature, 172,* 451–452.

Kiger, L., Tilleman, L., Geuens, E., Hoogewijs, D., Lechauve, C., Moens, L., et al. (2011). Electron transfer function versus oxygen delivery: A comparative study for several hexacoordinated globins across the animal kingdom. *PLoS One, 6,* e20478.

Kranz, R. G., Richard-Fogal, C., Taylor, J.-S., & Frawley, E. R. (2009). Cytochrome *c* biogenesis: Mechanisms for covalent modifications and trafficking of heme and for heme-iron redox control. *Microbiology and Molecular Biology Reviews, 73,* 510–528.

Kraulis, P. (1991). MOLSCRIPT: A program to produce both detailed and schematic plots of protein structures. *Journal of Applied Crystallography, 24,* 946–950.

Krieger-Liszkay, A. (2005). Singlet oxygen production in photosynthesis. *Journal of Experimental Botany, 56,* 337–346.

Kugelstadt, D., Haberkamp, M., Hankeln, T., & Burmester, T. (2004). Neuroglobin, cytoglobin, and a novel, eye-specific globin from chicken. *Biochemical and Biophysical Research Communications, 325,* 719–725.

Kundu, S., Premer, S. A., Hoy, J. A., Trent, J. T., III., & Hargrove, M. S. (2003). Direct measurement of equilibrium constants for high-affinity hemoglobins. *Biophysical Journal, 84,* 3931–3940.

La Mar, G. N., Satterlee, J. D., & de Ropp, J. S. (2000). Nuclear magnetic resonance of hemoproteins. In K. M. Smith, K. Kadish, & R. Guilard (Eds.), *The porphyrin handbook, Vol. 5,* (pp. 185–298). Burlington, MA: Academic Press.

Landfried, D. A. (2010). *Binding and folding of intrinsically disordered regions of proteins: Analysis of recognition elements in hemeproteins and peptides* (Ph.D. Thesis). Johns Hopkins University, Biophysics, Baltimore, MD.

Lechtreck, K. F., Brown, J. M., Sampaio, J. L., Craft, J. M., Shevchenko, A., Evans, J. E., et al. (2013). Cycling of the signaling protein phospholipase D through cilia requires the BBSome only for the export phase. *The Journal of Cell Biology, 201,* 249–261.

Lechtreck, K. F., Johnson, E. C., Sakai, T., Cochran, D., Ballif, B. A., Rush, J., et al. (2009). The *Chlamydomonas reinhardtii* BBSome is an IFT cargo required for export of specific signaling proteins from flagella. *The Journal of Cell Biology, 187,* 1117–1132.

Lecomte, J. T. J., Scott, N. L., Vu, B. C., & Falzone, C. J. (2001). Binding of ferric heme by the recombinant globin from the cyanobacterium *Synechocystis* sp. PCC 6803. *Biochemistry, 40,* 6541–6552.

Lecomte, J. T. J., Vu, B. C., & Falzone, C. J. (2005). Structural and dynamic properties of *Synechocystis* sp. PCC 6803 Hb revealed by reconstitution with Zn-protoporphyrin IX. *Journal of Inorganic Biochemistry, 99,* 1585–1592.

Lecomte, J. T. J., Vuletich, D. A., & Lesk, A. M. (2005). Structural divergence and distant relationships in proteins: Evolution of the globins. *Current Opinion in Structural Biology, 15,* 290–301.

Li, H., Hemann, C., Abdelghany, T. M., El-Mahdy, M. A., & Zweier, J. L. (2012). Characterization of the mechanism and magnitude of cytoglobin-mediated nitrite reduction and nitric oxide generation under anaerobic conditions. *The Journal of Biological Chemistry, 287,* 36623–36633.

Mallick, N., Rai, L. C., Mohn, F. H., & Soeder, C. J. (1999). Studies on nitric oxide (NO) formation by the green alga *Scenedesmus obliquus* and the diazotrophic cyanobacterium *Anabaena doliolum*. *Chemosphere, 39,* 1601–1610.

Markovic, D., Proll, S., Bubenzer, C., & Scheer, H. (2007). Myoglobin with chlorophyllous chromophores: Influence on protein stability. *Biochimica et Biophysica Acta, 1767,* 897–904.

Meeks, J. C., Elhai, J., Thiel, T., Potts, M., Larimer, F., Lamerdin, J., et al. (2001). An overview of the genome of *Nostoc punctiforme*, a multicellular, symbiotic cyanobacterium. *Photosynthesis Research, 70,* 85–106.

Meyer, M. T., Genkov, T., Skepper, J. N., Jouhet, J., Mitchell, M. C., Spreitzer, R. J., et al. (2012). Rubisco small-subunit α-helices control pyrenoid formation in *Chlamydomonas*. *Proceedings of the National Academy of Sciences of the United States of America, 109,* 19474–19479.

Milani, M., Ouellet, Y., Ouellet, H., Guertin, M., Boffi, A., Antonini, G., et al. (2004). Cyanide binding to truncated hemoglobins: A crystallographic and kinetic study. *Biochemistry, 43,* 5213–5221.

Milani, M., Pesce, A., Nardini, M., Ouellet, H., Ouellet, Y., Dewilde, S., et al. (2005). Structural bases for heme binding and diatomic ligand recognition in truncated hemoglobins. *Journal of Inorganic Biochemistry, 99,* 97–109.

Milani, M., Pesce, A., Ouellet, Y., Ascenzi, P., Guertin, M., & Bolognesi, M. (2001). *Mycobacterium tuberculosis* hemoglobin N displays a protein tunnel suited for O_2 diffusion to the heme. *The EMBO Journal, 20,* 3902–3909.

Milani, M., Pesce, A., Ouellet, Y., Dewilde, S., Friedman, J., Ascenzi, P., et al. (2004). Heme-ligand tunneling in group I truncated hemoglobins. *The Journal of Biological Chemistry*, *279*, 21520–21525.

Milani, M., Savard, P. Y., Ouellet, H., Ascenzi, P., Guertin, M., & Bolognesi, M. (2003). A TyrCD1/TrpG8 hydrogen bond network and a TyrB10-TyrCD1 covalent link shape the heme distal site of *Mycobacterium tuberculosis* hemoglobin O. *Proceedings of the National Academy of Sciences of the United States of America*, *100*, 5766–5771.

Mowat, C. G., Gazur, B., Campbell, L. P., & Chapman, S. K. (2010). Flavin-containing heme enzymes. *Archives of Biochemistry and Biophysics*, *493*, 37–52.

Mukai, M., Mills, C. E., Poole, R. K., & Yeh, S. R. (2001). Flavohemoglobin, a globin with a peroxidase-like catalytic site. *The Journal of Biological Chemistry*, *276*, 7272–7277.

Murray, J. W., Delumeau, O., & Lewis, R. J. (2005). Structure of a nonheme globin in environmental stress signaling. *Proceedings of the National Academy of Sciences of the United States of America*, *102*, 17320–17325.

Nadra, A. D., Marti, M. A., Pesce, A., Bolognesi, M., & Estrin, D. A. (2008). Exploring the molecular basis of heme coordination in human neuroglobin. *Proteins*, *71*, 695–705.

Nardini, M., Pesce, A., Milani, M., & Bolognesi, M. (2007). Protein fold and structure in the truncated (2/2) globin family. *Gene*, *398*, 2–11.

Neilson, J. A. D., & Durnford, D. G. (2010). Structural and functional diversification of the light-harvesting complexes in photosynthetic eukaryotes. *Photosynthesis Research*, *106*, 57–71.

Nothnagel, H. J., Preimesberger, M. R., Pond, M. P., Winer, B. Y., Adney, E. M., & Lecomte, J. T. J. (2011). Chemical reactivity of *Synechococcus* sp. PCC 7002 and *Synechocystis* sp. PCC 6803 hemoglobins: Covalent heme attachment and bishistidine coordination. *Journal of Biological Inorganic Chemistry*, *16*, 539–552.

Olson, J. S. (1981). Stopped-flow, rapid mixing measurements of ligand binding to hemoglobin and red cells. In E. Antonini, L. Rossi-Bernardi, & E. Chiancone (Eds.), *Methods in Enzymology*, Vol. 76, (pp. 631–651). New York: Academic Press.

Olson, J. S., & Phillips, G. N., Jr. (1996). Kinetic pathways and barriers for ligand binding to myoglobin. *The Journal of Biological Chemistry*, *271*, 17593–17596.

Olson, J. S., & Phillips, G. N., Jr. (1997). Myoglobin discriminates between O_2, NO, and CO by electrostatic interactions with the bound ligand. *Journal of Biological Inorganic Chemistry*, *2*, 544–552.

Olson, J. S., Soman, J., & Phillips, G. N., Jr. (2007). Ligand pathways in myoglobin: A review of Trp cavity mutations. *IUBMB Life*, *59*, 552–562.

Pastore, A., & Lesk, A. M. (1990). Comparison of the structures of globins and phycocyanins: Evidence for evolutionary relationship. *Proteins*, *8*, 133–155.

Pesce, A., Couture, M., Dewilde, S., Guertin, M., Yamauchi, K., Ascenzi, P., et al. (2000). A novel two-over-two α-helical sandwich fold is characteristic of the truncated hemoglobin family. *The EMBO Journal*, *19*, 2424–2434.

Petersen, M. G., Dewilde, S., & Fago, A. (2008). Reactions of ferrous neuroglobin and cytoglobin with nitrite under anaerobic conditions. *Journal of Inorganic Biochemistry*, *102*, 1777–1782.

Petrek, M., Otyepka, M., Banas, P., Kosinova, P., Koca, J., & Damborsky, J. (2006). CAVER: A new tool to explore routes from protein clefts, pockets and cavities. *BMC Bioinformatics*, *7*, 316.

Phillips, S. E. (1980). Structure and refinement of oxymyoglobin at 1.6 Å resolution. *Journal of Molecular Biology*, *142*, 531–554.

Pisciotta, J. M., Zou, Y., & Baskakov, I. V. (2010). Light-dependent electrogenic activity of cyanobacteria. *PLoS One*, *5*, e10821.

Pond, M. P., Majumdar, A., & Lecomte, J. T. J. (2012). Influence of heme post-translational modification and distal ligation on the backbone dynamics of a monomeric hemoglobin. *Biochemistry*, *51*, 5733–5747.

Potts, M., Angeloni, S. V., Ebel, R. E., & Bassam, D. (1992). Myoglobin in a cyanobacterium. *Science*, *256*, 1690–1691.

Preimesberger, M. R., Pond, M. P., Majumdar, A., & Lecomte, J. T. J. (2012). Electron self-exchange and self-amplified posttranslational modification in the hemoglobins from *Synechocystis* sp. PCC 6803 and *Synechococcus* sp. PCC 7002. *Journal of Biological Inorganic Chemistry*, *17*, 599–609.

Preimesberger, M. R., Wenke, B. B., Gilevicius, L., Pond, M. P., & Lecomte, J. T. J. (2013). Facile heme vinyl posttranslational modification in a hemoglobin. *Biochemistry*, *52*, 3478–3488.

Pröll, S., Wilhelm, B., Robert, B., & Scheer, H. (2006). Myoglobin with modified tetrapyrrole chromophores: Binding specificity and photochemistry. *Biochimica et Biophysica Acta*, *1757*, 750–763.

Rachmilewitz, E. A., & Harari, E. (1972). Intermediate hemichrome formation after oxidation of three unstable hemoglobins (Freiburg, Riverdale-Bronx and Koln). *Hämatologie und Bluttransfusion*, *10*, 241–250.

Rachmilewitz, E. A., & White, J. M. (1973). Hemichrome formation during *in vitro* oxidation of Hb Köln. *Nature: New Biology*, *241*, 115–117.

Reyes-Prieto, A., Weber, A. P. M., & Bhattacharya, D. (2007). The origin and establishment of the plastid in algae and plants. *Annual Review of Plant Genetics*, *41*, 147–168.

Richards, T. A., & van der Giezen, M. (2006). Evolution of the Isd11-IscS complex reveals a single α-proteobacterial endosymbiosis for all eukaryotes. *Molecular Biology and Evolution*, *23*, 1341–1344.

Rodriguez-Ezpeleta, N., Brinkmann, H., Burey, S. C., Roure, B., Burger, G., Loffelhardt, W., et al. (2005). Monophyly of primary photosynthetic eukaryotes: Green plants, red algae, and glaucophytes. *Current Biology*, *15*, 1325–1330.

Rosenbaum, J. L., & Witman, G. B. (2002). Intraflagellar transport. *Nature Reviews. Molecular Cell Biology*, *3*, 813–825.

Sakihama, Y., Nakamura, S., & Yamasaki, H. (2002). Nitric oxide production mediated by nitrate reductase in the green alga *Chlamydomonas reinhardtii*: An alternative NO production pathway in photosynthetic organisms. *Plant & Cell Physiology*, *43*, 290–297.

Samuni, U., Ouellet, Y., Guertin, M., Friedman, J. M., & Yeh, S. R. (2004). The absence of proximal strain in the truncated hemoglobins from *Mycobacterium tuberculosis*. *Journal of the American Chemical Society*, *126*, 2682–2683.

Sato, T., & Tamiya, H. (1937). Ueber die Atmungsfarbstoffe von *Paramecium*. *Cytologia (Tokyo), Fujii Jubilee Volume*, pp. 1133–1138.

Savard, P. Y., Daigle, R., Morin, S., Sebilo, A., Meindre, F., Lague, P., et al. (2011). Structure and dynamics of *Mycobacterium tuberculosis* truncated hemoglobin N: Insights from NMR spectroscopy and molecular dynamics simulations. *Biochemistry*, *50*, 11121–11130.

Scott, N. L., Falzone, C. J., Vuletich, D. A., Zhao, J., Bryant, D. A., & Lecomte, J. T. J. (2002). The hemoglobin of the cyanobacterium *Synechococcus* sp. PCC 7002: Evidence for hexacoordination and covalent adduct formation in the ferric recombinant protein. *Biochemistry*, *41*, 6902–6910.

Scott, E. E., Gibson, Q. H., & Olson, J. S. (2001). Mapping the pathways for O_2 entry into and exit from myoglobin. *The Journal of Biological Chemistry*, *276*, 5177–5188.

Scott, N. L., & Lecomte, J. T. J. (2000). Cloning, expression, purification, and preliminary characterization of a putative hemoglobin from the cyanobacterium *Synechocystis* sp. PCC 6803. *Protein Science*, *9*, 587–597.

Scott, N. L., Xu, Y., Shen, G., Vuletich, D. A., Falzone, C. J., Li, Z., et al. (2010). Functional and structural characterization of the 2/2 hemoglobin from *Synechococcus* sp. PCC 7002. *Biochemistry*, *49*, 7000–7011.

Shelnutt, J. A., Song, X. Z., Ma, J. G., Jia, S. L., Jentzen, W., & Medforth, C. J. (1998). Nonplanar porphyrins and their significance in proteins. *Chemical Society Reviews*, *27*, 31–41.

Shibata, T., Matsumoto, D., Nishimura, R., Tai, H., Matsuoka, A., Nagao, S., et al. (2012). Relationship between oxygen affinity and autoxidation of myoglobin. *Inorganic Chemistry*, *51*, 11955–11960.

Shikama, K. (2006). Nature of the FeO_2 bonding in myoglobin and hemoglobin: A new molecular paradigm. *Progress in Biophysics and Molecular Biology*, *91*, 83–162.

Shikama, K., & Matsuoka, A. (2004). Structure-function relationships in unusual nonvertebrate globins. *Critical Reviews in Biochemistry and Molecular Biology*, *39*, 217–259.

Smagghe, B. J., Sarath, G., Ross, E., Hilbert, J. L., & Hargrove, M. S. (2006). Slow ligand binding kinetics dominate ferrous hexacoordinate hemoglobin reactivities and reveal differences between plants and other species. *Biochemistry*, *45*, 561–570.

Smagghe, B. J., Trent, J. T., III., & Hargrove, M. S. (2008). NO dioxygenase activity in hemoglobins is ubiquitous in vitro, but limited by reduction in vivo. *PLoS One*, *3*, e2039.

Soule, T., Gao, Q. J., Stout, V., & Garcia-Pichel, F. (2013). The global response of *Nostoc punctiforme* ATCC 29133 to UVA stress, assessed in a Temporal DNA microarray study. *Photochemistry and Photobiology*, *89*, 415–423.

Spiro, T. G., & Soldatova, A. V. (2012). Ambidentate H-bonding of NO and O_2 in heme proteins. *Journal of Inorganic Biochemistry*, *115*, 204–210.

Springer, B. A., Sligar, S. G., Olson, J. S., & Philips, G. N., Jr. (1994). Mechanisms of ligand recognition in myoglobin. *Chemical Reviews*, *94*, 699–714.

Stewart, J. J., & Coyne, K. J. (2011). Analysis of raphidophyte assimilatory nitrate reductase reveals unique domain architecture incorporating a 2/2 hemoglobin. *Plant Molecular Biology*, *77*, 565–575.

Sturms, R., DiSpirito, A. A., Fulton, D. B., & Hargrove, M. S. (2011). Hydroxylamine reduction to ammonium by plant and cyanobacterial hemoglobins. *Biochemistry*, *50*, 10829–10835.

Sturms, R., DiSpirito, A. A., & Hargrove, M. S. (2011). Plant and cyanobacterial hemoglobins reduce nitrite to nitric oxide under anoxic conditions. *Biochemistry*, *50*, 3873–3878.

Sugawara, Y., Kadono, E., Suzuki, A., Yukuta, Y., Shibasaki, Y., Nishimura, N., et al. (2003). Hemichrome formation observed in human haemoglobin A under various buffer conditions. *Acta Physiologica Scandinavica*, *179*, 49–59.

Tardif, M., Atteia, A., Specht, M., Cogne, G., Rolland, N., Brugiere, S., et al. (2012). PredAlgo: A new subcellular localization prediction tool dedicated to green algae. *Molecular Biology and Evolution*, *29*, 3625–3639.

Teh, A.-H., Saito, J. A., Baharuddin, A., Tuckerman, J. R., Newhouse, J. S., Kanbe, M., et al. (2011). Hell's Gate globin I: An acid and thermostable bacterial hemoglobin resembling mammalian neuroglobin. *FEBS Letters*, *585*, 3250–3258.

Thorsteinsson, M. V., Bevan, D. R., Ebel, R. E., Weber, R. E., & Potts, M. (1996). Spectroscopical and functional characterization of the hemoglobin of *Nostoc commune* UTEX 584 (Cyanobacteria). *Biochimica et Biophysica Acta*, *1292*, 133–139.

Thorsteinsson, M. V., Bevan, D. R., Potts, M., Dou, Y., Eich, R. F., Hargrove, M. S., et al. (1999). A cyanobacterial hemoglobin with unusual ligand binding kinetics and stability properties. *Biochemistry*, *38*, 2117–2126.

Tiso, M., Tejero, J., Kenney, C., Frizzell, S., & Gladwin, M. T. (2012). Nitrite reductase activity of nonsymbiotic hemoglobins from *Arabidopsis thaliana*. *Biochemistry*, *51*, 5285–5292.

Tobin, J. L., & Beales, P. L. (2007). Bardet-Biedl syndrome: Beyond the cilium. *Pediatric Nephrology*, *22*, 926–936.

Tomita, T., Gonzalez, G., Chang, A. L., Ikeda-Saito, M., & Gilles-Gonzalez, M. A. (2002). A comparative resonance Raman analysis of heme-binding PAS domains: Heme iron coordination structures of the *Bj*FixL, *Ax*PDEA1, *Ec*Dos, and *Mt*Dos proteins. *Biochemistry*, *41*, 4819–4826.

Traylor, T. G., Mitchell, M. J., Tsuchiya, S., Campbell, D. H., Stynes, D. V., & Koga, N. (1981). Cyclophane hemes. 4. Steric effects on dioxygen and carbon monoxide binding to hemes and heme proteins. *Journal of the American Chemical Society*, *103*, 5234–5236.

Trent, J. T., III., & Hargrove, M. S. (2002). A ubiquitously expressed human hexacoordinate hemoglobin. *The Journal of Biological Chemistry*, *277*, 19538–19545.

Trent, J. T., III., Kundu, S., Hoy, J. A., & Hargrove, M. S. (2004). Crystallographic analysis of *Synechocystis* cyanoglobin reveals the structural changes accompanying ligand binding in a hexacoordinate hemoglobin. *Journal of Molecular Biology*, *341*, 1097–1108.

Tsai, A. L., Berka, V., Martin, E., & Olson, J. S. (2012). A "sliding scale rule" for selectivity among NO, CO, and O_2 by heme protein sensors. *Biochemistry*, *51*, 172–186.

Tsai, A. L., Martin, E., Berka, V., & Olson, J. S. (2012). How do heme-protein sensors exclude oxygen? Lessons learned from cytochrome *c'*, *Nostoc puntiforme* heme nitric oxide/oxygen-binding domain, and soluble guanylyl cyclase. *Antioxidants & Redox Signaling*, *17*, 1246–1263.

Vandelle, E., & Delledonne, M. (2011). Peroxynitrite formation and function in plants. *Plant Science*, *181*, 534–539.

Vazquez-Limon, C., Hoogewijs, D., Vinogradov, S. N., & Arredondo-Peter, R. (2012). The evolution of land plant hemoglobins. *Plant Science*, *191*, 71–81.

Vinogradov, S. N., Fernandez, I., Hoogewijs, D., & Arredondo-Peter, R. (2011). Phylogenetic relationships of 3/3 and 2/2 hemoglobins in Archaeplastida genomes to bacterial and other eukaryote hemoglobins. *Molecular Plant*, *4*, 42–58.

Vinogradov, S. N., Hoogewijs, D., & Arredondo-Peter, R. (2011). What are the origins and phylogeny of plant hemoglobins? *Communicative & Integrative Biology*, *4*, 443–445.

Vinogradov, S. N., Hoogewijs, D., Bailly, X., Arredondo-Peter, R., Guertin, M., Gough, J., et al. (2005). Three globin lineages belonging to two structural classes in genomes from the three kingdoms of life. *Proceedings of the National Academy of Sciences of the United States of America*, *102*, 11385–11389.

Vinogradov, S. N., Hoogewijs, D., Bailly, X., Mizuguchi, K., Dewilde, S., Moens, L., et al. (2007). A model of globin evolution. *Gene*, *398*, 132–142.

Vinogradov, S. N., & Moens, L. (2008). Diversity of globin function: Enzymatic, transport, storage, and sensing. *The Journal of Biological Chemistry*, *283*, 8773–8777.

Vinogradov, S. N., Tinajero-Trejo, M., Poole, R. K., & Hoogewijs, D. (2013). Bacterial and archaeal globins—A revised perspective. *Biochimica et Biophysica Acta–Proteins and proteomics*, *1834*, 1789–1800.

Virtanen, A. I., & Laine, T. (1946). Red, brown and green pigments in leguminous root nodules. *Nature*, *1157*, 25–26.

Vos, M. H., Battistoni, A., Lechauve, C., Marden, M. C., Kiger, L., Desbois, A., et al. (2008). Ultrafast heme-residue bond formation in six-coordinate heme proteins: Implications for functional ligand exchange. *Biochemistry*, *47*, 5718–5723.

Vothknecht, U. C., & Westhoff, P. (2001). Biogenesis and origin of thylakoid membranes. *Biochimica et Biophysica Acta–Molecular and Cell Biology*, *1541*, 91–101.

Vu, B. C., Jones, A. D., & Lecomte, J. T. J. (2002). Novel histidine-heme covalent linkage in a hemoglobin. *Journal of the American Chemical Society*, *124*, 8544–8545.

Vu, B. C., Vuletich, D. A., Kuriakose, S. A., Falzone, C. J., & Lecomte, J. T. J. (2004). Characterization of the heme-histidine cross-link in cyanobacterial hemoglobins from

Synechocystis sp. PCC 6803 and *Synechococcus* sp. PCC 7002. *Journal of Biological Inorganic Chemistry, 9,* 183–194.

Vuletich, D. A., & Lecomte, J. T. J. (2006). A phylogenetic and structural analysis of truncated hemoglobins. *Journal of Molecular Evolution, 62,* 196–210.

Wakabayashi, K., & King, S. M. (2006). Modulation of *Chlamydomonas reinhardtii* flagellar motility by redox poise. *The Journal of Cell Biology, 173,* 743–754.

Wakabayashi, S., Matsubara, H., & Webster, D. A. (1986). Primary sequence of a dimeric bacterial haemoglobin from *Vitreoscilla. Nature, 322,* 481–483.

Wardman, P. (2007). Fluorescent and luminescent probes for measurement of oxidative and nitrosative species in cells and tissues: progress, pitfalls, and prospects. *Free Radical Biology & Medicine, 43,* 995–1022.

Weiland, T. R., Kundu, S., Trent, J. T., III., Hoy, J. A., & Hargrove, M. S. (2004). Bis-histidyl hexacoordination in hemoglobins facilitates heme reduction kinetics. *Journal of the American Chemical Society, 126,* 11930–11935.

Wever, R., Oudega, B., & Van Gelder, B. F. (1973). Generation of superoxide radicals during the autoxidation of mammalian oxyhemoglobin. *Biochimica et Biophysica Acta—Enzymology, 302,* 475–478.

Wittenberg, J. B., Bolognesi, M., Wittenberg, B. A., & Guertin, M. (2002). Truncated hemoglobins: A new family of hemoglobins widely distributed in bacteria, unicellular eukaryotes and plants. *The Journal of Biological Chemistry, 277,* 871–874.

Wright, K. A., & Boxer, S. G. (1981). Solution properties of synthetic chlorophyllide– and bacteriochlorophyllide–apomyoglobin complexes. *Biochemistry, 20,* 7546–7556.

Wu, G. H., Wainwright, L. M., & Poole, R. K. (2003). Microbial globins. *Advances in Microbial Physiology, 47,* 255–310.

Yeh, D. C., Thorsteinsson, M. V., Bevan, D. R., Potts, M., & La Mar, G. N. (2000). Solution ^{1}H NMR study of the heme cavity and folding topology of the abbreviated chain 118-residue globin from the cyanobacterium *Nostoc commune. Biochemistry, 39,* 1389–1399.

Yoon, H. S., Hackett, J. D., Ciniglia, C., Pinto, G., & Bhattacharya, D. (2004). A molecular timeline for the origin of photosynthetic eukaryotes. *Molecular Biology and Evolution, 21,* 809–818.

Yoon, J., Herzik, M. A., Winter, M. B., Tran, R., Olea, C., & Marletta, M. A. (2010). Structure and properties of a bis-histidyl ligated globin from *Caenorhabditis elegans. Biochemistry, 13,* 5662–5670.

Zaghloul, N. A., & Katsanis, N. (2009). Mechanistic insights into Bardet-Biedl syndrome, a model ciliopathy. *The Journal of Clinical Investigation, 119,* 428–437.

CHAPTER SEVEN

The Dos Family of Globin-Related Sensors Using PAS Domains to Accommodate Haem Acting as the Active Site for Sensing External Signals

Shigetoshi Aono[1]

Okazaki Institute for Integrative Bioscience & Institute for Molecular Science, National Institutes of Natural Sciences, Okazaki, Japan
[1]Corresponding author: e-mail address: aono@ims.ac.jp

Contents

1.	Introduction	275
2.	General Properties of PAS Domains	276
3.	Single-Component Systems for Transcriptional Regulation	276
	3.1 CO oxidation operon	276
	3.2 Regulator of CO metabolism	278
	3.3 Neuronal PAS domain 2 protein	280
4.	Two-Component Systems for Transcriptional Regulation	281
	4.1 FixL/FixJ two-component system	281
	4.2 A haem-based redox sensor in NtrY/NtrX two-component system	294
	4.3 DosS and DosT using a haem-containing GAF domain as a sensor module	295
5.	Haem-Containing PAS Family for the Regulation of Diguanylate Cyclases and Phosphodiesterases	298
	5.1 Regulation of diguanylate cyclases activity through interactions between haem and external ligand	298
	5.2 PDEs consisting of a haem-containing PAS and EAL domains	301
	5.3 A haem-sensing PAS domain regulates the PDE activity of YybT	305
6.	Haem-Containing PAS Family for Chemotaxis Regulation	307
	6.1 Bacterial chemotaxis system	307
	6.2 MCP containing a PAS domain with b-type haem	308
	6.3 MCP containing a PAS domain with c-type haem	311
7.	Conclusion	316
	References	317

Advances in Microbial Physiology, Volume 63
ISSN 0065-2911
http://dx.doi.org/10.1016/B978-0-12-407693-8.00007-8

Abstract

Sensor proteins play crucial roles in maintaining homeostasis of cells by sensing changes in extra- and intracellular chemical and physical conditions to trigger biological responses. It has recently become clear that gas molecules function as signalling molecules in these biological regulatory systems responsible for transcription, chemotaxis, synthesis/hydrolysis of nucleotide second messengers, and other complex physiological processes. Haem-containing sensor proteins are widely used to sense gas molecules because haem can bind gas molecules reversibly. Ligand binding to the haem in the sensor proteins triggers conformational changes around the haem, which results in their functional regulation. Spectroscopic and crystallographic studies are essential to understand how these sensor proteins function in these biological regulatory systems. In this chapter, I discuss structural and functional relationships of haem-containing PAS and PAS-related families of the sensor proteins.

ABBREVIATIONS

AxPDEA1 *Acetobacter xylinum* phosphodiesterase A1
BjFixL FixL protein from *Bradyrhizobium japonicum*
BjFixLH haem-containing PAS domain of *Bj*FixL
BMAL1 brain and muscle Arnt-like protein
CooA transcriptional regulator for the CO oxidation operon
CRP cyclic AMP receptor protein
Cyclic di-GMP bis-(3′,5′)-cyclic dimeric guanosine monophosphate
DGC diguanylate cyclase
EAL phosphodiesterase having an amino acid sequence motif of Glu–Ala–Leu
EcDos direct oxygen sensor from *Escherichia coli*
EPR electron paramagnetic resonance
EXAFS extended X-ray absorption fine structure
FNR regulatory factor for fumarate and nitrate reduction
GAF acronyms of cGMP-specific phosphodiesterases, adenylyl cyclases, and FhlA
H-NOX haem-nitric oxide/oxygen binding
NPAS2 neuronal PAS domain protein 2
PAS acronyms of Period, Arnt, and Sim
PDE phosphodiesterase
RcoM regulator of CO metabolism
sh shoulder peak
SmFixL FixL protein from *Sinorhizobium meliloti*
SmFixLH haem-containing PAS domain of *Sm*FixL
SwMb sperm whale myoglobin
UV–vis ultraviolet/visible

1. INTRODUCTION

Haem is one of the most widely used prosthetic groups in biological systems in both prokaryotes and eukaryotes because haem can provide a variety of biological functions including oxygen transport/storage, electron transfer, enzymatic oxidation and oxygenation of various substrates, enzymatic dehydration of substrates, and signal transduction. Though the basic structure of haem is identical among these haem proteins, the chemical and physical properties of haem can be tuned diversely by changing in the coordination structure of haem and in haem environmental structures including interactions between haem and surrounding amino acid residues. For example, electronic properties of haem including redox potentials can be tuned by changing in the coordination structures (coordination number and axial ligands) of haem and/or of hydrophobicity/hydrophilicity of haem pocket. External ligand-binding properties can also be controlled by changing in the coordination and haem environmental structures. Thus, specific interactions between haem and protein matrix in different ways provide diversities of biological functions for haem proteins.

In the haem-based sensor proteins, haem acts as the active site for sensing external signals such as diatomic gas molecules or redox changes (Aono, 2003, 2011; Gilles-Gonzalez & Gonzalez, 2005; Roberts, Kerby, Youn, & Conrad, 2005). It makes sense that gas sensor proteins adopt haem as the active site for sensing O_2, CO, and NO because haem can bind these gas molecules reversibly. These sensor proteins are involved in the regulation of gene expressions, chemotaxis, and synthesis/degradation of second messenger molecules, where these gas molecules act as physiological effectors. They are usually multi-domain proteins consisting of regulatory domain(s) along with a sensor domain in the same molecule, while sensor proteins consisting of only a single sensor domain also exist. The sensor domains/proteins regulate biological functions of regulatory domains/proteins in response to their physiological effectors, for which intra- and/or intermolecular signal transduction play important roles.

The binding of a physiological gas molecule to haem is the first step for sensing of the cognate physiological effector and of functional regulation of the haem-based sensor proteins. Upon binding of O_2, CO, or NO to haem, a conformational change around the haem will be induced by interactions between the haem-bound ligand and surrounding amino acid residues, which will be a trigger of a conformational change in a whole protein. Conformational changes of the protein play an important role for the regulation of the physiological activity of the protein. Analyses of ligand binding-induced conformational changes in detail are required to understand the molecular mechanisms of sensing and signal transductions for the haem-based sensor proteins. As haem is a very good spectroscopic probe, many spectroscopic techniques including laser flash photolysis, ultraviolet/visible (UV–vis), resonance Raman, electron paramagnetic resonance (EPR), and extended X-ray absorption fine structure (EXAFS) have been used to study the structure and function relationships of the haem-based sensor proteins along with X-ray crystallography.

The haem-based sensor proteins can be categorised into several types based on the differences of the domains accommodating haem. PAS, GAF, H-NOX, CRP/FNR, and globin domains are known to be adopted as a sensor domain accommodating a haem in the haem-based sensor domains. This chapter focuses on the recent developments of the studies on the sensor proteins containing a haem-containing PAS domain or several related domain.

2. GENERAL PROPERTIES OF PAS DOMAINS

PAS domain consists of ca. 100–120 residues, which is widely distributed among Bacteria, Archaea, and Eukarya (Möglich, Ayers, & Moffat, 2009). Though the amino acid sequence homologies are not high among PAS domains, their three-dimensional structures are similar. In most cases, PAS domain(s) is a part of multi-domain proteins, where it plays a role for biological signal transductions by signal sensing and/or by protein–protein interactions. The canonical PAS fold shows the α/β structure consisting of five antiparalleled β strands (A_β, B_β, G_β, H_β, and I_β) and four α helices (C_α, D_α, E_α, and F_α) (Möglich et al., 2009). Some PAS domains accommodate a prosthetic group such as haem, flavin, or para–coumaric acid, which are used for sensing external signals such as diatomic gas molecules and light (Henry & Crosson, 2011).

3. SINGLE-COMPONENT SYSTEMS FOR TRANSCRIPTIONAL REGULATION

3.1. CO oxidation operon

There are two major systems to regulate biological functions in response to external chemical or physical signals: single-component and two–component signal transduction systems. In the single-component signal transduction system, the same protein shows the functions of not only sensing external signals but also regulating biological functions. For example, transcriptional regulators that consist of a sensor and DNA-binding domains are typical for the single-component system. In this system, conformational changes are induced upon sensing the cognate effector molecule/signal by the sensor domain, which ends up with regulation of the DNA-binding activity.

CO oxidation operon (CooA) protein is a single-component transcriptional regulator, which is one of the most widely studied haem-based sensors (Aono, 2003, 2011; Roberts et al., 2005). CooA consists of the N-terminal CRP/FNR-like sensor domain and the C-terminal DNA-binding domain (Komori, Inagaki, Yoshioka, Aono, & Higuchi, 2007; Lanzilotta et al., 2000). A b-type haem in the sensor domain acts as the active site for sensing CO (Aono, Nakajima, Saito, & Okada, 1996). The haem in CooA shows a dynamic change in the coordination structures of the haem. While the ferric haem in CooA from *Rhodospirillum rubrum* (Rr-CooA) is coordinated with Cys75 and Pro2, a ligand exchange takes place between Cys75 and His77 upon reduction of the haem (Fig. 7.1; Aono, 2003, 2011; Aono, Ohkubo, Matsuo, & Nakajima, 1998). This redox-dependent ligand exchange is fully reversible reaction. Though the ferrous haem in Rr-CooA is coordinationally saturated with two endogenous axial ligands, it can bind CO to form CO-bound haem under physiological conditions, during which the ligand exchange takes place between Pro2 and CO (Aono, 2003; Nakajima et al., 2001; Yamamoto et al., 2001).

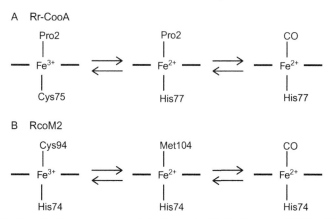

Figure 7.1 Coordination structures of the haem in (A) Rr-CooA and (B) RcoM2.

The binding of an external ligand to a 6-coordinated haem with ligand exchange will induce a large conformational change around the protein main chain at which the residue dissociated from the haem, which can be a trigger for conformational changes responsible for the functional regulation of proteins. In the case of Rr-CooA, Pro2 bound to the ferrous haem is dissociated from the haem iron upon CO binding (Aono, 2003; Nakajima et al., 2001; Yamamoto et al., 2001). As the N-terminus (Pro2) is bound to the haem in the ferric and ferrous Rr-CooA, the conformation of the N-terminal region in Rr-CooA is fixed. Upon CO binding to Rr-CooA, relocation of the N-terminal region is induced by the dissociation of Pro2 from the haem, which causes conformational changes of the DNA-binding domain, resulting in the activation of Rr-CooA for DNA binding (Aono, 2003, 2011).

3.2. Regulator of CO metabolism

Though CO concentration present in the natural environment is less than 1 ppm, a number of microbes express distinct CO-regulated systems including CO-dependent transcriptional regulators and proteins responsible for CO metabolism (King & Weber, 2007; Ragsdale, 2004). Two homologous proteins, RcoM1 and RcoM2 from *Burkholderia xenovorans*, function as CO-dependent transcriptional regulators, which consist of an N-terminal PAS domain and a C-terminal LytTR family DNA-binding domain (Kerby, Youn, & Roberts, 2008). The *rcoM1* gene in *B. xenovorans* is adjacent to the *coxMSL* genes encoding an aerobic CO oxidation system, suggesting that RcoM1 is a regulator of the *cox* expression. *In vivo* assays using *lacZ* fused to the promoter of the *coxM* as a reporter gene show that RcoM1 and RcoM2 activate the expression of the reporter gene only in the presence of CO (Kerby et al., 2008).

The N-terminal PAS domain binds a b-type haem in RcoM1 and RcoM2 (Smith et al., 2012). Amino acid sequence alignments reveal that the proximal His (His74 in RcoM) is conserved among regulator of CO metabolism (RcoM), FixL, and EcDos. RcoM1 shows the Soret, α, and β bands at 420, 574, and 539 nm; 426, 562, and 531 nm; 423, 570, and 540 nm; and 421, 578, and 545 nm in the ferric, ferrous, ferrous-CO, and ferrous-NO forms, respectively (Kerby et al., 2008). RcoM2 shows similar UV–vis spectra to those of RcoM1, and its detailed spectroscopic properties are studied by EPR, MCD, and resonance Raman spectroscopy (Marvin, Kerby, Youn, Roberts, & Burstyn, 2008). Ferric RcoM2 shows the EPR signals of $g = 2.52$, 2.28, and 1.88, which are similar to those of a thiolate-bound, low-spin haem such as Rr-CooA (Aono et al., 1998).

The MCD spectrum of ferric RcoM2 shows an intense temperature-dependent, derivative-shaped C term in the Soret region with a crossover at 419 nm, and another temperature-dependent C term is also present in the Q-bands region with a crossover at 570 nm (Marvin et al., 2008). Ferrous RcoM2 shows a temperature-independent A term in the Q-bands region with a crossover at 560 nm and much less intense temperature-independent feature in the Soret region with a crossover at 421 nm in its MCD spectrum (Marvin et al., 2008).

In the resonance Raman spectra, the $\nu 2$, $\nu 3$, $\nu 4$, and $\nu 10$ bands are observed at 1579, 1500, 1371, and 1634 cm^{-1}; 1580, 1490, 1359, and 1580 cm^{-1}; and 1575, 1494, 1368, and 1630 cm^{-1}, for ferric, ferrous, and CO-bound RcoM2, respectively (Marvin et al., 2008). The correlation between ν(Fe–CO) (485 cm^{-1}) and ν(C–O) (1965 cm^{-1}) bands indicates that the proximal ligand *trans* to CO is a neutral His in CO-bound RcoM2.

While the C94S variant of RcoM2 shows the identical spectroscopic features to the wild type in the ferrous and CO-bound forms, the ferric RcoM2 shows different UV–vis, EPR, and resonance Raman spectra from those of the wild type. Though the wild-type RcoM shows the δ-band at 354 nm typical for a thiolate-bound haem, a clear δ-band is not observed in the C94S variant. The C94S variant shows an axial signals of $g = 5.82$ and 2.00 instead of a set of rhombic signals observed in the wild-type RcoM2 (Smith et al., 2012). Resonance Raman spectrum reveals that the ferric haem in the C94S variant is a mixture of a low-spin and high-spin states, where the two bands are observed for both $\nu 2$ and $\nu 3$ at 1562 cm^{-1} (high spin)/1577 cm^{-1} (low spin) and 1493 cm^{-1} (high spin)/1503 cm^{-1} (low spin), respectively.

Based on these spectral features, the coordination structures of the haem in RcoM2 are proposed as shown in Fig. 7.1 (Smith et al., 2012). RcoM2 shows a ligand exchange upon the change in the oxidation state of the haem and upon binding CO. Though His74 is retained as the proximal ligand in the ferric, ferrous, and CO-bound haems, the distal ligand exchange takes place between the ferric and ferrous haems and between the ferrous and CO-bound haems. Cys94 coordinated to the ferric haem is replaced by Met104 upon reduction of the haem to form a 6-coordinate ferrous haem with His74 and Met104 as the axial ligands. Met104 is replaced by CO upon reaction of ferrous RcoM2 with CO to form the CO-bound RcoM2. Amino acid sequence alignments suggest that Met104 is located in the

FG loop in the PAS domain. Thus, the dissociation of Met104 upon CO binding will cause conformational changes at the distal haem pocket as is the case of FixL and EcDos, which is responsible for the activation of RcoM2 by CO.

3.3. Neuronal PAS domain 2 protein

Neuronal PAS domain 2 protein (NPAS2) is a mammalian transcriptional factor that binds DNA as a heterodimer with brain and muscle Arnt-like protein (BMAL1), which is thought to be involved in the regulation of circadian rhythm. NPAS2 knockout mice fail to exhibit rhythmic *per2* gene expression (Reick, Garcia, Dudley, & McKnight, 2001). NPAS2/BMAL1 heterodimer activates the expression of *per* and *cry* genes that are negative regulatory components of the circadian clock (King et al., 1997).

NPAS2 is a member of the basic helix–loop–helix (bHLH)-PAS family, which consists of the N-terminal bHLH domain and two PAS domains (PAS-A and PAS-B). Though both of the recombinant PAS-A and PAS-B bind a b-type haem, DNA-binding activity of NPAS2 is independent of presence or absence of haem (Dioum et al., 2002). NPAS2/BMAL1 heterodimer can bind to DNA regardless of whether NPAS2 is apo- or holo-forms (Dioum et al., 2002). On the other hand, CO regulates the DNA-binding activity only for holo-NPAS2, but not for apo-NAPS2 (Dioum et al., 2002). CO-dependent regulation of DNA binding for NPAS2 will be caused by the inhibition of the NPAS2/BMAL1 heterodimer formation, which will be induced by CO-binding to the haem in NPAS2. Haem oxygenase 2 (HO-2), which produces CO as a reaction product of haem degradation, is expressed in the same region where NPAS2 is expressed. These results suggest that CO generated by HO-2 acts as a physiological effector of NPAS2 (Dioum et al., 2002).

The resonance Raman spectra of the ferric and ferrous haems in PAS-A domain consist of a mixture of 6-coordinate and 5-coordinate states (Uchida et al., 2005). The coordination structures of the haem in the predominant species are proposed as shown in Fig. 7.2, which are elucidated by site-directed mutagenesis and resonance Raman spectroscopy. Mutation of His119 or Cys170 to Ala significantly increases the intensity of the $\nu 3$ band at 1490 cm^{-1} for a 5-coordinate, high-spin haem compared with that of the $\nu 3$ band at 1504 cm^{-1} for a 6-coordinate, low-spin haem, whereas the mutation of His148 or His171 induces a moderate increase (Uchida et al., 2005). These results suggest that His119 and Cys170 are axial ligands of

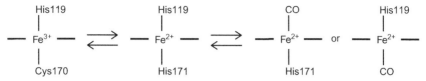

Figure 7.2 Coordination structures of the haem in NPAS2.

the ferric haem in the dominant species of the PAS-A domain. The bHLH-PAS-A shows different spectroscopic properties from those of the isolated PAS-A domain (Mukaiyama et al., 2006; Uchida, Sagami, Shimizu, Ishimori, & Kitagawa, 2012). The ferric haem in the bHLH-PAS-A is proposed to consist of His119 and His171 as the axial ligands instead of His119 and Cys170 (Uchida et al., 2012). These differences may be caused by an artificial effect by domain truncation or may suggest the existence of the equilibrium of the coordination structures between bis-His and Cys–His coordinations. Though it is proposed that ligand exchange takes place between Cys170 and His171 upon the change in the oxidation state of the haem (Fig. 7.2), its physiological role is not obvious at present. Upon CO binding to the haem, another ligand exchange takes place between the haem-bound His and CO, though it is not known which His is replaced by CO.

The mutation His119 or His171 to Ala results in remarkable reduced transcriptional activity, which is caused by impaired heterodimer formation with BMAL1 (Ishida, Ueha, & Sagami, 2008). The C170A variant shows the same transcriptional activity as the wild type (Ishida et al., 2008). These results suggest that the ligand exchange between the haem-bound His and CO plays an important role for the functional regulation of NPAS2, which may cause conformational changes in the haem pocket triggered by the dissociation of the haem-bound His.

4. TWO-COMPONENT SYSTEMS FOR TRANSCRIPTIONAL REGULATION

4.1. FixL/FixJ two-component system

Two-component signal transduction system consists of two separate proteins, a sensor kinase and a response regulator, which is one of the most utilised biological signal transduction systems (Jung, Fried, Behr, & Heermann, 2011; Kirby, 2009). In this system, sequential phosphoryl group transfer reactions play important role for signal transductions. The sensor kinases sense external signals, by which the autokinase activities of the sensor kinases are regulated. Once the sensor kinase is activated to take place its autophosphorylation, the phosphoryl group added in the sensor kinase transfers to the cognate response regulator to form a phosphorylated response regulator, which results in the regulation of the biological function of response regulator.

FixL is a sensor kinase in the FixL/FixJ two-component system that regulates the expression of nitrogen fixation genes in response to O_2 concentration in nodule bacteria (Bauer, Elsen, & Bird, 1999; Fischer, 1994; Sciotti, Chanfon, Hennecke, & Fischer, 2003). *Bradirhizobium japonicum* and *Sinorhizobium meliloti* are symbionts that form nodule upon infecting to soybean and alfalfa roots, respectively, FixL proteins from which have been studied extensively. In nodules, a series of nitrogen fixation genes are expressed as O_2 concentration decrease <50 µM (Gilles-Gonzalez, 2001; Soupène, Foussard, Boistard, Truchet, & Batut, 1995). As nitrogenase, a key enzyme for nitrogen fixation to form ammonia from N_2, is sensitive to O_2, its O_2-dependent expression is strictly regulated to prevent nitrogenase from inactivation by O_2. In *S. meliloti*, the FixL/FixJ system regulates the expression of *fixK* and *nifA* genes, while only *fixK_2* gene is regulated by the FixL/FixJ system in *B. japonicum* (David et al., 1998; Fischer, 1994). As FixK, NifA, and $FixK_2$ are transcriptional regulators, the regulatory cascades for gene expression proceed by a cue triggered with the FixL/FixJ two-component system.

FixL consists of the sensor and kinase domains, which is autophosphorylated with ATP, and then phosphoryl group transfer takes place between FixL and FixJ (Gilles-González, González, & Perutz, 1995; Gilles-Gonzalez et al., 1994). Unphosphorylated FixL preferably exists in the form of quaternary complex of $(FixL)_2(FixJ)_2$ (Miyatake et al., 1999a,b). Phosphorylation of FixL/FixJ takes place in the $(FixL)_2(FixJ)_2$ complex, in which ATP bound in one subunit is used to phosphorylate the conserved His (His285 in *Sm*FixL) in the other subunit (Gilles-Gonzalez & Gonzalez, 1993; Monson, Ditta, & Helinski, 1995; Nakamura, Kumita, Imai, Iizuka, & Shiro, 2004; Reyrat, David, Batut, & Boistard, 1994; Tuckerman, Gonzalez, & Gilles-Gonzalez, 2001). When phosphoryl group transfer has proceeded from the phosphorylated FixL to FixJ, phosphorylated FixJ is released from the complex to form a homodimer of phosphorylated FixJ. Upon phosphorylation, FixJ forms a homodimer that is the active form for the transcriptional regulator with specific DNA-binding activity (Agron, Ditta, & Helinski, 1993; Da Re et al., 1999; Reyrat, David, Blonski, Boistard, & Batut, 1993). O_2 inhibits the phosphorylation reaction of the $(FixL)_2(FixJ)_2$ complex with ATP but affects neither ATP binding (K_d = ca. 100 μM) nor the $(FixL)_2(FixJ)_2$ complex formation (K_d = ca. 4 μM) (Sousa, Gonzalez, & Gilles-Gonzalez, 2005).

ADP produced from ATP bound in the kinase domain of *Sm*FixL acts as an allosteric effector, reducing O_2-binding affinity of the sensor domain in *Sm*FixL (Nakamura et al., 2004). The K_d (O_2) values of the $(FixL)_2(FixJ)_2$ complex are 56 and 218 μM in the absence and presence of ADP, respectively (Nakamura et al., 2004). O_2-binding/dissociation kinetics measurements by stopped flow method reveal that decrease in O_2-binding affinity is mainly caused by an increase of k_{off} (rate constant of O_2 dissociation) (Nakamura et al., 2004). Nakamura et al. (2004) propose "a two-cylinder reciprocating engine model" of the ADP-dependent acceleration of kinase reaction, where ADP produced at the ATP-binding site in the FixL kinase domain of one subunit in the $(FixL)_2(FixJ)_2$ complex reduces the O_2-binding affinity of the other subunit of FixL in *trans*-acting manner.

4.1.1 O_2 regulates the activity of FixL

FixL contains a b-type haem and its kinase activity of FixL is regulated by O_2 binding to the haem (Gilles-Gonzalez, Ditta, & Helinski, 1991). Activities of FixL can be evaluated by autophosphorylation of FixL and/or phosphoryl group transfer to FixJ. Tuckerman, Gonzalez, Dioum, and Gilles-Gonzalez (2002) report that O_2 binding to the haem in deoxy *Sm*FixL (an external ligand-free Fe^{2+} form) causes >100-fold decrease in the activity of *Sm*FixJ phosphorylation (phosphoryl group transfer from *Sm*FixL to *Sm*FixJ), while autophosphorylation activity of FixL decreases by 15-fold. They also report that the activity of *Sm*FixJ phosphorylation activity decreases by 100-fold upon the oxidation of deoxy *Sm*FixL to met *Sm*FixL, though met *Sm*FixL shows the same activity for autophosphorylation of *Sm*FixL as does deoxy *Sm*FixL (Tuckerman et al., 2002). Akimoto, Tanaka, Nakamura, Shiro, and Nakamura (2003), however, reported that the met and deoxy forms of *Sm*FixL show comparable activities for both autophosphorylation of *Sm*FixL and *Sm*FixJ phosphorylation. They demonstrate that S—S bond formation at Cys301 causes aberrant kinase inactivation in met *Sm*FixL (Akimoto et al., 2003). The contents of the aberrant inactive form with the S—S bond will vary in preparation to preparation. Once the S—S bond is formed, it remains intact also in deoxy *Sm*FixL produced by reduction of met *Sm*FixL with sodium dithionite. On the other hand, DTT treatment of met *Sm*FixL to produce deoxy *Sm*FixL cleaves the S—S bond, if it is present, to restore kinase activity. Tuckerman et al. (2002) and Akimoto et al. (2003) use dithionite and DTT, respectively, to reduce met SmFixL, which may be a reason of the above discrepancies.

Though CO, NO, CN$^-$, F$^-$, and imidazole (Im) can be bound to the haem in FixL as is O$_2$, their regulatory effects on the FixL activity are not identical in all ligands. CO, NO, and F$^-$ do not significantly inhibit the kinase activity of FixL, while CN$^-$ and Im inhibit the ferric FixL activity as effective as O$_2$ (Akimoto et al., 2003; Tuckerman et al., 2002). These results clearly indicate that FixL discriminates O$_2$ from CO and NO. Spectroscopic and structural analyses of the ligand-bound FixLs with these ligands give useful information to elucidate the molecular mechanisms of selective O$_2$ sensing by FixL and of O$_2$-dependent signal transduction, which are described in the following sections.

4.1.2 X-ray crystal structures of the PAS domain in FixL

The X-ray crystal structures of the PAS domain have been determined in the met and deoxy forms for SmFixL (Miyatake et al., 2000). The overall structures of the met and deoxy forms are almost same, and there is no functionally important structural difference between these forms. A protohaem (b-type haem) is accommodated between the F helix and five antiparalleled β strands (A$_β$, B$_β$, H$_β$, I$_β$, and G$_β$ strands). His194 on the F helix in SmFixL (His200 in BjFixL) is the proximal ligand of the haem. The ferric haem is 5-coordinate with a vacant site at the sixth position, which is out-of-plane configuration with a doming of haem plane. The iron is displaced by 0.49 Å from the pyrrole nitrogen (Miyatake et al., 2000). The Fe–N distance between the haem iron and the proximal Im is 2.14 Å, which is consistent with the distance determined by EXAFS (2.11 Å) (Miyatake et al., 1999a,b). The orientation of His194 is fixed by a hydrogen-bond network among His194, carboxylate of Asp195, the amide group of Asn181, and a water molecule in the proximal haem pocket (Miyatake et al., 2000).

There are several hydrophobic residues (Ile209, Leu230, and Val232 in SmFixL) in the vicinity of the distal side on the haem. The distal haem pocket is packed so densely with these hydrophobic residues that O$_2$ could not bind to the haem iron without collisions with these residues, indicating that a conformational change in the distal haem pocket should be required to bind O$_2$ (Miyatake et al., 2000; Mukai, Nakamura, Nakamura, Iizuka, & Shiro, 2000). Based on these structural properties, the "hydrophobic triad model" is proposed for the signal transduction mechanism of FixL. In this model, it is proposed that conformational changes in these hydrophobic residues upon O_2 binding trigger a conformational change in the distal haem pocket, which results in conformational changes in the whole molecule of FixL for the regulation of its autokinase activity.

The X-ray crystal structures of the PAS domain have also been determined for met, deoxy, O_2-bound, CO-bound, NO-bound, Im-bound, MeIm-bound, and CN^--bound BjFixLH (haem-containing PAS domain of BjFixL) (Dunham et al., 2003; Gong, Hao, & Chan, 2000; Gong et al., 1998; Hao, Isaza, Arndt, Soltis, & Chan, 2002; Key & Moffat, 2005). The overall structures of the core PAS domains are same between haem-containing PAS domain of SmFixL (SmFixLH) and BjFixLH, though SmFixLH and BjFixLH are a homodimer and a monomer, respectively. Comparing the structures of deoxy and O_2-bound BjFixL reveal that a conformation of the haem itself is altered upon O_2 binding to the haem (Dunham et al., 2003; Gong et al., 2000, 1998; Hao et al., 2002; Key & Moffat, 2005). Though the met and deoxy haems are ruffling in BjFixLH, O_2 binding to the haem causes a flattening of the haem. A similar flattening of the haem is also observed in CO-, NO-, and CN^--bound haems for BjFixLH (Fig. 7.3). While O_2 and CN^- inhibit the FixL activity, CO and NO do not (Akimoto et al., 2003; Gilles-Gonzalez et al., 2006; Sousa, Tuckerman, Gondim, Gonzalez, & Gilles-Gonzalez, 2013). The degree of flattening upon CO or NO binding is smaller compared with that upon O_2 or CN^- binding. Thus, it partly correlates with the functional regulation of FixL whether or not the flattening of the haem takes place.

4.1.3 Functional roles of amino acid residues in the haem pocket

Electrostatic interactions network among the haem propionate groups and surrounding amino acid residues is rearranged upon O_2 binding. A salt bridge between Arg220, which is located at the C-terminal end of the FG loop, and the haem propionate in deoxy BjFixLH is lost in oxy BjFixLH. A new salt bridge is formed between Arg206 and the haem propionate 6 in oxy BjFixLH, while the side chain of Arg206 is interacting with the main-chain carbonyl oxygen of Asp212 without interaction with the haem propionate 6 in deoxy BjFixL (Hao et al., 2002; Key & Moffat, 2005). Resonance Raman spectroscopy reveals the reconstruction of salt bridge with the haem propionate (Balland et al., 2006). The R200A SmFixL variant (R206A for BjFixL) shows the same O_2-dependent regulation for autophosphorylation, indicating that the salt bridge interaction between the Arg residue and the haem propionate

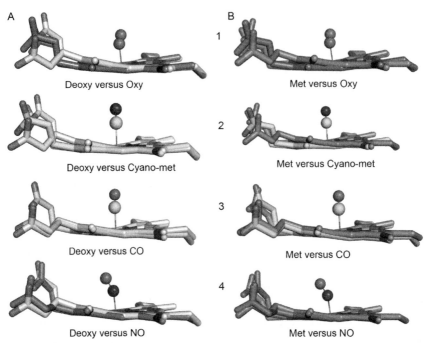

Figure 7.3 Comparison of haem ruffling for *Bj*FixL. (A) The deoxy haem (grey) is compared to (1) the oxy, (2) cyano-met, (3) CO-bound, and (4) NO-bound haems, respectively. (B) The met haem (orange) is compared to (1) the oxy, (2) cyano-met, (3) CO-bound, and (4) NO-bound haems, respectively. (For interpretation of the references to colour in this figure legend, the reader is referred to the online version of this chapter.)

6 is not essential for the functional regulation of FixL (Tanaka, Nakamura, Shiro, & Fujii, 2006).

The structure of R206A variant of *Bj*FixLH resembles that of met *Bj*FixLH and of CN⁻-bound *Bj*FixLH with the r.m.s.d. for all C_α atoms of 0.28 and 0.55 Å, respectively (Gilles-Gonzalez et al., 2006). Though Arg206 in wild-type *Bj*FixLH interacts with His214 and the haem propionate groups, these interactions are absent in R206A variant. The R206A variant shows a slightly lower O_2-binding affinity ($K_d = 350$ μM) compared with the wild-type *Bj*FixLH ($K_d = 142$ μM), which is caused by a faster rate of O_2 dissociation (Gilles-Gonzalez et al., 2006). Though the absence of Arg206 interactions also increases slightly the ruffling of the porphyrin ring, it is not responsible for the functional regulation of FixL. CN⁻ binding inhibits FixJ phosphorylation in FixL/FixJ complex as it does in wild-type *Bj*FixL (Gilles-Gonzalez et al., 2006).

X-ray structural analyses of *Bj*FixLH reveal that O_2 binding to the haem induces a shift of the FG loop (Thr209 to Arg220) close to the haem pocket (Fig. 7.4). This shift results in the formation of a hydrogen bond between the haem-bound O_2 and Arg220. O_2-binding affinity decreases by about 10-fold when Arg220 is replaced by Ala, indicating that the hydrogen-bonding interaction of Arg220 with the haem-bound O_2 plays an important role for the regulation of O_2-binding affinity (Table 7.1).

Though Arg220 forms a salt bridge with the haem propionate 7 in deoxy *Bj*FixLH, this salt bridge is dissociated upon O_2 binding and the guanidium side chain of Arg220 rotates into the distal haem pocket to form the hydrogen bond with the haem-bound O_2. The C_α—C_β bond of Arg220 is rotated with ca. 170° (Gong et al., 2000). There is a hydrogen-bond network among Arg220, O_2, a water molecule, and the carbonyl oxygen of Ile218, which will be stabilised by the proper orientation of Arg220 in oxy *Bj*FixLH.

While CN^- binding causes a similar FG loop shift, such an FG loop shift does not take place upon CO or NO binding (Gong et al., 2000; Hao et al., 2002). In CN^--bound *Bj*FixLH, Arg220 moves into the haem pocket to form an electrostatic interaction with the haem-bound CN^-, while the salt bridge between Arg220 and the haem propionate retains in CO- and NO-bound *Bj*FixLH. The structure of Im-bound *Bj*FixLH reveals changes in the FG loop that are similar to those observed in oxy and CN^--bound *Bj*FixLH, though Arg220 adopts a position outside of the haem pocket (Gong et al., 2000). Both CN^- and Im can regulate the FixL activity as does O_2. These results indicate that the interaction between the side chain of Arg220 and the haem-bound ligand is not a mandatory requirement for the regulation of FixL activity.

Based on these results, "the FG loop mechanism" is proposed for the regulation of FixL activity. In this mechanism, it is proposed that O_2 binding to

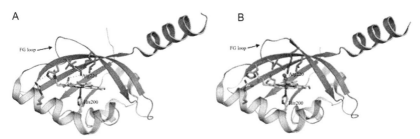

Figure 7.4 The structures of (A) met *Bj*FixLH (PDB 1DRM) and (B) oxy *Bj*FixLH (PDB 1DP6). (For colour version of this figure, the reader is referred to the online version of this chapter.)

Table 7.1 Ligand-binding kinetics parameters for FixL and Dos

	O₂			CO			CN⁻			
	k_{on} ($\mu M^{-1}\,s^{-1}$)	k_{off} (s^{-1})	K_d (μM)	k_{on} ($\mu M^{-1}\,s^{-1}$)	k_{off} (s^{-1})	K_d (μM)	k_{on} ($\mu M^{-1}\,s^{-1}$)	k_{off} (s^{-1})	K_d (μM)	References
BjFixL	0.14	20	142	0.08	0.04	4.8	1.1×10^{-4}	1.2×10^{-4}	0.94	Gilles–Gonzalez et al. (2006)
BjFixLH	0.3	10	33	—	—	—	—	—	—	Balland et al. (2005)
BjFixLH R220H	1.3	12	9	—	—	—	—	—	—	Balland et al. (2005)
BjFixLH R220A	1.4	1750	1250	—	—	—	—	—	—	Balland et al. (2005)
SmFixL	—	—	33	—	—	—	2.6×10^{-5}	3.1×10^{-4}	12	Tanaka et al. (2006)
SmFixLH	0.22	6.8	31	0.017	0.083	4.9	2.7×10^{-5}	1.0×10^{-4}	3.7	Gilles Gonzalez et al. (2006), Winkler et al. (1996)
SmFixLT	0.22	11	50	0.012	—	—	3.1×10^{-5}	1.5×10^{-4}	4.8	Gilles Gonzalez et al. (2006), Winkler et al. (1996)
AxPDEA1H	6.7	77	12	0.21	0.058	0.28	—	—	—	Chang et al. (2001)
EdDosH	0.0026	0.034	13	0.0011	0.011	10	—	—	—	Chang et al. (2001)
MtDosH	6	12	2	—	—	—	—	—	—	Tomita, Gonzalez, Chang, Ikeda-Saito, and Gilles-Gonzalez (2002)
SuMb	14	12	0.86	0.51	0.019	0.037	—	—	—	Winkler et al. (1996)

BjFixLH (residues 142–270); SmFixL (residues 127–505); SmFixLH (residues 127–260); SmFixLT (residues 78–464).

the haem induces the reconstruction of hydrogen-bond network around the haem with the conformational change of the FG loop, which results in triggering the conformational changes of FixL for the functional regulation of FixL.

4.1.4 Ligand-binding properties of the haem in FixL

Resonance Raman and UV–vis spectroscopy reveal that the ferric and ferrous haems are 5-coordinate and high-spin state (Balland et al., 2006; Lukat-Rodgers, Rexine, & Rodgers, 1998; Tomita et al., 2002). The ferric and ferrous haem in FixL can bind an external ligand at the vacant site of the haem as an axial ligand. Rate constants of ligand binding/dissociation for FixL are summarised in Table 7.1. O_2-binding affinity of FixL is lower compared with a typical O_2-binding protein, sperm whale myoglobin (SwMb), which is due to lower k_{on} by about two orders of magnitude.

O_2-binding affinity of BjFixL increases by a truncation of the C-terminal kinase domain by ca. fourfold, where the K_d values for O_2 are 142 and 33 μM for BjFixL with and without the kinase domain, respectively. Truncation of the kinase domain results in the increase and decrease of k_{on} and k_{off}, respectively, in BjFixL. These differences may be caused by conformational changes of the PAS domain induced by the truncation of the kinase domain.

4.1.5 Ligand-binding and protein dynamics revealed by time-resolved spectroscopy

Time-resolved spectroscopy is a useful tool to elucidate the dynamic properties of ligand binding and protein conformational changes. Time-resolved resonance Raman spectroscopy with subpicosecond time resolution reveals that no sizeable doming of the haem is observed in photolysis of O_2-bound BjFixLH, suggesting that efficient ultrafast O_2 rebinding to the haem occurs on the femtosecond timescale (Kruglik et al., 2007). This property is different from that for ligand-bound globins, in which haem doming occurs faster than 1 ps after photolysis of the haem-bound ligand. The dissociated O_2 cannot move away substantially from the haem because of steric constraints and remain in a favourable position for rebinding in BjFixLH (Kruglik et al., 2007). On the other hand, the recombination of CO to the haem does not occur on the time scale up to 4 ns (Jasaitis et al., 2006; Liebl, Bouzhir-Sima, Negrerie, Martin, & Vos, 2002). Time-resolved infrared spectroscopy reveals that photolyzed CO is trapped in the docking site, in which CO is oriented at a large angle with respect to the haem normal (Nuernberger et al., 2011). Escape of CO from the docking site to the solvent occurs on the tens or hundreds of nanosecond timescale (van Wilderen, Key, Van Stokkum, van Grondelle, & Groot, 2009). Anisotropy measurements reveal that the haem-bound CO makes a substantial angle with the haem normal (ca. 30°), indicating that the protein environments in FixL impose strain on the haem-bound CO (Nuernberger et al., 2011). This strain, which will be caused by the steric interaction among Ile215, Ile236, and CO, causes CO to tilt away from the perpendicular orientation and would be a reason of a lower CO-binding affinity for FixL compared with globins (Nuernberger et al., 2011).

Hiruma, Kikuchi, Tanaka, Shiro, and Mizutani (2007) have studied the structural changes of SmFixL by means of time-resolved resonance Raman spectroscopy. They reveal that changes in the time-resolved resonance Raman spectra occur in three steps. In the step 1, the band intensities of the ν(Fe–His), $\gamma 7$, and $\nu 8$ modes change in a timescale of 0.2–2 μs. In step 2, the intensity of the $\delta(C_\beta C_c C_d)$ band changes in a timescale of 0.8–9 μs. In step 3, the $\gamma 7$, $\nu 8$, and $\delta(C_\beta C_c C_d)$ modes change after >1 ms.

The $\delta(C_\beta C_c C_d)$ band is sensitive to the strength of the hydrogen bond of the haem propionates (Gottfried et al., 1996). The frequency of this band is down-shifted as the strength of the hydrogen bond decreases. The intensity of the $\nu 8$ mode is correlated with disorder in the orientation of the haem propionate groups (Peterson, Friedman, Chien, & Sligar, 1998). The $\gamma 7$ mode is associated with an out-of-plane motion of the methine carbons of haem, and this mode is observed in the high-spin haems, but not in the low-spin haems (Hu, Smith, & Spiro, 1996). Thus, the intensity change of the $\gamma 7$ mode is indicative of the extent of haem doming.

In the timescale of 10 ns to 10 μs, different relaxations are observed between O_2-dissociated and CO-dissociated species (Hiruma et al., 2007). In this timescale, fast intensity changes of $\gamma 7$ and $\nu 8$ modes are observed only for O_2-dissociated species. The intensity change of the $\delta(C_\beta C_c C_d)$ band is faster for O_2-dissociated species than for CO-dissociated species. The ν(Fe–His) band shows the intensity change for the O_2-dissociated species but not for CO-dissociated species in SmFixLH.

The relaxation of the ν(Fe–His) band after ligand photo-dissociation is different between SmFixLH and BjFixLH especially for O_2-dissociated species. In picosecond time-resolved resonance Raman spectroscopy with BjFixLH, the ν(Fe–His) band is not observed even at a time delay as short as 0.5 ps after photolysis of O_2-bound BjFixLH, which is consistent with ultrafast O_2 rebinding described earlier (Kruglik et al., 2007).

Hiruma et al. (2007) propose the following scheme of structural changes upon ligand dissociation. In the step 1, the interaction between Arg214 (Arg220 in BjFixL) and O_2 is lost upon O_2 dissociation, resulting in the reorientation of Arg214 towards haem propionate 7 to form a salt bridge with it. The increase in the $\nu 8$ mode intensity corresponds to the formation of the salt bridge between Arg214 and haem propionate 7. The formation of this salt bridge will intensify haem doming and thus intensifies the $\gamma 7$ band, which results in the intensity changes of the $\gamma 7$, $\nu 8$, and $\delta(C_\beta C_c C_d)$ modes. In the step 2, spectral changes are observed for the $\delta(C_\beta C_c C_d)$ band in a timescale of 1–10 μs, which is caused by the movement of Ile209 and Ile210 in the distal haem pocket. The main-chain amide of Ile209 is hydrogen bonded to haem propionate 6 in the ligand-bound forms, but not in the deoxy form. The spectral changes in the step 2 will correspond to reconstruction of this hydrogen-bond interaction. For both CO and O_2 dissociation, relaxations to the equilibrium deoxy form are not complete after 2 ms (Hiruma et al., 2007). The changes in the $\gamma 7$, $\nu 8$, and $\delta(C_\beta C_c C_d)$ modes will correspond to changes in the extent of haem doming and interactions between the amino acid residues in the FG loop and the haem propionates in the step 3.

4.1.6 A structural model of FixL/FixJ complex

Though the structures of the PAS domain of FixL are reported, the structure of full-length FixL is not solved yet. The structures of FixL in the full-length and FixL/FixJ complex are required to unveil the detailed molecular mechanisms of signal transductions in the FixL/FixJ two-component system. Yamada et al. (2009) report the structure of the complex of the PAS-containing sensory histidine kinase (ThkA) and its cognate response regulator (TrrA) in the two-component signal transduction system of *Thermotoga maritima*, which would be a good model of the FixL/FixJ system. They construct a structural model of ThkA/TrrA complex by fitting the structures of the isolated domains of ThkA and TrrA that are determined with high resolution (1.5–1.9 Å) into the electron density map of the ThkA/TrrA complex at a resolution of 3.6 Å (Fig. 7.5).

ThkA consists of the PAS, dimerization (DHp), and kinase catalytic (CA) domains. The domain organisation is conserved between FixL and ThkA, though the PAS domain in ThkA contains no prosthetic group. The $\beta 3$ in the ThkA PAS domain forms an interdomain, antiparallel β sheet with the $\beta 6$ in the CA domain of ThkA via four main-chain hydrogen bonds (Yamada et al., 2009). The $\beta 3$ in the ThkA PAS domain is located in the

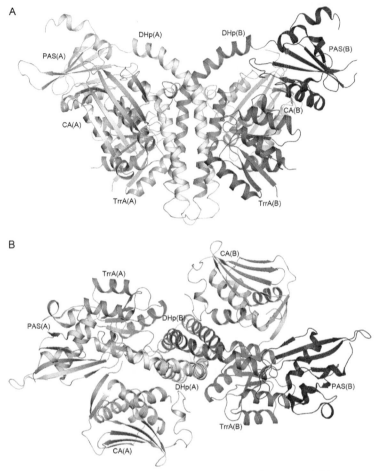

Figure 7.5 The structure of ThkA/TrrA complex from (A) side view and (B) top view. The complex forms 2:2 dimer with protomers A and B. (For colour version of this figure, the reader is referred to the online version of this chapter.)

region corresponding to the FG loop in FixL (Fig. 7.6), which suggests that a conformational change in the FG loop may induce directly conformational changes in the kinase catalytic domain of FixL through reconstructions of the interdomain interactions.

ThkA forms a homodimer, in which the four-helix bundle formed by two DHp domains is responsible for dimerization of two protomers (Yamada et al., 2009). His547 that is phosphorylated in ThkA is located at the C-terminal region of the α6 helix in the DHp domain. In the

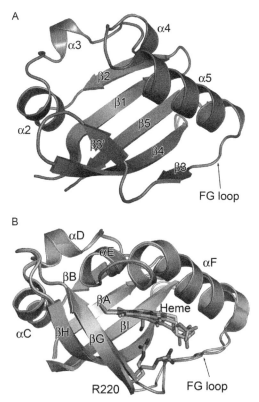

Figure 7.6 Structural comparison of the PAS domain of ThkA and the haem-containing PAS domain of *Bj*FixL. (A) Structure of the PAS domain of ThkA. (B) Superposition of the *Bj*FixL PAS domains in the oxy (red) and deoxy (grey) states. *Reproduced with permission from Yamada et al. (2009).* (For interpretation of the references to colour in this figure legend, the reader is referred to the online version of this chapter.)

ThkA/TrrA complex, the phosphoacceptor Asp residue in the TrrA faces the phosphodonor His547 in ThkA (Yamada et al., 2009; Fig. 7.7). The interaction between the PAS domain and TrrA is also present in the ThkA/TrrA complex, in which the β3–β4 loop and the α5 in the PAS domain in ThkA interact with the α4 in TrrA (Yamada et al., 2009). The formation of FixL/FixJ complex changes O_2-binding affinity of FixL (Nakamura et al., 2004), which may be caused by a similar interaction between the FixL PAS domain and FixJ.

The position of the CA domain relative to the DHp domain reveals that the ATP-binding site in the CA domain is far away from His547 in the DHp

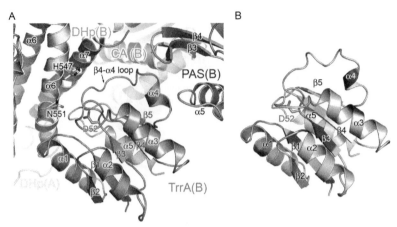

Figure 7.7 Interaction of TrrA with ThkA. (A) Close-up view of the interaction of TrrA with ThkA. (B) The structure of TrrA monomer. The regions interacting with ThkA, which are evaluated by NMR, are shown in blue and orange. Chemical shift perturbations (blue) and disappearance of the cross-peak (orange) are observed upon addition of an excess amount of the (DHp + CA) domain of ThkA to TrrA. *Reproduced with permission from Yamada et al. (2009).* (For interpretation of the references to colour in this figure legend, the reader is referred to the online version of this chapter.)

α7 helix (Yamada et al., 2009), indicating that phosphorylation of His547 cannot take place in this structure. Marina, Waldburger, and Hendrickson (2005) propose that the CA domain of histidine kinase should move significantly towards the phosphoacceptor His in the DHp domain. In the ThkA/TrrA complex, the interdomain β sheet with the PAS and CA domains will prevent the CA domain from moving towards the DHp domain (Yamada et al., 2009). Thus, the structure of ThkA might be an inactive form for autokinase activity, which will correspond to the O_2-bound form of FixL. If O_2 dissociation weakens the interdomain interaction by inducing a conformational change in the G_β, the CA domain would be permitted to freely move for autokinase reaction (Yamada et al., 2009).

4.2. A haem-based redox sensor in NtrY/NtrX two-component system

Genome sequence analyses reveal that *Brucella abortus* genome encodes 10 proteins with predicted PAS domains, 8 are associated with histidine kinase domains and 2 are associated with GGDEF/EAL domains, among which 5 proteins (BAB1_0640, 1139, 1621, 2101, and 0220) are predicted to contain a haem as a prosthetic group though it is not proved experimentally

(Carrica Mdel, Fernandez, Martí, Paris, & Goldbaum, 2012). NtrY (BAB1_1139) is the histidine kinase in the NtrY/NtrX two-component system found in *Brucella* spp. that are facultative intracellular gram-negative bacteria that belong to the α2 proteobacteria group. The expression of all the denitrification enzymes is increased in low O_2 tension in these bacteria, which is partly triggered by the NtrY/NtrX two-component system (Al Dahouk et al., 2009; Roop & Caswell, 2012). NtrY from *B. abortus* is a homologue to NtrY that is involved in nitrogen metabolism and/or fixation in *Azorhizobium caulinodans* and *Azospirillum brasilense* (Ishida et al., 2002; Pawlowski, Klosse, & de Bruijn, 1991), which contains predicted transmembrane regions at the N-terminus followed by a HAMP, PAS, and histidine kinase domains (Carrica Mdel et al., 2012). The PAS domain binds a haem, which shows the Soret, α, and β bands at 408, 558, and 533 nm, and 423, 555, and 527 nm in the ferric and ferrous states, respectively (Carrica Mdel et al., 2012). While a stable O_2-bound NtrY does not form due to an autoxidation to produce the ferric form, NO and CO can bind to the ferrous haem in NtrY to form a 5-coordinate nitrosyl haem and 6-coordinate CO-bound haem, respectively (Carrica Mdel et al., 2012). Ferrous NtrY possesses much higher autokinase activity than ferric NtrY. The autokinase activity of CO-bound and NO-bound NtrY is similar to that of ferrous NtrY (Carrica Mdel et al., 2012), though the coordination structures of the CO- and NO-bound haems are different from each other. The reason of this unique property is not obvious at present. Based on these results, it is proposed that NtrY acts as a redox sensor, not a gas sensor (Carrica Mdel et al., 2012). Alignments of amino acid sequences among NtrY and other haem-containing PAS domains reveal that NtrY does not show a conserved His as the proximal haem ligand within its PAS domain, which suggests that the haem binds to the PAS domain in NtrY in a different manner to that described for other haem-containing PAS domains. However, more detailed characterisation with spectroscopic and structural methods is required to prove this hypothesis with identification of the coordination structure of the haem in NtrY.

4.3. DosS and DosT using a haem-containing GAF domain as a sensor module

4.3.1 Regulation of kinase activities of DosS and DosT from Mycobacterium tuberculosis

GAF domains represent one of the largest and most widespread domain families, which provide a variety of functions including binding of small molecules, protein–protein interactions, and other processes (Heikaus, Pandit, & Klevit, 2009). They are distantly related to PAS domains, another superfamily with the same basic fold (Anantharaman, Koonin, & Aravind, 2001), both of which are involved in many signal transduction pathways and protein regulatory and sensory systems (Anantharaman et al., 2001; Aravind & Ponting, 1997). Several gas sensor proteins are identified adopting a GAF domain that binds a haem, Fe–S cluster (Nakajima et al., 2010) or non-haem iron (Bush, Ghosh, Tucker, Zhang, & Dixon, 2011; D'Autréaux, Tucker, Dixon, & Spiro, 2005) as the active site for gas sensing. DosS and DosT from *M. tuberculosis* are most extensively studied among them.

The sensor kinases DosS (also known as DevS) and DosT activate the transcriptional regulator DosR (also known as DevR) in response to both hypoxia and non-lethal level of NO. The DosS/DosR and DosT/DosR two-component signal transduction systems regulate the expression of the *dosR* regulon that is responsible for early adaptation to these stimuli as well as for initiating entrance of *M. tuberculosis* into non-replicating persistent state. DosS and DosT each contain two tandem GAF domains and a histidine kinase domain in their N-terminal and C-terminal regions, respectively. Both in DosS and DosT, the first GAF domain (GAF-A) contains a haem, while the second GAF domain (GAF-B) does not (Sardiwal et al., 2005; Sousa, Tuckerman, Gonzalez, & Gilles-Gonzalez, 2007). The second GAF domain (GAF-B) is suggested to help in the formation of a better-defined distal haem pocket of the GAF-A domain through inter-domain interactions within DosS (Lee et al., 2008). The activation of the response regulator DosR occurs through autophosphorylation of either DosS or DosT followed by transfer of the phosphate to DosR (Honaker, Dhiman, Narayanasamy, Crick, & Voskuil, 2010; Saini et al., 2004).

The kinase activities of DosS and DosT are dependent of the coordination state of the haem. In the both cases of DosS and DosT, kinase activity is observed when the haem is a 5-coordinate ferrous (Fe^{2+}) form, CO- or NO-bound forms (Fe^{2+}-CO or Fe^{2+}-NO) but is strongly inhibited in an oxy (Fe^{2+}-O_2) form (Sousa et al., 2007; Yukl, Ioanoviciu, Nakano, Ortiz de Montellano, & Moënne-Loccoz, 2008). The activities of DosS and DosT are inhibited by about 84% and 98%, respectively, upon O_2 binding to the ferrous haem (Sousa et al., 2007). These results indicate that both of DosS and DosT are direct O_2 sensor proteins.

Kim, Park, Ko, Kim, and Oh (2010) have proposed that the presence of both DosS and DosT paralogues in *M. tuberculosis* is not a functional redundancy but that they play distinct roles in sensing changing O_2 tension. They indicate that DosT appears to first respond to a decline in O_2 tension when *M. tuberculosis* is gradually transited from aerobic to anaerobic conditions (Kim et al., 2010). DosS and DosT show the equilibrium dissociation constants for O_2 of 3 and 26 mM, respectively (Sousa et al., 2007), which is consistent with the above results.

4.3.2 Crystal structures of the GAF-A domains of DosS and DosT

The crystal structures of the GAF-A domains of DosS (residues 63–210) and DosT (residues 61–208) have been reported (Cho, Cho, Kim, Oh, & Kang, 2009; Podust, Ioanoviciu, & Ortiz de Montellano, 2008). They show a similar fold each other, as expected from amino acid sequences homology, which consist of one five-stranded antiparallel β-sheet and four α-helices. A haem is located in the cavity between the β-sheet and the loop region covering the sheet. The haem is tethered to the protein by a histidine (H149 and H147 for DosS and DosT, respectively) from a long loop connecting the β3- and β4-strands at the proximal position of the haem. The haem plane is roughly perpendicular to the sheet and shielded by the α2 helix and α2–α3 loop from solvent (Cho et al., 2009; Podust et al., 2008).

The crystal structure of DosT reveals that Tyr169 forms a hydrogen bond with the haem-bound O_2 in DosT (Podust et al., 2008). Though the structure of the oxy form of DosS is not determined, Tyr171 in DosS (corresponding to Tyr169 in DosT) is thought to form a similar hydrogen bond with the haem-bound O_2 judging from the resonance Raman studies combined with site-directed mutagenesis (Ioanoviciu, Meharenna, Poulos, & Ortiz de Montellano, 2009; Yukl, Ioanoviciu, Ortiz de Montellano, & Moënne-Loccoz, 2007).

4.3.3 Ligand binding to the haem in DosS and DosT

DosS and DosT contain a 5-coordinated, high-spin state of ferrous haem that binds an exogenous ligand such as O_2, CO, or NO as an axial ligand of the haem (Ioanoviciu, Yukl, Moënne-Loccoz, & Ortiz de Montellano, 2007; Kumar, Toledo, Patel, Lancaster, & Steyn, 2007; Yukl et al., 2008, 2007, 2011). The resonance Raman spectra of the oxy DosS show the $v(Fe–O_2)$ band at 563 cm^{-1} that is perturbed by H_2O/D_2O exchange, supporting the presence of a hydrogen bond from a distal residue to the haem-bound O_2 (Yukl et al., 2007). There are two conformers for the haem-ligand complexes in the NO- and CO-bound forms of DosS. While one conformer is a major population in which the haem-bound CO or NO is free of hydrogen bonds, another one shows electrostatic and/or hydrogen-bond interaction within the distal haem pocket (Ioanoviciu et al., 2007; Yukl et al., 2007). In the resonance Raman spectra of Y171F DosS, these hydrogen-bonded CO and NO conformers are not observed (Yukl et al., 2008).

Y171F variant of DosS loses ability of ligand discrimination for CO, NO, and O_2. While only the oxy form is inactive for kinase activity in wild-type DosS, all of CO, NO, and O_2 forms are inactive complexes in the Y171F variant (Yukl et al., 2008). In contrast, the ferrous form of Y171F DosS exhibits kinase activity comparable to that of wild type (Yukl et al., 2008). Yukl et al. (2008) have proposed that interactions between Tyr171 and distal diatomic ligands turn the kinase activity on or off, and that mutation of Tyr171 to Phe disrupts the on–off switch, leading to an inactive kinase in all states.

5. HAEM-CONTAINING PAS FAMILY FOR THE REGULATION OF DIGUANYLATE CYCLASES AND PHOSPHODIESTERASES

5.1. Regulation of diguanylate cyclases activity through interactions between haem and external ligand

Bis-(3', 5')-cyclic dimeric guanosine monophosphate (cyclic di-GMP) is a ubiquitous bacterial second messenger involved in the regulation of cell motility, differentiation, development, virulence, biofilm formation, and factor-stimulated proliferation in human colon cancer cells (Cotter & Stibitz, 2007; Jenal, 2004; Jenal & Malone, 2006; Pesavento & Hengge, 2009; Römling & Amikam, 2006; Römling, Gomelsky, & Galperin, 2005; Tamayo, Pratt, & Camilli, 2007). While different biochemical processes seem to be controlled by cyclic di-GMP in response to different extracellular signals, this second messenger generally regulates transitions between the free-living, motile lifestyle and the sessile lifestyle (Cotter & Stibitz, 2007; Jenal, 2004; Jenal & Malone, 2006; Pesavento & Hengge, 2009; Römling & Amikam, 2006; Römling et al., 2005; Tamayo et al., 2007). Low concentrations of cyclic di-GMP promote motile growth, while high concentrations promote sessile growth with biofilm formation. The intracellular concentrations of cyclic di-GMP are controlled by the balance between synthesis and hydrolysis of cyclic di-GMP, which are catalysed by the enzymes, diguanylate cyclases (DGCs) and phosphodiesterases (PDEs), respectively (Cotter & Stibitz, 2007; Jenal, 2004; Jenal & Malone, 2006;

Pesavento & Hengge, 2009; Römling & Amikam, 2006; Römling et al., 2005; Tamayo et al., 2007). DGC contains the GGDEF domain, named from the conserved sequence motif (Gly-Gly-Asp-Glu-Phe) that constitutes part of the active site of the enzymes (Karatan & Watnick, 2009). Formation of cyclic di-GMP is catalysed by two GGDEF domains with two GTP (guanosine triphosphate) molecules. PDE catalyses hydrolysis of cyclic di-GMP to the linear form $5'$-pGpG.

The GGDEF domain is typically found coupled to a variety of other sensor and/or regulator domains including PAS and globin domains within multi-domain proteins (Jenal, 2004; Römling & Amikam, 2006). It is expected that these sensor and regulator domains will be responsible for receiving directly or indirectly the extracellular signal to regulate the DGC activity. A haem-containing sensor domain is adopted in several DGC and PDE including HemDGC, *Bpe*GReg, YddV (*Ec*DosC), *Ec*Dos, and *Ax*PDEA.

HemDGC, *Bpe*GReg, and YddV belong to the globin-coupled sensor proteins, which consists of globin and GGDEF domains. HemDGC from *Desulfotalea psychrophila* shows the DGC activity only when O_2 is bound to the haem (Sawai et al., 2009). Though HemDGC also binds CO and NO, CO- and NO-bound HemDGCs show no enzymatic activity (Sawai et al., 2009). Neither ferric nor ferrous HaemDGC shows the DGC activity. Mutagenesis and resonance Raman studies reveal that the ligand discrimination is achieved by changing a hydrogen-bonding network between the haem-bound ligand and surrounding amino acid residues in the distal haem pocket. In the O_2-bound form, the hydrogen-bonding network is formed among the haem-bound O_2, Tyr55, and Gln81, while only Gln81 is interacting with the haem-bound CO in the CO-bound form (Sawai et al., 2009).

The DGC activity of *Bpe*GReg (globin-coupled DGC from *Bordetella pertussis*) is also activated by 10-fold upon O_2 binding to the haem compared with that of the ferrous form (Wan et al., 2009). Unlike HemDGC, *Bpe*GReg shows the DGC activity in the NO- and CO-bound forms though their activity is lower than that of the O_2-bound form (Wan et al., 2009). Phenotype analyses reveal that *Bpe*GReg is involved in the regulation of biofilm formation (Wan et al., 2009). A homologue of *Bpe*GReg found in *Azotobacter vinelandii*, *Av*GReg, is purified and studied by resonance Raman spectroscopy for the haem environmental structure (Thijs et al., 2007).

The *yddV* (*dosC*) gene is a part of the operon encoding the *Ecdos* gene, and these two genes are co-transcribed (Tuckerman et al., 2009). The gene products, YddV (DosC) and *Ec*Dos, are suggested to form an associated complex for cyclic di-GMP signalling in *Escherichia coli* (Tuckerman et al., 2009). As the DGC activity of YddV (DosC) is lower compared with the PDE activity of *Ec*Dos, the formation of cyclic di-GMP by YddV (DosC) is the rate-determining step in the YddV (DosC)/*Ec*Dos system. The DGC activity of YddV (DosC) from *E. coli* is observed in the ferric (0.066 min^{-1}), O_2-bound (0.022 min^{-1}), and CO-bound forms (0.022 min^{-1}), while the ferrous and NO-bound forms show no significant activity (Kitanishi et al., 2010). Though resonance Raman studies suggest that Tyr43 forms the hydrogen bond with both O_2 and CO, detail mechanisms for ligand discrimination remain to be elucidated (Kitanishi et al., 2010).

There is also an intermolecular system in which the DGC activity is regulated by a separate haem–containing sensor protein such as bacterial H-NOX domains. H-NOX consists of seven α-helices and a four-stranded antiparallel β-sheet. A haem is sandwiched between a small α-helical subdomain at its distal side and a large mixed-α/β subdomain at its proximal side. The haem is tethered by His103 (numbering for H-NOX from *Shewanella oneidensis*) (Erbil, Price, Wemmer, & Marletta, 2009; Karow et al., 2004; Nioche et al., 2004; Olea, Boon, Pellicena, Kuriyan, & Marletta, 2008; Olea, Herzik, Kuriyan, & Marletta, 2010). The haem propionate shows hydrogen-bonding interaction with Tyr131, Ser133, and Arg135 that form an "YxSxR" motif (Karow et al., 2004; Nioche et al., 2004). These structural features including an "YxSxR" motif are conserved among H-NOX domains.

Stand-alone H-NOX domains are most often found in a predicted operon with a histidine kinase, a receiver domain/protein of a two-component system, a DGC containing a GGDEF domain, a phosphatase, or a cyclic diguanylate phosphodiesterases (Iyer, Anantharaman, & Aravind, 2003). The gene encoding *Lp* H-NOX1 in *Legionella pneumophila* is adjacent to a gene encoding a GGDEF–EAL protein (Lpg1057) that shows DGC activity *in vitro* (Carlson, Vance, & Marletta, 2010). *Lp* H-NOX1 contains a b-type haem that can bind NO, but not O_2 (Boon et al., 2006; Carlson et al., 2010). Upon reaction with NO, *Lp* H-NOX1 forms a 5-coorinated nitrosyl haem. NO-bound *Lp* H-NOX1(Fe^{2+}-NO) inhibits the DGC activity of Lpg1057 by about 20%, while an external ligand-free *Lp* H-NOX1(Fe^{2+}) does not (Carlson et al., 2010). Though complete inhibition by *Lp* H-NOX1(Fe^{2+}-NO) is not observed *in vitro*, deletion of the *h-nox 1* gene results in a hyper-biofilm phenotype that shows 40% more biofilm production than the wild type. This phenotype is caused by a lack of inhibition of the DGC activity of Lpg1057 by *Lp* H-NOX1 because increasing concentrations of cyclic di-GMP, a second messenger produced by

DGC, induces biofilm formation (Pesavento & Hengge, 2009; Seshasayee, Fraser, & Luscombe, 2010; Tamayo et al., 2007). These results suggest that *Lp* H-NOX1 is a NO sensor protein that regulates the DGC activity of Lpg1057 encoded in the same operon with *Lp* H-NOX1.

5.2. PDEs consisting of a haem-containing PAS and EAL domains

5.2.1 Spectroscopic properties of EcDos

*Ec*Dos contains 807 amino acid residues, which consists of three domains, two N-terminal PAS and C-terminal EAL domains. It contains one proto-haem per monomer and the haem is bound to the first PAS domain. Ferric and ferrous *Ec*Dos show the Soret and Q-bands at 416, 539, and 564 nm, and 427, 532, and 563 nm, respectively (Delgado-Nixon, Gonzalez, & Gilles-Gonzalez, 2000). Resonance Raman spectra show the $\nu2$, $\nu3$, and $\nu4$ bands at 1576, 1505, and 1370 cm^{-1}, and 1580, 1493, and 1361 cm^{-1} in ferric and ferrous *Ec*Dos, respectively (Tomita et al., 2002). These spectra indicate that the ferric and ferrous haems in *Ec*Dos are 6-coordinate, low-spin state.

Diatomic gas molecules such as O_2, CO, and NO can be bound to the haem in *Ec*Dos. O_2-, and CO-bound *Ec*Dos show the Soret and Q-bands at 417, 541, and 579 nm, and 423, 540, and 570 nm, respectively (Delgado-Nixon et al., 2000). The $\nu(Fe-CO)$ and $\nu(C-O)$ bands are observed at 487 and 1969 cm^{-1}, suggesting that the proximal ligand *trans* to CO is a neutral His in CO-bound *Ec*Dos (Tomita et al., 2002).

The frequency of the $\nu(Fe-O_2)$ band is shifted by the change in the pattern of hydrogen-bonding interactions to the haem-bound O_2. When the hydrogen bond to the proximal oxygen atom of the haem-bound O_2 exists, the lower frequency of the $\nu(Fe-O_2)$ band is observed compared with the O_2-bound haem having the hydrogen bond to the distal oxygen atom (Das, Couture, Ouellet, Guertin, & Rousseau, 2001; Mukai, Savard, Ouellet, Guertin, & Yeh, 2002; Yeh, Couture, Ouellet, Guertin, & Rousseau, 2000). For example, while the $\nu(Fe-O_2)$ band appears at ca. 570 cm^{-1} in Mb and Hb where a hydrogen bond exists between the distal oxygen atom of the haem-bound O_2 and the distal His, it downshifts to 562 cm^{-1} in the case of *M. tuberculosis* HbN where the hydrogen bond exists at the proximal oxygen atom (Ouellet, Milani, Couture, Bolognesi, & Guertin, 2006). In O_2-bound *Ec*Dos, the $\nu(Fe-O_2)$ band is observed at 562 cm^{-1} (Tomita et al., 2002), suggesting that a hydrogen bond to the proximal oxygen atom exists in *Ec*Dos.

5.2.2 Dynamic change in coordination structure of haem in EcDos

Site-directed mutagenesis and X-ray crystal structural analyses reveal that the ferric and ferrous haems in EcDos are coordinated with His77 and H_2O (or OH^-), and His77 and Met95, respectively, as the axial ligands. The haem in EcDos shows a redox-dependent ligand exchange between H_2O (or OH^-) and Met95. Though the ferrous haem coordinated with His77 and Met95 does not have a vacant site for an external ligand, it can bind an external diatomic gas molecule such as O_2, CO, and NO. When these gas molecules are bound to the haem, another ligand exchange takes place between the endogenous ligand, Met95, and the external ligand.

EcDos shows dynamic changes in its coordination structure of the haem in response to the change in the oxidation state of haem or to the external ligand. A similar dynamic ligand exchange was observed previously in cytochrome cd1 and CooA. Cytochrome cd1 from *Paracoccus pantotrophus* (formerly *Thiosphaera pantotropha*) is nitrite reductase, which has a c-type and d1-type haems that act as an electron carrier and the enzymatic active site, respectively. While the ferric haem c is coordinated with His17 and His69, His17 is replaced by Met106 to form a His/Met-ligated haem upon the reduction of the haem c (Baker et al., 1997; Fülöp, Moir, Ferguson, & Hajdu, 1995; Williams et al., 1997). The haem d1 also shows a redox-dependent ligand exchange. While the ferric d1 haem is coordinated with His200 and Tyr25, Tyr25 is dissociated from the haem iron to form a 5-coordinate haem upon the reduction of the d1 haem (Baker et al., 1997; Fülöp et al., 1995; Williams et al., 1997). These ligand exchanges will be responsible for the fine tuning of the redox potentials of the haem c and haem d1 to regulate the enzymatic activity. For the haem d1, this ligand exchange is also responsible for the regulation of substrate (nitrite) binding affinity. Though nitrite cannot bind to the 6-coordinate haem with His and Tyr as the axial ligands, it can bind to the 5-coordinate ferrous d1 haem.

5.2.3 Structures of the haem-containing PAS domain in EcDos

The crystal structures of the haem–containing PAS domain of EcDos (EcDosH) have been reported for the ferric, ferrous, and oxy forms (Kurokawa et al., 2004; Park, Suquet, Satterlee, & Kang, 2004), which show a typical PAS fold consisting of five distinct α helices and five-stranded β sheet. The haem is accommodated between the F helix and two β strands (G_β and H_β). The global structure of EcDos is similar to that of FixL (Fig. 7.8).

While the FG loop (residues 86–97) is disordered in the ferric *Ec*DosH, it is significantly rigidified upon reduction of the haem iron (Kurokawa et al., 2004). Electron densities of the FG loop are clearly observed in the ferrous *Ec*Dos, suggesting that the FG loop is rigidified upon reduction of the haem. As Met95 is located in the FG loop, tethering Met95 to the haem iron will restrict a flexibility of the FG loop. The distal water molecule (or OH^-) bound to the ferric haem is replaced by Met95 to form a 6-coordinated ferrous haem with His77 and Met95 as the axial ligands, which plays an important role for rigidifying the FG loop (Kurokawa et al., 2004; Park et al., 2004).

The crystal structure of O_2-bound *Ec*Dos clearly shows that another ligand-exchange reaction occurs upon O_2 binding between Met95 and O_2, where Met95 is replaced by O_2 to form the O_2-bound *Ec*Dos. Met95 tethered to the haem iron is dissociated from it upon O_2 binding, by which conformational changes in the FG loop are caused. The conformational changes of the FG loop will be involved in the signalling process to regulate the PDE activity of *Ec*Dos (Sasakura, Yoshimura-Suzuki, Kurokawa, & Shimizu, 2006).

O_2 binding to the haem also causes rearrangement of hydrogen-bonding networks around the haem, as observed in FixL. While Arg97 is out of the distal haem pocket in ferrous *Ec*Dos, the side chain of Arg97 is rotated by about $180°$ upon O_2 binding to form a hydrogen bond to the haem-bound O_2 (Park et al., 2004). This hydrogen bond is essential for a stable O_2 binding to the haem. In Arg97 variants, the stable O_2-bound *Ec*Dos is not formed

A   B

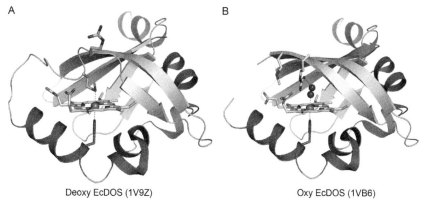

Deoxy EcDOS (1V9Z)   Oxy EcDOS (1VB6)

Figure 7.8 The structures of the haem-containing PAS domain of EcDos in (A) the deoxy (PDB 1V9Z) and (B) oxy (PDB 1VB6) states. (For colour version of this figure, the reader is referred to the online version of this chapter.)

due to rapid autoxidation and/or low affinity for O_2 (Tanaka, Takahashi, & Shimizu, 2007). The rotation of the propionate groups of the haem also occurs upon O_2 binding, which may trigger the formation of the hydrogen bond between Asn84 and Tyr126 through the haem propionate 6 hydrogen-bond network (El-Mashtoly, Takahashi, Shimizu, & Kitagawa, 2007; Sato et al., 2002). These hydrogen-bond networks are proposed to be responsible for the intramolecular signal transduction from the sensor to enzymatic domains (El-Mashtoly et al., 2007).

5.2.4 Regulation of PDEs activity of EcDos

The EAL domain, which contains a conserved sequence motif of Glu-Ala-Leu as the enzymatic active site, shows PDE activity with a substrate of cyclic di-GMP (Delgado-Nixon et al., 2000; Tuckerman et al., 2009). The physiological function of EcDos is regulated by exogenous ligand binding to the haem. Thus, the PDE activity is enhanced upon O_2 binding to the Fe^{2+} haem (Tanaka et al., 2007; Tuckerman et al., 2009). CO and NO binding to the haem causes similar enhancement of the PDE activity (Tanaka et al., 2007; Tuckerman et al., 2009), indicating that EcDos cannot discriminate these gas molecules. Positive cooperativity in O_2 binding is observed for EcDos (Lechauve et al., 2009; Tuckerman et al., 2009). Lechauve et al. (2009) have reported that the second O_2 molecule binds to EcDos with a sixfold higher intrinsic affinity than the first one. They propose that the replacement of Met95 by O_2 and its swapping out of the haem pocket in one subunit leads to a modification at the dimer interface, presumably leading to decrease Met95 affinity for haem, which will increase O_2-binding affinity, in the other subunit. The PDE activity also shows a non-linear dependence on the fraction of the O_2-saturated haem, which enables to activate EcDos in narrow range of O_2 concentrations (Kobayashi et al., 2010; Tuckerman et al., 2009).

The basal activity of the ferrous EcDos increases to a similar value for O_2-, CO-, or NO-bound form by mutation of Met95 to Ala or Leu (Tanaka et al., 2007). These results suggest that the dissociation of Fe–Met bond upon ligand binding results in a conformational change of the FG loop to activate the PDE activity of EcDos. In addition to Met95, several amino acid residues are identified by mutagenesis and resonance Raman studies as those responsible for controlling ligand-binding properties and/or enzymatic activity (El-Mashtoly, Nakashima, Tanaka, Shimizu, & Kitagawa, 2008; Ishitsuka et al., 2008; Ito, Araki, et al., 2009; Ito, Igarashi, & Shimizu, 2009; Tanaka & Shimizu, 2008).

5.2.5 Acetobacter xylinum *phosphodiesterase A1*

The phosphodiesterase A1 (*Ax*PDEA1) of *Gluconacetobacter xylinus* (formerly *Acetobacter xylinum*) plays an important role for the regulation of bacterial cellulose synthesis, which consists of the N-terminal haem-containing PAS and C-terminal PDE domains (Chang et al., 2001). *Ax*PDEA1 catalyses hydrolysis of cyclic di-GMP to $5'$-pGpG. The enzymatic activity of O_2-bound *Ax*PDEA1 is about one-third compared with that of ferrous *Ax*PDEA1, indicating that the haem-containing PAS domain regulates the biological function of *Ax*PDEA1 via O_2 binding to the haem. The PAS domains of *Ax*PDEA1, FixL, and *Ec*Dos are >30% identical in amino acid sequence and the proximal His to the haem is conserved at the corresponding positions among these PAS domains (Chang et al., 2001).

Resonance Raman spectroscopy reveals that the ferric haem in the PAS domain of *Ax*PDEA1 is a mixture of 6-coordinate, low-spin and 5-coordinate, high-spin states and that the ferrous haem is a 5-coordinate, high-spin state (Tomita et al., 2002). While the spectral properties of ferric *Ax*PDEA1 are similar to those of ferric *Ec*Dos, ferrous *Ax*PDEA1 shows different properties from those of *Ec*Dos. The ferrous haem in *Ax*PDEA1 is a 5-coordinate state, but it is a 6-coordinate state in EcDos. Thus, O_2 is bound to a vacant distal site of the ferrous haem in *Ax*PDEA1 without any ligand exchange unlike *Ec*Dos. These differences suggest different mechanisms of O_2-dependent signal transductions between *Ax*PDEA1 and *Ec*Dos.

The PAS domain of *Ax*PDEA1 (*Ax*PDEA1H) shows higher k_{on} ($6.6\ \mu M^{-1}\ s^{-1}$) and k_{off} ($77\ s^{-1}$) values for O_2 binding and dissociation, respectively, compared with FixL and *Ec*Dos. The autoxidation rate of *Ax*PDEA1 is slowest (>12 h of the half-life for the oxy form) among haem-containing PAS (Chang et al., 2001). These properties of *Ax*PDEA1 would be caused by structural differences around the haem or/and differences in flexibility of the haem pocket, but detail mechanisms remain to be elucidated.

5.3. A haem-sensing PAS domain regulates the PDE activity of YybT

YybT family proteins are widely distributed among *Staphylococcus aureus*, *Streptococcus mutans*, and *Listeria monocytogenes*, which contain two N-terminal transmembrane helices and three predicted protein domains, a putative PAS domain, a highly degenerate GGDEF domain, and a DHH/DHHA1 domain (Rao, Ji, Soehano, & Liang, 2011). The C-terminal DHH/DHHA1 domain exhibits PDE activity towards the cyclic dinucleotides, cyclic di-AMP, and cyclic di-GMP, while the GGDEF domain possesses a weak ATPase activity instead of the DGC activity (Rao et al., 2010). The DHH/DHHA1 domain hydrolyzes cyclic di-AMP and cyclic di-GMP to generate the linear dinucleotides, 5'-pApA and 5'-pGpG, respectively.

YybT from *Bacillus subtilis* shows the Soret and Q-bands at 412, 533, and 566 nm, and 423, 528, and 557 nm in the ferric and ferrous forms, respectively, though the haem contents of the recombinant YybT are low (ca. 5%) (Rao et al., 2011). These spectral features suggest that the ferric and ferrous haems in YybT are in a 6-coordinate and low-spin state, though it remains unknown what residues are the axial ligands of the haem in YybT. As autoxidation proceeds to produce the ferric form upon the reaction of ferrous YybT with O_2, no stable O_2-bound form of YybT is formed. NO and CO can be bound to the ferrous YybT to form a 5-coordinate nitrosyl haem and a 6-coordinate CO-bound haem, respectively (Rao et al., 2011).

YybT binds a b-type haem as described earlier. However, amino acid sequence alignment of the PAS domain of YybT family proteins with the haem-binding PAS domain of FixL and *Ec*Dos reveals that there is no candidate of the axial ligands of the haem in YybT. Secondary structure prediction and structural modelling suggest that the F_α helix and G_β strand are shorter than those of FixL and *Ec*Dos and the FG loop is absent in YybT, though the same topology is conserved in the PAS domain in YybT (Rao et al., 2011). Very recently, the solution structure of the PAS domain of YybT from *Geobacillus thermodenitrificans* (*Gt*Yyb) is determined by NMR to indicate that it adopts the characteristic PAS fold composed of a five-stranded antiparallel β sheet and a few short α helices (Tan et al., 2013). It forms a homodimer with the interaction of the central β-sheet in an antiparallel fashion (Tan et al., 2013). As three of the β strands and one of the helices in the PAS domain of *Gt*YybT are shorter than those found in other typical PAS domains, the PAS domain of *Gt*YybT consists of only 80 residues. Despite the small size of the domain, it forms a hydrophobic cavity by backbone and side chains of predominantly non-polar residues, Leu88, Leu103, Ala104, Phe107, Glu109, Leu112, Leu117, Leu120, Ser121, Leu124, Leu145, and Phe158 (Tan et al., 2013). As expected from the sequence analyses, there is no residue that can act as the axial ligands of the haem in this cavity. These results suggest that the haem is bound in this cavity by hydrophobic interactions.

The catalytic activity of YybT with a substrate of cyclic di-AMP is regulated by whether or not YybT binds a haem. Apo-YybT shows a higher activity by 15-fold compared with holo-YybT. The k_{cat} and the K_m for holo-YybT is 23-fold lower and 12-fold higher than those of apo-YybT, respectively (Rao et al., 2011). Thus, a catalytic efficiency (k_{cat}/K_m) for holo-YybT is 276-fold lower than that of the apo-YybT. The catalytic activity increases upon binding of NO to ferrous YybT by a threefold increase. CO has a negligible effect on the catalytic activity. The redox change in the haem also shows no effect on the enzymatic activity.

Based on these results, Rao et al. (2011) propose that YybT is a haem sensor (a sensor protein for haem), not a haem-based sensor such as the proteins described in the other sections. Haem binding to YybT would initiate cellular response through the cyclic di-AMP signalling (Tan et al., 2013). Though a TetR family haem sensor, HrtR from *Lactococcus lactis*, is well characterised (Sawai, Yamanaka, Sugimoto, Shiro, & Aono, 2012), no PAS family haem sensor is known to date. YybT is the first example of haem sensors that adopt a PAS domain to sense a haem molecule. An *in vivo* study for *S. aureus* suggests that YybT family proteins may respond to cellular haem levels to regulate hemolysin secretion, which is required for acquiring nutrients including haem iron from the host by lysing the host cells. The knockout of the SA0013 gene encoding YybT disrupts hemolysin secretion in *S. aureus* (Burnside et al., 2010). The physiological function of YybT is also studied by disruption of the *yybT* gene in *L. lactis*. *L. lactis* mutant strain lacking YybT shows greater sensitivity towards haem treatment, which will be caused by modulation of cyclic di-AMP concentration (Tan et al., 2013).

6. HAEM-CONTAINING PAS FAMILY FOR CHEMOTAXIS REGULATION

6.1. Bacterial chemotaxis system

Bacterial chemotaxis has extensively been studied with enteric bacteria (Blair, 1995; Szurmant & Ordal, 2004). The chemotaxis regulation system consists of a signal transducer protein (sometimes called *m*ethyl-accepting *c*hemotaxis *p*rotein, MCP), CheA, CheW, CheY, and some other Che proteins that are responsible for adaptation. MCP acts as a sensor for the external signal and regulates the self-kinase activity of CheA that is a component of the CheA/CheY signal transduction system (Blair, 1995; Szurmant & Ordal, 2004). CheA is the phosphodonor for the cognate response regulator CheY that binds to the flagellar motor to control its rotational direction for the regulation of the swimming behaviour of bacteria (Blair, 1995; Stock, Lukat, & Stock, 1991; Szurmant & Ordal, 2004).

MCPs are usually membrane-bound proteins in which the periplasmic and cytoplasmic domains are responsible for sensing the external signal and signalling to CheA, respectively. The binding of a repellant or attractant to the periplasmic sensor domain of MCPs is a cue of a conformational change of the cytoplasmic signalling domain. As CheA forms a complex with MCP and CheW, the conformational change of the cytoplasmic signalling domain of MCP causes a conformational change of CheA, which results in the regulation of the autokinase activity of CheA. As O_2 can be freely transmitted across cell membranes, several MCPs for aerotaxis are soluble proteins located in cytoplasm.

6.2. MCP containing a PAS domain with b-type haem

Two types of signal transducer, HemAT and Aer, are known to act as the sensor protein in microbial aerotaxis control system. HemAT is a soluble signal transducer protein consisting of globin and MCP domains as the sensor and regulatory domains, respectively, which senses molecular O_2 by the haem in the globin domain (Aono, 2011; Hou et al., 2000). On the other hand, Aer is a membrane-bound signal transducer protein, which adopts a flavin-containing PAS domain as the sensor domain to sense O_2 indirectly through the redox change of the flavin (Taylor, Rebbapragada, & Johnson, 2001).

A homologue of Aer, Aer2, is recently found in *Pseudomonas aeruginosa*, which is a soluble protein consisting of a haem–containing PAS domain. Aer2 homologues are present in some other bacteria (Baraquet, Théraulaz, Iobbi-Nivol, Méjean, & Jourlin-Castelli, 2009; Osterberg, Skärfstad, & Shingler, 2010; Sarand et al., 2008). Physiological function of Aer2 is not clear at present. Though Hong et al. propose that Aer2 from *P. aeruginosa* is a signal transducer responsible for aerotaxis (Hong, Kuroda, Takiguchi, Ohtake, & Kato, 2005; Hong et al., 2004), Watts, Taylor, and Johnson (2011) argue against it. *E. coli* cells expressing Aer2 tumble constantly in the presence of air and swim smoothly when O_2 is replaced by N_2 (Watts et al., 2011). This response is opposite to the classic Aer-mediated response in which *E. coli* cells tumble in response to a decrease in O_2 concentration. The Aer2-mediated tumbling O_2 is dependent on the extent of receptor methylation, but the direction of the response is the same whether or not Aer2 is methylated (Watts et al., 2011). *E. coli* cells expressing Aer2 also tumble in response to CO or NO, indicating that Aer2 is able to interact with the chemotaxis system of *E. coli* to mediate repellent responses to O_2, CO, and NO (Watts et al., 2011).

Amino acid sequence analyses suggest that Aer2 consists of the N-terminal poly-HAMP (the first to third HAMP), PAS, di-HAMP, and MCP domains. Aer2 shows the unusual domain arrangement compared to most well-studied chemotaxis signal transducer proteins, which contain a single HAMP domain immediately following a membrane-spanning segment. The N-terminal poly HAMP domain of Aer2 adopts a four-helix bundle (Fig. 7.9). Deleting the second HAMP and/or the third HAMP results in a kinase-off phenotype, whereas deleting the fourth HAMP and/or the fifth HAMP results in a kinase-on phenotype that is unresponsive to changes in O_2 concentration (Watts et al., 2011).

The isolated PAS domain (residues 173–289) shows typical UV–vis spectra of a 5-coordinate and high-spin haem with the Soret band at 395 and 433 nm in the ferric and ferrous states, respectively (Watts et al., 2011). On the other hand, the full-length Aer2 show the Soret bands at 394 (sh)/412 nm and 425/434(sh) nm in the ferric and ferrous states, respectively, indicating the existence of a 6-coordinate haem as a predominant species (Sawai, Sugimoto, et al., 2012). These results suggest that a truncation of the HAMP domains affects a conformation of the haem pocket, which causes changes in the coordination structure of the haem.

Sawai, Sugimoto, et al. (2012) report the crystal structure of a truncated Aer2, PH-Aer2 consisting of the PAS and di-HAMP domains (the residues 178–384) in cyano-met (Fe^{3+}–CN) form, at 2.4 Å resolution (Fig. 7.10). Though the HAMP domain in PH-Aer2 is disordered for the most part (residues 308–384), the PAS domain is well ordered. The residues 184–285 constitutes a core of the PAS domain that is composed of two α-helices (α2 and α3), one 3_{10} helix (η1), and a single antiparallel, five-stranded β-sheet (β1–β5) (Sawai, Sugimoto, et al., 2012). The α1 helix is a part of the linker between the N-terminal poly-HAMP and PAS domains. Though the overall structure of the PAS domain in Aer2 is similar to that in FixL and EcDos, it shows differences in haem pocket region and haem ligation. In FixL and EcDos, the proximal His is located in the F_α helix, but it is not the case in Aer2. His234 in the 3_{10} helix provides the proximal ligand of the haem in Aer2 (Sawai, Sugimoto, et al., 2012).

A b-type haem is accommodated in a hydrophobic pocket composed of non-polar side chains. The 7-propionate forms a salt bridge with His251, and the 6-propionate has hydrogen bonds with a water molecule, the main-chain N atom of Lys235, and NE2 atom of Gln240 (Sawai, Sugimoto, et al., 2012). The refined model indicates that the Fe—C—N bond angle is 159° and the Fe—C distance is 2.0 Å (Sawai, Sugimoto,

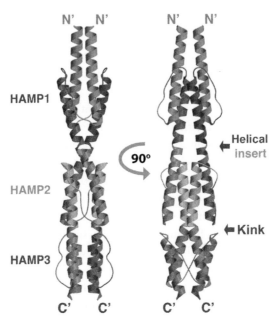

Figure 7.9 Crystal structure of the Aer2 N-terminal domain contains three successive and interwoven HAMP domains. Ribbon representation of the Aer2 (1–172) dimer with HAMP1 (AS1, light blue; AS2, blue), HAMP2 (AS1, orange; AS2, yellow), and HAMP3 (AS1, light purple; AS2, purple). HAMP2/3 forms a concatenated structure. AS2 of HAMP1 is contiguous with AS1 of HAMP2, and AS2 of HAMP2 is contiguous with AS1 of HAMP3. HAMP3 is rotated roughly 90° relative to HAMP1 and HAMP2. *Reproduced with permission from Airola et al. (2010).* (For interpretation of the references to colour in this figure legend, the reader is referred to the online version of this chapter.)

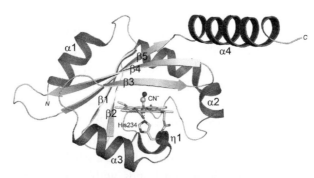

Figure 7.10 Crystal structure of PH-Aer2 consisting of the PAS and di-HAMP (residues 173–384) domains in the CN^--bound form. The region from 307 to 384 is disordered. (For colour version of this figure, the reader is referred to the online version of this chapter.)

et al., 2012). The bond distance between the N atom of CN^- and the NE2 atom of Trp283 is 2.9 Å, suggesting that the haem-bound CN^- interacts with the side chain of Trp283. Given that CN^- is a model of O_2, Trp283 will form a hydrogen bond with the haem-bound O_2, which is the first example of O_2-binding haem proteins in which a Trp residue is involved for the stabilisation of the oxy state as a hydrogen bond donor.

The formation of the hydrogen bond between Trp283 and the haem-bound O_2 will cause a conformational change of the β5 strand upon O_2 binding because Trp283 is located in the C-terminal region of the β5 strand. Once the PAS domain in Aer2 senses O_2 by binding it to the haem, the signal of O_2 sensing should be transduced to the C-terminal MCP domain. As the β5 strand is directly linked to the α4 helix composing a linker between the PAS and di-HAMP domains, the conformational change of the β5 strand could cause a change in the conformation of the di-HAMP domain, which will result in a conformational change of the MCP domain. This signal transduction would play an important role for the functional regulation of Aer2 in response to O_2.

Recently, Airola et al. (2013) report the crystal structure of the PAS domain (residues 173–289) of Aer2 in the ferric state. Comparing the structures of the CN^--bound and ligand-free forms reveals some structural differences. The indole ring of Trp283 rotates ca. 90° in the absence of the external ligand and Leu264 contacts towards the haem iron to occupy the position where CN^- binds. These changes are accompanied with the movement of β3, β4, and β5 strands. Haem itself also shifts up towards Leu264 by 1.5–2.0 Å upon ligand binding (Airola et al., 2013). These changes will be involved in the intramolecular signal transductions upon O_2 binding to the haem.

6.3. MCP containing a PAS domain with c-type haem
6.3.1 GSU0935 and GSU0582

Geobacter sulfurreducens has the ability to oxidise organic compounds to CO_2 with Fe^{3+} or other metal ions as the electron acceptor. Genome sequence analyses reveal that there are 10 sequences containing at least one c-type haem-binding motif ($Cys-X_{2-4}-Cys-His$), which are parts of proteins annotated as signal transduction/chemotaxis proteins: five are annotated as being sensor histidine kinases (GSU0059, GS2916, GSU1302, GSU2816, and GSU2314), two as chemotaxis signal transducer proteins (GSU0935 and GSU0582), two as cytochrome *c* family proteins (GSU0303 and GSU0591), and one as a HAMP/GAF/HD-GYP protein (GSU2622)

(Londer, Dementieva, D'Ausilio, Pokkuluri, & Schiffer, 2006). Among five proteins (GSU2816, GSU2314, GSU0935, GSU0582, and GSU0303) containing a PAS domain in their sequences, GSU0935 and GSU0582 are characterised spectroscopically and structurally.

Both GSU0935 and GSU0582 have similar predicted topologies: a cytoplasmic N-terminal tail followed by a transmembrane helix, a periplasmic domain (ca. 135 residues), another transmembrane helix, and cytoplasmic domains consisting of a HAMP domain followed by a MCP domain. X-ray crystal structures of the periplasmic PAS domains of GSU0582 and 0935 reveal that these PAS domains form a swapped dimmer (Fig. 7.11; Pokkuluri et al., 2008). In the swapped dimers, the N-terminal two helices from one protomer associate with the β sheet of the other (Pokkuluri et al., 2008). The haem is covalently bound in the loop between H_β and I_β strands and is in proximity to the N-terminal two helices of the other protomer (Pokkuluri et al., 2008). The location of the haem in the PAS domain is different from that in the b-type haem-containing PAS domains such as FixL and EcDos, suggesting that these c-type haem-containing PAS domains adopt the different mechanisms for intramolecular signal transductions triggered by the haem upon sensing the external signal. The PAS domains in GSU0582 and 0935 lack the FG loop that plays an important role for the signal transductions in FixL and EcDos, which supports the above hypothesis.

Though the crystal structures reveal the homodimeric form of the PAS domains for GSU0935 and GSU0582, they are monomeric in diluted solution (Pokkuluri et al., 2008; Silva, Lucas, Salgueiro, & Gomes, 2012). Dimer formation is observed in concentrated solution, suggesting the existence of equilibrium between the monomer and dimer forms in solution. If this is the case in cells, significant folding changes and conformational arrangement should be required to the formation/dissociation of the swapped dimer. The stability of the PAS fold would modulate domain swapping and dimerization. Intriguingly, the two PAS domains for GSU0935 and GSU0582 have distinct levels of intrinsic disordered region, which are responsible for conformational stability and signalling properties of these proteins (Silva et al., 2012).

The coordination structure of the haem in GSU0935 and GSU0582 is similar to that in EcDos (Fig. 7.12). The ferric haem is coordinated by His143 in the Cys-x-x-Cys-His motif and a water molecule in GSU0582. The haem-bound water molecule forms a hydrogen bond to the main-chain carbonyl oxygen atom of Met60 of the other protomer.

A GSU0582 (PDB 3B47)

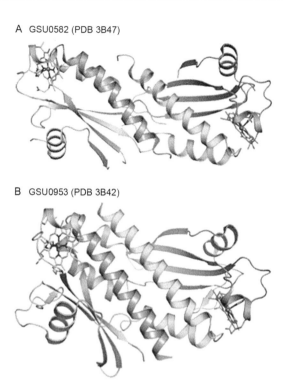

B GSU0953 (PDB 3B42)

Figure 7.11 Crystal structures of the periplasmic PAS domain of (A) GSU0582 (PDB 3B47) and GSU0953 (PDB 3B42). (For colour version of this figure, the reader is referred to the online version of this chapter.)

The ligand exchange takes place upon the reduction between the haem-bound water molecule and Met60 to form the ferrous haem with two endogenous residues as the axial ligands. In the structure of GSU0935, the coordination structures of the haems in two protomers are different from each other. For the haem of chain B, it is in high-spin state with His144 and a water molecule as the axial ligands. The haem-bound water molecule has a hydrogen-bonding pattern similar to that observed in GSU0582. The haem of chain A is in 6-coorinate and low-spin state with His144 and Met60.

GSU0582 and GSU0935 show EPR spectra with $g = 5.90$ and 5.93, respectively, characteristic of high-spin haems with $S = 5/2$. While the high-spin signals are predominant in both cases, a trace amount of low-spin haem ($S = 1/2$) is also observed at g values of 2.28 and 2.75, and 2.28 and 2.81 for GSU0582 and GSU0935, respectively, which suggest an equilibrium between low-spin and high-spin states (Pokkuluri et al., 2008). In the resonance Raman spectra, ferric GSU0582 and GSU0935 show two

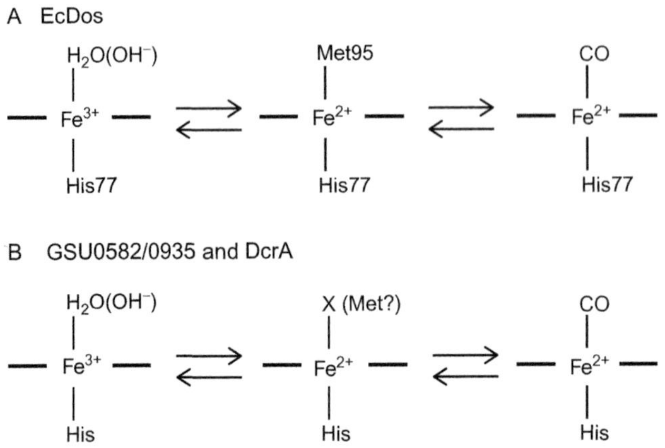

A EcDos

B GSU0582/0935 and DcrA

Figure 7.12 Coordination structures of the haem in (A) GSU0582/GSU0953 and (B) EcDos.

v2 and v3 bands at 1567/1588 cm^{-1} and 1483/1508 cm^{-1} for GSU0582 and at 1570/1590 cm^{-1} and 1479/1508 cm^{-1} for GSU0935, respectively (Catarino et al., 2010), which indicate that the ferric haem is in a mixture of a 6-coordinate and low-spin and a 6-coordinate and high-spin states.

Though the ferrous haem in GSU0582 and GSU0935 is 6-coordinated, it can bind CO and NO to form the CO- and NO-bound haems as are the cases of EcDos, CooA, and DcrA. Formation of a stable O_2-bound form is not reported for GSU0582 and GSU0935. Resonance Raman spectroscopy reveals formation of a mixture of a 6-coordinate and low-spin nitrosyl haem, and a 5-coordinate and high-spin nitrosyl haem upon the reaction of ferrous GSU0582 or GSU0935 with NO (Catarino et al., 2010). Upon the reaction of ferric GSU0582 and GSU0935 with NO, a 6-coordinate Fe^{3+}-nitrosyl haem is formed along with a 5-coordinate Fe^{2+}-nitrosyl haem produced by reductive nitrosylation (Catarino et al., 2010). The K_d values of GSU0582 and GSU0935 for CO to the ferrous haem, NO to the ferrous haem, and NO to the ferric haem are 0.05, 0.08, and 0.34 μM, and 0.08, 0.04, and 17 μM, respectively. MD calculations reveal that the difference of K_d(NO) for the ferric haems between GSU0582 and GSU0935 can be rationalised in terms of distal haem pocket accessibility (Catarino et al., 2010).

6.3.2 DcrA

The amino acid sequence of DcrA from *Desulfovibrio vulgaris* Hildenborough indicates homology with the methyl-accepting chemotaxis proteins from enteric bacteria (Deckers & Voordouw, 1996; Dolla, Fu, Brumlik, &

Voordouw, 1992). DcrA from *D. vulgaris* consists of an N-terminal transmembrane helix, periplasmic PAS, another transmembrane helix, HAMP, cytoplasmic PAS, and MCP domains (Deckers & Voordouw, 1994a, 1994b). Though its physiological effector signal is not obvious at present, it is supposed to act as a chemotaxis signal transducer protein sensing oxygen concentration and/or redox potential (Fu & Voordouw, 1997; Fu, Wall, & Voordouw, 1994; Yoshioka et al., 2005).

Though the sequence identity of the periplasmic domain of DcrA (DcrA-N) from *D. vulgaris* with the PAS domains in GSU0935 and GSU0582 is low, the c-type haem-binding motif (Cys-x-x-Cys-His) is conserved among them (Fig. 7.13). DcrA contains a c-type haem covalently bound to Cys in this motif with a thioether bond (Fu et al., 1994; Yoshioka et al., 2005). In the resonance Raman spectra, DcrA-N shows the $\delta(C_\beta C_a C_b)$ and $\nu(C_a\text{–}S)$ bands that are characteristic bands for c-type haems containing covalent thioether bonds (Desbois, 1994; Hu, Morris, Singh, Smith, & Spiro, 1993; Yoshioka et al., 2005).

The ferric haem in DcrA-N is a mixture of a 6-coordinate, low-spin and 6-coordinate, high-spin states, which is revealed by resonance Raman spectrum showing two $\nu2$ (1568 and 1580 cm^{-1}) and $\nu3$ (1481 and 1509 cm^{-1}) bands (Yoshioka et al., 2005). Ferric DcrA-N shows a similar UV–vis spectrum to that of M80A variant of cytochrome c with the Soret band at 400 nm (Brem & Gray, 1993). Ferrous DcrA-N shows the $\nu2$ and $\nu3$ bands at 1593 and 1495 cm^{-1}, respectively, indicating that the ferrous haem in DcrA-N is a 6-coordinate, low-spin state (Yoshioka et al., 2005). Resonance Raman and UV–vis spectra of DcrA-N reveal the coordination structures of the haem as shown in Fig. 7.12B. The His in the Cys-x-x-Cys-His motif is the proximal ligand of the haem in DcrA. Given that DcrA is a redox sensor, the ligand exchange between H_2O (or OH^-) and Met upon the change in the oxidation state of the haem will play an important role for intramolecular signal transductions because the coordination/dissociation of the sixth ligand may cause conformational changes in the distal haem pocket. Though the sixth axial ligand in the ferrous haem in DcrA-N is not identified, Met61 will be a candidate of the axial ligand of the ferrous haem as are the cases of GSU0935 and GSU0582 because Met 61 is conserved at the corresponding positions of Met60 in GSU0935 and GSU0582 (Fig. 7.13).

While a stable O_2-bound form is not produced because an autoxidation takes place upon the reaction of ferrous DcrA with O_2, CO is bound to the haem in DcrA to form the CO-bound DcrA. CO-bound DcrA-N shows

```
GSU0935    RSSLDLQLKNARNLAGLIIHDIDGYMMKG-DSSEVDRFISAVK----SKNFIMDLR 85
GSU0582    NAIMDLQTRNTRGLSTLVVRDIGELMMAG-DMAVIERYVADVR----GKGAVLDLR 85
DcrA-N     QSALTLIDSGAVRASELLLDAIADPMSKGNDTGTTEKFDAIARRYADIRAYMTDFR 91
                                   *

GSU0935    ---VFDEQAKEVSPTPSQTPNA-----KIQQAIAAGRTLEFKETLDGKRTLSLVLP 133
GSU0582    ---IYDAAGRPAG-KKQDAPDG-----EVQAALTSGATAEKRHKVDGRHVLSFIVP 132
DcrA-N     GNITYSTSKDVLRRDLADVVQAPALLDKFNAALKTDSREEQLATIDGMAFYATIRS 147

GSU0935    FPNEQRCQSCHDAGAAYLGGLLVTTSIEEGYEGARH-------- 169
GSU0582    LANEVRCQSCHEQGARFNGAMLLTTSLEEGYAGARN-------- 168
DcrA-N     IPNAPECHHCHGRSQPILGTLVMLQDVTPAMSELRLHQYETVGL 191
```

Figure 7.13 Amino acid sequence alignments among the periplasmic haem-binding PAS domains of GSU0582, GSU0935, and DcrA-N. The conserved Cys in the Cys-x-x-Cys-His motif and Met are shown in bold.

the $\nu(Fe–CO)$ and $\nu(C–O)$ bands at 493 and 1955 cm^{-1}, respectively. The correlation between the $\nu(Fe–CO)$ and $\nu(C–O)$ bands reveals that a neutral His is the proximal ligand of the CO-bound haem (Yoshioka et al., 2005). Thus, the ligand exchange between the endogenous axial ligand and CO takes place upon CO binding. These properties for the coordination structures of the haem and ligand exchange upon CO binding are same as those of GSU0935 and GSU0582. However, the CO-binding affinity ($K_d = 138$ μM) is lower by 1700- and 2700-fold compared with GSU0935 ($K_d = 0.08$ μM) and GSU0582 ($K_d = 0.05$ μM). *D. vulgaris* possesses the *coo* operon encoding the proteins for CO metabolism (Voordouw, 2002), suggesting that it can gain energy for growth by CO metabolism. If it is the case, it makes sense there is a chemotaxis regulation system sensing suitable concentrations of CO. DcrA would be a possible candidate of a CO sensor with a low CO-binding affinity to sense the existence of enough concentrations of CO for gaining energy by CO metabolism. This is an attractive hypothesis, but more experiments are required to prove whether or not it is the case.

7. CONCLUSION

Haem-containing PAS domains are widely used as a sensory module in sensor proteins that sense various environmental cues. Haem acts as the active site for sensing the cognate physiological signals in these sensor proteins. Binding of a gas molecule to the haem is the first step of sensing the physiological effector in gas sensor proteins, while the change in the oxidation state of the haem is the first cue in redox sensors. As various external ligands can be bound to the haem, haem-based sensor proteins should discriminate its physiological effector among them. There will be two strategies in principle for discriminating a physiological gas molecule from others. First, if only the physiological effector is able to bind to the haem, the discrimination can be achieved. Second, if there are specific interactions between the haem-bound ligand and surrounding residues only upon binding of a physiological effector, it can be used for discrimination of the cognate physiological effector. The haem-containing PAS family of sensor proteins adopts the second strategy with some modifications in individual proteins. In these systems, the binding of the cognate physiological effector to the haem results in the reconstruction of the hydrogen-bonding network among the haem, haem-bound ligand, and surrounding residues along with conformational changes in the haem pocket. Dynamic exchange of the axial ligand upon redox change of the haem or upon binding of an external ligand can also be a trigger for conformational changes of several haem-containing PAS domains. Conformational changes induced by sensing the cognate effector are crucial processes for the regulation of the sensor proteins. As described in this chapter, spectroscopic and structural studies of the isolated PAS domains give the detailed information of what and how conformational changes occur in the sensor domains upon sensing the effector. However, structural studies of sensor proteins in full-length remain to be achieved. To fully understand the structural and functional relationships of the haem-containing PAS family, it is required to determine the structures at "on-state" and "off-state" in the full-length protein.

REFERENCES

Agron, P. G., Ditta, G. S., & Helinski, D. R. (1993). Oxygen regulation of *nifA* transcription in vitro. *Proceedings of the National Academy of Sciences of the United States of America, 90,* 3506–3510.

Airola, M. V., Huh, D., Sukomon, N., Widom, J., Sircar, R., Borbat, P. P., et al. (2013). Architecture of the soluble receptor Aer2 indicates an in-line mechanism for PAS and HAMP domain signaling. *Journal of Molecular Biology, 425,* 886–901.

Airola, M. V., Watts, K. J., Bilwes, A. M., & Crane, B. R. (2010). Structure of concatenated HAMP domains provides a mechanism for signal transduction. *Structure, 18,* 436–448.

Akimoto, S., Tanaka, A., Nakamura, K., Shiro, Y., & Nakamura, H. (2003). O_2-specific regulation of the ferrous heme-based sensor-kinase FixL from *Sinorhizobium meliloti* and its aberrant inactivation in the ferric form. *Biochemical and Biophysical Research Communications, 304,* 136–142.

Al Dahouk, S., Loisel-Meyer, S., Scholz, H. C., Tomaso, H., Kersten, M., Harder, A., et al. (2009). Proteomic analysis of *Brucella suis* under oxygen deficiency reveals flexibility in adaptive expression of various pathways. *Proteomics, 9,* 3011–3021.

Anantharaman, V., Koonin, E. V., & Aravind, L. (2001). Regulatory potential, phyletic distribution and evolution of ancient, intracellular small-molecule-binding domains. *Journal of Molecular Biology, 307,* 1271–1292.

Aono, S. (2003). Biochemical and biophysical properties of the CO-sensing transcriptional activator CooA. *Accounts of Chemical Research, 36*, 825–831.

Aono, S. (2011). Metal-containing sensor proteins sensing diatomic gas molecules. *Dalton Transactions*, (24), 3137–3146.

Aono, S., Nakajima, H., Saito, K., & Okada, M. (1996). A novel heme protein that acts as a carbon monoxide-dependent transcriptional activator in *Rhodospirillum rubrum*. *Biochemical and Biophysical Research Communications, 228*, 752–756.

Aono, S., Ohkubo, K., Matsuo, T., & Nakajima, H. (1998). Redox-controlled ligand exchange of the heme in the CO-sensing transcriptional activator CooA. *The Journal of Biological Chemistry, 273*, 25757–25764.

Aravind, L., & Ponting, C. P. (1997). The GAF domain: An evolutionary link between diverse phototransducing proteins. *Trends in Biochemical Sciences, 22*, 458–459.

Baker, S. C., Saunders, N. F., Willis, A. C., Ferguson, S. J., Hajdu, J., & Fülöp, V. (1997). Cytochrome cd1 structure: Unusual haem environments in a nitrite reductase and analysis of factors contributing to beta-propeller folds. *Journal of Molecular Biology, 269*, 440–455.

Balland, V., Bouzhir-Sima, L., Anxolabéhère-Mallart, E., Boussac, A., Vos, M. H., Liebl, U., et al. (2006). Functional implications of the propionate 7-arginine 220 interaction in the FixLH oxygen sensor from *Bradyrhizobium japonicum*. *Biochemistry, 45*, 2072–2084.

Balland, V., Bouzhir-Sima, L., Kiger, L., Marden, M. C., Vos, M. H., Liebl, U., et al. (2005). Role of arginine 220 in the oxygen sensor FixL from *Bradyrhizobium japonicum*. *The Journal of Biological Chemistry, 280*, 15279–15288.

Baraquet, C., Théraulaz, L., Iobbi-Nivol, C., Méjean, V., & Jourlin-Castelli, C. (2009). Unexpected chemoreceptors mediate energy taxis towards electron acceptors in *Shewanella oneidensis*. *Molecular Microbiology, 73*, 278–290.

Bauer, C. E., Elsen, S., & Bird, T. H. (1999). Mechanisms for redox control of gene expression. *Annual Review of Microbiology, 53*, 495–523.

Blair, D. F. (1995). How bacteria sense and swim. *Annual Review of Microbiology, 49*, 489–522.

Boon, E. M., Davis, J. H., Tran, R., Karow, D. S., Huang, S. H., Pan, D., et al. (2006). Nitric oxide binding to prokaryotic homologs of the soluble guanylate cyclase beta1 H-NOX domain. *The Journal of Biological Chemistry, 281*, 21892–21902.

Brem, K. L., & Gray, H. B. (1993). Structurally engineered cytochromes with novel ligand-binding sites: Oxy and carbonmonoxy derivatives of semisynthetic horse heart Ala80 cytochrome c. *Journal of the American Chemical Society, 115*, 10382–103803.

Burnside, K., Lembo, A., de Los Reyes, M., Iliuk, A., Binhtran, N. T., Connelly, J. E., et al. (2010). Regulation of hemolysin expression and virulence of *Staphylococcus aureus* by a serine/threonine kinase and phosphatase. *PLoS One, 5*, e11071. http://dx.doi.org/10.1371/journal.pone.0011071.

Bush, M., Ghosh, T., Tucker, N., Zhang, X., & Dixon, R. (2011). Transcriptional regulation by the dedicated nitric oxide sensor, NorR: A route toward NO detoxification. *Biochemical Society Transactions, 39*, 289–293.

Carlson, H. K., Vance, R. E., & Marletta, M. A. (2010). H-NOX regulation of c-di-GMP metabolism and biofilm formation in *Legionella pneumophila*. *Molecular Microbiology, 77*, 930–942.

Carrica Mdel, C., Fernandez, I., Martí, M. A., Paris, G., & Goldbaum, F. A. (2012). The NtrY/X two-component system of *Brucella* spp. acts as a redox sensor and regulates the expression of nitrogen respiration enzymes. *Molecular Microbiology, 85*, 39–50.

Catarino, T., Pessanha, M., De Candia, A. G., Gouveia, Z., Fernandes, A. P., Pokkuluri, P. R., et al. (2010). Probing the chemotaxis periplasmic sensor domains from Geobacter sulfurreducens by combined resonance Raman and molecular dynamic approaches: NO and CO sensing. *The Journal of Physical Chemistry. B, 114*, 11251–11260.

Chang, A. L., Tuckerman, J. R., Gonzalez, G., Mayer, R., Weinhouse, H., Volman, G., et al. (2001). Phosphodiesterase A1, a regulator of cellulose synthesis in *Acetobacter xylinum*, is a heme-based sensor. *Biochemistry, 40*, 3420–3426.

Cho, H. Y., Cho, H. J., Kim, Y. M., Oh, J. I., & Kang, B. S. (2009). Structural insight into the heme-based redox sensing by DosS from *Mycobacterium tuberculosis*. *The Journal of Biological Chemistry, 284*, 13057–13067.

Cotter, P. A., & Stibitz, S. (2007). c-di-GMP-mediated regulation of virulence and biofilm formation. *Current Opinion in Microbiology, 10*, 17–23.

Da Re, S., Schumacher, J., Rousseau, P., Fourment, J., Ebel, C., & Kahn, D. (1999). Phosphorylation-induced dimerization of the FixJ receiver domain. *Molecular Microbiology, 34*, 504–511.

Das, T. K., Couture, M., Ouellet, Y., Guertin, M., & Rousseau, D. L. (2001). Simultaneous observation of the O-O and Fe-O_2 stretching modes in oxyhemoglobins. *Proceedings of the National Academy of Sciences of the United States of America, 98*, 479–484.

D'Autréaux, B., Tucker, N. P., Dixon, R., & Spiro, S. (2005). A non-haem iron centre in the transcription factor NorR senses nitric oxide. *Nature, 437*, 769–772.

David, M., Daveran, M. L., Batut, J., Dedieu, A., Domergue, O., Ghai, J., et al. (1998). Cascade regulation of nif gene expression in *Rhizobium meliloti*. *Cell, 54*, 671–683.

Deckers, H. M., & Voordouw, G. (1994a). Identification of a large family of genes for putative chemoreceptor proteins in an ordered library of the *Desulfovibrio vulgaris* Hildenborough genome. *Journal of Bacteriology, 176*, 351–358.

Deckers, H. M., & Voordouw, G. (1994b). Membrane topology of the methyl-accepting chemotaxis protein DcrA from *Desulfovibrio vulgaris* Hildenborough. *Antonie Van Leeuwenhoek, 65*, 7–12.

Deckers, H. M., & Voordouw, G. (1996). The dcr gene family of *Desulfovibrio*: Implications from the sequence of dcrH and phylogenetic comparison with other mcp genes. *Antonie Van Leeuwenhoek, 70*, 21–29.

Delgado-Nixon, V. M., Gonzalez, G., & Gilles-Gonzalez, M. A. (2000). Dos, a heme-binding PAS protein from *Escherichia coli*, is a direct oxygen sensor. *Biochemistry, 39*, 2685–2691.

Desbois, A. (1994). Resonance Raman spectroscopy of c-type cytochromes. *Biochimie, 76*, 693–707.

Dioum, E. M., Rutter, J., Tuckerman, J. R., Gonzalez, G., Gilles-Gonzalez, M. A., & McKnight, S. L. (2002). NPAS2: A gas-responsive transcription factor. *Science, 298*, 2385–2397.

Dolla, A., Fu, R., Brumlik, M. J., & Voordouw, G. (1992). Nucleotide sequence of dcrA, a *Desulfovibrio vulgaris* Hildenborough chemoreceptor gene, and its expression in *Escherichia coli*. *Journal of Bacteriology, 174*, 1726–1733.

Dunham, C. M., Dioum, E. M., Tuckerman, J. R., Gonzalez, G., Scott, W. G., & Gilles-Gonzalez, M. A. (2003). A distal arginine in oxygen-sensing heme-PAS domains is essential to ligand binding, signal transduction, and structure. *Biochemistry, 42*, 7701–7708.

El-Mashtoly, S. F., Nakashima, S., Tanaka, A., Shimizu, T., & Kitagawa, T. (2008). Roles of Arg-97 and Phe-113 in regulation of distal ligand binding to heme in the sensor domain of Ec DOS protein. Resonance Raman and mutation study. *The Journal of Biological Chemistry, 283*, 19000–19100.

El-Mashtoly, S. F., Takahashi, H., Shimizu, T., & Kitagawa, T. (2007). Ultraviolet resonance Raman evidence for utilization of the heme 6-propionate hydrogen-bond network in signal transmission from heme to protein in Ec DOS protein. *Journal of the American Chemical Society, 129*, 3556–3563.

Erbil, W. K., Price, M. S., Wemmer, D. E., & Marletta, M. A. (2009). A structural basis for H-NOX signaling in *Shewanella oneidensis* by trapping a histidine kinase inhibitory conformation. *Proceedings of the National Academy of Sciences of the United States of the Amer-ica, 106,* 19753–19760.

Fischer, H. M. (1994). Genetic regulation of nitrogen fixation in rhizobia. *Microbiological Reviews, 58,* 352–386.

Fu, R., & Voordouw, G. (1997). Targeted gene-replacement mutagenesis of dcrA, encoding an oxygen sensor of the sulfate-reducing bacterium *Desulfovibrio vulgaris* Hildenborough. *Microbiology (Reading, England), 143,* 1815–1826.

Fu, R., Wall, J. D., & Voordouw, G. (1994). DcrA, a c-type heme-containing methyl-accepting protein from *Desulfovibrio vulgaris* Hildenborough, senses the oxygen concen-tration or redox potential of the environment. *Journal of Bacteriology, 176,* 344–350.

Fülöp, V., Moir, J. W., Ferguson, S. J., & Hajdu, J. (1995). The anatomy of a bifunctional enzyme: Structural basis for reduction of oxygen to water and synthesis of nitric oxide by cytochrome cd1. *Cell, 81,* 369–377.

Gilles-Gonzalez, M. A. (2001). Oxygen signal transduction. *IUBMB Life, 51,* 165–173.

Gilles-Gonzalez, M. A., Caceres, A. I., Sousa, E. H., Tomchick, D. R., Brautigam, C., Gonzalez, C., et al. (2006). A proximal arginine R206 participates in switching of the *Bradyrhizobium japonicum* FixL oxygen sensor. *Journal of Molecular Biology, 360,* 80–89.

Gilles-Gonzalez, M. A., Ditta, G. S., & Helinski, D. R. (1991). A haemoprotein with kinase activity encoded by the oxygen sensor of *Rhizobium meliloti*. *Nature, 350,* 170–172.

Gilles-Gonzalez, M. A., & Gonzalez, G. (1993). Regulation of the kinase activity of heme protein FixL from the two-component system FixL/FixJ of *Rhizobium meliloti*. *The Journal of Biological Chemistry, 268,* 16293–16297.

Gilles-Gonzalez, M. A., & Gonzalez, G. (2005). Heme-based sensors: Defining characteris-tics, recent developments, and regulatory hypotheses. *Journal of Inorganic Biochemistry, 99,* 1–22.

Gilles-González, M. A., González, G., & Perutz, M. F. (1995). Kinase activity of oxygen sensor FixL depends on the spin state of its heme iron. *Biochemistry, 34,* 232–236.

Gilles-Gonzalez, M. A., Gonzalez, G., Perutz, M. F., Kiger, L., Marden, M. C., & Poyart, C. (1994). Heme-based sensors, exemplified by the kinase FixL, are a new class of heme protein with distinctive ligand binding and autoxidation. *Biochemistry, 33,* 8067–8073.

Gong, W., Hao, B., & Chan, M. K. (2000). New mechanistic insights from structural studies of the oxygen-sensing domain of *Bradyrhizobium japonicum* FixL. *Biochemistry, 39,* 3955–3962.

Gong, W., Hao, B., Mansy, S. S., Gonzalez, G., Gilles-Gonzalez, M. A., & Chan, M. K. (1998). Structure of a biological oxygen sensor: A new mechanism for heme-driven signal transduction. *Proceedings of the National Academy of Sciences of the United States of America, 95,* 15177–15182.

Gottfried, D. S., Peterson, E. S., Sheikh, A. G., Wang, J., Yang, M., & Friedman, J. M. (1996). Evidence for damped hemoglobin dynamics in a room temperature trehalose glass. *The Journal of Physical Chemistry, 100,* 12034–12042.

Hao, B., Isaza, C., Arndt, J., Soltis, M., & Chan, M. K. (2002). Structure-based mechanism of O_2 sensing and ligand discrimination by the FixL heme domain of *Bradyrhizobium japonicum*. *Biochemistry, 41,* 12952–12958.

Heikaus, C. C., Pandit, J., & Klevit, R. E. (2009). Cyclic nucleotide binding GAF domains from phosphodiesterases: Structural and mechanistic insights. *Structure, 17,* 1551–1557.

Henry, J. T., & Crosson, S. (2011). Ligand-binding PAS domains in a genomic, cellular, and structural context. *Annual Review of Microbiology, 65,* 261–286.

Hiruma, Y., Kikuchi, A., Tanaka, A., Shiro, Y., & Mizutani, Y. (2007). Resonance Raman observation of the structural dynamics of FixL on signal transduction and ligand discrim-ination. *Biochemistry, 46,* 6086–6096.

Honaker, R. W., Dhiman, R. K., Narayanasamy, P., Crick, D. C., & Voskuil, M. I. (2010). DosS responds to a reduced electron transport system to induce the *Mycobacterium tuberculosis* DosR regulon. *Journal of Bacteriology*, *192*, 6447–6455.

Hong, C. S., Kuroda, A., Takiguchi, N., Ohtake, H., & Kato, J. (2005). Expression of *Pseudomonas aeruginosa aer-2*, one of two aerotaxis transducer genes, is controlled by RpoS. *Journal of Bacteriology*, *187*, 1533–1535.

Hong, C. S., Shitashiro, M., Kuroda, A., Ikeda, T., Takiguchi, N., Ohtake, H., et al. (2004). Chemotaxis proteins and transducers for aerotaxis in *Pseudomonas aeruginosa*. *FEMS Microbiology Letters*, *231*, 247–252.

Hou, S., Larsen, R. W., Boudko, D., Riley, C. W., Karatan, E., Zimmer, M., et al. (2000). Myoglobin-like aerotaxis transducers in Archaea and Bacteria. *Nature*, *403*, 540–544.

Hu, S., Morris, I. K., Singh, J. P., Smith, K. M., & Spiro, T. G. (1993). Complete assignment of cytochrome-c resonance Raman spectra via enzymatic reconstitution with isotopically-labeled hemes. *Journal of the American Chemical Society*, *115*, 12446–12458.

Hu, S. Z., Smith, K. M., & Spiro, T. G. (1996). Assignment of protoheme resonance Raman spectrum by heme labeling in myoglobin. *Journal of the American Chemical Society*, *118*, 12638–12646.

Ioanoviciu, A., Meharenna, Y. T., Poulos, T. L., & Ortiz de Montellano, P. R. (2009). DevS oxy complex stability identifies this heme protein as a gas sensor in *Mycobacterium tuberculosis* dormancy. *Biochemistry*, *48*, 5839–5848.

Ioanoviciu, A., Yukl, E. T., Moënne-Loccoz, P., & Ortiz de Montellano, P. R. (2007). DevS, a heme-containing two-component oxygen sensor of *Mycobacterium tuberculosis*. *Biochemistry*, *46*, 4250–4260.

Ishida, M. L., Assumpção, M. C., Machado, H. B., Benelli, E. M., Souza, E. M., & Pedrosa, F. O. (2002). Identification and characterization of the two-component NtrY/NtrX regulatory system in *Azospirillum brasilense*. *Brazilian Journal of Medical and Biological Research*, *35*, 651–661.

Ishida, M., Ueha, T., & Sagami, I. (2008). Effects of mutations in the heme domain on the transcriptional activity and DNA-binding activity of NPAS2. *Biochemical and Biophysical Research Communications*, *368*, 292–297.

Ishitsuka, Y., Araki, Y., Tanaka, A., Igarashi, J., Ito, O., & Shimizu, T. (2008). Arg97 at the heme-distal side of the isolated heme-bound PAS domain of a heme-based oxygen sensor from *Escherichia coli* (Ec DOS) plays critical roles in autoxidation and binding to gases, particularly O_2. *Biochemistry*, *47*, 8874–8884.

Ito, S., Araki, Y., Tanaka, A., Igarashi, J., Wada, T., & Shimizu, T. (2009). Role of Phe113 at the distal side of the heme domain of an oxygen-sensor (Ec DOS) in the characterization of the heme environment. *Journal of Inorganic Biochemistry*, *103*, 989–996.

Ito, S., Igarashi, J., & Shimizu, T. (2009). The FG loop of a heme-based gas sensor enzyme, Ec DOS, functions in heme binding, autoxidation and catalysis. *Journal of Inorganic Biochemistry*, *103*, 1380–1385.

Iyer, L. M., Anantharaman, V., & Aravind, L. (2003). Ancient conserved domains shared by animal soluble guanylyl cyclases and bacterial signaling proteins. *BMC Genomics*, *4*, 5. http://dx.doi.org/10.1186/1471-2164-4-5.

Jasaitis, A., Hola, K., Bouzhir-Sima, L., Lambry, J. C., Balland, V., Vos, M. H., et al. (2006). Role of distal arginine in early sensing intermediates in the heme domain of the oxygen sensor FixL. *Biochemistry*, *45*, 6018–6026.

Jenal, U. (2004). Cyclic di-guanosine-monophosphate comes of age: A novel secondary messenger involved in modulating cell surface structures in bacteria? *Current Opinion in Microbiology*, *7*, 185–191.

Jenal, U., & Malone, J. (2006). Mechanisms of cyclic-di-GMP signaling in bacteria. *Annual Review of Genetics*, *40*, 385–407.

Jung, K., Fried, L., Behr, S., & Heermann, R. (2011). Histidine kinases and response regulators in networks. *Current Opinion in Microbiology, 15,* 118–124.

Karatan, E., & Watnick, P. (2009). Signals, regulatory networks, and materials that build and break bacterial biofilms. *Microbiology and Molecular Biology Review, 73,* 310–347.

Karow, D. S., Pan, D., Tran, R., Pellicena, P., Presley, A., Mathies, R. A., et al. (2004). Spectroscopic characterization of the soluble guanylate cyclase-like heme domains from *Vibrio cholerae* and *Thermoanaerobacter tengcongensis. Biochemistry, 43,* 10203–10211.

Kerby, R. L., Youn, H., & Roberts, G. P. (2008). RcoM: A new single-component transcriptional regulator of CO metabolism in bacteria. *Journal of Bacteriology, 190,* 3336–3343.

Key, J., & Moffat, K. (2005). Crystal structures of deoxy and CO-bound *bj*FixLH reveal details of ligand recognition and signaling. *Biochemistry, 44,* 4627–4635.

Kim, M. J., Park, K. J., Ko, I. J., Kim, Y. M., & Oh, J. I. (2010). Different roles of DosS and DosT in the hypoxic adaptation of Mycobacteria. *Journal of Bacteriology, 192,* 4868–4875.

King, G. M., & Weber, C. F. (2007). Distribution, diversity and ecology of aerobic CO-oxidizing bacteria. *Nature Reviews. Microbiology, 5,* 107–118.

King, D. P., Zhao, Y., Sangoram, A. M., Wilsbacher, L. D., Tanaka, M., Antoch, M. P., et al. (1997). Positional cloning of the mouse circadian clock gene. *Cell, 89,* 641–653.

Kirby, J. R. (2009). Chemotaxis-like regulatory systems: Unique roles in diverse bacteria. *Annual Review of Microbiology, 63,* 45–59.

Kitanishi, K., Kobayashi, K., Kawamura, Y., Ishigami, I., Ogura, T., Nakajima, K., et al. (2010). Important roles of Tyr43 at the putative heme distal side in the oxygen recognition and stability of the Fe(II)-O_2 complex of YddV, a globin-coupled heme-based oxygen sensor diguanylate cyclase. *Biochemistry, 49,* 10381–10393.

Kobayashi, K., Tanaka, A., Takahashi, H., Igarashi, J., Ishitsuka, Y., Yokota, N., et al. (2010). Catalysis and oxygen binding of Ec DOS: A haem-based oxygen-sensor enzyme from *Escherichia coli. Journal of Biochemistry, 148,* 693–703.

Komori, H., Inagaki, S., Yoshioka, S., Aono, S., & Higuchi, Y. (2007). Crystal structure of CO-sensing transcription activator CooA bound to exogenous ligand imidazole. *Journal of Molecular Biology, 367,* 864–871.

Kruglik, S. G., Jasaitis, A., Hola, K., Yamashita, T., Liebl, U., Martin, J. L., et al. (2007). Subpicosecond oxygen trapping in the heme pocket of the oxygen sensor FixL observed by time-resolved resonance Raman spectroscopy. *Proceedings of the National Academy of Sciences of the United States of America, 104,* 7408–7413.

Kumar, A., Toledo, J. C., Patel, R. P., Lancaster, J. R., Jr., & Steyn, A. J. (2007). *Mycobacterium tuberculosis* DosS is a redox sensor and DosT is a hypoxia sensor. *Proceedings of the National Academy of Sciences of the United States of America, 104,* 11568–11573.

Kurokawa, H., Lee, D. S., Watanabe, M., Sagami, I., Mikami, B., Raman, C. S., et al. (2004). A redox-controlled molecular switch revealed by the crystal structure of a bacterial heme PAS sensor. *The Journal of Biological Chemistry, 279,* 20186–20193.

Lanzilotta, W. N., Schuller, D. J., Thorsteinsson, M. V., Kerby, R. L., Roberts, G. P., & Poulos, T. L. (2000). Structure of the CO sensing transcription activator CooA. *Natural Structural Biology, 7,* 876–880.

Lechauve, C., Bouzhir-Sima, L., Yamashita, T., Marden, M. C., Vos, M. H., Liebl, U., et al. (2009). Heme ligand binding properties and intradimer interactions in the full-length sensor protein dos from *Escherichia coli* and its isolated heme domain. *The Journal of Biological Chemistry, 284,* 36146–36159.

Lee, J. M., Cho, H. Y., Cho, H. J., Ko, I. J., Park, S. W., Baik, H. S., et al. (2008). O_2- and NO-sensing mechanism through the DevSR two-component system in *Mycobacterium smegmatis. Journal of Bacteriology, 190,* 6795–6804.

Liebl, U., Bouzhir-Sima, L., Negrerie, M., Martin, J. L., & Vos, M. H. (2002). Ultrafast ligand rebinding in the heme domain of the oxygen sensors FixL and Dos: General regulatory implications for heme-based sensors. *Proceedings of the National Academy of Sciences of the United States of America, 99*, 12771–12776.

Londer, Y. Y., Dementieva, I. S., D'Ausilio, C. A., Pokkuluri, P. R., & Schiffer, M. (2006). Characterization of a c-type heme-containing PAS sensor domain from *Geobacter sulfurreducens* representing a novel family of periplasmic sensors in Geobacteraceae and other bacteria. *FEMS Microbiology Letters, 258*, 173–181.

Lukat-Rodgers, G. S., Rexine, J. L., & Rodgers, K. R. (1998). Heme speciation in alkaline ferric FixL and possible tyrosine involvement in the signal transduction pathway for regulation of nitrogen fixation. *Biochemistry, 37*, 13543–13552.

Marina, A., Waldburger, C. D., & Hendrickson, W. A. (2005). Structure of the entire cytoplasmic portion of a sensor histidine-kinase protein. *EMBO Journal, 24*, 4247–4259.

Marvin, K. A., Kerby, R. L., Youn, H., Roberts, G. P., & Burstyn, J. N. (2008). The transcription regulator RcoM-2 from *Burkholderia xenovorans* is a cysteine-ligated hemoprotein that undergoes a redox-mediated ligand switch. *Biochemistry, 47*, 9016–9028.

Miyatake, H., Kanai, M., Adachi, S., Nakamura, H., Tamura, K., Tanida, H., et al. (1999a). Dynamic light-scattering and preliminary crystallographic studies of the sensor domain of the haem-based oxygen sensor FixL from *Rhizobium meliloti. Acta Crystallographica. Section D, Biological Crystallography, 55*, 1215–1218.

Miyatake, H., Mukai, M., Adachi, S., Nakamura, H., Tamura, K., Iizuka, T., et al. (1999b). Iron coordination structures of oxygen sensor FixL characterized by Fe K-edge extended X-ray absorption fine structure and resonance Raman spectroscopy. *The Journal of Biological Chemistry, 274*, 23176–23184.

Miyatake, H., Mukai, M., Park, S. Y., Adachi, S., Tamura, K., Nakamura, H., et al. (2000). Sensory mechanism of oxygen sensor FixL from *Rhizobium meliloti*: Crystallographic, mutagenesis and resonance Raman spectroscopic studies. *Journal of Molecular Biology, 301*, 415–431.

Möglich, A., Ayers, R. A., & Moffat, K. (2009). Structure and signaling mechanism of Per-ARNT-Sim domains. *Structure, 17*, 1282–1294.

Monson, E. K., Ditta, G. S., & Helinski, D. R. (1995). The oxygen sensor protein, FixL, of *Rhizobium meliloti*. Role of histidine residues in heme binding, phosphorylation, and signal transduction. *The Journal of Biological Chemistry, 270*, 5243–5250.

Mukai, M., Nakamura, K., Nakamura, H., Iizuka, T., & Shiro, Y. (2000). Roles of Ile209 and Ile210 on the heme pocket structure and regulation of histidine kinase activity of oxygen sensor FixL from *Rhizobium meliloti. Biochemistry, 39*, 13810–13816.

Mukai, M., Savard, P. Y., Ouellet, H., Guertin, M., & Yeh, S. R. (2002). Unique ligand-protein interactions in a new truncated hemoglobin from *Mycobacterium tuberculosis. Biochemistry, 41*, 3897–3905.

Mukaiyama, Y., Uchida, T., Sato, E., Sasaki, A., Sato, Y., Igarashi, J., et al. (2006). Spectroscopic and DNA-binding characterization of the isolated heme-bound basic helix-loop-helix-PAS-A domain of neuronal PAS protein 2 (NPAS2), a transcription activator protein associated with circadian rhythms. *FEBS Journal, 273*, 2528–2539.

Nakajima, H., Honma, Y., Tawara, T., Kato, T., Park, S. Y., Miyatake, H., et al. (2001). Redox properties and coordination structure of the heme in the CO-sensing transcriptional activator CooA. *The Journal of Biological Chemistry, 276*, 7055–7061.

Nakajima, H., Takatani, N., Yoshimitsu, K., Itoh, M., Aono, S., Takahashi, Y., et al. (2010). The role of the Fe-S cluster in the sensory domain of nitrogenase transcriptional activator VnfA from *Azotobacter vinelandii. FEBS Journal, 277*, 817–832.

Nakamura, H., Kumita, H., Imai, K., Iizuka, T., & Shiro, Y. (2004). ADP reduces the oxygen-binding affinity of a sensory histidine kinase, FixL: The possibility of an enhanced reciprocating kinase reaction. *Proceedings of the National Academy of Sciences of the United States of America, 101*, 2742–2746.

Nioche, P., Berka, V., Vipond, J., Minton, N., Tsai, A. L., & Raman, C. S. (2004). Femtomolar sensitivity of a NO sensor from *Clostridium botulinum*. *Science, 306*, 1550–1553.

Nuernberger, P., Lee, K. F., Bonvalet, A., Bouzhir-Sima, L., Lambry, J. C., Liebl, U., et al. (2011). Strong ligand-protein interactions revealed by ultrafast infrared spectroscopy of CO in the heme pocket of the oxygen sensor FixL. *Journal of the American Chemical Society, 133*, 17110–17113.

Olea, C., Boon, E. M., Pellicena, P., Kuriyan, J., & Marletta, M. A. (2008). Probing the function of heme distortion in the H-NOX family. *ACS Chemical Biology, 3*, 703–710.

Olea, C., Jr., Herzik, M. A., Jr., Kuriyan, J., & Marletta, M. A. (2010). Structural insights into the molecular mechanism of H-NOX activation. *Protein Science, 19*, 881–887.

Osterberg, S., Skärfstad, E., & Shingler, V. (2010). The sigma-factor FliA, ppGpp and DksA coordinate transcriptional control of the aer2 gene of *Pseudomonas putida*. *Environmental Microbiology, 12*, 1439–1451.

Ouellet, Y., Milani, M., Couture, M., Bolognesi, M., & Guertin, M. (2006). Ligand interactions in the distal heme pocket of *Mycobacterium tuberculosis* truncated hemoglobin N: Roles of TyrB10 and GlnE11 residues. *Biochemistry, 45*, 8770–8781.

Park, H., Suquet, C., Satterlee, J. D., & Kang, C. (2004). Insights into signal transduction involving PAS domain oxygen-sensing heme proteins from the X-ray crystal structure of *Escherichia coli* Dos heme domain (Ec DosH). *Biochemistry, 43*, 2738–2746.

Pawlowski, K., Klosse, U., & de Bruijn, F. J. (1991). Characterization of a novel *Azorhizobium caulinodans* ORS571 two-component regulatory system, NtrY/NtrX, involved in nitrogen fixation and metabolism. *Molecular and General Genetics, 231*, 124–138.

Pesavento, C., & Hengge, R. (2009). Bacterial nucleotide-based second messengers. *Current Opinion in Microbiology, 12*, 170–176.

Peterson, E. S., Friedman, J. M., Chien, E. Y., & Sligar, S. G. (1998). Functional implications of the proximal hydrogen-bonding network in myoglobin: A resonance Raman and kinetic study of Leu89, Ser92, His97, and F-helix swap mutants. *Biochemistry, 37*, 12301–12319.

Podust, L. M., Ioanoviciu, A., & Ortiz de Montellano, P. R. (2008). 2.3 Å X-ray structure of the heme-bound GAF domain of sensory histidine kinase DosT of *Mycobacterium tuberculosis*. *Biochemistry, 47*, 12523–12531.

Pokkuluri, P. R., Pessanha, M., Londer, Y. Y., Wood, S. J., Duke, N. E., Wilton, R., et al. (2008). Structures and solution properties of two novel periplasmic sensor domains with c-type heme from chemotaxis proteins of *Geobacter sulfurreducens*: Implications for signal transduction. *Journal of Molecular Biology, 377*, 1498–1517.

Ragsdale, S. W. (2004). Life with carbon monoxide. *Critical Reviews in Biochemistry and Molecular Biology, 39*, 165–195.

Rao, F., Ji, Q., Soehano, I., & Liang, Z. X. (2011). Unusual heme-binding PAS domain from YybT family proteins. *Journal of Bacteriology, 193*, 1543–1551.

Rao, F., See, R. Y., Zhang, D., Toh, D. C., Ji, Q., & Liang, Z. X. (2010). YybT is a signaling protein that contains a cyclic dinucleotide phosphodiesterase domain and a GGDEF domain with ATPase activity. *The Journal of Biological Chemistry, 285*, 473–482.

Reick, M., Garcia, J. A., Dudley, C., & McKnight, S. L. (2001). NPAS2: An analog of clock operative in the mammalian forebrain. *Science, 293*, 506–509.

Reyrat, J. M., David, M., Batut, J., & Boistard, P. (1994). FixL of *Rhizobium meliloti* enhances the transcriptional activity of a mutant FixJD54N protein by phosphorylation of an alternate residue. *Journal of Bacteriology, 176*, 1969–1976.

Reyrat, J. M., David, M., Blonski, C., Boistard, P., & Batut, J. (1993). Oxygen-regulated in vitro transcription of *Rhizobium meliloti* nifA and fixK genes. *Journal of Bacteriology, 175*, 6867–6872.

Roberts, G. P., Kerby, R. L., Youn, H., & Conrad, M. (2005). CooA, a paradigm for gas sensing regulatory proteins. *Journal of Inorganic Biochemistry*, *99*, 280–292.

Römling, U., & Amikam, D. (2006). Cyclic di-GMP as a second messenger. *Current Opinion in Microbiology*, *9*, 218–228.

Römling, U., Gomelsky, M., & Galperin, M. (2005). C-di-GMP: The dawning of a novel bacterial signaling system. *Molecular Microbiology*, *57*, 629–639.

Roop, R. M., & Caswell, C. C. (2012). Redox-responsive regulation of denitrification genes in *Brucella*. *Molecular Microbiology*, *85*, 5–7.

Saini, D. K., Malhotra, V., Dey, D., Pant, N., Das, T. K., & Tyagi, J. S. (2004). DevR-DevS is a bona fide two-component system of *Mycobacterium tuberculosis* that is hypoxia-responsive in the absence of the DNA-binding domain of DevR. *Microbiology (Reading, England)*, *150*, 865–875.

Sarand, I., Osterberg, S., Holmqvist, S., Holmfeldt, P., Skärfstad, E., Parales, R. E., et al. (2008). Metabolism-dependent taxis towards (methyl)phenols is coupled through the most abundant of three polar localized Aer-like proteins of *Pseudomonas putida*. *Environmental Microbiology*, *10*, 1320–1334.

Sardiwal, S., Kendall, S. L., Movahedzadeh, F., Rison, S. C., Stoker, N. G., & Djordjevic, S. A. (2005). GAF domain in the hypoxia/NO-inducible *Mycobacterium tuberculosis* DosS protein binds haem. *Journal of Molecular Biology*, *353*, 929–936.

Sasakura, Y., Yoshimura-Suzuki, T., Kurokawa, H., & Shimizu, T. (2006). Structure-function relationships of EcDOS, a heme-regulated phosphodiesterase from *Escherichia coli*. *Accounts of Chemical Research*, *39*, 37–43.

Sato, A., Sasakura, Y., Sugiyama, S., Sagami, I., Shimizu, T., Mizutani, Y., et al. (2002). Stationary and time-resolved resonance Raman spectra of His77 and Met95 mutants of the isolated heme domain of a direct oxygen sensor from *Escherichia coli*. *The Journal of Biological Chemistry*, *277*, 32650–32658.

Sawai, H., Sugimoto, H., Shiro, Y., Ishikawa, H., Mizutani, Y., & Aono, S. (2012). Structural basis for oxygen sensing and signal transduction of the heme-based sensor protein Aer2 from *Pseudomonas aeruginosa*. *Chemical Communications*, *48*, 6523–6525.

Sawai, H., Yamanaka, M., Sugimoto, H., Shiro, Y., & Aono, S. (2012). Structural basis for the transcriptional regulation of heme homeostasis in *Lactococcus lactis*. *The Journal of Biological Chemistry*, *287*, 30755–30768.

Sawai, H., Yoshioka, S., Uchida, T., Hyodo, M., Hayakawa, Y., Ishimori, K., et al. (2009). Molecular oxygen regulates the enzymatic activity of a heme-containing diguanylate cyclase (HemDGC) for the synthesis of cyclic di-GMP. *Biochimica et Biophysica Acta*, *1804*, 166–172.

Sciotti, M. A., Chanfon, A., Hennecke, H., & Fischer, H. M. (2003). Disparate oxygen responsiveness of two regulatory cascades that control expression of symbiotic genes in *Bradyrhizobium japonicum*. *Journal of Bacteriology*, *185*, 5639–5642.

Seshasayee, A. S., Fraser, G. M., & Luscombe, N. M. (2010). Comparative genomics of cyclic-di-GMP signaling in bacteria: Post-translational regulation and catalytic activity. *Nucleic Acids Research*, *38*, 5970–5981.

Silva, M. A., Lucas, T. G., Salgueiro, C. A., & Gomes, C. M. (2012). Protein folding modulates the swapped dimerization mechanism of methyl-accepting chemotaxis heme sensors. *PLoS One*, *7*, e46328. http://dx.doi.org/10.1371/journal.pone.0046328.

Smith, A. T., Marvin, K. A., Freeman, K. M., Kerby, R. L., Roberts, G. P., & Burstyn, J. N. (2012). Identification of Cys94 as the distal ligand to the Fe(III) heme in the transcriptional regulator RcoM-2 from *Burkholderia xenovorans*. *Journal of Biological Inorganic Chemistry*, *17*, 1071–1082.

Soupène, E., Foussard, M., Boistard, P., Truchet, G., & Batut, J. (1995). Oxygen as a key developmental regulator of *Rhizobium meliloti* N_2-fixation gene expression within the alfalfa root nodule. *Proceedings of the National Academy of Sciences of the United States of America*, *92*, 3759–3763.

Sousa, E. H., Gonzalez, G., & Gilles-Gonzalez, M. A. (2005). Oxygen blocks the reaction of the FixL-FixJ complex with ATP but does not influence binding of FixJ or ATP to FixL. *Biochemistry*, *44*, 15359–15365.

Sousa, E. H., Tuckerman, J. R., Gondim, A. C., Gonzalez, G., & Gilles-Gonzalez, M. A. (2013). Signal transduction and phosphoryl transfer by a FixL hybrid kinase with low oxygen affinity: Importance of the vicinal PAS domain and receiver aspartate. *Biochemistry*, *52*, 456–465.

Sousa, E. H., Tuckerman, J. R., Gonzalez, G., & Gilles-Gonzalez, M. A. (2007). DosT and DevS are oxygen-switched kinases in *Mycobacterium tuberculosis*. *Protein Science*, *16*, 1708–1719.

Stock, J. B., Lukat, G. S., & Stock, A. M. (1991). Bacterial chemotaxis and the molecular logic of intracellular signal transduction networks. *Annual Review of Biophysics and Biophysical Chemistry*, *20*, 109–136.

Szurmant, H., & Ordal, G. W. (2004). Diversity in chemotaxis mechanisms among the bacteria and archaea. *Microbiology and Molecular Biology Reviews*, *68*, 301–319.

Tamayo, R., Pratt, J. T., & Camilli, A. (2007). Roles of cyclic diguanylate in the regulation of bacterial pathogenesis. *Annual Review of Microbiology*, *61*, 131–148.

Tan, E., Rao, F., Pasunooti, S., Pham, T. H., Soehano, I., Turner, M. S., et al. (2013). Solution structure of the PAS domain of a thermophilic YybT homolog reveals a potential ligand-binding site. *The Journal of Biological Chemistry*, *288*, 11949–11959.

Tanaka, A., Nakamura, H., Shiro, Y., & Fujii, H. (2006). Roles of the heme distal residues of FixL in O_2 sensing: A single convergent structure of the heme moiety is relevant to the downregulation of kinase activity. *Biochemistry*, *45*, 2515–2523.

Tanaka, A., & Shimizu, T. (2008). Ligand binding to the Fe(III)-protoporphyrin IX complex of phosphodiesterase from *Escherichia coli* (Ec DOS) markedly enhances catalysis of cyclic di-GMP: Roles of Met95, Arg97, and Phe113 of the putative heme distal side in catalytic regulation and ligand binding. *Biochemistry*, *47*, 13438–13446.

Tanaka, A., Takahashi, H., & Shimizu, T. (2007). Critical role of the heme axial ligand, Met95, in locking catalysis of the phosphodiesterase from *Escherichia coli* (Ec DOS) toward cyclic diGMP. *The Journal of Biological Chemistry*, *282*, 21301–21307.

Taylor, B. L., Rebbapragada, A., & Johnson, M. S. (2001). The FAD-PAS domain as a sensor for behavioral responses in *Escherichia coli*. *Antioxidants & Redox Signaling*, *3*, 867–879.

Thijs, L., Vinck, E., Bolli, A., Trandafir, F., Wan, X., Hoogewijs, D., et al. (2007). Characterization of a globin-coupled oxygen sensor with a gene-regulating function. *The Journal of Biological Chemistry*, *282*, 37325–37340.

Tomita, T., Gonzalez, G., Chang, A. L., Ikeda-Saito, M., & Gilles-Gonzalez, M. A. (2002). A comparative resonance Raman analysis of heme-binding PAS domains: Heme iron coordination structures of the *Bj*FixL, AxPDEA1, EcDos, and MtDos proteins. *Biochemistry*, *41*, 4819–4826.

Tuckerman, J. R., Gonzalez, G., Dioum, E. M., & Gilles-Gonzalez, M. A. (2002). Ligand and oxidation-state specific regulation of the heme-based oxygen sensor FixL from *Sinorhizobium meliloti*. *Biochemistry*, *41*, 6170–6177.

Tuckerman, J. R., Gonzalez, G., & Gilles-Gonzalez, M. A. (2001). Complexation precedes phosphorylation for two-component regulatory system FixL/FixJ of *Sinorhizobium meliloti*. *Journal of Molecular Biology*, *308*, 449–455.

Tuckerman, J. R., Gonzalez, G., Sousa, E. H., Wan, X., Saito, J. A., Alam, M., et al. (2009). An oxygen-sensing diguanylate cyclase and phosphodiesterase couple for c-di-GMP control. *Biochemistry*, *48*, 9764–9774.

Uchida, T., Sagami, I., Shimizu, T., Ishimori, K., & Kitagawa, T. (2012). Effects of the bHLH domain on axial coordination of heme in the PAS-A domain of neuronal PAS domain protein 2 (NPAS2): Conversion from His119/Cys170 coordination to His119/His171 coordination. *Journal of Inorganic Biochemistry, 108*, 188–195.

Uchida, T., Sato, E., Sato, A., Sagami, I., Shimizu, T., & Kitagawa, T. (2005). CO-dependent activity-controlling mechanism of heme-containing CO-sensor protein, neuronal PAS domain protein 2. *The Journal of Biological Chemistry, 280*, 21358–21368.

van Wilderen, L. J., Key, J. M., Van Stokkum, I. H., van Grondelle, R., & Groot, M. L. (2009). Dynamics of carbon monoxide photodissociation in *Bradyrhizobium japonicum* FixL probed by picosecond midinfrared spectroscopy. *The Journal of Physical Chemistry. B, 113*, 3292–3297.

Voordouw, G. (2002). Carbon monoxide cycling by *Desulfovibrio vulgaris* Hildenborough. *Journal of Bacteriology, 184*, 5903–5911.

Wan, X., Tuckerman, J. R., Saito, J. A., Freitas, T. A., Newhouse, J. S., Denery, J. R., et al. (2009). Globins synthesize the second messenger bis-(3'-5')-cyclic diguanosine monophosphate in bacteria. *Journal of Molecular Biology, 388*, 262–270.

Watts, K. J., Taylor, B. L., & Johnson, M. S. (2011). PAS/poly-HAMP signaling in Aer-2, a soluble haem-based sensor. *Molecular Microbiology, 79*, 686–699.

Williams, P. A., Fülöp, V., Garman, E. F., Saunders, N. F., Ferguson, S. J., & Hajdu, J. (1997). Haem-ligand switching during catalysis in crystals of a nitrogen-cycle enzyme. *Nature, 389*, 406–412.

Winkler, W. C., Gonzalez, G., Wittenberg, J. B., Hille, R., Dakappagari, N., Jacob, A., et al. (1996). Nonsteric factors dominate binding of nitric oxide, azide, imidazole, cyanide, and fluoride to the rhizobial heme-based oxygen sensor FixL. *Chemistry & Biology, 3*, 841–850.

Yamada, S., Sugimoto, H., Kobayashi, M., Ohno, A., Nakamura, H., & Shiro, Y. (2009). Structure of PAS-linked histidine kinase and the response regulator complex. *Structure, 17*, 1333–1344.

Yamamoto, K., Ishikawa, H., Takahashi, S., Ishimori, K., Morishima, I., Nakajima, H., et al. (2001). Binding of CO at the Pro2 side is crucial for the activation of CO-sensing transcriptional activator CooA. ^{1}H NMR spectroscopic studies. *The Journal of Biological Chemistry, 276*, 11473–11476.

Yeh, S. R., Couture, M., Ouellet, Y., Guertin, M., & Rousseau, D. L. (2000). A cooperative oxygen binding hemoglobin from *Mycobacterium tuberculosis*. Stabilization of heme ligands by a distal tyrosine residue. *The Journal of Biological Chemistry, 275*, 1679–1684.

Yoshioka, S., Kobayashi, K., Yoshimura, H., Uchida, T., Kitagawa, T., & Aono, S. (2005). Biophysical properties of a c-type heme in chemotaxis signal transducer protein DcrA. *Biochemistry, 44*, 15406–15413.

Yukl, E. T., Ioanoviciu, A., Nakano, M. M., Ortiz de Montellano, P. R., & Moënne-Loccoz, P. (2008). A distal tyrosine residue is required for ligand discrimination in DevS from *Mycobacterium tuberculosis*. *Biochemistry, 47*, 12532–12539.

Yukl, E. T., Ioanoviciu, A., Ortiz de Montellano, P. R., & Moënne-Loccoz, P. (2007). Interdomain interactions within the two-component heme-based sensor DevS from *Mycobacterium tuberculosis*. *Biochemistry, 46*, 9728–9736.

Yukl, E. T., Ioanoviciu, A., Sivaramakrishnan, S., Nakano, M. M., Ortiz de Montellano, P. R., & Moënne-Loccoz, P. (2011). Nitric oxide dioxygenation reaction in DevS and the initial response to nitric oxide in *Mycobacterium tuberculosis*. *Biochemistry, 50*, 1023–1028.

CHAPTER EIGHT

The Globins of Cold-Adapted *Pseudoalteromonas haloplanktis* TAC125: From the Structure to the Physiological Functions

Daniela Giordano[*], Daniela Coppola[*], Roberta Russo[*],
Mariana Tinajero-Trejo[†], Guido di Prisco[*], Federico Lauro[‡],
Paolo Ascenzi[*,§], Cinzia Verde[*,¶,1]

[*]Institute of Protein Biochemistry, CNR, Naples, Italy
[†]Department of Molecular Biology & Biotechnology, The University of Sheffield, Sheffield, United Kingdom
[‡]School of Biotechnology & Biomolecular Sciences, The University of New South Wales, Sydney, New South Wales, Australia
[§]Interdepartmental Laboratory for Electron Microscopy, University Roma 3, Rome, Italy
[¶]Department of Biology, University Roma 3, Rome, Italy
[1]Corresponding author: e-mail address: c.verde@ibp.cnr.it

Contents

1. The Polar Environments   331
2. Phylogeny and Biogeography of Cold-Adapted Marine Microorganisms   334
3. The Role of Temperature in Evolutionary Adaptations   337
4. Bacterial Globins   340
    4.1 Flavohaemoglobins   341
    4.2 Single-domain globins   344
    4.3 Truncated haemoglobins   345
5. The Antarctic Marine Bacterium *Pseudoalteromonas haloplanktis* TAC125: A Case Study   348
    5.1 General aspects   348
    5.2 Genomic and post-genomic insights   350
    5.3 Excess of O_2 and metabolic constraints   355
    5.4 Biotechnological applications   357
6. *P. haloplanktis* TAC125 Globins   358
    6.1 General aspects   358
    6.2 Structure–function relationships of *Ph*-2/2HbO   360
7. Conclusion and Perspectives   372
Acknowledgements   374
References   374

Advances in Microbial Physiology, Volume 63
ISSN 0065-2911
http://dx.doi.org/10.1016/B978-0-12-407693-8.00008-X

329

Abstract

Evolution allowed Antarctic microorganisms to grow successfully under extreme conditions (low temperature and high O_2 content), through a variety of structural and physiological adjustments in their genomes and development of programmed responses to strong oxidative and nitrosative stress. The availability of genomic sequences from an increasing number of cold-adapted species is providing insights to understand the molecular mechanisms underlying crucial physiological processes in polar organisms. The genome of *Pseudoalteromonas haloplanktis* TAC125 contains multiple genes encoding three distinct truncated globins exhibiting the 2/2 α-helical fold. One of these globins has been extensively characterised by spectroscopic analysis, kinetic measurements and computer simulation. The results indicate unique adaptive structural properties that enhance the overall flexibility of the protein, so that the structure appears to be resistant to pressure-induced stress. Recent results on a genomic mutant strain highlight the involvement of the cold-adapted globin in the protection against the stress induced by high O_2 concentration. Moreover, the protein was shown to catalyse peroxynitrite isomerisation *in vitro*. In this review, we first summarise how cold temperatures affect the physiology of microorganisms and focus on the molecular mechanisms of cold adaptation revealed by recent biochemical and genetic studies. Next, since only in a very few cases the physiological role of truncated globins has been demonstrated, we also discuss the structural and functional features of the cold-adapted globin in an attempt to put into perspective what has been learnt about these proteins and their potential role in the biology of cold-adapted microorganisms.

ABBREVIATIONS

CAP cold-adapted protein
CSP cold-shock protein
Cygb cytoglobin
FHb flavohaemoglobin
Hb haemoglobin
Hmp FHb from *E. coli*
HSP heat-shock protein
Mb myoglobin
Ngb neuroglobin
***Ph*TAC125** *Pseudoalteromonas haloplanktis* TAC125
PTS phosphoenolpyruvate-dependent phosphotransferase system
RNS reactive nitrogen species
ROS reactive oxygen species
SDgb single-domain globin
TBDTs TonB-dependent transport systems
TF trigger factor

1. THE POLAR ENVIRONMENTS

Although many cold-ocean species have been studied, we still have limited knowledge about adaptation to low temperatures in sea water. In the light of the ongoing climate change, there is growing interest in polar marine organisms and how they have evolved at constantly cold temperatures. The rate of the impact of current climate change in relation to the capacity of species to acclimate or adapt is a crucial study area for managing polar ecosystems in the future.

While the early biological works focussed on specific aspects of molecular adaptation of single genes and proteins to cold, progress in genomics and postgenomics, as well as the availability of genomic sequences of several model species, allowed highlighting the adaptation mechanisms permitting species evolution in polar regions (Peck, 2011; Somero, 2010).

The planet is currently losing sea ice, most notably in the Arctic region, because of warming trends over the last century (Moritz, Bitz, & Steig, 2002). The dramatic sea-ice decrease is progressing from the Barents and Bering Seas to the central Arctic Ocean.

Isolation of the two polar oceans has occurred to a different extent. The land masses surrounding the Arctic Ocean have partially limited water exchanges with other oceans for the last 60 million years or so. The Southern Ocean was isolated much more thoroughly by the Antarctic Circumpolar Current, the strongest current system in the world, since 20–40 million years ago (Eastman, 2005).

Geography, oceanography and biology of species inhabiting Arctic and Antarctic polar regions have often been intercompared (see Dayton, Mordida, & Bacon, 1994) to detect and outline differences between the two ecosystems. The northern polar region is characterised by extensive, shallow shelf sea areas of the land masses that surround a partially land-locked ocean; in contrast, the Antarctic region comprises a dynamic open ocean that surrounds the continent, and a continental shelf (Smetacek & Nicol, 2005) that is very deep because of the enormous weight exerted on the continent by the covering ice sheet, which has a thickness of 2–4000 m. Although the climate drivers acting on the biota are relatively similar, the two polar environments are quite different from each other. One of the main differences is the freshwater supply. Arctic surface waters are modified by the input of large rivers that influence the nutrient regimes and their differences. In the Southern Ocean, the influence of freshwater, mostly from glaciers, is much smaller.

The Arctic is currently experiencing some of the most rapid and severe climate changes on Earth. Some effects of increasing temperature on marine ecosystems are already evident (Rosenzweig et al., 2008). Over the next 100 years, warming is expected to accelerate, contributing to major physical, ecological, social and economic changes. In fact, all models forecast reduction of cold areas and expansion of the warmer ones, with consequent threat for cold-adapted organisms. However, there is still little understanding regarding how the loss of a species or groups of species will affect ecosystem services.

The most important Arctic characters include seasonality in light, cold temperatures with winter extremes, and extensive shelf seas around a deep central ocean basin. The Arctic comprises a vast ocean surrounded by the northern coasts of three continents, open to influx of warm water from the Atlantic and, to a lesser extent, from the Pacific. The "permanent cap" of ice, composed of multi- and first-year ice that forms annually and extends and retreats seasonally, is probably the most important feature of Arctic marine systems (Polyak et al., 2010). The major decline in sea ice that began to take place in the Arctic since 2000 is the most important climate-change signal (Comiso, Parkinson, Gersten, & Stock, 2008). Expectations are that summer sea ice will continue to decline in the future. Climate models have indicated that the Arctic Ocean might essentially be ice-free during summer by the later half of the twenty-first century (Overpeck et al., 2006; Wang & Overland, 2009), with dramatic and potentially devastating effects on a number of species associated with the sea ice (Moline et al., 2008) and significant biological consequences (Clarke et al., 2007).

In terms of constantly low temperatures, the southern polar environment is considered the most extreme on our planet. Antarctica has the capacity to influence the Earth's climate and ocean-ecosystem function, and from this standpoint, it is the world's most important continent. As such, its palaeo- and current geological and climatic history, physical and biological oceanography, as well as marine and terrestrial ecosystems have been— and are—the target of a wealth of studies. As a frozen deep mass of ice, most of Antarctica reflects the sun's radiation, buffering global warming trends. The current lack of warming is also due to the shielding effect produced by the stratospheric winds driven by the human-induced Ozone Hole. These winds generate a Polar Vortex extending to the surface that acts as a strong barrier, keeping warm and moist air away (Turner et al., 2009). When and if the Ozone Hole closes, such a protection will no longer be efficient, and warming caused by greenhouse gases, another anthropogenic contribution, will take the lead. However, shielding by the Polar Vortex is not taking place in the Antarctic Peninsula, especially on the western side. Similar to the Arctic, the Peninsula is thus experiencing one of the fastest rates of warming on the planet (Convey et al., 2009; Turner et al., 2009). The consequences are already being seen on land (Convey, 2006, 2010), where ice loss leads to new land becoming available for rapidly occurring colonisation. Reduction of sea ice causes displacement of key invertebrate and fish species, whose reproductive processes, closely associated with sea ice, are upset (Moline et al., 2008). Such migrations have consequences on the whole food chain.

The Southern Ocean is the planet's fourth largest ocean. Over geological time, environmental conditions and habitats in the Antarctic have changed dramatically. Two key events allowed the establishment of the powerful Antarctic Circumpolar Current: (i) the opening of the Tasman Seaway, which occurred approximately 32–35 million years ago, according to tectonics and marine geology (Kennett, 1977; Lawver & Gahagan, 2003); and (ii) the opening of the Drake Passage, between South America and the Antarctic Peninsula, dated between 40 and 17 million years ago (Scher & Martin, 2006; Thomson, 2004). The Antarctic Polar Front, a roughly circular oceanic feature running between 50°S and 60°S, is the northern boundary of the Circumpolar Current. Along the Front, the surface layers of the north-moving Antarctic waters sink beneath the less cold and less dense sub-Antarctic waters, generating almost permanent turbulence. Just north of the Front, the surface water temperature is ca. 2–3 °C warmer. Separating warm northern waters from cold southern waters, the Front acts as a barrier for migration of marine organisms between Antarctica and the lower latitudes.

Thus, Antarctica is a closed system, shielded by the Antarctic Circumpolar Current from the influence of waters from latitudes lower than 60° (Eastman, 2005). These conditions promoted adaptive evolution to low temperature and extreme seasonality to develop in isolation, and led to the current composition of the Antarctic marine biota (Eastman, 2005). The earliest cold-climate marine fauna is thought to date from the late Eocene–Oligocene (about 35 million years ago). The water column south of the Front is close to O_2 saturation at all depths. O_2 solubility increases with temperature decrease, thus the cold seas are an O_2-rich habitat.

The dominant feature of the modern continent is its ice sheet, covering most of the land area. In the sea, during wintertime, ice coverage currently extends from the Antarctic coastline northward to approximately 60°S.

Antarctic marine environments are thus considered the most extreme on Earth in combining the globally lowest and most stable temperatures with the highest O_2 content, and at the same time great variability in light intensity, ice cover and phytoplankton productivity (Peck, Convey, & Barnes, 2006). Antarctic marine habitats include sea water and sediments at near -1 °C and the sea ice, where internal fluids remain liquid to -35 °C during winter.

2. PHYLOGENY AND BIOGEOGRAPHY OF COLD-ADAPTED MARINE MICROORGANISMS

Global warming is expected to increase the microbial activity and decrease the availability of energy and food for organisms that are at higher levels in the food chain (Kirchman, Morán, & Ducklow, 2009). The role of microorganisms in polar waters is essential. Microbial processes in polar ecosystems are highly sensitive to small environmental changes and influence ecosystem functioning. Cold marine environments are colonised by a wide diversity of microorganisms including bacteria, archaea, yeasts, fungi and algae (Margesin & Miteva, 2011; Murray & Grzymski, 2007). Sea-ice microbial communities at the two poles display closely related organisms (Brinkmeyer et al., 2003) and are continually seeded by alien microorganisms, including mesophilic species that contribute to the potential environmental pool of DNA (Cowan et al., 2011).

In the past, understanding bacterial, archaeal and viral diversity in polar marine environments has been somewhat impaired by the difficulty of accessing sampling locations year-round. In comparison, rapid advances in microbial ecological theory have been achieved through results from temperate oceans, particularly, with respect to long-term microbial observatories such as BATS (Bermuda Atlantic Time Series) (Steinberg et al., 2001) and Station HOT (Hawaii Ocean Time Series) (DeLong et al., 2006).

Some of the drawbacks in data acquisition have been overcome with the advent of culture-independent molecular tools and large-scale community sequencing. Studies based on denaturing gradient gel electrophoresis (Abell & Bowman, 2005; Giebel, Brinkhoff, Zwisler, Selje, & Simon, 2009) have revolutionised our views of diversity in Antarctic waters, making the Southern Ocean one of the focal regions of microbial ecology.

However, the applicability of these investigations to the peculiar characteristics of the polar microbiota is largely unknown. Deeper understanding is necessary in light of the important role that polar waters play in global carbon cycling. The microbial component represents up to 90% of cellular DNA (Paul, Jeffrey, & DeFlaun, 1985) and is estimated to be responsible for up to 80% of the primary carbon production (Douglas, 1984; Ducklow, 1999; Li et al., 1983) and for most of the carbon flux between the sea water and the atmosphere (Azam, 1998; Azam & Malfatti, 2007).

Environmental genomics revealed that heterotrophic bacteria play a key role in controlling carbon fluxes within oceans. These bacteria dominate biogeochemical cycles and are part of the microbial loop which, at least in part, causes the response of oceanic ecosystems to climate change (Kirchman et al., 2009). Recent diversity studies that employed sequencing of ribosomal RNA genes (Galand, Casamayor, Kirchman, & Lovejoy, 2009; Ghiglione & Murray, 2012; Kirchman, Cottrell, & Lovejoy, 2010), and metagenomics and metaproteomics (Grzymski et al., 2012; Wilkins et al., 2013; Williams et al., 2012, 2013) have clarified some of the aspects of the interactions between microorganisms and the polar environment, which is unique in terms of environmental parameters such as temperature, day length and trophic interactions.

Antarctic marine waters harbour taxa of heterotrophic microbes similar to those found in temperate and tropical waters. Among these, the most dominant are α-Proteobacteria and, in particular, specific phylotypes of SAR11 (Brown et al., 2012), γ-Proteobacteria, Flavobacteria and ammonia-oxidising Marine Group I Crenarchaeota (Grzymski et al., 2012; Wilkins et al., 2013; Williams et al., 2012). However, the emerging view is that, while the taxa present might be distributed worldwide, there are clear signatures of allopatric speciation, which are only evident at a finer phylogenetic scale (Brown et al., 2012).

The geographic separation necessary for such evolutionary events is provided by sharp transitions in chemicophysical parameters that mark and isolate water masses (Agogué, Lamy, Neal, Sogin, & Herndl, 2011). The Antarctic Polar Front provides one of the most dramatic examples of such transitions. Here, the water drops ∼3 °C in temperature over a space of less than 30 miles which results in abrupt shifts in the microbial community composition and functional gene distribution (Wilkins et al., 2013).

Moreover, certain taxa become transiently dominant in response to particular seasonal changes in environmental parameters such as the Marine Group I Crenarchaeota which show a dramatic increase in relative abundance and activity during the winter in the Antarctic Peninsula (Grzymski et al., 2012; Williams et al., 2012). Similarly, bacterial clades within the Rhodobacteraceae, uncultivated γ-Proteobacteria and Bacteriodetes show large seasonal variations between samples from summer and winter from both the Antarctic Peninsula and the sub-Antarctic Kerguelen Islands (Ghiglione & Murray, 2012). On top of these oscillations, over shorter time and spatial scales, Flavobacteria can become dominant in response to algal blooms (Ghiglione & Murray, 2012; Grzymski et al., 2012; Williams et al., 2013).

The composition of the sea-ice microbiota is also unique as a result of its seasonal nature and physicochemical environment (Bowman et al., 2012; Brown & Bowman, 2001). It has an important role in providing the "seed populations" for the productive springtime microbial communities. It is still unclear whether the selective pressure within the winter sea ice generates significant genetic bottlenecks on different microbial species (Connelly, Tilburg, & Yager, 2006; Junge, Imhoff, Staley, & Deming, 2002). What is clear is that, when compared to surrounding sea water, the species richness in sea ice is lower than in surrounding waters (Bowman et al., 2012), which in turn has been shown to decrease when moving from lower to higher latitudes (Sul, Oliver, Ducklow, Amaral-Zettler, & Sogin, 2013). This lower richness might not provide enough resilience in case of future climatic changes.

Compared to the heterotrophic community, the latitudinal distribution and temporal variation of primary producers are even more extreme. Cyanobacteria, for example, *Synechococcus* sp. and *Prochlorococcus* sp., are fundamental in carbon fixation and responsible for more than half of primary production in oligotrophic ocean waters (Liu et al., 1998; Liu, Nolla, & Campbell, 1997). A consistent trend is the progressive disappearance of *Prochlorococcus* populations south of the Polar Front and the appearance of specific clades of *Synechococcus* which dominate at higher latitudes (Scanlan et al., 2009). This trend holds true at both poles. In fact, bipolar distribution of organisms is the rule rather than the exception amongst microbial taxa (Sul et al., 2013). In microeukaryotes, the observation of pheromone cross signalling amongst Arctic and Antarctic strains of the polar protozoan ciliate *Euplotes nobilii* suggests mechanisms for recent genetic exchange (Di Giuseppe et al., 2011). If associated with the strong bipolar biogeographical patterns, this could be true for all classes of organisms living at low Reynolds numbers, with the caveat that deep-sea currents, in particular those associated with thermohaline circulation, are allowing an ongoing genetic exchange between the poles (Lauro, Chastain, Blankenship, Yayanos, & Bartlett, 2007).

Complementing these culture-independent studies, the last few years have seen an increase in genomic sequences of cultured isolates. The study of individual genomes facilitates the characterisation of physiological adaptations to the specific polar conditions. Nevertheless, in contrast with the large diversity observed with molecular techniques, the phylogenetic breadth of the taxa with at least one representative genome sequence is limited to a few genera (Fig. 8.1).

In view of the high degree of temporal and spatial variability observed in polar environments, which positively correlates with changes in microbial community structure and function, there is a pressing need for increasing culturing efforts and single-cell genomic analysis targeted at under-represented phyla. These should be integrated within the larger framework of global organismal biogeography and ocean models.

3. THE ROLE OF TEMPERATURE IN EVOLUTIONARY ADAPTATIONS

The bulk of the Earth's biosphere is cold (e.g. 90% of the ocean is below 5 °C), sustaining a broad diversity of microbial life. Evolution under extreme conditions has been marked by a suite of adaptations (evolutionary gains) including the development of proteins that function optimally in the cold. A commonly accepted view for protein cold adaptation is the activity/stability/flexibility relationships. Although active sites are generally highly conserved among homologous proteins, adaptive changes may occur at recognition site(s). These alterations in the strength of subunit interactions may affect thermal stability and energy changes associated with conformational transitions due to ligand binding (D'Amico, Collins, Marx, Feller, & Gerday, 2006).

Comparative genome analysis indicates that the cold-adapted lifestyle is generally conferred by a collection of changes in the overall genome content and composition. The flexible structures of enzymes from cold-adapted bacteria compensate for the environment's low kinetic energy.

In cold environments, challenges to cellular function and structural integrity include low rates of transcription, translation and cell division, inappropriate protein folding and cold denaturation, as well as intracellular ice formation (D'Amico et al., 2006). The ability of an organism to survive and grow in the cold is dependent on a number of adaptive strategies (Table 8.1) to maintain vital cellular functions at cold temperatures (Rodrigues & Tiedje, 2008). These strategies include the synthesis of

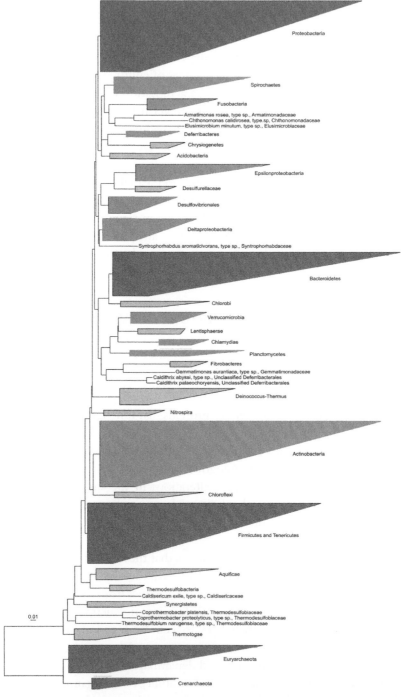

Figure 8.1 Bacterial and Archaeal phylogenetic tree adapted from the Silva comprehensive ribosomal RNA database (http://www.arb-silva.de/). Blue: phyla containing at least one sequenced polar genome. Green: phyla containing polar isolates or numerically relevant in culture-independent surveys of polar regions. (For interpretation of the references to colour in this figure legend, the reader is referred to the online version of this chapter.)

Table 8.1 Molecular adaptations in cold-adapted bacteria

Molecular adaptations	Explanation or consequence	Reference
Protection against reactive oxygen species (ROS): lower frequency of oxidisable amino acids; oxidoreductases, superoxide dismutases, catalases, peroxidases	Due to increased solubility of O_2 at low temperatures forming increased ROS	Rabus et al. (2004), Médigue et al. (2005), Methé et al. (2005), Bakermans et al. (2007), Duchaud et al. (2007), Ayub, Tribelli, and Lopez (2009) and Piette et al. (2010)
Enzymes	Maintain catalytic efficiency at low temperatures	Georlette et al. (2004)
Membranes: increased unsaturation and decreased chain length of fatty acids, carotenoids, desaturases	Increase the fluidity of membranes	Jagannadham, Rao, and Shivaji (1991), Chauhan and Shivaji (1994) and Ray et al. (1998)
Synthesis of specific elements: cold-shock proteins, molecular chaperones, compatible solutes	Maintain vital cellular functions at cold temperatures	Motohashi, Watanabe, Yohda, and Yoshida (1999), Cavicchioli, Thomas, and Curmi (2000), Watanabe and Yoshida (2004) and Pegg (2007)
Molecular mechanisms involved in protein flexibility	**Consequence**	**Reference**
Decreased number of H bonds and salt bridges	Increased flexibility	Feller and Gerday (1997)
Reduced proline and arginine content	Increased molecular entropy	Ray et al. (1998), Russell (2000), D'Amico et al. (2002), Cavicchioli, Siddiqui, Andrews, and Sowers (2002) and Rodrigues and Tiedje (2008)
Reduced frequency of surface, inter-domain and inter-subunit ionic linkages and ion-network	Increased conformational flexibility and reduced enthalpic contribution to stability	D'Amico et al. (2006)

Adapted from Casanueva, Tuffin, Cary, and Cowan (2010).

cold-shock proteins (CSPs) (Cavicchioli et al., 2000) and molecular chaperones (Motohashi et al., 1999), solutes (Pegg, 2007) and structural modifications for maintaining membrane fluidity (Chintalapati, Kiran, & Shivaji, 2004; Russell, 1998). Moreover, cold-adapted organisms must develop an effective and intricate network of defence mechanisms against oxidative stress: an increasing number of oxidoreductases, superoxide dismutases, catalases and peroxidases can be seen in this perspective (Ayub et al., 2009; Bakermans et al., 2007; Duchaud et al., 2007; Médigue et al., 2005; Methé et al., 2005; Piette et al., 2010; Rabus et al., 2004).

In addition to adaptations at the cellular level, a key adaptive strategy is the maintenance of adequate reaction rates at thermal extremes; therefore, adequate features of catalytic processes become crucial. Enzyme catalytic rates at low temperatures depend on increased protein flexibility and concomitant increase in thermolability (Georlette et al., 2004).

In this respect, a suite of factors contribute to maintaining enzyme molecular flexibility in all cold-adapted organisms. In bacterial enzymes, Ser, Asp, Thr and Ala are over-represented in the coil regions of secondary structures. On the other hand, in the helical regions, aliphatic, basic, aromatic and hydrophilic residues are generally under-represented (Cavicchioli et al., 2002; D'Amico et al., 2002; Ray et al., 1998; Rodrigues & Tiedje, 2008; Russell, 2000). Moreover, a reduction of surface, inter-domain, inter-subunit ionic linkages and a decreased number of hydrogen bonds and salt bridges are key mechanisms to induce an increasing of conformational flexibility of psychrophilic enzymes (D'Amico et al., 2006; Feller & Gerday, 1997).

Cold adaptation is also strongly linked to the capacity of the organism to sense temperature changes, perhaps by virtue of mechanisms linked with the lipid composition of the cell membrane and alterations in the DNA and RNA topology. The latter may enhance (or halt) the replication, transcription and translation processes (Eriksson, Hurme, & Rhen, 2002). Although increased unsaturation and decreased chain length of fatty acids are the major modifications of cell membranes, other membrane-associated components may well play important roles in adaptation to low temperatures (Jagannadham et al., 1991; Ray et al., 1998). Studies of Antarctic psychrotrophic bacteria *in vitro* have shown that carotenoids may have a function in buffering membrane fluidity (Jagannadham et al., 1991).

4. BACTERIAL GLOBINS

The traditional view of the exclusive role of haemoglobin (Hb) as O_2 carrier in vertebrates is obsolete. The discovery of globin genes in prokaryotic and eukaryotic microorganisms, including bacteria, yeasts, algae, protozoa and fungi, suggests that the globin superfamily is exceptionally flexible in terms of biological roles and possible functions. The number of "globin-like" proteins is currently increasing as different genomes from microorganisms are sequenced and annotated (Vinogradov et al., 2005; Vinogradov & Moens, 2008; Vinogradov, Tinajero-Trejo, Poole, & Hoogewijs, 2013).

Non-vertebrate globins display high variability in primary and tertiary structures, which probably indicates their adaptations to additional functions with respect to their vertebrate homologues (Vinogradov et al., 2005; Vinogradov & Moens, 2008; Vinogradov et al., 2013).

The most recent bioinformatic survey of globin-like sequences in prokaryotic genomes revealed that over half of the more than 2200 bacterial genomes sequenced so far contain putative globins (Vinogradov et al., 2013). A new global nomenclature including prokaryotic and eukaryotic globins has been proposed, and globins have been classified within three families: (i) myoglobin (Mb)-like family (M) (displaying the classical three-on-three (3/3) α-helical sandwich motif) containing flavohaemoglobins (FHbs) and single-domain globins (SDgb); (ii) sensor globins family (S); and (iii) truncated haemoglobins family (T), showing the two-on-two (2/2) α-helical sandwich motif (Vinogradov et al., 2013).

Although there are still some uncertainties about the evolutionary relationship between these three classes of microbial globins, it has been proposed that prokaryotic and eukaryotic globins, including vertebrate α/β globins, Mb, neuroglobin (Ngb), cytoglobin (Cygb), and invertebrate and plant Hbs, emerged from a common ancestor (Vinogradov et al., 2005).

4.1. Flavohaemoglobins

Chimeric globins seem to have kept their original enzymatic functions in prokaryotes, plants and some unicellular eukaryotes. Therefore, the FHb sub-family has been the only one able to adapt to different functions more extensively than the other two globin families. Moreover, the presence of Hbs in unicellular organisms suggests that O_2 transport in metazoans is a relatively recent evolutionary acquisition and that the early Hb functions have been enzymatic and O_2 sensing (Vinogradov & Moens, 2008).

FHbs are widely present in bacteria, yeasts and fungi and belong to the ferredoxin reductase-like protein family. They consist of an N-terminal haem-globin domain fused with C-terminal reductase domain binding NAD(P)H and FAD (Bolognesi, Bordo, Rizzi, Tarricone, & Ascenzi, 1997; Bonamore & Boffi, 2008).

Sequence alignments in Gram-negative and Gram-positive bacteria and unicellular eukaryotes indicate that the FHbs family is a very homogeneous group of proteins that share highly conserved active sites in both domains. Amino acid residues building up the haem domain and the flavin-binding region are widely conserved, and the architecture of FHb domains is closely similar to those of globins and flavodoxin-reductase proteins. This finding suggests that the haem domain displays globin-like functional properties and that the flavin moiety acts as an electron-transfer module from NADH to the haem. However, sequence alignments on separate domains strongly diverge towards the homologous proteins, suggesting that the rise of FHbs comes from fusion of a protoglobin ancestor and a flavin-binding domain (Bonamore & Boffi, 2008).

Details of the structure, function and reaction mechanism of purified native or recombinant FHbs from bacteria and yeast are available (Lewis, Corker, Gollan, & Poole, 2008). FHb from *Escherichia coli* (Hmp) is the best characterised member of the family. Hmp is distributed into both the cytoplasmic and periplasmic space, although the biochemically active protein is exclusively found in the cytoplasmic fraction (Vasudevan, Tang, Dixon, & Poole, 1995). It is subject to complex control (reviewed by Spiro, 2007; Poole, 2008), being dramatically up-regulated in response to NO and nitrosating agents (Membrillo-Hernández et al., 1999; Membrillo-Hernández, Coopamah, Channa, Hughes, & Poole, 1998; Poole et al., 1996). In particular, the *hmp* gene is predominantly regulated at the transcriptional level by NO-sensitive transcription factors, especially NsrR and Fnr (Spiro, 2007). Remarkably, the fine-tuning of Hmp synthesis appears critical for *E. coli* survival. In fact, the constitutive expression of Hmp in the absence of NO generates oxidative stress because of partial O_2 oxidation by the haem to superoxide and peroxide anion; accumulation of O_2 radicals has been stressed both in kinetic studies on the purified protein (Orii, Ioannidis, & Poole, 1992; Poole, Rogers, D'mello, Hughes, & Orii, 1997; Wu, Corker, Orii, & Poole, 2004) and by detecting the superoxide-generating activity of Hmp *in vivo* (Membrillo-Hernández, Ioannidis, & Poole, 1996). Similar results have also been obtained in *Salmonella enterica* where the constitutive expression of Hmp makes cells hypersensitive to paraquat and H_2O_2 (Gilberthorpe, Lee, Stevanin, Read, & Poole, 2007), as well as to peroxynitrite ($ONOO^-$) (McLean, Bowman, & Poole, 2010).

FHbs display a pivotal role in NO detoxification (Poole & Hughes, 2000). NO, involved in many beneficial and/or dangerous physiological and pathological processes, is a signalling and defence molecule of great importance in biological systems, and nowadays has an important role in contemporary medicine, physiology, biochemistry and microbiology. The behaviour of NO is made particularly complex by its ability to be oxidised to the nitrosonium cation (NO^+) or reduced to the nitroxyl anion (NO^-) and to react with O_2 to form nitrite (NO_2^-). Moreover, the reactions of NO led to production of reactive nitrogen species (RNS) (reviewed by Bowman, McLean, Poole, & Fukuto, 2011; Poole & Hughes, 2000), such as $ONOO^-$, formed by the reaction of NO with superoxide anion.

Extensive literature deals with NO and related species, especially considering that NO plays vital anti-microbial roles in innate immunity (Granger, Perfect, & Durack, 1986; Green, Meltzer, Hibbs, & Nacy, 1990; Iyengar, Stuehr, & Marletta, 1987; Liew, Millott, Parkinson, Palmer, & Moncada, 1990; Marletta, Yoon, Iyengar, Leaf, & Wishnok, 1988; Stuehr, Gross, Sakuma, Levi, & Nathan, 1989) and that microorganisms have evolved a large number of NO-sensitive targets and defence mechanisms against its toxic effects.

FHbs catalyse reaction of NO with O_2 to yield the relatively innocuous NO_3^- (Gardner, 2005; Mowat, Gazur, Campbell, & Chapman, 2010; Poole & Hughes, 2000) by a dioxygenase (Gardner et al., 2006, 2000; Gardner, Gardner, Martin, & Salzman, 1998) or denitrosylase (Hausladen, Gow, & Stamler, 1998, 2001). Anaerobically, Hmp shows low NO-reductase activity, converting NO to N_2O (Kim, Orii, Lloyd, Hughes, & Poole, 1999; Liu, Zeng, Hausladen, Heitman, & Stamler, 2000; Poole & Hughes, 2000; Vinogradov & Moens, 2008).

Deletion of the *hmp* gene alone abolishes the NO-consuming activity (Liu et al., 2000) and is sufficient to render bacteria hypersensitive to NO and related compounds, not only *in vitro* (Membrillo-Hernández et al., 1999) but also *in vivo* (Stevanin, Poole, Demoncheaux, & Read, 2002).

Similar to *E. coli* (Stevanin, Read, & Poole, 2007), the *S. enterica* serovar Typhimurium mutant defective in FHb shows enhanced sensitivity to mouse and human-macrophage microbicidal activity (Gilberthorpe et al., 2007; Stevanin et al., 2002), suggesting that the globin contributes to protection from NO-mediated toxicity in macrophages.

FHbs from several other bacteria show protective functions against RNS, such as those from *Ralstonia eutropha*, *Bacillus subtilis*, *Pseudomonas aeruginosa*, *Deinococcus radiodurans*, *S. enterica* and *Klebsiella pneumoniae* (reviewed by Frey, Farres, Bollinger, & Kallio, 2002).

Although FHbs protect pathogenic microorganisms from the host immune systems, they also defend non-pathogenic organisms from endogenous NO generated by nitrate and nitrite reduction, under anaerobic conditions (Rogstam, Larsson, Kjelgaard, & von Wachenfeldt, 2007). Vinogradov and Moens (2008) also propose that FHb may be involved in the coordination of intracellular NO concentration with extracellular O_2 levels, as suggested by the mitochondrial generation of NO in *Saccharomyces* exposed to hypoxia, with concomitant localisation of FHb in the pro-mitochondrial matrix (Castello, David, McClure, Crook, & Poyton, 2006).

Although FHbs are involved in NO detoxification, other physiological functions have been identified; for instance, Hmp has alkyl hydroperoxide reductase activity under anaerobic conditions, suggesting that, together with the unique lipid-binding properties, this globin is capable of catalysing the reduction of lipid-membrane hydroperoxides into the corresponding alcohols using NADH as electron donor and may thus be involved in the repair of the lipid-membrane oxidative damage generated during oxidative/nitrosative stress (Bonamore & Boffi, 2008; D'Angelo et al., 2004).

4.2. Single-domain globins

The first identified and sequenced SDgb was the Hb of *Vitreoscilla* (Vgb), whose expression is significantly increased under microaerobic conditions. It comprises a single domain, unmistakably globin-like, and the protohaem; it lacks the reductase domain seen in FHbs. Despite evidence that its expression in heterologous hosts can provide some protection from nitrosative stress, the generally accepted view is that Vgb facilitates O_2 utilisation, perhaps by directly interacting with a terminal oxidase (Wu, Wainwright, Membrillo-Hernández, & Poole, 2004, Wu, Wainwright, & Poole, 2003).

Another example of SDgb is offered by the microaerophilic, foodborne, pathogenic bacterium *Campylobacter jejuni*, exposed to NO and other nitrosating species during host infection. This globin (Cgb) is dramatically up-regulated in response to nitrosative stress (Elvers et al., 2005, Elvers, Wu, Gilberthorpe, Poole, & Park, 2004; Monk, Pearson, Mulholland, Smith, & Poole, 2008; Smith, Shepherd, Monk, Green, & Poole, 2011) and provides a specific and inducible defence against NO and nitrosating agents (Poole, 2005); it detoxifies NO and presents a peroxidase-like haem-binding cleft. In contrast to Vgb, there is no evidence that Cgb functions in O_2 delivery.

Since a detailed overview of SDgbs is beyond the goal of this contribution, the reader is directed elsewhere (Bowman et al., 2011; Frey et al., 2002;

Frey & Kallio, 2003, Frey, Shepherd, Jokipii-Lukkari, Haggman, & Kallio, 2011; Vinogradov et al., 2005, 2013).

4.3. Truncated haemoglobins

Members of the T family are found in eubacteria, cyanobacteria, protozoa and plants but not in animals (Milani et al., 2005; Wittenberg, Bolognesi, Wittenberg, & Guertin, 2002). The N-termini of these globins are 20–40-residue shorter; these globins display the 2/2 α-helical sandwich fold (composed of helices B, E, G and H), which has been recognised as a subset of the classical 3/3 α-helical sandwich (Milani et al., 2005; Wittenberg et al., 2002). The 2/2 α-helical sandwich fold results in four α-helices (B/E and G/H) arranged in antiparallel pairs connected by an extended polypeptide loop that replaces the α-helix F (Fig. 8.2).

It is noteworthy that the haem-proximal helix F is replaced by a poly-peptide segment (Milani et al., 2001; Pesce et al., 2000). The residues com-prising the F loop affect the orientation of the proximal HisF8 modulating the O_2-binding properties (Milani et al., 2005; Pathania, Navani, Gardner, Gardner, & Dikshit, 2002). Helix E is very close to the haem distal site; therefore, residues at positions B10, CD1, E7, E11, E15 and/or G8 (Table 8.2) modulate ligand binding (Milani et al., 2005).

On the basis of phylogenetic analysis, the T family can be further divided into three distinct sub-families: TrHbI (or N), TrHbII (or O) and TrHbIII (or P); specific structural features that depend on residues of the distal haem pocket distinguish each group (Table 8.2). In group I, the hydrogen bond network stabilising the haem-bound ligand involves the B10, E7 and E11 residues. Strongly conserved Tyr B10 plays a key role in ligand stabilisation through OH pointing directly to the heam-bound ligand. Normally, com-plete stabilisation by the H-bond network is provided by Glu located at E7 or E11, or at both positions. In group II, TrpG8 is fully conserved, contrib-uting to ligand stabilisation by the H bond linking the indole nitrogen atom and the ligand.

Further, Tyr at CD1 in some TrHbIIs drastically modifies the interaction network (Pesce et al., 2000). In group III, for example, *C. jejuni* TrHbIII (Ctb), the hydrogen bond network stabilising the haem-bound ligand involves TyrB10 and TrpG8 (Nardini et al., 2006). Interestingly, the affinity of cyanide for the ferrous derivative of *C. jejuni* Ctb is higher than that reported for any known (in)vertebrate ferrous globin and is reminiscent

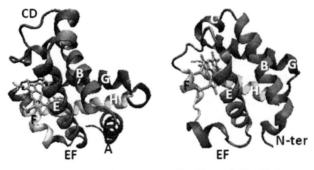

Sperm whale myoglobin   *Bacillus subtilis* HbO

Figure 8.2 A stereo view of sperm whale Mb (3/3) (pdb:1JP6, Urayama, Gruner, & Phillips, 2002) and *B. subtilis* TrHbII (HbO) (pdb:1UX8, Ilari, Giangiacomo, Boffi, & Chiancone, 2005) tertiary structures, including the haem group and helices. Modifications of the conventional 3/3 Mb fold occur particularly at helix A and in the CD-D regions, which are virtually absent, and in the EF-F regions of TrHbs. A very short segment linking helices C and E forces the haem-distal helix E very close to the haem distal side. (For colour version of this figure, the reader is referred to the online version of this chapter.)

of that of ferrous horseradish peroxidase, suggesting that this globin may participate in cyanide detoxification (Bolli et al., 2008).

Strikingly, TrHbs belonging to group III (or N) host a protein matrix tunnel system offering a potential path for ligand diffusion to and from the haem distal site. The apolar tunnel/cavity system, extending for approximately 28 Å through the protein matrix, is conserved in TrHbs belonging to group N, although with modulation of its size and/or structure (Milani et al., 2001; Pesce, Milani, Nardini, & Bolognesi, 2008). It has been proposed that in *Mycobacterium tuberculosis* HbN, the haem–Fe/O_2 stereochemistry and the protein matrix tunnel may promote O_2/NO chemistry *in vivo*, as a *M. tuberculosis* defence mechanism against macrophage nitrosative stress (Milani et al., 2001).

Unlike HbN, *M. tuberculosis* HbO does not host the protein matrix tunnel but two topologically equivalent matrix cavities. Moreover, the small apolar Ala E7 residue leaves room for ligand access to the haem distal site through the conventional E7 path (Pesce et al., 2008), as proposed for Mb.

In contrast to TrHbs I and II (Milani et al., 2004, 2001; Nardini et al., 2006; Pesce et al., 2008), Ctb does not display protein matrix tunnel/cavity systems at all (Nardini et al., 2006). Although the gating role of HisE7 in the modulation of ligand access into and out of the heam pocket is debated

Table 8.2 Amino acid residues building up the proximal (F8) and distal (B10, CD1, E7, E11, E15 and G8) haem pocket of sperm whale Mb and TrHbs belonging to groups I, II and III

Protein	Group	B10	CD1	E7	E11	E15	F8	G8
Sperm whale Mb		L	F	H	V	L	H	I
Nostoc commune		H	F	Q	L	L	H	I
Paramecium caudatum		Y	F	Q	T	L	H	V
Chlamydomonas eugametos	I	Y	F	Q	Q	L	H	V
Synechocystis		Y	F	Q	Q	L	H	V
Mycobacterium tuberculosis N		Y	F	L	Q	F	H	V
Mycobacterium avium N		Y	F	L	Q	F	H	V
Mycobacterium avium O		Y	Y	A	L	L	H	W
Mycobacterium tuberculosis O		Y	Y	A	L	L	H	W
Mycobacterium smegmatis O		Y	Y	A	L	L	H	W
Mycobacterium leprae	II	Y	Y	A	L	L	H	W
Thermobifida fusca		Y	Y	A	L	L	H	W
Bacillus subtilis		Y	F	T	Q	L	H	W
Arabidopsis thaliana		Y	F	A	Q	F	H	W
Thiobacillus ferrooxidans		Y	F	H	L	W	H	W
Mycobacterium avium P		Y	F	H	M	W	H	W
Campylobacter jejuni	III	Y	F	H	I	W	H	W
Bordetella pertussis 1		Y	F	H	L	W	H	W

Adapted from Milani et al. (2005).

(Lu, Egawa, Wainwright, Poole, & Yeh, 2007; Nardini et al., 2006), this mechanism appears to be operative in the *C. jejuni* TrHbIII (Nardini et al., 2006).

The sequence identity between TrHbs belonging to the three phylogenetic groups is very low (<20%) (Nardini, Pesce, Milani, & Bolognesi, 2007; Vuletich & Lecomte, 2006; Wittenberg et al., 2002), but may be higher than 80% within a given group. Analysis of the distribution of TrHbs suggests a scenario for the evolution of the different groups where the group II gene is ancestral and group-I and group-III genes are the results of duplications and transfer events (Vuletich & Lecomte, 2006).

TrHbs belonging to the three groups may coexist in some bacteria, suggesting distinct functions. These globins have been hypothesised to store ligands and/or substrates, to facilitate NO detoxification, to sense O_2/NO, to display (pseudo)enzymatic activities and to deliver O_2 under hypoxic conditions (Vinogradov & Moens, 2008; Wittenberg et al., 2002). The high O_2 affinity suggests that some TrHbs may function as O_2 scavengers rather than O_2 transporters (Ouellet et al., 2003; Wittenberg et al., 2002).

5. THE ANTARCTIC MARINE BACTERIUM *PSEUDOALTEROMONAS HALOPLANKTIS* TAC125: A CASE STUDY

5.1. General aspects

Despite the fact that the Antarctic marine environment is characterised by permanent low temperatures, the surface water and the sea-ice zones host a surprisingly high level of microbial activity.

A typical representative of γ-Proteobacteria found in the Antarctic is the marine cold-adapted psychrophile *P. haloplanktis* TAC125 (*Ph*TAC125), a Gram-negative bacterium, isolated in Antarctic coastal sea water in the vicinity of the French station Dumont d'Urville, Terre Adélie (66°40′S; 140°01′E). As in many marine γ-Proteobacteria, its genome is made up of two chromosomes (Médigue et al., 2005). This strain thrives between −2 °C and 4 °C, but is also able to survive long-term frozen conditions when entrapped in the winter sea ice. *Ph*TAC125 can grow in a wide temperature range (4–25 °C) (Fig. 8.3A) and achieve very high cell density even under uncontrolled laboratory conditions (Fig. 8.3B). In a marine broth, *Ph*TAC125 displays a doubling time of about 4 h at 4 °C and 5 h 15 min at 0 °C. At higher temperatures, the bacterium divides actively and the generation time decreases moderately (e.g. 1 h 40 min at 18 °C), with increase in the biomass produced at the stationary phase (Piette et al., 2011). In contrast, higher temperatures cause a drastic reduction in cell density at the stationary phase, suggesting that the heat-induced stress affects the growth (Piette et al., 2011).

The doubling time of *Ph*TAC125 at 16 °C is approximately 2 h, almost three times faster than that of *E. coli* under similar growth conditions (Piette et al., 2010). Consistent with the high growth rate, at room temperature, *Ph*TAC125 shows a very efficient chemotactic response, 10 times faster than that of *E. coli*, allowing it to exploit nutrient patches in the marine

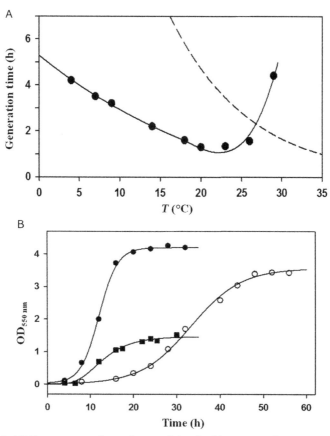

Figure 8.3 (A) Temperature dependence of the doubling time of Antarctic bacterium *Ph*TAC125, grown in marine broth (solid line and circles), compared to a typical growth curve of *E. coli* RR1 strain, obtained in LB broth (dashed line). (B) Growth curves of *Ph*TAC125, performed at 4 °C (open circles), 18 °C (filled circles) and 26 °C (filled squares). *Adapted from Piette et al. (2011).*

environment before they dissipate (Stocker, Seymour, Samadani, Hunt, & Polz, 2008). *Ph*TAC125 grows with extremely high growth rates in defined sea water medium, with peptone as the only carbon and nitrogen source, suggesting that these growth conditions resemble the favoured natural environment of the marine bacterium, which can be easily isolated from damaged tissues of fishes or molluscs, where such substrates are available (Wilmes et al., 2011). *Ph*TAC125 lacks a cyclic AMP (cAMP)–catabolite activator protein complex, that regulates carbon availability in related organisms, and a phosphoenolpyruvate-dependent phosphotransferase system (PTS) for the transport and first metabolic step of so-called PTS sugars (Médigue et al., 2005; Wilmes et al., 2011), making the bacterium unable to grow on glucose.

Approximately 13% of the identified proteins in the periplasmic protein fraction of PhTAC125 are transport-related proteins, mostly belonging to TonB-dependent transport systems (TBDTs) (Wilmes et al., 2011). The high amount of TBDTs in the genome (Médigue et al., 2005) and in the periplasmic proteome (Wilmes et al., 2011) support the idea that these transporters permit efficient use and scavenge the large variety of substrates found in the marine environment, probably representing an important prerequisite for fast growth under nutrient-rich conditions. Besides TBDTs, three ABC transporters, four porins and the transporter TolB are the other detected putative substrate-transport-related proteins (Médigue et al., 2005; Wilmes et al., 2011).

PhTAC125 is also able to grow in anaerobiosis, although with lower yields (Médigue et al., 2005). It is worth noting that lower duplication rate and poor growth of PhTAC125 in micro-aerobiosis have been observed (Parrilli, Giuliani, Giordano, et al., 2010). Due to lower O_2 solubility at 15 °C than at 4 °C, $OD_{600, max}$ is approximately 7.2 at 15 °C, and 4.3 at 4 °C, in extreme aerobiosis; moreover, $OD_{600, max}$ is approximately 1.25 at 4 °C and 0.38 at 15 °C, in micro-aerobiosis.

5.2. Genomic and post-genomic insights

Over the last decade, several genomes from psychrophilic bacteria and Archaea have been sequenced (Casanueva et al., 2010). Some of these (Allen et al., 2009; Ayala-del-Rio et al., 2010; Duchaud et al., 2007; Médigue et al., 2005; Methé et al., 2005; Rabus et al., 2004; Riley et al., 2008; Rodrigues et al., 2008; Saunders et al., 2003) have been analysed with respect to cold adaptation through proteomic and transcriptomic approaches (Bakermans et al., 2007, Bergholz, Bakermans, & Tiedje, 2009; Campanaro et al., 2011; Goodchild, Raftery, Saunders, Guilhaus, & Cavicchioli, 2005; Goodchild et al., 2004; Kawamoto, Kurihara, Kitagawa, Kato, & Esaki, 2007; Piette et al., 2011, 2010; Qiu, Kathariou, & Lubman, 2006; Ting et al., 2010; Williams et al., 2010; Wilmes et al., 2011; Zheng et al., 2007).

Through the MaGe annotation platform (http://www.genoscope.cns.fr/agc/mage/wwwpkgdb/Login/log.php?pid=7#ancreLogin), and by *in silico* and *in vivo* analyses, several exceptional genomic and metabolic features have been identified in PhTAC125 (Médigue et al., 2005).

*Ph*TAC125 is an interesting model for investigating strategies adopted to survive at low temperature (Médigue et al., 2005). The available genome sequence, combined with remarkable versatility and fast growth (Duilio, Tutino, & Marino, 2004), makes *Ph*TAC125 an attractive model to study protein-secretion mechanisms in marine environments, in addition to its use as a non-conventional host for recombinant production of thermal-labile and aggregation-prone proteins at low temperature (Cusano et al., 2006; Gasser et al., 2008; Parrilli, Duilio, & Tutino, 2008; Parrilli, Giuliani, Pezzella, et al., 2010; Vigentini, Merico, Tutino, Compagno, & Marino, 2006). Actually, low temperatures improve the quality of the products, removing the negative effects of high temperatures on protein folding, due to the strong temperature dependence of hydrophobic interactions that mainly drive the aggregation (Kiefhaber, Rudolph, Kohler, & Buchner, 1991). The growth of *E. coli* at lower temperatures to minimise aggregation has not been successful, probably because sub-optimal temperatures act negatively on cell performance (Gasser et al., 2008).

The efficiency of the cold-adapted *Ph*TAC125 expression system was demonstrated by the production of biologically active soluble products, for example, a yeast α-glucosidase, the mature human nerve growth factor and a cold-adapted lipase (de Pascale et al., 2008; Parrilli et al., 2008).

At the genome level, a relatively large number of rRNA genes (nine rRNA gene clusters) and tRNA genes (106 genes, organised in long runs of repeated sequences) have been observed in *Ph*TAC125 (Médigue et al., 2005), similar to *Colwellia psychrerythraea* (Methé et al., 2005) and *Psychromonas ingrahamii* (Riley et al., 2008). This finding may be explained as a response to the limited speed of transcription/translation at low temperature, allowing fast growth in the cold. However, it has recently been speculated that a high number of rRNA genes may reflect an ecological bacterial strategy to improve the response to perturbations in nutrient resources (Klappenbach, Dunbar, & Schmidt, 2000). Moreover, *Ph*TAC125 contains 19 genes presumably encoding known RNA-binding proteins or RNA chaperones. An unexpected feature is the prominent absence of *hns*, an RNA/nucleoid-associated cold-shock gene found in all γ-Proteobacteria. In contrast, the presence of many RNA helicases (three copies of *rhlE*, and probably a fourth one, *PSHAa0641*, and two copies of *srmB*) instead of the single one in *E. coli*, suggests that the control of RNA folding and degradation is important at low temperature (Médigue et al., 2005). In fact, RNA helicases have been found to be over-expressed at low temperature in many other psychrophilic microorganisms, such as *Methanococcoides burtonii*

(Lim, Thomas, & Cavicchioli, 2000), *Exiguobacterium sibiricum* (Rodrigues et al., 2008), *Sphingopyxis alaskensis* (Ting et al., 2010) and *Psychrobacter arcticus* (Bergholz et al., 2009; Zheng et al., 2007). This feature may reflect the need for help to unwind the RNA secondary structures for highly efficient translation in the cold (Cartier, Lorieux, Allemand, Dreyfus, & Bizebard, 2010).

Sequence analyses using genomic and metagenomic data clearly show that different mechanisms of adaptation to cold, including a bias towards specific residues, occur. The analysis of the amino acid composition of γ-Proteobacteria from different biotopes reveals a similar trend in the various genomes, Leu being most abundant, and Trp, Cys, His and Met most unusual. The proteome of thermophilic microorganisms shows several differences when compared to those of mesophilic and psychrophilic species, for example, Gln is poorly represented in thermophilic species, whereas mesophilic and psychrophilic species prefer Ala (except in *Oceanobacillus iheyensis*). A few remarkable differences have been identified between mesophilic and psychrophilic species, in particular, that some Asn and Gln are pivotal for bacteria growth in cold environments.

Specifically, proteins from *Ph*TAC125 are rich in Asn (Médigue et al., 2005) and contain few hydrophilic uncharged residues bearing an often thermolabile amide group (Stratton et al., 2001; Weintraub & Manson, 2004; Zhou, Cocco, Russ, Brunger, & Engelman, 2000), which undergo deamidating cyclisation, a process extremely sensitive to temperature (Daniel, Dines, & Petach, 1996). The richness in Asn in the *Ph*TAC125 proteome makes this Antarctic bacterium an organism of choice for foreign protein production when deamidation ought to be at a minimum (Weintraub & Manson, 2004).

Other aspects of cold-adapted proteins (CAPs) are (i) the significantly high level of non-charged polar Gln and Thr, and the low content of hydrophobic residues (particularly Leu) in the archaeal psychrophiles *Methanogenium frigidum* and *M. burtonii* (Saunders et al., 2003); (ii) the low content of polar residues such as Ser, the replacement of Asp with Glu, and the general decrease in charged residues in the proteins of *C. psychrerythraea* (Methé et al., 2005); (iii) the reduction of Pro and Arg codons in the *P. arcticus* genome, in particular, in the cell-growth and reproduction genes (Ayala-del-Rio et al., 2010); and (iv) decrease in Ala, Pro and Arg in *Shewanella halifaxensis* and *S. sediminis* (Zhao, Deng, Manno, & Hawari, 2010). These findings support the hypothesis of the increased flexibility and reduced thermostability of CAPs (Methé et al., 2005). However, no substitution promoting cold adaptation has been found in the *Desulfotalea psychrophila* genome (Rabus et al., 2004).

The proteomes expressed by *Ph*TAC125 at 4 and 18 °C were compared (Piette et al., 2010, 2011) to identify the cold-acclimation proteins, that is, those continuously over-expressed at high level at low temperatures, and to highlight the numerous down-regulated cellular functions (Piette et al., 2011). Interestingly, three proteins (Pnp, TypA and Tig, involved in distinct functions such as degradosome, membrane integrity and protein folding, respectively) have been identified as CAPs in *Ph*TAC125 (Piette et al., 2010) and as CSPs in mesophiles. Moreover, several CSP homologues have been reported in other cold-adapted bacteria (Bakermans et al., 2007; Bergholz et al., 2009; Kawamoto et al., 2007), suggesting striking similarities between CSPs in mesophiles and CAPs in psychrophiles. In agreement with these results, it has been proposed that "from an evolutionary point of view, one of the adaptive mechanisms for growth in the cold is to regulate the cold-shock response, shifting from a transient expression of CSPs to a continuous synthesis of at least some of them" (Piette et al., 2010, 2011).

The proteomic analyses of *Ph*TAC125 revealed that 30% of the identified CAPs are directly related to protein synthesis, covering all essential steps, from transcription (including RNA polymerase RpoB) to translation (i.e. methionyl-tRNA synthetase MetG that can be connected to the need of an increased pool of initiation tRNA to promote protein synthesis) and folding (i.e. the trigger factor—TF—Tig that acts on proteins synthesised by the ribosome; and PpiD, involved in the folding of outer membrane proteins). The genes *pnp* and *rpsA* also encode components of the degradosome that regulate transcript lifetimes and two putative proteases (*PSHAa2492* and *PSHAa2260*), identified as CAPs, likely to be involved in the proteolysis of misfolded proteins (Piette et al., 2010). Other identified CAPs are both components and regulators of the outer membrane architecture (Piette et al., 2010). In particular, the TonB-dependent receptor is indicative of sensing and exchanges with the external medium, and TypA (involved in lipopolysaccharides core synthesis) and Pal (a peptidoglycan-associated protein) are involved in the outer membrane stability and integrity (Abergel, Walburger, Chenivesse, & Lazdunski, 2001).

These results strongly suggest that low temperatures impair protein synthesis and folding, resulting in up-regulation of the associated functions and indicating that both cellular processes are limiting factors for bacterial development in cold environments (Piette et al., 2010).

Similar amounts of ribosomal and translation-specific proteins have also been revealed in mesophilic fast-growing bacteria, such as *B. subtilis* or *B. licheniformis*, under optimal growth conditions (Buttner et al., 2001; Voigt et al., 2004; Wilmes et al., 2011), suggesting that also the expression of these proteins, directly related to protein synthesis, is likely growth-rate dependent (Klumpp, Zhang, & Hwa, 2009).

Based on the cytoplasmic proteome and the available genome sequence, the analysis of the amino acid degradation pathways showed that all the common degradation routes are present in *Ph*TAC125, with the exception of those involved in Trp and Lys catabolism (Wilmes et al., 2011). Since the Antarctic genome contains coding sequences of biosynthetic enzymes for all 20 proteinogenic amino acids, it is likely that the degradation of Trp and Lys occurs by reversal of the biosynthetic routes. For instance, an alternative way to use Lys may be decarboxylation to cadaverine via *PSHAa1094* (annotated as a putative basic amino acid decarboxylase) (Wilmes et al., 2011). Further, a relatively high abundance of the tricarboxylic acid cycle enzymes in the cyto-plasmic proteome analysed at 16 °C, needed for efficient catabolism of the peptone-based amino acids, is in line with the extremely high growth rate (maximal rate being 0.35 h^{-1}) of the Antarctic bacterium (Wilmes et al., 2011).

The major CAP, 37-fold over-expressed at 4 °C (Piette et al., 2010), is the TF Tig, a CSP in *E. coli* (Kandror & Goldberg, 1997). TF is the first molecular chaperone interacting with virtually all newly synthesised polypeptides on the ribosome; it assists folding by delaying premature chain compaction and maintaining the elongating polypeptide in a non-aggregated state until adequate structural information for correct folding is available, and later promotes protein folding (Hartl & Hayer-Hartl, 2009; Martinez-Hackert & Hendrickson, 2009; Merz et al., 2008). TF also possesses a peptidyl-prolyl *cis–trans* isomerase activity (Kramer et al., 2004), the rate-limiting step in the folding of a wide range of proteins (Baldwin, 2008). In *Ph*TAC125, the peptidyl-prolyl *cis–trans* isomerase involved in protein folding is up-regulated at low temperature (Piette et al., 2010).

The major heat-shock proteins (HSPs), such as the chaperone DnaK, the chaperonin GroEL/ES and the chaperone Hsp90, are cold-repressed in the proteome of *Ph*TAC125. However, their expression is up-regulated when the bacterium is grown at higher temperature, indicating heat-induced cel-lular stress (Goodchild et al., 2005; Rosen & Ron, 2002). Synthesis of HSPs is also repressed during growth at low temperatures (Kandror & Goldberg, 1997) in *E. coli*. Accordingly, the observed cold repression of HSPs would be beneficial not only to *Ph*TAC125 but also to *E. coli*.

TF is transiently expressed in mesophilic bacteria but continuously over-expressed in psychrophiles to achieve cold adaptation, rescuing the chaperone function at low temperatures (Piette et al., 2010, 2011).

Either PPiases or TF act as potential CAPs in the proteome of most cold-adapted microorganisms analysed so far (Goodchild et al., 2005, 2004; Kawamoto et al., 2007; Qiu et al., 2006; Suzuki, Haruki, Takano, Morikawa, & Kanaya, 2004; Ting et al., 2010), suggesting that the constraint imposed by protein folding at low temperature and the cellular responses are common traits in most psychrophiles (Piette et al., 2010). In contrast, an almost inverse regulation was found in *P. arcticus* where GroEL/ES chaperonins and repression of TF are up-regulated under cold conditions (Bergholz et al., 2009; Zheng et al., 2007). Increased synthesis of chaperonins has also been reported in *S. alaskensis* (Ting et al., 2010) possessing two sets of *dnaK–dnaJ–grpE* gene clusters; proteomic analysis suggests that one of these sets functions as a low-temperature chaperone system whereas the other functions at higher temperatures (Ting et al., 2010).

At low temperature, in accordance with reduced biomass, almost half of the down-regulated proteins are involved in general bacterial metabolism. Most of these proteins are involved in oxidative metabolism, including glycolysis, the pentose phosphate pathway, the Kreb's cycle and the electron chain transporters (Piette et al., 2011; Wilmes et al., 2011).

The *Ph*TAC125 genome contains genes putatively involved in NO metabolism, such as NO reductase, *PSHAa2417*, and NO_2^- reductase, *PSHAa1477* (Médigue et al., 2005). In this context, the presence of multiple genes in distinct positions on chromosome I encoding three TrHbs (annotated as *PSHAa0030*, *PSHAa0458*, *PSHAa2217*) and a FHb (*PSHAa2880*) (Giordano et al., 2007; Médigue et al., 2005) may be pivotal for cell protection (see Section 6).

5.3. Excess of O_2 and metabolic constraints

Gases (e.g. O_2) and radicals (e.g. NO) are highly soluble and stable at low temperature with visible consequences in genome annotations in cold-adapted bacteria, having developed responses to strong oxidative stress (see Casanueva et al., 2010).

The apparent benefits of easier O_2 supply are contrasted by the adverse effects of low temperature on (macro)molecular functions and on the increased production of RNS and reactive oxygen species (ROS) (Casanueva et al., 2010; D'Amico et al., 2006). In fact, although RNS and

ROS could act as signalling molecules during cell differentiation and cell cycle progression, and in response to extracellular stimuli (Sauer, Wartenberg, & Hescheler, 2001), they are potentially toxic for cells (Finkel, 2003), being involved in a large number of pathological mechanisms.

Several mechanisms to cope with RNS and ROS have been proposed for cold-adapted bacteria. They include slightly lower frequency of oxidisable residues in protein sequences, occurrence of specific reductases, presence of dioxygenases and deletion of RNS- and ROS-producing metabolic pathways (see Casanueva et al., 2010). Interestingly, acyl desaturases (that introduce a double bond into fatty-acyl chains, using O_2 as substrate) combine the elimination of toxic O_2 with the improvement of membrane fluidity (Zhang & Rock, 2008). Therefore, the augmented capacity in antioxidant defence is likely an important component of evolutionary adaptation to a cold and O_2-rich environment (Ayub et al., 2009; Bakermans et al., 2007; Duchaud et al., 2007; Médigue et al., 2005; Methé et al., 2005; Piette et al., 2010; Rabus et al., 2004).

The cold environment raises the question of how PhTAC125 can cope with RNS and ROS. We have evidence proving that, in order to prevent significant damage to cellular structures, PhTAC125 improves the redox buffering capacity of the cytoplasm, and glutathione synthetase is strongly up-regulated at low temperature (Piette et al., 2010). These adjustments in antioxidant defences are needed to maintain the steady-state concentration of ROS and may be important components in evolutionary adaptations in cold and O_2-rich environments (Chen et al., 2008).

The main adaptive strategy used by PhTAC125, exposed to permanent oxidative stress, is expected to be increased production of enzymes active against hydrogen peroxide and superoxide. Surprisingly, in the genome of PhTAC125, only two genes, encoding an iron superoxide dismutase (sodB; PSHAa1215) and a catalase (katB, with the possible homologue PSHAa1737), have been identified. This catalase has very high similarity to catalases from other α-, β- and γ-Proteobacteria, for example, Psychrobacter, Mannheimia, Haemophilus and Neisseria (Médigue et al., 2005). Moreover, while the O_2-responding OxyR control has been found in PhTAC125, SoxR regulation is absent (Médigue et al., 2005).

Proteomic analyses of PhTAC125 reveal that oxidative stress-related proteins, such as catalase, glutathione reductase and peroxiredoxin (Piette et al., 2011), are repressed at 4 °C. However, because PhTAC125 metabolism is stimulated at 18 °C, it should be mentioned that, although these proteins would be repressed at 4 °C, they would most likely be induced at 18 °C (Piette et al., 2011).

In contrast, the Arctic bacterium *C. psychrerythraea* has developed an enhanced antioxidant capacity owing to the presence of three copies of catalase genes as well as two superoxide-dismutase genes, one of which codes for a nickel-containing superoxide-dismutase, never reported before in proteobacteria (Methé et al., 2005).

*Ph*TAC125 copes with increased O_2 solubility by deleting entire metabolic pathways that generate ROS as side products. It is worth noting that, despite the availability of molybdate in sea water (Hille, 2002), *Ph*TAC125 not only lacks the molybdate biosynthetic and transport genes but also genes encoding enzymes using molybdate as cofactor, for example, trimethylamine N-oxide reductase, xanthine oxidase, biotin sulphoxide reductase and oxido-reductase YedY (Loschi et al., 2004).

The *Ph*TAC125 genome also contains all genes required for the pentose phosphate pathway (Médigue et al., 2005); moreover, glucose-6-phosphate dehydrogenase (Zwf; *PSHAa1140*), transketolase (TktA; *PSHAa0671*) and transaldolase (TalB; *PSHAa2559*) have been found in the cytoplasmic proteome (Wilmes et al., 2011). This feature increases the concentration of NADPH, which in turn provides high levels of reduced thioredoxin that can help to protect against the toxic effects of O_2. Therefore, to develop better oxidative stress adaptation in the cold, *Ph*TAC125 can use the pentose phosphate pathway for carbohydrate inter-conversion.

In order to cope with the improved stability of ROS at low temperatures, iron-related proteins are down-regulated at 4 °C (Piette et al., 2011), presumably to avoid oxidative cell damage induced by the deleterious Fenton reaction (Valko, Morris, & Cronin, 2005). Cell protection may be achieved by dioxygenases that are coded in large number in both chromosomes (Médigue et al., 2005). Moreover, O_2-consuming lipid desaturases protect against toxic O_2 by increasing the membrane fluidity at low temperature (Médigue et al., 2005).

A further tool, possibly related to the peculiar features of the Antarctic habitat, may be the synthesis of globins facilitating several biological functions, including protection from nitrosative and oxidative stress (see Section 6.2.4).

5.4. Biotechnological applications

Microorganisms are an interesting source of cold-active enzymes endowed with biotechnological potential. Enzymes from psychrophiles have recently received increasing attention, because they offer novel opportunities in several industrial processes where high enzymatic activity or peculiar stereo-specificity at low temperature is required. The high specific activity of cold-adapted enzymes is due to the lack of a number of non-covalent stabilising interactions, providing improved flexibility of the conformation (Feller, 2010; Feller & Gerday, 2003; Siddiqui & Cavicchioli, 2006); this is a key adaptation to compensate for the exponential decrease in chemical reaction rates at lower temperatures. For instance, cold-active esterases and lipases found in the genome of *Ph*TAC125 (Aurilia, Parracino, Saviano, Rossi, & D'Auria, 2007; de Pascale et al., 2008; Médigue et al., 2005) can be added to detergents for use at low temperatures and to biocatalysts for biotransformation of heat-labile compounds (Margesin & Schinner, 1994). Cold-active Lip1 lipase (de Pascale et al., 2008), encoded by the *PSHAa0051* gene, was functionally over-expressed in *Ph*TAC125 at 4 °C (Duilio et al., 2004). In contrast, in mesophilic *E. coli*, the recombinant production was always found associated with the inclusion bodies and refolding was unsuccessful (de Pascale et al., 2008).

Engineered *Ph*TAC125 expressing a toluene-*o*-xylene mono-oxygenase from mesophilic *Pseudomonas* sp. OX1 (Bertoni, Bolognese, Galli, & Barbieri, 1996), combined with the endogenous laccase-like protein induced by copper, can convert several aromatic compounds into non-toxic metabolites (Papa, Parrilli, & Sannia, 2009; Parrilli, Papa, Tutino, & Sannia, 2010; Siani, Papa, Di Donato, & Sannia, 2006). This strategy endows *Ph*TAC125 with degrading capabilities and wide potentiality in bioremediation applications, for example, removal of organic pollutants from chemically contaminated marine environments and cold effluents of industrial processes (Parrilli, Papa, et al., 2010).

6. *P. haloplanktis* TAC125 GLOBINS
6.1. General aspects

Three TrHbs were identified in *Ph*TAC125: one TrHbI (encoded by the *PSHAa0458* gene) and two TrHbsII (encoded by *PSHAa0030* and *PSHAa2217*) (Giordano et al., 2007). The sequence identity between TrHbs from different groups is generally low but may be higher within a given group. The identity between the two TrHbs belonging to group II is 24%, suggesting that these proteins may have different function(s) in bacterial cellular metabolism. Moreover, a FHb, annotated as *PSHAa2880*, has been found in *Ph*TAC125 (Giordano et al., 2007).

The distribution of globins in polar bacteria is very different (Table 8.3).

Table 8.3 Distribution of FHb and TrHbs in polar bacteria

Polar bacteria	Strain origin	FHb	TrHbs
Colwellia psychrerythraea 34H	Arctic marine sediments	1	–
Shewanella frigidimarina NCMB400	Sea ice, sea water, Antarctica	1	2
Psychrobacter arcticus 273-4	Siberian permafrost	–	–
Psychrobacter cryohalolentis K5	Siberian permafrost	1	–
Oleispira antarctica RB-8	Rod Bay, Ross Sea Antarctica	–	–
Pseudoalteromonas haloplanktis TAC125	Coastal Antarctic sea water, Terre Adélie	1	3
Desulfotalea psychrophila LSv54	Arctic marine sediments, Svalbard	–	–
Exiguobacterium sibiricum 255-15	Siberian permafrost	1	1
Psychroflexus torquis ATCC 700755	Sea-ice algal assemblage Prydz Bay, Antarctica	–	–
Polaribacter filamentous 215	Surface sea water, north of Deadhorse, Alaska	–	–
Polaribacter irgensii 23-P	Nearshore marine waters off Antarctic Peninsula	–	–
Psychromonas ingrahamii 37	Sea ice, off Point Barrow, Northern Alaska	1	1
Marine actinobacterium PHSC20C1	Nearshore marine waters of Antarctic Peninsula	1	1

Interestingly, the *C. psychrerythraea* and *P. cryohalolentis* K5 genomes do not possess genes encoding TrHbs, but contain a gene for a single FHb; *E. sibiricum* 255-15 (Rodrigues et al., 2008), *P. ingrahamii* 37 (Riley et al., 2008) and *Marine actinobacterium* PHSC20C1 genomes contain one gene encoding a TrHb and one encoding an FHb; *S. frigidimarina* NCMB400 contains two genes encoding TrHbs and one encoding a FHb; *P. arcticus* 273-4, *Oleispira antarctica* RB-8, *D. psychrophila* LSv54, *Psychroflexus torquis* ATCC 700755, *Polaribacter filamentous* 215 and *Polaribacter irgensii* 23-P do not possess globin genes (Table 8.3).

The presence of multiple genes encoding globins may be considered a mechanism of defence against oxidative and nitrosative stress also in mesophilic organisms.

M. tuberculosis carries both a TrHbI (HbN) and a TrHbII (HbO) (Milani et al., 2005). It is worth noting that HbN efficiently protects *M. tuberculosis* from nitrosative damage, contributing to its survival in the host macrophage (Bidon-Chanal et al., 2006), whereas HbO may have an oxidation/reduction function because it has peroxidase activity with formation of ferryl intermediates (Ouellet et al., 2007). On the other hand, *Mycobacterium leprae* only displays HbO, which is capable of protecting against NO as well as oxidative stress (Ascenzi, De Marinis, Coletta, & Visca, 2008). *B. subtilis* (Ouellet et al., 2007) and *Thermobifida fusca* (Bonamore et al., 2005) encode both a TrHbII and an FHb. To our knowledge, *Ph*TAC125 is the first example of coexistence of genes encoding a FHb and three TrHbs (see Section 6.2.2).

A transcriptional analysis of the *PSHAa0030, PSHAa0458* and *PSHAa2217* genes encoding the TrHbs and the FHb-encoding gene was carried out on *Ph*TAC125 wild type and on a mutant strain in which *PSHAa0030* was inactivated. In *Ph*TAC125 wild-type cells, *PHSAa0030* is expressed at 4 °C and 15 °C. *PSHAa0458* and *PSHAa2217* encoding the other TrHbs are expressed in both strains under all conditions, whereas transcription of the FHb-encoding gene is detectable only in mutant cells grown at 4 °C in micro-aerobiosis (Parrilli, Giuliani, Giordano, et al., 2010) (see Section 6.2.4).

To date, only group II of the TrHbs encoded by *PSHAa0030* (hereafter named *Ph*-2/2HbO) has been thoroughly investigated from the structural and functional viewpoints (Coppola et al., 2013; Giordano et al., 2007, 2011; Howes et al., 2011; Parrilli, Giuliani, Giordano, et al., 2010; Russo et al., 2013). The gene was selected as the first of the three because its position on chromosome I is very close to the origin of replication of the bacterium, indicating an important physiological role (Giordano et al., 2007).

Since transcription of FHb-encoding genes is usually directly or indirectly induced by NO (Hausladen et al., 1998; Spiro, 2007), the observed FHb-gene expression only in a *Ph*TAC125 mutant strain is suggestive of occurrence of an NO-induced stress related to the absence of the TrHb encoded by *PSHAa0030* (Parrilli, Giuliani, Giordano, et al., 2010; see Section 6.2.4).

6.2. Structure–function relationships of *Ph*-2/2HbO

6.2.1 Structure
Group II of TrHbs is by far the most populated of the three and is characterised by specific residues building up the haem cavity (Vuletich &

Lecomte, 2006; Wittenberg et al., 2002). The crystal structures of TrHbsII from *B. subtilis* (Giangiacomo, Ilari, Boffi, Morea, & Chiancone, 2005), *T. fusca* (Bonamore et al., 2005), *Geobacillus stearothermophilus* (Ilari et al., 2007) and *M. tuberculosis* (Milani et al., 2005, 2003) show a network of interactions between polar residues and the haem-Fe atom that may explain the high O_2 affinity of these globins (Bonamore et al., 2005; Giangiacomo et al., 2005; Ilari et al., 2007; Milani et al., 2005; Mukai, Savard, Ouellet, Guertin, & Yeh, 2002; Ouellet et al., 2003).

Ph-2/2HbO displays structural features typical of TrHbII (Giordano et al., 2007). In particular, *Ph*-2/2HbO has Trp at G8, and Tyr at both CD1 and B10 (Fig. 8.4; Howes et al., 2011). These three positions are pivotal for the stabilisation of the haem-bound O_2 in TrHbsII (Milani et al., 2005). It is worth noting that CD1 Phe, that wedges the haem into its pocket, is considered a conserved residue among globins, unlike members of TrHbsII from *M. tuberculosis*, *Mycobacterium avium*, *M. leprae*, *Mycobacterium smegmatis*, *Streptomyces coelicolor*, *Corynebacterium diphtheriae* and *T. fusca*, which host Tyr instead (Table 8.2; Bonamore et al., 2005; Milani et al., 2005).

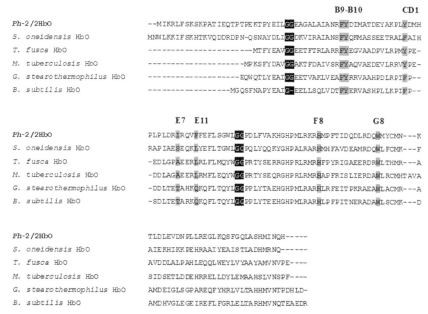

Figure 8.4 Sequence alignment of some representative TrHbs of group II. Identical functionally important residues of the distal haem pocket (B9, B10, CD1, E7, E11 and G8) and the proximal His F8 are highlighted in grey. The Gly-Gly motifs typical of TrHbs are highlighted in black. *Adapted from Howes et al. (2011).*

Ph-2/2HbO shows structural differences with respect to other TrHbsII. In particular, the insertion of three residues in the CD loop (Howes et al., 2011; Fig. 8.4) confers higher flexibility that may facilitate its action at low temperature, providing greater freedom for the correct positioning of ligand(s) (Feller & Gerday, 2003; Siddiqui & Cavicchioli, 2006). In contrast to TrHbsI, the E7 and E11 positions are occupied by non-polar residues, Ile and Phe, respectively, precluding haem-bound ligand stabilisation. On the proximal side, HisF8, conserved in all members of the globin superfamily (Howes et al., 2011), is coordinated to the haem-Fe atom (Table 8.2 and Fig. 8.4).

Ph-2/2HbO shows an unusual extension of 15 residues at the N-terminus (pre-helix A), similar to *M. tuberculosis* HbN and to many slow-growing species of *Mycobacterium*, such as *M. bovis*, *M. avium*, *M. microti*, *M. marinum* (Lama et al., 2009) and *Shewanella oneidensis* (Vuletich & Lecomte, 2006). The pre-A motif of *M. tuberculosis* HbN does not significantly contribute to the structural integrity of the protein, protruding out of the compact globin fold (Milani et al., 2001). However, the deletion of this motif reduces the ability of *M. tuberculosis* to scavenge NO (Lama et al., 2009). Unlike in *M. tuberculosis* HbN, the deletion of the N-terminal extension of *Ph*-2/2HbO does not seem to reduce the NO scavenging activity (Coppola et al., 2013) (see Section 6.2.4).

6.2.2 Hexacoordination

Ph-2/2HbO displays hexacoordination of the ferric and ferrous haem-Fe atom (Giordano et al., 2011; Howes et al., 2011). Hydrostatic pressure enhances hexacoordination in both oxidation states of the haem-Fe atom, as previously shown in other haem proteins (Hamdane et al., 2005), indicating that a flexible protein allows structural changes (Russo et al., 2013).

Binding of O_2 to Mb and Hb occurs on the distal side of the pentacoordinated haem-Fe atom, where O_2 establishes a sixth coordination bond to the Fe atom, whereas the fifth coordination position is occupied by invariant HisF8 (Fig. 8.5). The haem-Fe-bound O_2 is generally stabilised by interaction(s) with distal residues. The main O_2 stabilising interaction is usually provided by an H bond donated by HisE7 (Fig. 8.5).

In hexacoordinated globins, where, in the absence of external ligands, the sixth position is taken by an internal residue, exogenous ligands (e.g. O_2, CO and NO) compete with the internal ligand to bind Fe, this behaviour being the basis of the control of Fe reactivity (Smagghe, Trent, & Hargrove, 2008; Trent, Hvitved, & Hargrove, 2001).

Haem-Fe hexacoordination is widespread in globins, having been found in unicellular eukaryotes (Wittenberg et al., 2002), plants (Watts et al.,

Hexacoordinated ⇄ **Pentacoordinated** ⇄ Oxygenated

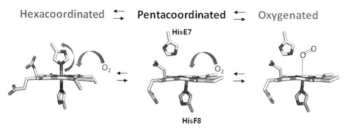

Figure 8.5 Schematical representation of the iron coordination in hexacoordinated, pentacoordinated and oxygenated forms. The protein is in hexacoordinated conformation when, in the absence of external ligands, there is an amino acid residue as internal ligand. The sixth ligand is usually provided by His. Upon addition of gaseous ligands, for example, O_2, there is competition between the external ligand and the sixth ligand, with replacement of the internal ligand with O_2. (For colour version of this figure, the reader is referred to the online version of this chapter.)

2001), invertebrates (Dewilde et al., 2006), but only a few cases have been reported in bacterial TrHbs (Falzone, Christie Vu, Scott, & Lecomte, 2002; Razzera et al., 2008; Scott et al., 2002; Vinogradov & Moens, 2008; Visca et al., 2002). Hexacoordination has also been found in higher vertebrates, for example, in ferric β-chains of tetrameric Antarctic fish Hbs (Riccio, Vitagliano, di Prisco, Zagari, & Mazzarella, 2002; Vergara et al., 2007; Vergara, Vitagliano, Verde, di Prisco, & Mazzarella, 2008; Vitagliano et al., 2004, 2008) and in the ferric and ferrous states of mammalian (Pesce et al., 2003; Vallone, Nienhaus, Brunori, & Nienhaus, 2004) and Antarctic fish (Giordano et al., 2012) Ngbs and Cygbs (de Sanctis et al., 2004; Alessia Riccio et al., unpublished results). The occurrence of ferrous (haemochrome) and ferric (haemichrome) oxidation states in members of the Hb superfamily is not uniform, suggesting that the functional roles of these states are multiple, possibly being a tool for modulating ligand-binding or redox properties (Vergara et al., 2008; Vitagliano et al., 2008). Exchange between haemichrome and pentacoordinated forms may play a physiological role in Antarctic fish due to higher peroxidase activity (Vergara et al., 2008; Vitagliano et al., 2008).

Over the years, haemichromes in tetramers have been considered precursors of Hb denaturation (Rifkind, Abugo, Levy, & Heim, 1994); however, haemichromes can be obtained under non-denaturing as well as physiological conditions (Vergara et al., 2008). It has also been suggested that haemichromes can be involved in Hb protection from peroxide attack (Feng et al., 2005), given that the haemichrome species of human α-subunits complexed with the α-helix-stabilising protein do not exhibit peroxidase activity (Feng et al., 2005).

Hexacoordination of the haem–Fe atom may suggest a common physiological mechanism for protecting cells against oxidative chemistry in response to high O_2 concentration. Several roles have been hypothesised for hexacoordinated Ngb and Cygb, for example, as O_2 scavengers under hypoxic conditions (Burmester, Ebner, Weich, & Hankeln, 2002; Burmester, Weich, Reinhardt, & Hankeln, 2000), as terminal oxidases (Sowa, Guy, Sowa, & Hill, 1999), as O_2-sensor proteins (Kriegl et al., 2002), and in NO metabolism (Smagghe et al., 2008). It was recently reported that Ngb over-expression and intracellular localisation confer protection to neurons, both *in vitro* and *in vivo*, against oxidative stress and enhance cell survival under anoxia and ischaemic conditions (Fiocchetti, De Marinis, Ascenzi, & Marino, 2013). However, their physiological role is still a matter of debate.

Hexacoordinated globins are characterised by specific electronic absorption bands in the UV–visible spectra, clearly indicating the electronic structure of the Fe atom and its axial ligands (Dewilde et al., 2001).

The electronic-absorption spectrum of ferric *Ph*-2/2HbO is characterised by hexacoordinated high-spin (bands at 503 nm and charge-transfer transition at 635 nm) and low-spin forms (bands at 533 and 570 nm) (Fig. 8.6A), the latter being characteristic of a Tyr coordinated

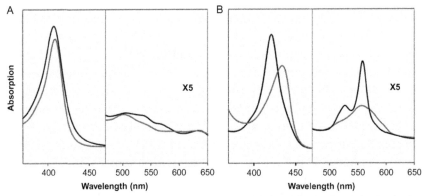

Figure 8.6 Overlay of absorption spectra of (A) ferric hexacoordinated *Ph*-2/2HbO (black line) with ferric Mb (red line) and (B) deoxy ferrous hexacoordinated *Ph*-2/2HbO (black line) with ferrous Mb (red line). All measurements were at pH 7.6 and 25 °C. The ferrous samples were prepared by adding 2 μl of sodium dithionite (10 mg ml^{-1}) to 600 μl of deoxygenated buffered solution of ferric globins, obtained flushing the ferric forms with nitrogen. The protein concentration was 10 μM on a haem basis. (For interpretation of the references to colour in this figure legend, the reader is referred to the online version of this chapter.)

to the Fe atom (Howes et al., 2011). The ferrous state shows a mixture of a predominant hexacoordinated low-spin state (Soret band at 421 nm and Q bands at 528 and 559 nm) and a pentacoordinated high-spin state (shoulder at 440 nm) (Fig. 8.6B; Giordano et al., 2011). These spectra are in marked contrast to those of monomeric Mb, in which the Fe atom is pentacoordinated. In fact, the deoxygenated ferrous form has a broad peak at 556 nm (Fig. 8.6B), whereas the ferric form exhibits two peaks at 504 and 632 nm (Fig. 8.6A; Antonini & Brunori, 1971).

Based on the spectroscopic data and molecular-dynamics simulation (Howes et al., 2011), it has been shown that either TyrCD1 or TyrB10 can coordinate the ferrous atom. Although His is the most common residue that coordinates the Fe atom, Tyr coordinates Fe of ferrous *Herbaspirillum seropedicae* Hb (Razzera et al., 2008) and of ferrous and ferric *Chlamydomonas* Hb (Couture et al., 1999; Das et al., 1999; Milani et al., 2005).

6.2.3 Reactivity

Reversible hexa- to pentacoordination of the haem-Fe atom modulates the reactivity of *Ph*-2/2HbO; in fact, the cleavage of the haem distal Fe-TyrCD1 or Fe-TyrB10 bonds is the rate-limiting step for the association of exogenous ligands (e.g. O_2, CO and NO) and (pseudo)enzymatic activities (Russo et al., 2013).

CO binding to *Ph*-2/2HbO displays a rapid spectroscopic phase independent of CO concentration, followed by standard bimolecular recombination. CO-rebinding kinetics show an unusually slow geminate phase, which becomes dominant at low temperature. While geminate recombination usually occurs on the ns timescale, *Ph*-2/2HbO displays a component of about 1 μs that accounts for half of the geminate phase at 8 °C, indicative of a relatively slow internal ligand binding (Russo et al., 2013).

After ligand escape, bimolecular recombination takes place. Second-order rebinding indicates two major conformations at 25 °C, characterised by CO-association rates that differ by a factor of 20, with pH-dependent relative fractions. A dynamic equilibrium was found between a predominant hexacoordinated low-spin state and a pentacoordinated high-spin state. A shift in the equilibrium between the two conformations may also provide a large change in the ligand affinity. The second-order rate constant of the fast phase (Russo et al., 2013) is of the order of 10^7 M^{-1} s^{-1} and closely similar to that of human Ngb (Uzan et al., 2004), whereas the second-order rate constant of the slow process (Russo et al., 2013) is compatible with that of Mb (Table 8.4), being in the range of 10^5 M^{-1} s^{-1} (Springer, Sligar,

Table 8.4 Values of kinetic parameters for O_2 and CO binding to penta and hexacoordinated Hbs.

haem protein	$k_{on}O_2$ ($\mu M^{-1} s^{-1}$)	$k_{off}O_2$ (s^{-1})	$k_{on}CO$ ($\mu M^{-1} s^{-1}$)	$k_{off}CO$ (s^{-1})	$P_{50}O_2$ (Torr)	Reference
Hexacoordinated						
*Ph-2/2*HbO (8 °C)	0.9	1.0	0.35	–	1	Russo et al. (2013)
*Ph-2/2*HbO (25 °C)		4.2	0.69 (slow) 12.0 (fast)	–	–	Giordano et al. (2011)
Ngb	170	0.7	40	–	6.8 (0.9 S–S)	Uzan et al. (2004)
Pentacoordinated						
Sperm whale Mb	14	12	0.51	0.019	0.51	Springer et al. (1994)
M. tuberculosis HbO*	0.11	0.0014	0.014	0.004	–	Ouellet et al. (2003)
	(80%) 0.85 (20%)	(78%) 0.0058 (22%)	(79%) 0.18 (21%)	(60%) 0.0015 (40%)	–	

*The relative percentage of the two rate constants are reported in parentheses.

Olson, & Phillips, 1994). The relatively fast CO-dependent kinetic process is unusual for a TrHbs of group II; the second-order rate constant of the fast phase for carbonylation of *M. tuberculosis* HbO is in the range of $10^5 M^{-1} s^{-1}$, whereas the second-order rate constant of the slow phase is in the range of $10^4 M^{-1} s^{-1}$ (Ouellet et al., 2003). Thus the proteins displays two conformations that greatly differ in the ligand association rate, suggesting that they may switch between two distinct functional levels.

At 8 °C, 85% of the CO bimolecular recombination occurs on the ms timescale at a rate similar to that of a 3/3 Mb ($3.5 \times 10^5 M^{-1} s^{-1}$), whereas the remaining kinetic component is faster ($10^7 M^{-1} s^{-1}$) (Russo et al., 2013), as observed at 25 °C (Giordano et al., 2011; Table 8.4).

Hexacoordination of the haem-Fe atom of *Ph-2/2*HbO via distal Tyr is only partial (the Tyr equilibrium affinity is close to 1), indicating a weak interaction between Tyr and the Fe atom under atmospheric pressure (Russo et al., 2013). The fast binding and dissociation of Tyr from the Fe atom can be a molecular event that triggers the shift of the globin between two conformations (penta- vs. hexacoordinated haem) with different redox potentials. This behaviour may be considered as an ancestral mechanism for modulating a conformational switch between two functional species (Russo et al., 2013).

Ph-2/2HbO is quickly oxidised in the presence of O_2, probably due to the superoxide character of the haem-Fe-O_2 adduct, affected by the presence of the surrounding hydrogen-bond donor residues (Milani et al., 2001, 2003).

The O_2 affinity, poorly affected by competition with Tyr, is about 1 Torr at 8 °C, pH 7.0 (Table 8.4). The O_2 affinity of *Ph*-2/2HbO is compatible with the *in vivo* conditions (Fig. 8.7A), considering that the *Ph*TAC125 bacterial metabolism must cope with high O_2 concentration and high-salinity conditions at low temperature. However, Mb-like functions do not seem to be possible for *Ph*-2/2HbO, requiring a still unknown efficient reducing system, and a local high globin concentration (Russo et al., 2013).

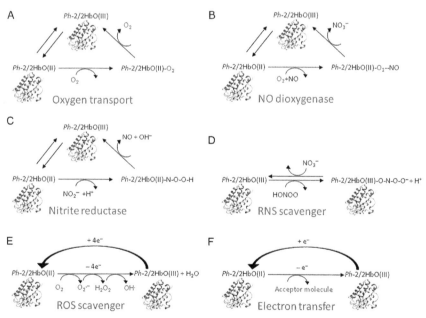

Figure 8.7 Some postulated functions of *Ph*-2/2HbO. (A) O_2 carrier even though this function requires the presence of a still unknown coupled reductase system for high turnover reaction; (B) *Ph*-2/2HbO may convert NO to nitrate and (C) nitrite to NO; (D) it may detoxify RNS and (E) ROS; (F) it may function as electron transfer coupled to an acceptor molecule. *The model of* Ph-2/2HbO *was kindly provided by L. Boechi.* (For colour version of this figure, the reader is referred to the online version of this chapter.)

Thus Ph-2/2HbO is likely to be involved in a redox reaction(s) associating diatomic ligands and their derived oxidative species. Such a reaction is not unusual, since globins generally display NO dioxygenase activity leading to nitrate synthesis (Flögel, Merx, Gödecke, Decking, & Schrader, 2001; Gardner et al., 2006; Fig. 8.7B). It is noteworthy that Ph-2/2HbO provides protection against NO and related reactive species (Fig. 8.7D), under aerobic conditions (Coppola et al., 2013; see Section 6.2.4). Moreover, Ph-2/2HbO exhibits a twofold higher nitrite-reductase activity than horse Mb at pH 7.0, 25 °C. This evidence suggests (Fig. 8.7C) that, during an anaerobic phase, Ph-2/2HbO may supply NO via nitrite reduction (Russo et al., 2013).

Other reactions (Fig. 8.7E) may involve complex ROS chemistry (Flögel, Gödecke, Klotz, & Schrader, 2004). O_2 is necessary for bacterial metabolism, but can become poisonous if it is responsible for oxidative-stress burst. A number of reactions take place between ROS and the haem-Fe atom since O_2 is reduced by four e^- before yielding a water molecule with three ROS intermediates, whereas the haem-Fe atom is susceptible to oxidation from +2 to +4. In general, pentacoordinated globins are more prone to ROS oxidation than hexacoordinated forms (Herold, Kalinga, Matsui, & Watanabe, 2004; Lardinois, Tomer, Mason, & Deterding, 2008). However, in Ph-2/2HbO, due to the low affinity of the haem for distal Tyr (weak protection) in ferrous and ferric states, the protection against deleterious oxidation at high ROS concentration (autocatalytic oxidations leading to irreversible haem oxidation and globin degradation) is not expected (Russo et al., 2013).

The redox state of Ph-2/2HbO in vivo will depend on the presence of specific reductases and on the O_2 levels. The redox state could be involved in an electron-transfer reaction or in a regulatory mechanism with the protein acting as a redox sensor. In fact, at high O_2 concentration, in the presence of a reducing system that can compensate for autoxidation, the globin will be mainly ferrous, but under intermediate conditions it could be in the ferric form. This behaviour has been observed in pentacoordinated sperm whale and pig Mb (II), probably upon nucleophile attack such as that mediated by water (Brantley, Smerdon, Wilkinson, Singleton, & Olson, 1993), and in hexacoordinated bis-His form of the globin GLB-26 of the nematode worm Caenorhabditis elegans and of human Ngb, which promotes O_2 reduction (Kiger et al., 2011). By analogy, the hexacoordinated His-Fe-Tyr adduct of Ph-2/2HbO (Fig. 8.7F) may be involved in electron transfer (Russo et al., 2013).

6.2.4 NO detoxification

The remarkably high number of TrHbs in the *Ph*TAC125 genome strongly suggests that these globins fulfil important functional roles associated with the extreme features of the Antarctic habitat. The involvement of cold-adapted *Ph*-2/2HbO in detoxification of RNS and ROS may be a mechanism associated with high production of toxic species upon cold stress. A similar function has been reported in HbN of *M. tuberculosis* (Pathania et al., 2002) and *M. bovis* (Ouellett et al., 2002) and *M. leprae* HbO (Fabozzi, Ascenzi, Renzi, & Visca, 2006), which protect pathogenic bacteria from the toxic activity of macrophage-generated RNS and ROS.

The physiological role of *Ph*-2/2HbO has been investigated using a genomic approach, by the construction of a *Ph*TAC125 mutant strain in which the *PSHAa0030* gene was inactivated by insertional mutagenesis. The mutant strain was grown under controlled conditions and its growth behaviour was compared to that of wild-type cells, changing O_2 pressure in solution and growth temperature (4 and 15 °C). Regardless of temperature, growth of the mutant strain in extreme aerobiosis is lower than that of the wild type, also in terms of biomass. The presence of *Ph*-2/2HbO in wild-type cells is thus an advantage when cells are grown at high O_2 concentration. In micro-aerobiosis, both strains slow down their replication kinetics. At 4 °C, the wild-type cells appear better suited to the challenging conditions, reaching higher biomass than the mutant cells (Parrilli, Giuliani, Giordano, et al., 2010).

The inactivation of the *Ph*-2/2HbO gene makes the mutant strain sensitive to high O_2 levels, hydrogen peroxide and nitrosating agents (Parrilli, Giuliani, Giordano, et al., 2010), suggesting involvement of the protein in protection from oxidative and nitrosative stress. Moreover, the transcription of the FHb-encoding gene occurs only in the mutant in which *PSHAa0030* is inactivated, when grown in micro-aerobiosis at 4 °C, suggesting that the occurrence of the NO-induced stress is probably related to the absence of *Ph*-2/2HbO (Parrilli, Giuliani, Giordano, et al., 2010). In micro-aerobiosis, *Ph*TAC125 may endogenously produce NO, due to a gene encoding a nitrite reductase (*PSHAa1477*), as reported in other Gram-negative bacteria (Corker & Poole, 2003; Ji & Hollocher, 1988). Further, NO may accumulate when its spontaneous oxidation is limited by low O_2 availability. In micro-aerobiosis, O_2 is further reduced when the biomass is increased, that is, in the late exponential phase, and NO accumulation may become a serious threat for cell viability. Induction of the FHb gene may be viewed as a suitable strategy aimed at counteracting NO-induced stress due to the absence of *Ph*-2/2HbO (Parrilli, Giuliani, Giordano, et al., 2010).

An up-to-date approach set up by Coppola et al. (2013) to establish the participation of Ph-2/2HbO in RNS detoxification, tested the influence of heterologous expression of the $PSHAa0030$ gene $in\ vivo$ on protection from NO toxicity in a NO-sensitive $E.\ coli$ strain ($E.\ coli\ hmp$, defective in the FHb) (see Section 4.1).

The growth properties and O_2 uptake of $E.\ coli\ hmp$ having the $PSHAa0030$ gene was analysed in an attempt to demonstrate that Ph-2/2HbO offers resistance to nitrosative stress. Wild-type $E.\ coli$ and a hmp mutant, carrying the $PSHAa0030$ gene or not, were grown at 25 °C under aerobic conditions and treated with either the NO-releaser DETA-NONOate or the nitrosating agent GSNO. As expected, exposure to these sources of NO has no effect on the growth of wild-type $E.\ coli$ or the expression of Hmp from the complemented plasmid, and a comparable level of resistance is evident in cells expressing Ph-2/2HbO. In contrast, in the absence of Ph-2/2HbO, expression results in severe growth inhibition (Coppola et al., 2013).

Moreover, Coppola et al. (2013) demonstrated that upon addition of NO, $E.\ coli\ hmp$ not expressing Ph-2/2HbO shows prolonged inhibition of O_2 uptake (Fig. 8.8A) until the NO level falls steadily. In contrast, in the mutant strain carrying hmp^+, the addition of NO does not inhibit O_2 uptake (Fig. 8.8B), confirming that Hmp is able to detoxify NO, as reported previously (Membrillo-Hernández et al., 1999; Mills, Sedelnikova, Søballe, Hughes, & Poole, 2001; Stevanin et al., 2000). Upon addition of NO to the $E.\ coli$ mutant carrying Ph-2/2HbO, only very short periods of inhibition of respiration are observed and, again, the disappearance of NO is very fast (Fig. 8.8C). When NO reaches negligible levels, the O_2 uptake is brought back to a rate similar to that occurring before NO addition, unlike in the cells bearing the empty vector (Fig. 8.8A). Following exhaustion of O_2, further addition of NO results in a larger signal and a slower rate of consumption (Coppola et al., 2013).

Under aerobic conditions, over-expression of Ph-2/2HbO provides significant resistance to NO and nitrosating agents and distinct NO consumption ability to the NO-sensitive $E.\ coli\ hmp$ mutant. In contrast, growth curves and cellular respiration are strongly inhibited in $E.\ coli\ hmp$ not expressing the Antarctic globin gene. These results are clear evidence of a very important physiological role of Ph-2/2HbO in PhTAC125.

$In\ vitro$ kinetic studies of peroxynitrite isomerisation by the ferric protein support the NO and O_2^- detoxification activity as a possible functional role of the cold-adapted globin, thus confirming the involvement of Ph-2/2HbO in the protection from nitrosative stress. The high reactivity of

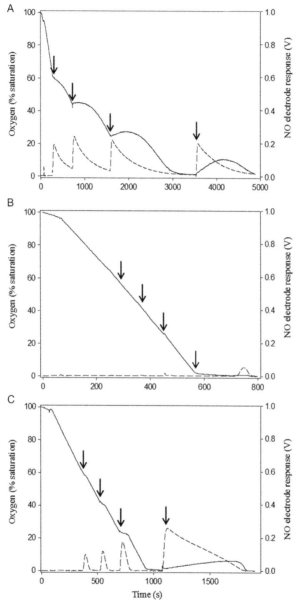

Figure 8.8 NO uptake and respiration of *E. coli hmp* cells (A) carrying the empty vector pBAD/HisA, or (B) expressing Hmp, or (C) *Ph*-2/2HbO. Respiration was followed in a Clark-type O$_2$ electrode (solid lines) upon addition of 1 μM Proli-NONOate (arrows). NO uptake was measured simultaneously with NO electrode (dashed lines). *Adapted from Coppola et al. (2013).*

ferric Ph-2/2HbO towards peroxynitrite at low temperature ($k_{on} = 3.5 \times 10^4\ M^{-1}\ s^{-1}$ at 5 °C and $2.9 \times 10^4\ M^{-1}\ s^{-1}$ at 20 °C) suggests that the protection of the psychrophile PhTAC125 against RNS and ROS may happen also in the cold Antarctic environment (Coppola et al., 2013). The k_{on} values for the ferric Ph-2/2HbO-mediated peroxynitrite isomerisation are similar to those reported for catalysis of the ferric equine-heart Mb ($2.9 \times 10^4\ M^{-1}\ s^{-1}$, Herold & Kalinga, 2003), sperm–whale Mb ($1.6 \times 10^4\ M^{-1}\ s^{-1}$, Herold et al., 2004) and human Hb ($1.2 \times 10^4\ M^{-1}\ s^{-1}$, Herold & Kalinga, 2003), whereas they are lower than those of horse-heart native and carboxymethylated cytochrome c in the presence of saturating cardiolipin (3.2×10^5 and $5.3 \times 10^5\ M^{-1}\ s^{-1}$, respectively), horse–heart car-boxymethylated cytochrome c ($6.8 \times 10^4\ M^{-1}\ s^{-1}$), and human serum haem–albumin ($4.1 \times 10^5\ M^{-1}\ s^{-1}$) (Ascenzi, Bolli, et al., 2011; Ascenzi, Bolli, Gullotta, Fanali, & Fasano, 2010; Ascenzi, Ciaccio, Sinibaldi, Santucci, & Coletta, 2011a, 2011b; Ascenzi et al., 2009). As reported for several haemoproteins, the acceleration of the peroxynitrite–isomerisation rate by ferric Ph-2/2HbO seems to be caused by reaction of peroxynitrite with the ferric penta-coordinated derivative only (Ascenzi, Bolli, et al., 2011, 2010; Ascenzi et al., 2011a, 2011b, 2009; Goldstein, Lind, & Merényi, 2005; Herold & Kalinga, 2003; Herold et al., 2004).

Taken together, the $in\ vivo$ and $in\ vitro$ evidence suggests that, under aerobic conditions, Ph-2/2HbO supplies cell protection against RNS and ROS, compensating for the defect in NO detoxification of $E.\ coli\ hmp$, which lacks the major NO-scavenging protein.

Finally, attempting to ascertain whether the N–terminal motif of Ph-2/2HbO is a requirement for efficient NO scavenging, similar to $M.\ tuberculosis$ HbN (Lama et al., 2009; see Section 6.2.1), Coppola et al. (2013) investigated the effect of deletion of the first 20 residues of the N-terminus of the protein on the ability of $E.\ coli\ hmp$ strain to deal with nitrosative stress; while full-length Ph-2/2HbO restores the ability to survive and grow under nitrosative-stress conditions, the protein without the N–terminal extension does not significantly contribute to NO detoxification, since its deletion does reduce the globin NO-scavenging ability in the heterologous host, unlike in $M.\ tuberculosis$ HbN.

7. CONCLUSION AND PERSPECTIVES

The Antarctic exhibits stable living conditions due to substantial isolation, also by virtue of the Polar Front. The evolutionary processes of

Antarctic organisms and the time scales in the context of geological and climatic changes have been extensively analysed and discussed (Peck, 2011).

The structure and function of proteins are the basis for understanding the evolutionary forces operating at sub-zero temperature, and—in this context—the knowledge gained at the molecular level is also crucial for predictions of the evolutionary consequences of global warming. In fact, at all analysed levels, the functional adaptation to permanently low temperature appears to require maintenance of flexibility of molecules in order to adequately support the cellular functioning. Proteins are one main factor of the ensuing mechanisms of adaptation.

Temperature is the prime driver that shaped the current structure and function of polar communities. Amongst abiotic factors influenced by temperature, O_2 and CO_2, and their concentrations, play an important role in life-sustaining processes. Due to low temperature, their concentrations are several-fold higher than in temperate and tropical marine habitats. The temperature-dependent balance between O_2 demand/supply and the associated functional capacity for specific functions of macromolecules shape the performance window in polar species (Pörtner et al., 2007). In polar environments, the benefits of O_2 levels (high by default) are indeed largely apparent, because they are counterbalanced by the kinetics of biological processes operating at low temperature (D'Amico et al., 2006) which decreases the rates, and by increased production of ROS. O_2 is obviously necessary for aerobic bacterial metabolism, but it can become poisonous in triggering oxidative-stress bursts.

In view of these considerations, in all Antarctic organisms, biological processes envisaging O_2 (respiration, transport/release, scavenging, reactive species, etc.) and other gases are bound to attract the interest of biologists.

In the realm of microbial life, the cold-adapted bacterial protein *Ph-2/2HbO* displays hexacoordination of the ferric and ferrous haem–Fe atom (Giordano et al., 2011; Howes et al., 2011). Investigating the features of this globin, in an attempt to shed light on possible multiplicity of functions, has been an important task. For instance, *Ph-2/2HbO* appears to exhibit a pseudo-enzymatic function in which O_2 is involved (Russo et al., 2013) and is available for reactions with NO to produce nitrate anions. In a single molecule, multiple conformations (penta- vs. hexacoordinated states) may account for multiple functions: under aerobic conditions, on one hand, *Ph-2/2HbO* provides cells with protection against NO and related RNS; on the other, during the anaerobic phase, *Ph-2/2HbO* may provide NO via nitrite reduction. The evidence summarised here indicates that

Ph-2/2HbO displays unique adaptive structural properties ensuring higher flexibility, thus facilitating its function at low temperatures, for example, by enhancing the capacity for correct positioning of ligand(s), which would be made more difficult by a rigid structure. In summary, this globin is a notable case study of relationship between molecular structure, cold adaptation and a wide range of equally important biological functions.

Hexacoordinated globins in Antarctic microorganisms call for efforts in shedding light on their place and role in the context of evolution. Knowledge of the range of their functions and physiological role *in vivo* is still incomplete and a matter of lively debate. However, modern concepts of biological sciences appear more and more to support the idea that the physiological role of a given molecule is not restricted to a single aspect, although one aspect may well be predominant. Based on this assumption, the current knowledge summarised in this review seems to be a useful starting point to achieve progress by further investigations aimed at increasing our albeit incomplete understanding of the biological function of a fundamentally important class of macromolecules such as globins, not only *Ph*-2/2HbO—a valuable case study—but also other globins, for example, Ngb and Cygb, whose biomedical significance is steadily growing.

ACKNOWLEDGEMENTS

The authors wish to thank the Centre de Ressources Biologiques de l'Institut Pasteur, Paris, France (http://www.crbip.pasteur.fr) for supplying the *P. haloplanktis* CIP 108707 strain. This study was carried out in the framework of the SCAR programme "Antarctic Thresholds–Ecosystem Resilience and Adaptation" (AnT-ERA), and of the "Coordination Action for Research Activities on Life in Extreme Environments" (CAREX), European Commission FP7 call ENV.2007.2.2.1.6. It was financially supported by the Italian National Programme for Antarctic Research (PNRA).

F. M. is supported by a fellowship from the Australian Research Council (DE120102610).

M. T.-T. is supported by Consejo Nacional de Ciencia y Tecnologia (CONACyT) (Mexico) through grant number 99171 and Consejo Estatal de Ciencia, Tecnología e Innovación de Michoacán (CECTI) through grant number 007. Thanks are due to Prof. Robert K. Poole for kindly inviting us to submit this review. We apologise in case we have neglected references relevant to the subject.

REFERENCES

Abell, G. C., & Bowman, J. P. (2005). Ecological and biogeographic relationships of class Flavobacteria in the Southern Ocean. *FEMS Microbiology Ecology*, *51*, 265–277.

Abergel, C., Walburger, A., Chenivesse, S., & Lazdunski, C. (2001). Crystallization and preliminary crystallographic study of the peptidoglycan-associated lipoprotein from *Escherichia coli*. *Acta Crystallographica. Section D, Biological Crystallography*, *57*, 317–319.

Agogué, H., Lamy, D., Neal, P. R., Sogin, M. L., & Herndl, G. J. (2011). Water mass-specificity of bacterial communities in the North Atlantic revealed by massively parallel sequencing. *Molecular Ecology, 20,* 258–274.

Allen, M. A., Lauro, F. M., Williams, T. J., Burg, D., Siddiqui, K. S., De Francisci, D., et al. (2009). The genome sequence of the psychrophilic archaeon, *Methanococcoides burtonii*: The role of genome evolution in cold adaptation. *The ISME Journal, 3,* 1012–1035.

Antonini, E., & Brunori, M. (1971). *Hemoglobin and myoglobin in their reactions with ligands.* Amsterdam: North-Holland Publishing Co.

Ascenzi, P., Bolli, A., di Masi, A., Tundo, G. R., Fanali, G., Coletta, M., et al. (2011). Isoniazid and rifampicin inhibit allosterically heme binding to albumin and peroxynitrite isomerization by heme-albumin. *Journal of Biological Inorganic Chemistry, 16,* 97–108.

Ascenzi, P., Bolli, A., Gullotta, F., Fanali, G., & Fasano, M. (2010). Drug binding to Sudlow's site I impairs allosterically human serum heme-albumin-catalyzed peroxynitrite detoxification. *IUBMB Life, 62,* 776–780.

Ascenzi, P., Ciaccio, C., Sinibaldi, F., Santucci, R., & Coletta, M. (2011a). Cardiolipin modulate allosterically peroxynitrite detoxification by horse heart cytochrome *c. Biochemical and Biophysical Research Communications, 404,* 190–194.

Ascenzi, P., Ciaccio, C., Sinibaldi, F., Santucci, R., & Coletta, M. (2011b). Peroxynitrite detoxification by horse heart carboxymethylated cytochrome *c* is allosterically modulated by cardiolipin. *Biochemical and Biophysical Research Communications, 415,* 463–467.

Ascenzi, P., De Marinis, E., Coletta, M., & Visca, P. (2008). H_2O_2 and NO scavenging by *Mycobacterium leprae* truncated hemoglobin O. *Biochemical and Biophysical Research Communications, 373,* 197–201.

Ascenzi, P., di Masi, A., Coletta, M., Ciaccio, C., Fanali, G., Nicoletti, F. P., et al. (2009). Ibuprofen impairs allosterically peroxynitrite isomerization by ferric human serum heme-albumin. *Journal of Biological Chemistry, 284,* 31006–31017.

Aurilia, V., Parracino, A., Saviano, M., Rossi, M., & D'Auria, S. (2007). The psychrophilic bacterium *Pseudoalteromonas haloplanktis* TAC125 possesses a gene coding for a cold-adapted feruloyl esterase activity that shares homology with esterase enzymes from γ-proteobacteria and yeast. *Gene, 397,* 51–57.

Ayala-del-Rio, H. L., Chain, P. S., Grzymski, J. J., Ponder, M. A., Ivanova, N., Bergholz, P. W., et al. (2010). The genome sequence of *Psychrobacter arcticus* 273-4, a psychroactive Siberian permafrost bacterium, reveals mechanisms for adaptation to low-temperature growth. *Applied and Environmental Microbiology, 76,* 2304–2312.

Ayub, N. D., Tribelli, P. M., & Lopez, N. I. (2009). Polyhydroxyalkanoates are essential for maintenance of redox state in the Antarctic bacterium *Pseudomonas* sp. 14-3 during low temperature adaptation. *Extremophiles, 13,* 59–66.

Azam, F. (1998). Microbial control of oceanic carbon flux: The plot thickens. *Science, 280,* 694–696.

Azam, F., & Malfatti, F. (2007). Microbial structuring of marine ecosystems. *Nature Reviews. Microbiology, 5,* 782–791.

Bakermans, C., Tollaksen, S. L., Giometti, C. S., Wilkerson, C., Tiedje, J. M., & Thomashow, M. F. (2007). Proteomic analysis of *Psychrobacter cryohalolentis* K5 during growth at subzero temperatures. *Extremophiles, 11,* 343–354.

Baldwin, R. L. (2008). The search for folding intermediates and the mechanism of protein folding. *Annual Review of Biophysics, 37,* 1–21.

Bergholz, P. W., Bakermans, C., & Tiedje, J. M. (2009). *Psychrobacter arcticus* 273-4 uses resource efficiency and molecular motion adaptations for subzero temperature growth. *Journal of Bacteriology, 191,* 2340–2352.

Bertoni, G., Bolognese, F., Galli, E., & Barbieri, P. (1996). Cloning of the genes for and characterization of the early stages of toluene and o-xylene catabolism in *Pseudomonas stutzeri* OX1. *Applied and Environmental Microbiology, 62,* 3704–3711.

Bidon-Chanal, A., Martí, M. A., Crespo, A., Milani, M., Orozco, M., Bolognesi, M., et al. (2006). Ligand-induced dynamical regulation of NO conversion in *Mycobacterium tuberculosis* truncated hemoglobin-N. *Proteins, 64*, 457–464.

Bolli, A., Ciaccio, C., Coletta, M., Nardini, M., Bolognesi, M., Pesce, A., et al. (2008). Ferrous *Campylobacter jejuni* truncated hemoglobin P displays an extremely high reactivity for cyanide—A comparative study. *The FEBS Journal, 275*, 633–645.

Bolognesi, M., Bordo, D., Rizzi, M., Tarricone, C., & Ascenzi, P. (1997). Nonvertebrate hemoglobins: Structural bases for reactivity. *Progress in Biophysics and Molecular Biology, 68*, 29–68.

Bonamore, A., & Boffi, A. (2008). Flavohemoglobin: Structure and reactivity. *IUBMB Life, 60*, 19–28.

Bonamore, A., Ilari, A., Giangiacomo, L., Bellelli, A., Morea, V., & Boffi, A. (2005). A novel thermostable hemoglobin from the actinobacterium *Thermobifida fusca*. *The FEBS Journal, 272*, 4189–4201.

Bowman, L. A., McLean, S., Poole, R. K., & Fukuto, J. M. (2011). The diversity of microbial responses to nitric oxide and agents of nitrosative stress close cousins but not identical twins. *Advances in Microbial Physiology, 59*, 135–219.

Bowman, J. S., Rasmussen, S., Blom, N., Deming, J. W., Rysgaard, S., & Sicheritz-Ponten, T. (2012). Microbial community structure of Arctic multiyear sea ice and surface seawater by 454 sequencing of the 16S RNA gene. *The ISME Journal, 6*, 11–20.

Brantley, R. E., Jr., Smerdon, S. J., Wilkinson, A. J., Singleton, E. W., & Olson, J. S. (1993). The mechanism of autooxidation of myoglobin. *Journal of Biological Chemistry, 268*, 6995–7010.

Brinkmeyer, R., Knittel, K., Jürgens, J., Weyland, H., Amann, R., & Helmke, E. (2003). Diversity and structure of bacterial communities in arctic versus Antarctic pack ice. *Applied and Environmental Microbiology, 69*, 6610–6619.

Brown, M. V., & Bowman, J. P. (2001). A molecular phylogenetic survey of sea-ice microbial communities (SIMCO). *FEMS Microbiology Ecology, 35*, 267–275.

Brown, M. V., Lauro, F. M., DeMaere, M. Z., Muir, L., Wilkins, D., Thomas, T., et al. (2012). Global biogeography of SAR11 marine bacteria. *Molecular Systems Biology, 8*, 595.

Burmester, T., Ebner, B., Weich, B., & Hankeln, T. (2002). Cytoglobin: A novel globin type ubiquitously expressed in vertebrate tissues. *Molecular Biology and Evolution, 19*, 416–421.

Burmester, T., Weich, B., Reinhardt, S., & Hankeln, T. (2000). A vertebrate globin expressed in the brain. *Nature, 407*, 520–523.

Buttner, K., Bernhardt, J., Scharf, C., Schmid, R., Mäder, U., Eymann, C., et al. (2001). A comprehensive two-dimensional map of cytosolic proteins of *Bacillus subtilis*. *Electrophoresis, 22*, 2908–2935.

Campanaro, S., Williams, T. J., Burg, D. W., De Francisci, D., Treu, L., Lauro, F. M., et al. (2011). Temperature-dependent global gene expression in the Antarctic archaeon *Methanococcoides burtonii*. *Environmental Microbiology, 13*, 2018–2038. http://dx.doi.org/10.1111/j.1462-2920.2010.02367.x.

Cartier, G., Lorieux, F., Allemand, F., Dreyfus, M., & Bizebard, T. (2010). Cold adaptation in DEAD-box proteins. *Biochemistry, 49*, 2636–2646.

Casanueva, A., Tuffin, M., Cary, C., & Cowan, D. A. (2010). Molecular adaptations to psychrophily: The impact of 'omic' technologies. *Trends in Microbiology, 18*, 374–381.

Castello, P., David, P., McClure, T., Crook, Z., & Poyton, R. (2006). Mitochondrial cytochrome oxidase produces nitric oxide under hypoxic conditions: Implications for oxygen sensing and hypoxic signaling in eukaryotes. *Cell Metabolism, 3*, 277–287.

Cavicchioli, R., Siddiqui, K., Andrews, D., & Sowers, K. R. (2002). Low temperature extremophiles and their applications. *Current Opinion in Biotechnology, 13*, 253–261.

Cavicchioli, R., Thomas, T., & Curmi, P. M. (2000). Cold stress response in Archaea. *Extremophiles, 4*, 321–331.

Chauhan, S., & Shivaji, S. (1994). Growth and pigmentation in *Sphingobacterium antarcticus*, a psychrotrophic bacterium from Antarctica. *Polar Biology, 1*, 31–36.

Chen, Z., Cheng, C. H. C., Zhang, J., Cao, L., Chen, L., Zhou, L., et al. (2008). Transcriptomic and genomic evolution under constant cold in Antarctic notothenioid fish. *Proceedings of the National Academy of Sciences of the United States of America, 105*, 12944–12949.

Chintalapati, S., Kiran, M. D., & Shivaji, S. (2004). Role of membrane lipid fatty acids in cold adaptation. *Cell and Molecular Biology (Noisy-le-Grand), 50*, 631–642.

Clarke, A., Murphy, E. J., Meredith, M. P., King, J. C., Peck, L. S., Barnes, D. K. A., et al. (2007). Climate change and the marine ecosystem of the western Antarctic Peninsula. *Philosophical Transactions of the Royal Society B: Biological Sciences, 362*, 149–166.

Comiso, J. C., Parkinson, C. L., Gersten, R., & Stock, L. (2008). Accelerated decline in the Arctic sea ice cover. *Geophysical Research Letters, 35*, L01703. http://dx.doi.org/10.1029/2007GL031972.

Connelly, T. L., Tilburg, C. M., & Yager, P. L. (2006). Evidence for psychrophiles outnumbering psychrotolerant marine bacteria in the springtime coastal Arctic. *Limnology and Oceanography, 51*, 1205–1210.

Convey, P. (2006). Antarctic climate change and its influences on terrestrial ecosystems. In D. M. Bergstrom, P. Convey, & A. H. L. Huiskes (Eds.), *Trends in Antarctic terrestrial and limnetic ecosystems: Antarctica as a global indicator* (pp. 253–272). Springer: Dordrecht.

Convey, P. (2010). Life history adaptations to polar and alpine environments. In D. L. Denlinger & R. E. Lee (Eds.), *Low temperature biology of insects* (pp. 297–321). Cambridge: Cambridge University Press.

Convey, P., Bindschadler, R., di Prisco, G., Fahrbach, E., Gutt, J., Hodgson, D. A., et al. (2009). Antarctic climate change and the environment. *Antarctic Science, 21*, 541–563.

Coppola, D., Giordano, D., Tinajero-Trejo, M., di Prisco, G., Ascenzi, P., Poole, R. K., et al. (2013). Antarctic bacterial hemoglobin and its role in the protection against nitrogen reactive species. *Biochimica et Biophysica Acta, 1834*(9), 1923–1931.

Corker, H., & Poole, R. K. (2003). Nitric oxide formation by *Escherichia coli*—Dependence on nitrite reductase, the NO-sensing regulator FNR, and flavohemoglobin Hmp. *Journal of Biological Chemistry, 278*, 31584–31592.

Couture, M., Das, T. K., Lee, H. C., Peisach, J., Rousseau, D. L., Wittenberg, B. A., et al. (1999). *Chlamydomonas* chloroplast ferrous hemoglobin. Heme pocket structure and reaction with ligands. *Journal of Biological Chemistry, 274*, 6898–6910.

Cowan, D. A., Chown, S. L., Convey, P., Tuffin, M., Hughes, K., Pointing, S., et al. (2011). Non-indigenous microorganisms in the Antarctic: Assessing the risks. *Trends in Microbiology, 19*, 540–548.

Cusano, A. M., Parrilli, E., Duilio, A., Sannia, G., Marino, G., & Tutino, M. L. (2006). Secretion of psychrophilic alpha-amylase deletion mutants in *Pseudoalteromonas haloplanktis* TAC125. *FEMS Microbiology Letters, 258*, 67–71.

D'Amico, S., Claverie, P., Collins, T., Georlette, D., Gratia, E., Hoyoux, A., et al. (2002). Molecular basis of cold adaptation. *Philosophical Transactions of the Royal Society B: Biological Sciences, 357*, 917–925.

D'Amico, S., Collins, T., Marx, J. C., Feller, G., & Gerday, C. (2006). Psychrophilic microorganisms: Challenges for life. *EMBO Reports, 7*, 385–389.

D'Angelo, P., Lucarelli, D., Della Longa, S., Benfatto, M., Hazemann, J. L., Feis, A., et al. (2004). Unusual heme iron-lipid acyl chain coordination in Escherichia coli flavohemoglobin. *Biophysical Journal, 86*, 3882–3892.

Daniel, R. M., Dines, M., & Petach, H. H. (1996). The denaturation and degradation of stable enzymes at high temperatures. *The Biochemical Journal, 317*, 1–11.

Das, T. K., Couture, M., Lee, H. C., Peisach, J., Rousseau, D. L., Wittenberg, B. A., et al. (1999). Identification of the ligands to the ferric heme of *Chlamydomonas* chloroplast hemoglobin: Evidence for ligation of tyrosine-63 (B10) to the heme. *Biochemistry, 38,* 15360–15368.

Dayton, P. K., Mordida, B. J., & Bacon, F. (1994). Polar marine communities. *American Zoology, 34,* 90–99.

DeLong, E. F., Preston, C. M., Mincer, T., Rich, V., Hallam, S. J., Frigaard, N. U., et al. (2006). Community genomics among stratified microbial assemblages in the ocean's interior. *Science, 311,* 496–503.

de Pascale, D., Cusano, A. M., Autore, F., Parrilli, E., di Prisco, G., Marino, G., et al. (2008). The cold-active Lip1 lipase from the Antarctic bacterium *Pseudoalteromonas haloplanktis* TAC125 is a member of a new bacterial lipolytic enzyme family. *Extremophiles, 12,* 311–323.

de Sanctis, D., Dewilde, S., Pesce, A., Moens, L., Ascenzi, P., Hankeln, T., et al. (2004). Crystal structure of cytoglobin: The fourth globin type discovered in man displays heme hexa-coordination. *Journal of Molecular Biology, 336,* 917–927.

Dewilde, S., Ebner, B., Vinck, E., Gilany, K., Hankeln, T., Burmester, T., et al. (2006). The nerve hemoglobin of the bivalve mollusc *Spisula solidissima*: Molecular cloning, ligand binding studies, and phylogenetic analysis. *Journal of Biological Chemistry, 281,* 5364–5372.

Dewilde, S., Kiger, L., Burmester, T., Hankeln, T., Baudin-Creuza, V., Aerts, T., et al. (2001). Biochemical characterization and ligand binding properties of neuroglobin, a novel member of the globin family. *Journal of Biological Chemistry, 276,* 38949–38955.

Di Giuseppe, G., Erra, F., Dini, F., Alimenti, C., Vallesi, A., Pedrini, B., et al. (2011). Antarctic and Arctic populations of the ciliate Euplotes nobilii show common pheromone-mediated cell-cell signaling and cross-mating. *Proceedings of the National Academy of Sciences of the United States of America, 108,* 3181–3186.

Douglas, D. J. (1984). Microautoradiography-based enumeration of photosynthetic picoplankton with estimates of carbon-specific growth rates. *Marine Ecology Progress Series, 14,* 223–228.

Duchaud, E., Boussaha, M., Loux, V., Bernardet, J. F., Michel, C., Kerouault, B., et al. (2007). Complete genome sequence of the fish pathogen *Flavobacterium psychrophilum*. *Nature Biotechnology, 25,* 763–769.

Ducklow, H. (1999). The bacterial component of the oceanic euphotic zone. *FEMS Microbiology Ecology, 30,* 1–10.

Duilio, A., Tutino, M. L., & Marino, G. (2004). Recombinant protein production in Antarctic Gram-negative bacteria. *Methods in Molecular Biology, 267,* 225–237.

Eastman, J. T. (2005). The nature of the diversity of Antarctic fishes. *Polar Biology, 28,* 93–107.

Elvers, K. T., Turner, S. M., Wainwright, L. M., Marsden, G., Hinds, J., Cole, J. A., et al. (2005). NssR, a member of the Crp-Fnr superfamily from *Campylobacter jejuni*, regulates a nitrosative stress-responsive regulon that includes both a single-domain and a truncated haemoglobin. *Molecular Microbiology, 57,* 735–750.

Elvers, K. T., Wu, G., Gilberthorpe, N. J., Poole, R. K., & Park, S. F. (2004). Role of an inducible single-domain hemoglobin in mediating resistance to nitric oxide and nitrosative stress in *Campylobacter jejuni* and *Campylobacter coli*. *The Biochemical Journal, 186,* 5332–5341.

Eriksson, S., Hurme, R., & Rhen, M. (2002). Low temperature sensors in bacteria. *Philosophical Transactions of the Royal Society B: Biological Sciences, 357,* 887–893.

Fabozzi, G., Ascenzi, P., Renzi, S. D., & Visca, P. (2006). Truncated hemoglobin GlbO from *Mycobacterium leprae* alleviates nitric oxide toxicity. *Microbial Pathogenesis, 40,* 211–220.

Falzone, C. J., Christie Vu, B., Scott, N. L., & Lecomte, J. T. (2002). The solution structure of the recombinant hemoglobin from the cyanobacterium Synechocystis sp. PCC 6803 in its hemichrome state. *Journal of Molecular Biology, 324,* 1015–1029.

Feller, G. (2010). Protein stability and enzyme activity at extreme biological temperatures. *Journal of Physics: Condensed Matter, 22*, 323101.

Feller, G., & Gerday, C. (1997). Psychrophilic enzymes: Molecular basis of cold adaptation. *Cellular and Molecular Life Sciences, 53*, 830–841.

Feller, G., & Gerday, C. (2003). Psychrophilic enzymes: Hot topics in cold adaptation. *Nature Reviews. Microbiology, 1*, 200–208.

Feng, L., Zhou, S., Gu, L., Gell, D., Mackay, J., Weiss, M., et al. (2005). Structure of oxidized α-haemoglobin bound to AHSP reveals a protective mechanism for heme. *Nature, 435*, 697–701.

Finkel, T. (2003). Oxidant signals and oxidative stress. *Current Opinion in Cell Biology, 15*, 247–254.

Fiocchetti, M., De Marinis, E., Ascenzi, P., & Marino, M. (2013). Neuroglobin and neuronal cell survival. *Biochimica et Biophysica Acta, 1834*(9), 1744–1749.

Flögel, U., Gödecke, A., Klotz, L. O., & Schrader, J. (2004). Role of myoglobin in the antioxidant defense of the heart. *The FASEB Journal, 18*, 1156–1158.

Flögel, U., Merx, M. W., Gödecke, A., Decking, U. K., & Schrader, J. (2001). Myoglobin: A scavenger of bioactive NO. *Proceedings of the National Academy of Sciences of the United States of America, 98*, 735–740.

Frey, A. D., Farres, J., Bollinger, C. J. T., & Kallio, P. T. (2002). Bacterial hemoglobins and flavohemoglobins for alleviation of nitrosative stress in *Escherichia coli*. *Applied and Environmental Microbiology, 68*, 4835–4840.

Frey, A. D., & Kallio, P. T. (2003). Bacterial hemoglobins and flavohemoglobins: Versatile proteins and their impact on microbiology and biotechnology. *FEMS Microbiology Reviews, 27*, 525–545.

Frey, A. D., Shepherd, M., Jokipii-Lukkari, S., Haggman, H., & Kallio, P. T. (2011). The single-domain globin of *Vitreoscilla* augmentation of aerobic metabolism for biotechnological applications. *Advances in Microbial Physiology, 58*, 81–139.

Galand, P. E., Casamayor, E. O., Kirchman, D. L., & Lovejoy, C. (2009). Ecology of the rare microbial biosphere of the Arctic Ocean. *Proceedings of the National Academy of Sciences of the United States of America, 106*, 22427–22432.

Gardner, P. R. (2005). Nitric oxide dioxygenase function and mechanism of flavohemoglobin, hemoglobin, myoglobin and their associated reductases. *Journal of Inorganic Biochemistry, 99*, 247–266.

Gardner, P. R., Gardner, A. M., Brashear, W. T., Suzuki, T., Hvitved, A. N., Setchell, K. D., et al. (2006). Hemoglobins dioxygenate nitric oxide with high fidelity. *Journal of Inorganic Biochemistry, 100*, 542–550.

Gardner, P. R., Gardner, A. M., Martin, L. A., Dou, Y., Li, T. S., Olson, J. S., et al. (2000). Nitric-oxide dioxygenase activity and function of flavohemoglobins-sensitivity to nitric oxide and carbon monoxide inhibition. *Journal of Biological Chemistry, 275*, 31581–31587.

Gardner, P. R., Gardner, A. M., Martin, L. A., & Salzman, A. L. (1998). Nitric oxide dioxygenase: An enzymic function for flavohemoglobin. *Proceedings of the National Academy of Sciences of the United States of America, 95*, 10378–10383.

Gasser, B., Saloheimo, M., Rinas, U., Dragosits, M., Rodríguez-Carmona, E., Baumann, K., et al. (2008). Protein folding and conformational stress in microbial cells producing recombinant proteins: A host comparative overview. *Microbial Cell Factories, 7*, 11. http://dx.doi.org/10.1186/1475-2859-7-11.

Georlette, D., Blaise, V., Collins, T., D'Amico, S., Gratia, E., Hoyoux, A., et al. (2004). Some like it cold: Biocatalysis at low temperatures. *FEMS Microbiology Reviews, 28*, 25–42.

Ghiglione, J. F., & Murray, A. E. (2012). Pronounced summer to winter differences and higher wintertime richness in coastal Antarctic marine bacterioplankton. *Environmental Microbiology, 14*, 617–629.

Giangiacomo, A., Ilari, L., Boffi, A., Morea, V., & Chiancone, E. (2005). The truncated oxygen-avid hemoglobin from *Bacillus subtilis*: X-ray structure and ligand binding properties. *Journal of Biological Chemistry*, *280*, 9192–9202.

Giebel, H. A., Brinkhoff, T., Zwisler, W., Selje, N., & Simon, M. (2009). Distribution of Roseobacter RCA and SAR11 lineages and distinct bacterial communities from the subtropics to the Southern Ocean. *Environmental Microbiology*, *11*, 2164–2178.

Gilberthorpe, N. J., Lee, M. E., Stevanin, T. M., Read, R. C., & Poole, R. K. (2007). NsrR: A key regulator circumventing *Salmonella enterica* serovar Typhimurium oxidative and nitrosative stress in vitro and in IFN-gamma-stimulated J774.2 macrophages. *Microbiology*, *153*, 1756–1771.

Giordano, D., Boron, I., Abbruzzetti, S., Van Leuven, W., Nicoletti, F. P., Forti, F., et al. (2012). Biophysical characterisation of neuroglobin of the icefish, a natural knockout for hemoglobin and myoglobin. Comparison with human neuroglobin. *PLoS One*, *7*, e44508.

Giordano, D., Parrilli, E., Dettaï, A., Russo, R., Barbiero, G., Marino, G., et al. (2007). The truncated hemoglobins in the Antarctic psychrophilic bacterium *Pseudoalteromonas haloplanktis* TAC125. *Gene*, *398*, 69–77.

Giordano, D., Russo, R., Ciaccio, C., Howes, B. D., di Prisco, G., Smulevich, G., et al. (2011). Ligand- and proton-linked conformational changes of the ferrous 2/2 hemoglobin of *Pseudoalteromonas haloplanktis* TAC125. *IUBMB Life*, *63*, 566–573.

Goldstein, S., Lind, J., & Merényi, G. (2005). Chemistry of peroxynitrites and peroxynitrates. *Chemical Reviews*, *105*, 2457–2470.

Goodchild, A., Raftery, M., Saunders, N. F., Guilhaus, M., & Cavicchioli, R. (2005). Cold adaptation of the Antarctic archaeon, *Methanococcoides burtonii* assessed by proteomics using ICAT. *Journal of Proteomic Research*, *4*, 473–480.

Goodchild, A., Saunders, N. F., Ertan, H., Raftery, M., Guilhaus, M., Curmi, P. M., et al. (2004). A proteomic determination of cold adaptation in the Antarctic archaeon, *Methanococcoides burtonii*. *Molecular Microbiology*, *53*, 309–321.

Granger, D. L., Perfect, J. R., & Durack, D. T. (1986). Macrophage-mediated fungistasis in vitro: Requirements for intracellular and extracellular cytotoxicity. *Journal of Immunology*, *136*, 672–680.

Green, S. J., Meltzer, M. S., Hibbs, J. B., Jr., & Nacy, C. A. (1990). Activated macrophages destroy intracellular *Leishmania major* amastigotes by an L-arginine-dependent killing mechanism. *Journal of Immunology*, *144*, 278–283.

Grzymski, J. J., Riesenfeld, C. S., Williams, T. J., Dussaq, A. M., Ducklow, H., Erickson, M., et al. (2012). A metagenomic assessment of winter and summer bacterioplankton from Antarctica Peninsula coastal surface waters. *The ISME Journal*, *6*, 1901–1915.

Hamdane, D., Kiger, L., Hui Bon Hoa, G., Dewilde, S., Uzan, J., Burmester, T., et al. (2005). High pressure enhances hexacoordination in neuroglobin and other globins. *Journal of Biological Chemistry*, *280*, 36809–36814.

Hartl, F. U., & Hayer-Hartl, M. (2009). Converging concepts of protein folding in vitro and in vivo. *Nature Structural & Molecular Biology*, *16*, 574–581.

Hausladen, A., Gow, A. J., & Stamler, J. S. (1998). Nitrosative stress: Metabolic pathway involving the flavohemoglobin. *Proceedings of the National Academy of Sciences of the United States of America*, *95*, 14100–14105.

Hausladen, A., Gow, A., & Stamler, J. S. (2001). Flavohemoglobin denitrosylase catalyzes the reaction of a nitroxyl equivalent with molecular oxygen. *Proceedings of the National Academy of Sciences of the United States of America*, *98*, 10108–10112.

Herold, S., & Kalinga, S. (2003). Metmyoglobin and methemoglobin catalyze the isomerization of peroxynitrite to nitrate. *Biochemistry*, *42*, 14036–14046.

Herold, S., Kalinga, S., Matsui, T., & Watanabe, Y. (2004). Mechanistic studies of the isomerization of peroxynitrite to nitrate catalyzed by distal histidine metmyoglobin mutants. *Journal of the American Chemical Society, 126,* 6945–6955.

Hille, R. (2002). Molybdenum and tungsten in biology. *Trends in Biochemical Sciences, 27,* 360–367.

Howes, B. D., Giordano, D., Boechi, L., Russo, R., Mucciacciaro, S., Ciaccio, C., et al. (2011). The peculiar heme pocket of the 2/2 hemoglobin of cold adapted *Pseudoalteromonas haloplanktis* TAC125. *Journal of Biological Inorganic Chemistry, 16,* 299–311.

Ilari, A., Giangiacomo, L., Boffi, A., & Chiancone, E. (2005). The truncated oxygen-avid hemoglobin from *Bacillus subtilis*: X-ray structure and ligand binding properties. *Journal of Biological Chemistry, 280,* 9192–9202.

Ilari, A., Kjelgaard, P., von Wachenfeldt, C., Catacchio, B., Chiancone, E., & Boffi, A. (2007). Crystal structure and ligand binding properties of the truncated hemoglobin from *Geobacillus stearothermophilus*. *Archives of Biochemistry and Biophysics, 457,* 85–94.

Iyengar, R., Stuehr, D. J., & Marletta, M. A. (1987). Macrophage synthesis of nitrite, nitrate and N-nitrosamines: Precursors and role of the respiratory burst. *Proceedings of the National Academy of Sciences of the United States of America, 84,* 6369–6373.

Jagannadham, M. V., Rao, V. J., & Shivaji, S. (1991). The major carotenoid pigment of a psychrotrophic *Micrococcus roseus*: Purifcation, structure, and interaction of the pigment with synthetic membranes. *Journal of Bacteriology, 173,* 7911–7917.

Ji, X. B., & Hollocher, T. C. (1988). Reduction of nitrite to nitric oxide by enteric bacteria. *Biochemical and Biophysical Research Communications, 157,* 106–108.

Junge, K., Imhoff, F., Staley, T., & Deming, W. (2002). Phylogenetic diversity of numerically important Arctic sea-ice bacteria cultured at subzero temperature. *Microbial Ecology, 43,* 315–328.

Kandror, O., & Goldberg, A. L. (1997). Trigger factor is induced upon cold shock and enhances viability of *Escherichia coli* at low temperatures. *Proceedings of the National Academy of Sciences of the United States of America, 94,* 4978–4981.

Kawamoto, J., Kurihara, T., Kitagawa, M., Kato, I., & Esaki, N. (2007). Proteomic studies of an Antarctic cold-adapted bacterium, *Shewanella livingstonensis* Ac10, for global identification of cold-inducible proteins. *Extremophiles, 11,* 819–826.

Kennett, J. P. (1977). Cenozoic evolution of Antarctic glaciation. The Circum-Antarctic Ocean, and the impact of global paleoceanography. *Journal of Geophysical Research, 82,* 3843–3860.

Kiefhaber, T., Rudolph, R., Kohler, H., & Buchner, J. (1991). Protein aggregation *in vitro* and *in vivo*: A quantitative model of the kinetic competition between folding and aggregation. *Biotechnology (NY), 9,* 825–829.

Kiger, L., Tilleman, L., Geuens, E., Hoogewijs, D., Lechauve, C., Moens, L., et al. (2011). Electron transfer function versus oxygen delivery: A comparative study for several hexacoordinated globins across the animal kingdom. *PLoS One, 6,* e20478.

Kim, S. O., Orii, Y., Lloyd, D., Hughes, M. N., & Poole, R. K. (1999). Anoxic function for the *Escherichia coli* flavohaemoglobin (Hmp): Reversible binding of nitric oxide and reduction to nitrous oxide. *FEBS Letters, 445,* 389–394.

Kirchman, D. L., Cottrell, M. T., & Lovejoy, C. (2010). The structure of bacterial communities in the western Arctic Ocean as revealed by pyrosequencing of 16S rRNA genes. *Environmental Microbiology, 12,* 1132–1143.

Kirchman, D. L., Morán, X. A. G., & Ducklow, H. (2009). Microbial growth in the polar oceans-role of temperature and potential impact of climate change. *Nature Reviews. Microbiology, 7,* 451–459.

Klappenbach, J. A., Dunbar, J. M., & Schmidt, T. M. (2000). rRNA operon copy number reflects ecological strategies of bacteria. *Applied and Environmental Microbiology, 66,* 1328–1333.

Klumpp, S., Zhang, Z., & Hwa, T. (2009). Growth rate-dependent global effects on gene expression in bacteria. *Cell, 139,* 1366–1375.

Kramer, G., Patzelt, H., Rauch, T., Kurz, T. A., Vorderwulbecke, S., Bukau, B., et al. (2004). Trigger factor peptidylprolyl cis/trans isomerase activity is not essential for the folding of cytosolic proteins in *Escherichia coli. Journal of Biological Chemistry, 279,* 14165–14170.

Kriegl, J. M., Bhattacharyya, A. J., Nienhaus, K., Deng, P., Minkow, O., & Nienhaus, G. U. (2002). Ligand binding and protein dynamics in neuroglobin. *Proceedings of the National Academy of Sciences of the United States of America, 99,* 7992–7997.

Lama, A., Pawaria, S., Bidon-Chanal, A., Anand, A., Gelpí, J. L., Arya, S., et al. (2009). Role of Pre-A motif in nitric oxide scavenging by truncated hemoglobin, HbN, of *Mycobacterium tuberculosis. Journal of Biological Chemistry, 284,* 14457–14468.

Lardinois, O. M., Tomer, K. B., Mason, R. P., & Deterding, L. J. (2008). Identification of protein radicals formed in the human neuroglobin-H_2O_2 reaction using immuno-spin trapping and mass spectrometry. *Biochemistry, 47,* 10440–10448.

Lauro, F. M., Chastain, R. A., Blankenship, L. E., Yayanos, A. A., & Bartlett, D. H. (2007). The unique 16S rRNA genes of piezophiles reflect both phylogeny and adaptation. *Applied and Environmental Microbiology, 73,* 838–845.

Lawver, L. A., & Gahagan, L. M. (2003). Evolution of Cenozoic seaways in the circum-Antarctic region. *Palaeogeography, Palaeoclimatology, Palaeoecology, 198,* 11–38.

Lewis, M. E. S., Corker, H. A., Gollan, B., & Poole, R. K. (2008). A survey of methods for the purification of microbial flavohemoglobins. *Methods in Enzymology, 436,* 169–186.

Li, W. K. W., Subba Rao, D. V., Harrison, W. G., Smith, J. C., Cullen, J. J., Irwin, B., et al. (1983). Autotrophic picoplankton in the tropical ocean. *Science (New York), 219,* 292–295.

Liew, F. Y., Millott, S., Parkinson, C., Palmer, R. M. J., & Moncada, S. (1990). Macrophage killing of *Leishmania* parasite *in vivo* is mediated by nitric oxide from L-arginine. *Journal of Immunology, 144,* 4794–4797.

Lim, J., Thomas, T., & Cavicchioli, R. (2000). Low temperature regulated DEAD-box RNA helicase from the Antarctic archaeon, *Methanococcoides burtonii. Journal of Molecular Biology, 297,* 553–567.

Liu, H., Campbell, L., Landry, M. R., Nolla, H. A., Brown, S. L., & Constantinou, J. (1998). *Prochlorococcus* and *Synechococcus* growth rates and contributions to production in the Arabian Sea during the 1995 Southwest and Northeast Monsoons. *Deep Sea Research Part II: Topical Studies in Oceanography, 45,* 2327–2352.

Liu, H., Nolla, H., & Campbell, L. (1997). *Prochlorococcus* growth rate and contribution to primary production in the equatorial and subtropical North Pacific Ocean. *Aquatic Microbial Ecology, 12,* 39–47.

Liu, L., Zeng, M., Hausladen, A., Heitman, J., & Stamler, J. S. (2000). Protection from nitrosative stress by yeast flavohemoglobin. *Proceedings of the National Academy of Sciences of the United States of America, 7,* 4672–4676.

Loschi, L., Brokx, S. J., Hills, T. L., Zhang, G., Bertero, M. G., Lovering, A. L., et al. (2004). Structural and biochemical identification of a novel bacterial oxidoreductase. *Journal of Biological Chemistry, 279,* 50391–50400.

Lu, C., Egawa, T., Wainwright, L. M., Poole, R. K., & Yeh, S. R. (2007). Structural and functional properties of a truncated hemoglobin from a food-borne pathogen *Campylobacter jejuni. Journal of Biological Chemistry, 282,* 13627–13636.

Margesin, R., & Miteva, V. (2011). Diversity and ecology of psychrophilic microorganisms. *Research in Microbiology, 162,* 346–361.

Margesin, R., & Schinner, F. (1994). Properties of cold adapted microorganisms and their role in biotechnology. *Journal of Biotechnology, 33,* 1–4.

Marletta, M. A., Yoon, P. S., Iyengar, R., Leaf, C. D., & Wishnok, J. S. (1988). Macrophage oxidation of L-arginine to nitrite and nitrate-nitric oxide is an intermediate. *Biochemistry, 27,* 8706–8711.

Martinez-Hackert, E., & Hendrickson, W. A. (2009). Promiscuous substrate recognition in folding and assembly activities of the trigger factor chaperone. *Cell, 138,* 923–934.

McLean, S., Bowman, L. A. H., & Poole, R. K. (2010). Peroxynitrite stress is exacerbated by flavohaemoglobin-derived oxidative stress in *Salmonella Typhimurium* and is relieved by nitric oxide. *Microbiology, 156,* 3556–3565.

Médigue, C., Krin, E., Pascal, G., Barbe, V., Bernsel, A., Bertin, P. N., et al. (2005). Coping with cold: The genome of the versatile marine Antarctica bacterium *Pseudoalteromonas haloplanktis* TAC125. *Genome Research, 15,* 1325–1335.

Membrillo-Hernández, J., Coopamah, M. D., Anjum, M. F., Stevanin, T. M., Kelly, A., Hughes, M. N., et al. (1999). The flavohemoglobin of Escherichia coli confers resistance to a nitrosating agent, a "nitric oxide releaser", and paraquat and is essential for transcriptional responses to oxidative stress. *Journal of Biological Chemistry, 274,* 748–754.

Membrillo-Hernández, J., Coopamah, M. D., Channa, A., Hughes, M. N., & Poole, R. K. (1998). A novel mechanism for upregulation of the *Escherichia coli* K-12 *hmp* (flavohaemoglobin) gene by the 'NO releaser', S-nitrosoglutathione: Nitrosation of homocysteine and modulation of MetR binding to the *glyA-hmp* intergenic region. *Molecular Microbiology, 29,* 1101–1112.

Membrillo-Hernández, J., Ioannidis, N., & Poole, R. K. (1996). The flavohaemoglobin (HMP) of *Escherichia coli* generates superoxide *in vitro* and causes oxidative stress *in vivo*. *FEBS Letters, 382,* 141–144.

Merz, F., Boehringer, D., Schaffitzel, C., Preissler, S., Hoffmann, A., Maier, T., et al. (2008). Molecular mechanism and structure of trigger factor bound to the translating ribosome. *The EMBO Journal, 27,* 1622–1632.

Methé, B. A., Nelson, K. E., Deming, J. W., Momen, B., Melamud, E., Zhang, X., et al. (2005). The psychrophilic lifestyle as revealed by the genome sequence of *Colwellia psychrerythraea* 34H through genomic and proteomic analyses. *Proceedings of the National Academy of Sciences of the United States of America, 102,* 10913–10918.

Milani, M., Ouellet, Y., Ouellet, H., Guertin, M., Boffi, A., Antonini, G., et al. (2004). Cyanide binding to truncated hemoglobins: A crystallographic and kinetic study. *Biochemistry, 43,* 5213–5221.

Milani, M., Pesce, A., Nardini, M., Ouellet, H., Ouellet, Y., Dewilde, S., et al. (2005). Structural bases for heme binding and diatomic ligand recognition in truncated hemoglobins. *Journal of Inorganic Biochemistry, 99,* 97–109.

Milani, M., Pesce, A., Ouellet, Y., Ascenzi, P., Guertin, M., & Bolognesi, M. (2001). *Mycobacterium tuberculosis* hemoglobin N displays a protein tunnel suited for O_2 diffusion to the heme. *The EMBO Journal, 20,* 3902–3909.

Milani, M., Savard, P. Y., Ouellet, H., Ascenzi, P., Guertin, M., & Bolognesi, M. (2003). A TyrCD1/TrpG8 hydrogen bond network and a TyrB10TyrCD1 covalent link shape the heme distal site of *Mycobacterium tuberculosis* hemoglobin O. *Proceedings of the National Academy of Sciences of the United States of America, 100,* 5766–5771.

Mills, C. E., Sedelnikova, S., Søballe, B., Hughes, M. N., & Poole, R. K. (2001). *Escherichia coli* flavohaemoglobin (Hmp) with equistoichiometric FAD and haem contents has a low affinity for dioxygen in the absence or presence of nitric oxide. *The Biochemical Journal, 353,* 207–213.

Moline, M. A., Karnovsky, N. J., Brown, Z., Divoky, G. J., Frazer, T. K., Jacoby, C. A., et al. (2008). High latitude changes in ice dynamics and their impact on polar marine ecosystems. *Annals of the New York Academy of Sciences, 1134,* 267–319.

Monk, C. E., Pearson, B. M., Mulholland, F., Smith, H. K., & Poole, R. K. (2008). Oxygen- and NssR-dependent globin expression and enhanced iron acquisition in the response of *Campylobacter* to nitrosative stress. *Journal of Biological Chemistry*, *283*, 28413–28425.

Moritz, R. E., Bitz, C. M., & Steig, E. J. (2002). Dynamics of recent climate change in the Arctic. *Science*, *297*, 1497–1502.

Motohashi, K., Watanabe, Y., Yohda, M., & Yoshida, M. (1999). Heat-inactivated proteins are rescued by the DnaK.J-GrpE set and ClpB chaperones. *Proceedings of the National Academy of Sciences of the United States of America*, *96*, 7184–7189.

Mowat, C. G., Gazur, B., Campbell, L. P., & Chapman, S. K. (2010). Flavin-containing heme enzymes. *Archives of Biochemistry and Biophysics*, *493*, 37–52.

Mukai, M., Savard, P. Y., Ouellet, H., Guertin, M., & Yeh, S. R. (2002). Unique ligand- protein interactions in a new truncated hemoglobin from *Mycobacterium tuberculosis*. *Biochemistry*, *41*, 3897–3905.

Murray, A. E., & Grzymski, J. J. (2007). Diversity and genomics of Antarctic marine microorganisms. *Philosophical Transactions of the Royal Society of London. Series B, Biological Sciences*, *362*, 2259–2271.

Nardini, M., Pesce, A., Labarre, M., Richard, C., Bolli, A., Ascenzi, P., et al. (2006). Structural determinants in the group III truncated hemoglobin from *Campylobacter jejuni*. *Journal of Biological Chemistry*, *281*, 37803–37812.

Nardini, M., Pesce, A., Milani, M., & Bolognesi, M. (2007). Protein fold and structure in the truncated (2/2) globin family. *Gene*, *398*, 2–11.

Orii, Y., Ioannidis, N., & Poole, R. K. (1992). The oxygenated flavohaemoglobin from *Escherichia coli*: Evidence from photodissociation and rapid-scan studies for two kinetic and spectral forms. *Biochemical and Biophysical Research Communications*, *187*, 94–100.

Ouellet, H., Juszczak, L., Dantsker, D., Samuni, U., Ouellet, Y. H., Savard, P. Y., et al. (2003). Reactions of *Mycobacterium tuberculosis* truncated hemoglobin O with ligands reveal a novel ligand-inclusive hydrogen bond network. *Biochemistry*, *42*, 5764–5774.

Ouellet, H., Ranguelova, K., Labarre, M., Wittenberg, J., Wittenberg, B., Magliozzo, R., et al. (2007). Reaction of *Mycobacterium tuberculosis* truncated hemoglobin O with hydro- gen peroxide. Evidence for peroxidatic activity and formation of protein-based radicals. *Journal of Biological Chemistry*, *282*, 7491–7503.

Ouellett, H., Ouellett, Y., Richard, C., Labarre, M., Wittenberg, B., Wittenberg, J., et al. (2002). Truncated hemoglobin HbN protects *Mycobacterium bovis* from nitric oxide. *Proceedings of the National Academy of Sciences of the United States of America*, *99*, 5902–5907.

Overpeck, J. T., Otto-Bliesner, B. L., Miller, G. H., Muhs, D. R., Alley, R. B., & Kiehl, J. T. (2006). Paleoclimatic evidence for future ice-sheet instability and rapid sea-level rise. *Science*, *311*, 1747–1750.

Papa, R., Parrilli, E., & Sannia, G. (2009). Engineered marine Antarctic bacterium *Pseudoalteromonas haloplanktis* TAC125: A promising microorganism for the bioremedi- ation of aromatic compounds. *Journal of Applied Microbiology*, *106*, 49–56.

Parrilli, E., Duilio, A., & Tutino, M. L. (2008). Heterologous protein expression in psychro- philic hosts. In R. Margesin, F. Schinner, J. C. Marx, & C. Gerday (Eds.), *Psychrophiles: From biodiversity to biotechnology* (pp. 365–379). Berlin/Heidelberg: Springer-Verlag.

Parrilli, E., Giuliani, M., Giordano, D., Russo, R., Marino, G., Verde, C., et al. (2010). The role of a 2-on-2 haemoglobin in oxidative and nitrosative stress resistance of Antarctic *Pseudoalteromonas haloplanktis* TAC125. *Biochimie*, *92*, 1003–1009.

Parrilli, E., Giuliani, M., Pezzella, C., Danchin, A., Marino, G., & Tutino, M. L. (2010). PssA is required for alpha-amylase secretion in Antarctic *Pseudoalteromonas haloplanktis*. *Micro- biology*, *156*, 211–219.

Parrilli, E., Papa, R., Tutino, M. L., & Sannia, G. (2010). Engineering of a psychrophilic bacterium for the bioremediation of aromatic compounds. *Bioengineered Bugs*, *1*, 213–216.

Pathania, R., Navani, N. K., Gardner, A. M., Gardner, P. R., & Dikshit, K. L. (2002). Nitric oxide scavenging and detoxification by the *Mycobacterium tuberculosis* haemoglobin, HbN in *Escherichia coli*. *Molecular Microbiology, 45*, 1303–1314.

Paul, J., Jeffrey, W. H., & DeFlaun, M. (1985). Particulate DNA in sub-tropical oceanic & estuarine planktonic environments. *Marine Biology, 90*, 95–101.

Peck, L. S. (2011). Organisms and responses to environmental change. *Marine Genomics, 4*, 237–243.

Peck, L. S., Convey, P., & Barnes, D. K. A. (2006). Environmental constraints on life histories in Antarctic ecosystems: Tempos, timings and predictability. *Biological Review, 81*, 75–109.

Pegg, D. E. (2007). Principles of cryopreservation. *Methods in Molecular Biology, 368*, 39–57.

Pesce, A., Couture, M., Dewilde, S., Guertin, M., Yamauchi, K., Ascenzi, P., et al. (2000). A novel two-over-two a-helical sandwich fold is characteristic of the truncated hemoglobin family. *The EMBO Journal, 19*, 2424–2434.

Pesce, A., Dewilde, S., Nardini, M., Moens, L., Ascenzi, P., Hankeln, T., et al. (2003). Human brain neuroglobin structure reveals a distinct mode of controlling oxygen affinity. *Structure, 11*, 1087–1095.

Pesce, A., Milani, M., Nardini, M., & Bolognesi, M. (2008). Mapping heme-ligand tunnels in group I truncated (2/2) hemoglobins. *Methods in Enzymology, 436*, 303–315.

Piette, F., D'Amico, S., Mazzucchelli, G., Danchin, A., Leprince, P., & Feller, G. (2011). Life in the cold: A proteomic study of cold-repressed proteins in the Antarctic bacterium *Pseudoalteromonas haloplanktis* TAC125. *Applied and Environmental Microbiology, 77*, 3881–3883.

Piette, F., D'Amico, S., Struvay, C., Mazzucchelli, G., Renaut, J., Tutino, M. L., et al. (2010). Proteomics of life at low temperatures: Trigger factor is the primary chaperone in the Antarctic bacterium *Pseudoalteromonas haloplanktis* TAC125. *Molecular Microbiology, 76*, 120–132.

Polyak, L., Alley, R. B., Andrews, J. T., Brigham-Grette, J., Cronin, T. M., Darby, D. A., et al. (2010). History of sea ice in the Arctic. *Quaternary Science Reviews, 29*, 1679–1715.

Poole, R. K. (2005). Nitric oxide and nitrosative stress tolerance in bacteria. *Biochemical Society Transactions, 33*, 176–180.

Poole, R. K. (2008). Microbial haemoglobins: Proteins at the crossroads of oxygen and nitric oxide metabolism. *Methods in Enzymology, 9*, 241–257.

Poole, R. K., Anjum, M. F., Membrillo-Hernández, J., Kim, S. O., Hughes, M. N., & Stewart, V. (1996). Nitric oxide, nitrite, and Fnr regulation of *hmp* (flavohemoglobin) gene expression in *Escherichia coli* K-12. *Journal of Bacteriology, 178*, 5487–5492.

Poole, R. K., & Hughes, M. N. (2000). New functions for the ancient globin family: Bacterial responses to nitric oxide and nitrosative stress. *Molecular Microbiology, 36*, 775–783.

Poole, R. K., Rogers, N. J., D'mello, R. A. M., Hughes, M. N., & Orii, Y. (1997). *Escherichia coli* flavohaemoglobin (Hmp) reduces cytochrome *c* and Fe(III)-hydroxamate K by electron transfer from NADH via FAD: Sensitivity of oxidoreductase activity to haem-bound dioxygen. *Microbiology, 143*, 1557–1565.

Pörtner, H. O., Peck, L. S., & Somero, G. N. (2007). Thermal limits and adaptation: An integrative view (Antarctic Ecology: From Genes to Ecosystems). *Philosophical Transactions of the Royal Society B, 362*, 2233–2258.

Qiu, Y., Kathariou, S., & Lubman, D. M. (2006). Proteomic analysis of cold adaptation in a Siberian permafrost bacterium *Exiguobacterium sibiricum* 255-15 by two-dimensional liquid separation coupled with mass spectrometry. *Proteomics, 6*, 5221–5233.

Rabus, R., Ruepp, A., Frickey, T., Rattei, T., Fartmann, B., Stark, M., et al. (2004). The genome of *Desulfotalea psychrophila*, a sulfate-reducing bacterium from permanently cold Arctic sediments. *Environmental Microbiology, 6*, 887–902.

Ray, M. K., Kumar, G. S., Janiyani, K., Kannan, K., Jagtap, P., Basu, M. K., et al. (1998). Adaptation to low temperature and regulation of gene expression in Antarctic psychrotrophic bacteria. *Journal of Biosciences, 23*, 423–435.

Razzera, G., Vernal, J., Baruh, D., Serpa, V. I., Tavares, C., Lara, F., et al. (2008). Spectroscopic characterization of a truncated hemoglobin from the nitrogen-fixing bacterium *Herbaspirillum seropedicae*. *Journal of Biological Inorganic Chemistry*, *13*, 1085–1096.

Riccio, A., Vitagliano, L., di Prisco, G., Zagari, A., & Mazzarella, L. (2002). The crystal structure of a tetrameric hemoglobin in a partial hemichrome state. *Proceedings of the National Academy of Sciences of the United States of America*, *99*, 9801–9806.

Rifkind, J. M., Abugo, O., Levy, A., & Heim, J. M. (1994). Detection, formation, and relevance of hemichrome and hemochrome. *Methods in Enzymology*, *231*, 449–480.

Riley, M., Staley, J. T., Danchin, A., Wang, T. Z., Brettin, T. S., Hauser, L. J., et al. (2008). Genomics of an extreme psychrophile, *Psychromonas ingrahamii*. *BMC Genomics*, *9*, 210. http://dx.doi.org/10.1186/1471-2164-1189-1210.

Rodrigues, D. F., Ivanova, N., He, Z., Huebner, M., Zhou, J., & Tiedje, J. M. (2008). Architecture of thermal adaptation in an *Exiguobacterium sibiricum* strain isolated from 3 million year old permafrost: A genome and transcriptome approach. *BMC Genomics*, *9*, 547. http://dx.doi.org/10.1186/1471-2164-1189-1547.

Rodrigues, D. F., & Tiedje, J. M. (2008). Coping with our cold planet. *Applied and Environmental Microbiology*, *74*, 1677–1686.

Rogstam, A., Larsson, J., Kjelgaard, P., & von Wachenfeldt, C. (2007). Mechanisms of adaptation to nitrosative stress in *Bacillus subtilis*. *Journal of Bacteriology*, *189*, 3063–3071.

Rosen, R., & Ron, E. Z. (2002). Proteome analysis in the study of the bacterial heat-shock response. *Mass Spectrometry Reviews*, *21*, 244–265.

Rosenzweig, C., Karoly, D., Vicarelli, M., Neofotis, P., Wu, Q., Casassa, G., et al. (2008). Attributing physical and biological impacts to anthropogenic climate change. *Nature*, *453*, 353–357.

Russell, N. J. (1998). Molecular adaptations in psychrophilic bacteria: Potential for biotechnological applications. *Advances in Biochemical Engineering/Biotechnology*, *61*, 1–21.

Russell, N. J. (2000). Towards a molecular understanding of cold activity of enzymes from psychrophiles. *Extremophiles*, *4*, 83–90.

Russo, R., Giordano, D., di Prisco, G., Hui Bon Hoa, G., Marden, M. C., Verde, C., et al. (2013). Ligand-rebinding kinetics of 2/2 hemoglobin from the Antarctic bacterium *Pseudoalteromonas haloplanktis* TAC125. *Biochimica et Biophysica Acta*, http://dx.doi.org/10.1016/j.bbapap.2013.02.013, pii: S1570-9639(13)00076-9.

Sauer, H., Wartenberg, M., & Hescheler, J. (2001). Reactive oxygen species as intracellular messengers during cell growth and differentiation. *Cellular Physiology and Biochemistry*, *11*, 173–186.

Saunders, N. F., Thomas, T., Curmi, P. M., Mattick, J. S., Kuczek, E., Slade, R., et al. (2003). Mechanisms of thermal adaptation revealed from the genomes of the Antarctic Archaea *Methanogenium frigidum* and *Methanococcoides burtonii*. *Genome Research*, *13*, 1580–1588.

Scanlan, D. J., Ostrowski, M., Mazard, S., Dufresne, A., Garczarek, L., Hess, W. R., et al. (2009). Ecological genomics of marine picocyanobacteria. *Microbiology and Molecular Biology Reviews*, *73*, 249–299.

Scher, H. D., & Martin, E. E. (2006). Timing and climatic consequences of the opening of Drake Passage. *Science*, *312*, 428–430.

Scott, N. L., Falzone, C. J., Vuletich, D. A., Zhao, J., Bryant, D. A., & Lecomte, J. T. (2002). Truncated hemoglobin from the cyanobacterium *Synechococcus* sp. PCC 7002: Evidence for hexacoordination and covalent adduct formation in the ferric recombinant protein. *Biochemistry*, *41*, 6902–6910.

Siani, L., Papa, R., Di Donato, A., & Sannia, G. (2006). Recombinant expression of toluene o-xylene monooxygenase (ToMO) from *Pseudomonas stutzeri* OX1 in the marine Antarctic bacterium *Pseudoalteromonas haloplanktis* TAC125. *Journal of Biotechnology*, *126*, 334–341.

Siddiqui, K. S., & Cavicchioli, R. (2006). Cold-adapted enzymes. *Annual Review of Biochemistry*, *75*, 403–433.

Smagghe, B. J., Trent, J. T., III., & Hargrove, M. S. (2008). NO dioxygenase activity in hemoglobins is ubiquitous *in vitro*, but limited by reduction *in vivo*. *PLoS One*, *3*, e2039.

Smetacek, V., & Nicol, S. (2005). Polar ocean ecosystems in a changing world. *Nature*, *437*, 362–368.

Smith, H. K., Shepherd, M., Monk, C., Green, J., & Poole, R. K. (2011). The NO-responsive hemoglobins of *Campylobacter jejuni*: Concerted responses of two globins to NO and evidence *in vitro* for globin regulation by the transcription factor NssR. *Nitric Oxide*, *25*, 234–241.

Somero, G. N. (2010). The physiology of climate change: How potentials for acclimatization and genetic adaptation will determine "winners" and "losers". *The Journal of Experimental Biology*, *213*, 912–920.

Sowa, A. W., Guy, P. A., Sowa, S., & Hill, R. D. (1999). Nonsymbiotic haemoglobins in plants. *Acta Biochimica Polonica*, *46*, 431–445.

Spiro, S. (2007). Regulators of bacterial responses to nitric oxide. *FEMS Microbiology Reviews*, *31*, 193–211.

Springer, B. A., Sligar, S. G., Olson, J. S., & Phillips, G. N., Jr. (1994). Mechanisms of ligand recognition in myoglobin. *Chemical Reviews*, *94*, 699–714.

Steinberg, D. K., Carlson, C. A., Bates, N. R., Johnson, R. J., Michaels, A. F., & Knap, A. H. (2001). Overview of the U.S. JGOFS Bermuda Atlantic Time-series Study (BATS): A decade-scale look at ocean biology and biogeochemistry. *Deep Sea Research II*, *48*, 1405–1447.

Stevanin, T. M., Ioannidis, N., Mills, C. E., Kim, S. O., Hughes, M. N., & Poole, R. K. (2000). Flavohemoglobin Hmp affords inducible protection for *Escherichia coli* respiration, catalyzed by cytochromes *bo'* or *bd*, from nitric oxide. *Journal of Biological Chemistry*, *275*, 35868–35875.

Stevanin, T. M., Poole, R. K., Demoncheaux, E. A. G., & Read, R. C. (2002). Flavohemoglobin Hmp protects *Salmonella enterica* serovar Typhimurium from nitric oxide-related killing by human macrophages. *Infection and Immunity*, *70*, 4399–4405.

Stevanin, T. A., Read, R. C., & Poole, R. K. (2007). The *hmp* gene encoding the NO-inducible flavohaemoglobin in *Escherichia coli* confers a protective advantage in resisting killing within macrophages, but not *in vitro*: Links with swarming motility. *Gene*, *398*, 62–68.

Stocker, R., Seymour, J. R., Samadani, A., Hunt, D. E., & Polz, M. F. (2008). Rapid chemotactic response enables marine bacteria to exploit ephemeral microscale nutrient patches. *Proceedings of the National Academy of Sciences of the United States of America*, *105*, 4209–4214.

Stratton, L. P., Kelly, R. M., Rowe, J., Shively, J. E., Smith, D. D., Carpenter, J. F., et al. (2001). Controlling deamidation rates in a model peptide: Effects of temperature, peptide concentration, and additives. *Journal of Pharmaceutical Sciences*, *90*, 2141–2148.

Stuehr, D. J., Gross, S. S., Sakuma, I., Levi, R., & Nathan, C. F. (1989). Activated murine macrophages secrete a metabolite of arginine with the bioactivity of endothelium-derived relaxing factor and the chemical reactivity of nitric oxide. *The Journal of Experimental Medicine*, *169*, 1011–1020.

Sul, W. J., Oliver, T. A., Ducklow, H. W., Amaral-Zettler, L. A., & Sogin, M. L. (2013). Marine bacteria exhibit a bipolar distribution. *Proceedings of the National Academy of Sciences of the United States of America*, *110*, 2342–2347.

Suzuki, Y., Haruki, M., Takano, K., Morikawa, M., & Kanaya, S. (2004). Possible involvement of an FKBP family member protein from a psychrotrophic bacterium *Shewanella* sp. SIB1 in cold-adaptation. *European Journal of Biochemistry*, *271*, 1372–1381.

Thomson, M. R. A. (2004). Geological and palaeoenvironmental history of the Scotia Sea region as a basis for biological interpretation. *Deep Sea Research Part II*, *51*, 1467–1487.

Ting, L., Williams, T. J., Cowley, M. J., Lauro, F. M., Guilhaus, M., Raftery, M. J., et al. (2010). Cold adaptation in the marine bacterium, *Sphingopyxis alaskensis*, assessed using quantitative proteomics. *Environmental Microbiology*, *12*, 2658–2676.

Trent, J. T., III., Hvitved, A. N., & Hargrove, M. S. (2001). A model for ligand binding to hexacoordinate hemoglobins. *Biochemistry*, *40*, 6155–6163.

Turner, J., Bindschadler, R., Convey, P., di Prisco, G., Fahrbach, E., Gutt, J., et al. (2009). *Antarctic climate change and the environment*. Cambridge, UK: SCAR Scott Polar Research Institute.p. 526. *www.scar.org*.

Urayama, P., Gruner, S. M., & Phillips, G. N., Jr. (2002). Probing substates in sperm whale myoglobin using high-pressure crystallography. *Structure*, *10*, 51–60.

Uzan, J., Dewilde, S., Burmester, T., Hankeln, T., Moens, L., Hamdane, D., et al. (2004). Neuroglobin and other hexacoordinated hemoglobins show a weak temperature dependence of oxygen binding. *Biophysical Journal*, *87*, 1196–1204.

Valko, M., Morris, H., & Cronin, M. T. (2005). Metals, toxicity and oxidative stress. *Current Medicinal Chemistry*, *12*, 1161–1208.

Vallone, B., Nienhaus, K., Brunori, M., & Nienhaus, G. U. (2004). The structure of murine neuroglobin: Novel pathways for ligand migration and binding. *Proteins*, *56*, 85–92.

Vasudevan, S. G., Tang, P., Dixon, N. E., & Poole, R. K. (1995). Distribution of the flavohaemoglobin, HMP, between periplasm and cytoplasm in *Escherichia coli*. *FEMS Microbiology Letters*, *125*, 219–224.

Vergara, A., Franzese, M., Merlino, A., Vitagliano, L., di Prisco, G., Verde, C., et al. (2007). Structural characterization of ferric hemoglobins from three antarctic fish species of the suborder notothenioidei. *Biophysical Journal*, *93*, 2822–2829.

Vergara, A., Vitagliano, L., Verde, C., di Prisco, G., & Mazzarella, L. (2008). Spectroscopic and crystallographic characterization of bis-histidyl adducts in tetrameric hemoglobins. *Methods in Enzymology*, *436*, 425–444.

Vigentini, I., Merico, A., Tutino, M. L., Compagno, C., & Marino, G. (2006). Optimization of recombinant human nerve growth factor production in the psychrophilic *Pseudoalteromonas haloplanktis*. *Journal of Biotechnology*, *127*, 141–150.

Vinogradov, S., Hoogewijs, D., Bailly, X., Arredondo-Peter, R., Gough, J., Dewilde, S., et al. (2005). Three globin lineages belonging to two structural classes in genomes from the three kingdoms of life. *Proceedings of the National Academy of Sciences of the United States of America*, *102*, 11385–11389.

Vinogradov, S., & Moens, L. (2008). Diversity of globin function: Enzymatic, transport, storage, and sensing. *Journal of Biological Chemistry*, *283*, 8773–8777.

Vinogradov, S., Tinajero-Trejo, M., Poole, R. K., & Hoogewijs, D. (2013). Bacterial and archaeal globins—A revised perspective. *Biochimica et Biophysica Acta*, *1834*(9), 1789–1800.

Visca, P., Fabozzi, G., Petrucca, A., Ciaccio, C., Coletta, M., De Sanctis, G., et al. (2002). The truncated hemoglobin from *Mycobacterium leprae*. *Biochemical and Biophysical Research Communications*, *294*, 1064–1070.

Vitagliano, L., Bonomi, G., Riccio, A., di Prisco, G., Smulevich, G., & Mazzarella, L. (2004). The oxidation process of Antarctic fish hemoglobins. *European Journal of Biochemistry*, *271*, 1651–1659.

Vitagliano, L., Vergara, A., Bonomi, G., Merlino, A., Verde, C., di Prisco, G., et al. (2008). Spectroscopic and crystallographic analysis of a tetrameric hemoglobin oxidation pathway reveals features of an intermediate R/T state. *Journal of the American Chemical Society*, *130*, 10527–10535.

Voigt, B., Schweder, T., Becher, D., Ehrenreich, A., Gottschalk, G., Feesche, J., et al. (2004). A proteomic view of cell physiology of *Bacillus licheniformis*. *Proteomics*, *4*, 1465–1490.

Vuletich, D. A., & Lecomte, J. T. (2006). A phylogenetic and structural analysis of truncated hemoglobins. *Journal of Molecular Evolution*, *62*, 196–210.

Wang, M., & Overland, J. E. (2009). A sea ice free summer Arctic within 30 years? *Geophysical Research Letters*, *36*, L07502.

Watanabe, Y. H., & Yoshida, M. (2004). Trigonal DnaK–DnaJ complex versus free DnaK and DnaJ: Heat stress converts the former to the latter, and only the latter can do disaggregation in cooperation with ClpB. *Journal of Biological Chemistry*, *279*, 15723–15727.

Watts, R. A., Hunt, P. W., Hvitved, A. N., Hargrove, M. S., Peacock, W. J., & Dennis, E. S. (2001). A hemoglobin from plants homologous to truncated hemoglobins of microorganisms. *Proceedings of the National Academy of Sciences of the United States of America*, *98*, 10119–10124.

Weintraub, S. J., & Manson, S. R. (2004). Asparagine deamidation: A regulatory hourglass. *Mechanisms of Ageing and Development*, *125*, 255–257.

Wilkins, D., Lauro, F. M., Williams, T. J., Demaere, M. Z., Brown, M. V., Hoffman, J. M., et al. (2013). Biogeographic partitioning of Southern Ocean microorganisms revealed by metagenomics. *Environmental Microbiology*, *15*, 1318–1333. http://dx.doi.org/10.1111/1462-2920.12035.

Williams, T. J., Burg, D. W., Raftery, M. J., Poljak, A., Guilhaus, M., Pilak, O., et al. (2010). Global proteomic analysis of the insoluble, soluble, and supernatant fractions of the psychrophilic archaeon *Methanococcoides burtonii*. Part I: The effect of growth temperature. *Journal of Proteomic Research*, *9*, 640–652.

Williams, T. J., Long, E., Evans, F., Demaere, M. Z., Lauro, F. M., Raftery, M. J., et al. (2012). A metaproteomic assessment of winter and summer bacterioplankton from Antarctic Peninsula coastal surface waters. *The ISME Journal*, *6*, 1883–1900.

Williams, T. J., Wilkins, D., Long, E., Evans, F., Demaere, M. Z., Raftery, M. J., et al. (2013). The role of planktonic Flavobacteria in processing algal organic matter in coastal East Antarctica revealed using metagenomics and metaproteomics. *Environmental Microbiology*, *15*, 1302–1317. http://dx.doi.org/10.1111/1462-2920.12017.

Wilmes, B., Kock, H., Glagla, S., Albrecht, D., Voigt, B., Markert, S., et al. (2011). Cytoplasmic and periplasmic proteomic signatures of exponentially growing cells of the psychrophilic bacterium *Pseudoalteromonas haloplanktis* TAC125. *Applied and Environmental Microbiology*, *77*, 1276–1283.

Wittenberg, J. B., Bolognesi, M., Wittenberg, B. A., & Guertin, M. (2002). Truncated hemoglobins: A new family of hemoglobins widely distributed in bacteria, unicellular eukaryotes, and plants. *Journal of Biological Chemistry*, *227*, 871–874.

Wu, G., Corker, H., Orii, Y., & Poole, R. K. (2004). *Escherichia coli* Hmp, an "oxygen-binding flavohaemoprotein," produces superoxide anion and self-destructs. *Archives of Microbiology*, *182*, 193–203.

Wu, G., Wainwright, L. M., Membrillo-Hernández, J., & Poole, R. K. (2004). Bacterial hemoglobins: old proteins with "new" functions? Roles in respiratory and nitric oxide metabolism. In D. Zannoni (Ed.), *Respiration in a Archaea and bacteria. Vol 1: Diversity of prokaryotic electron transport carriers*, Vol. 1, (pp. 251–286). Dordrecht: Kluwer Academic Publishers.

Wu, G., Wainwright, L. M., & Poole, R. K. (2003). Microbial globins. *Advances in Microbial Physiology*, *47*, 255–310.

Zhang, Y. M., & Rock, C. O. (2008). Membrane lipid homeostasis in bacteria. *Nature Reviews. Microbiology*, *6*, 222–233.

Zhao, J. S., Deng, Y., Manno, D., & Hawari, J. (2010). Shewanella spp. genomic evolution for a cold marine lifestyle and in-situ explosive biodegradation. *PLoS One*, *5*, e9109.

Zheng, S., Ponder, M. A., Shih, J. Y., Tiedje, J. M., Thomashow, M. F., & Lubman, D. M. (2007). A proteomic analysis of *Psychrobacter articus* 273-4 adaptation to low temperature and salinity using a 2-D liquid mapping approach. *Electrophoresis*, *28*, 467–488.

Zhou, F. X., Cocco, M. J., Russ, W. P., Brunger, A. T., & Engelman, D. M. (2000). Interhelical hydrogen bonding drives strong interactions in membrane proteins. *Nature Structural & Molecular Biology*, *7*, 154–160.

CHAPTER NINE

Microbial Eukaryote Globins

Serge N. Vinogradov*, Xavier Bailly†, David R. Smith‡, Mariana Tinajero-Trejo§, Robert K. Poole§, David Hoogewijs¶,1

*Department of Biochemistry and Molecular Biology, Wayne State University School of Medicine, Detroit, Michigan, USA
†Marine Plants and Biomolecules, Station Biologique de Roscoff, Roscoff, France
‡Department of Biology, Western University, London, Ontario, Canada
§Institute of Biology and Biotechnology, Department of Molecular Biology and Biotechnology, The University of Sheffield, Sheffield, United Kingdom
¶Institute of Physiology and Zürich Center for Integrative Human Physiology, University of Zürich, Zürich, Switzerland
1Corresponding author: e-mail address: david.hoogewijs@access.uzh.ch

Contents

1. Overview	393
2. Historical Perspective	394
3. Globin Nomenclature	395
4. Microbial Eukaryote Globins	397
4.1 Archaeplastida	397
4.2 SAR clade	402
4.3 Excavata	403
4.4 Amoebozoa	404
4.5 Other microbial eukaryotes	404
4.6 Opisthokonta	405
4.7 Fungi	406
5. Phylogenetic Relationships	406
5.1 Methods of globin identification and alignment	406
5.2 Multiple sequence alignments and phylogenetic analysis	409
5.3 Origin of protist globins and their relationship to higher eukaryote globins	411
6. Globin Function	420
6.1 Overview of globin functions	420
6.2 Yeasts	421
6.3 Other fungi	423
6.4 Protozoa	429
6.5 Algae	433
6.6 Perspectives	433
7. Conclusion	434
Acknowledgements	434
References	436

391

Abstract

A bioinformatics survey of about 120 protist and 240 fungal genomes and transcriptomes revealed a broad array of globins, representing five of the eight subfamilies identified in bacteria. Most conspicuous is the absence of protoglobins and globin-coupled sensors, except for a two-domain globin in *Leishmanias*, that comprises a nucleotidyl cyclase domain, and the virtual absence of truncated group 3 globins. In contrast to bacteria, co-occurrence of more than two globin subfamilies appears to be rare in protists. Although globins were lacking in the Apicomplexa and the Microsporidia intracellular pathogens, they occurred in the pathogenic Trypanosomatidae, Stramenopiles and certain fungi. Flavohaemoglobins (FHbs) and related single-domain globins occur across the protist groups. Fungi are unique in having FHbs co-occurring with sensor single-domain globins (SSDgbs). Obligately biotrophic fungi covered in our analysis lack globins. Furthermore, SSDgbs occur only in a heterolobosean amoeba, *Naegleria* and the stramenopile *Hyphochytrium*. Of the three subfamilies of truncated Mb-fold globins, TrHb1s appear to be the most widespread, occurring as multiple copies in chlorophyte and ciliophora genomes, many as multidomain proteins. Although the ciliates appear to have only TrHb1s, the chlorophytes have Mb-like globins and TrHb2s, both closely related to the corresponding plant globins. The presently available number of protist genomes is inadequate to provide a definitive census of their globins. Bayesian molecular analyses of single-domain 3/3 Mb-fold globins suggest a close relationship of chlorophyte and haptophyte globins, including choanoflagellate and *Capsaspora* globins to land plant symbiotic and non-symbiotic haemoglobins and to vertebrate neuroglobins.

ABBREVIATIONS

Adgb androglobin
AGN Agnathan, jawless vertebrates
Cygb cytoglobin
FHb flavohaemoglobin; N-terminal 3/3 globin linked covalently to a ferredoxin-NAD(P)H reductase-like domain
GbX globin X, found in fish, amphibians and gnathostomes
GbY globin Y, found in amphibians
GCS globin-coupled sensor; N-terminal 3/3 globin linked covalently to a gene-regulating or methyl-accepting chemotaxis domain
Hb haemoglobin
LECA last eukaryote common ancestor
Ma million years ago
Mb myoglobin
ML maximum likelihood
Ngb neuroglobin
NJ neighbour-joining
Pgb protoglobin; single–domain 3/3 globin related to the N-terminal of GCSs

SDgb single-domain 3/3 globin related to the N-terminal of FHbs
SSDgb sensor single-domain 3/3 globin related to the N-terminal of GCSs

1. OVERVIEW

Haemoglobins (Hbs) and related haemeproteins (e.g. myoglobin (Mb)), that we call globins for short, undoubtedly represent the most studied family of proteins. Furthermore, the determinations of the crystal and primary structures of horse heart haemoglobin and of sperm whale myoglobin and of the sequences of their polypeptide chains in the 1950s mark the beginning of protein molecular biology. Although it had become clear by the mid-1990s that globins occurred in a wide variety of metazoans, in microbial eukaryotes such as ciliates (Iwaasa, Takagi, & Shikama, 1989; Yamauchi, Tada, & Usuki, 1995) and fungi (Zhu & Riggs, 1992), in bacteria and were ubiquitous in plants (Hardison, 1996; Hunt et al., 2001), a detailed panorama of the globin phylogenomic distribution was not revealed until the arrival of bacterial genomic information. A bioinformatic survey of the then available (~420) bacterial genomes (Vinogradov et al., 2005) demonstrated the existence of three globin families, of which the flavohaemoglobin (FHb) and globin-coupled sensor (GCS) families have the canonical 3/3 Mb-fold, had been discovered by one of us (R. K. P.) (Vasudevan et al., 1991) and by Alam and his collaborators (Hou et al., 2001), respectively. The third family comprised globins with a novel, truncated 2/2 Mb-fold, characterized by an absent or vestigial helix A and a loop instead of helix E (Pesce et al., 2000). Although globins belonging to each of the three families had already been recognized, additional bacterial genomes demonstrated the presence of single-domain globins in the FHb and GCS families (Vinogradov et al., 2006, 2007), indicating eight globin subfamilies (Vinogradov, Hoogewijs, Vanfleteren, et al., 2011): FHbs and related single-domain globins (SDgbs), the GCSs and two related, but distinct single-domain sensor globins, protoglobins (Pgbs), discovered by Alam and his collaborators (Freitas, Saito, Hou, & Alam, 2005) and sensor single-domain globins (SSDgbs), and three classes of TrHbs, classes 1, 2 and 3 (Vuletich & Lecomte, 2006; Wittenberg, Bolognesi, Wittenberg, & Guertin, 2002). A recent update of the bacterial globin census based on over 2200 bacterial and about 140 archaeal genomes has fully confirmed and extended earlier findings (Vinogradov, Tinajero-Trejo, Poole, & Hoogewijs, 2013).

2. HISTORICAL PERSPECTIVE

In 1674, Antonie Van Leeuwenhoek, a draper from Delft and an amateur naturalist skilled in the polishing of lenses and construction of microscopes capable of up to $275 \times$ magnification (Ford, 1995), sent a letter to the Secretary of the Royal Society in London, describing 'many little animalcules' that he had observed in a sample from a local lake (Rothschild, 1989). The term 'Protozoa' (first animals) was introduced by the German naturalist Georg Goldfuss in 1820 for microscopic organisms within the Kingdom Animalia (Scamardella, 1999). By the middle of the nineteenth century, microscopic organisms were viewed as Protozoa (unicellular animals), Protophyta (primitive plants), Phytozoa (animal-like plants) and Bacteria (regarded primarily as plants) (Scamardella, 1999). The British naturalist John Hogg (1800–1869) used the term 'Protoctista' to denote the microorganisms that share characteristics of both plants and animals. The German naturalist Ernst Haeckel (1834–1919) introduced the term 'Protista' (primordial/first) to define organisms intermediate between plants and animals. Over the next century, progress in the understanding of microorganisms was based primarily on traditional studies of morphology, physiology and biochemistry. Concurrently, the distinction between prokaryote and eukaryotes based only on the absence or presence of nuclei was introduced by Chatton (1925), and the term protozoa was superseded by Protista. The advent of molecular phylogeny based on small-subunit ribosomal RNA (SSU rRNA) sequences, pioneered by Woese, Kandler, and Wheelis (1990) helped to demonstrate that protists comprise numerous independently evolved and widely diverse lineages (Sogin, Morrison, Hinkle, & Silberman, 1996; Sogin & Silberman, 1998). Based on molecular phylogeny, six clusters or supergroups of protists have been recognized (Simpson & Roger, 2002): (1) the Opisthokonta, comprising animals (metazoans), fungi, choanoflagellates and mesomycetozoa, (2) the Amoebozoa, grouping amoebas, amoeba-flagellates and amitochonriate amoebas, (3) the Excavata, encompassing oxymonads, parabasalids, jacobids, diplomonads, Euglenozoa and Heterolobosea, (4) the Rhizaria, grouping the Foraminifera, Radiolaria and Cercozoa, (5) the Archaeaplastida, comprising the Glaucophyte, red and green algae and plants, and (6), the Chromalveolata, grouping the Alveolata (ciliates, dinoflagellates and Apicomplexa), the Stramenopiles (brown algae, diatoms, zoosporic fungi and opalinids) together with the Haptophyta and Cryptophyceae (Adl et al., 2005). Protists are defined as eukaryotes with

Table 9.1 Estimates of known and potential number of species in the major protist groups

Group	Known	Potential
Amoebozoa	4110	~22,600
Opisthokonta (excluding animals)	407,370	~2,500,000
Rhizaria	ca. 11,900	$>2.5 \times 10^6$
Archaeplastida (excluding plants)	ca. 20,300	~220,000
Chromalveolata	ca. 216,553	$>2 \times 10^6$
Excavata	ca. 2300	~3200

From Table 1 of Adl et al. (2007).

unicellular, colonial, filamentous or parenchymatous organization, which lack vegetative tissue differentiation except for reproduction (Adl et al., 2007). Protists can be free living or parasitic and are widely distributed on land and water. Molecular barcoding using SSU rRNA gene sequences has been used to reveal an extensive genetic diversity of protists in a wide range of ecosystems (Bik et al., 2012; Del Campo & Ruiz-Trillo, 2013; Lopez-Garcia, Rodriguez-Valera, Pedros-Alio, & Moreira, 2001; Moon-van der Staay, De Wachter, & Vaulot, 2001; Pawlowski et al., 2011), including lineages only known by their genetic signatures (orphan environmental sequences), found everywhere. Furthermore, the rate of discovery of new species remains high. Table 9.1 provides estimates of known and potential number of species based on unknown DNA sequences found in environmental samples (Adl et al., 2007). The recent update of protist classification (Adl et al., 2012) recognizes the grouping of Opishokonta and Amoebozoa into the Unikonta (Roger & Simpson, 2009) and the SAR clade, encompassing the Stramenopiles, Alveolata and Rhizaria (Burki et al., 2007). Some 145 groups or genera incertae sedis remain with uncertain affiliation within protists (Adl et al., 2012).

3. GLOBIN NOMENCLATURE

Table 9.2 summarizes a recently proposed comprehensive nomenclature for all globins, including eukaryotes, based on the bacterial globin families (Vinogradov et al., 2013). Over the last decade, Burmester, Hankeln and their collaborators have discovered several new globins in vertebrate genomes (Burmester et al., 2004), including the ubiquitous neuroglobins

Table 9.2 Proposed global globin nomenclature (Vinogradov et al., 2013)

Family	M (Mb-like globins)		S (sensor globins)			T (truncated Mb-fold)		
Mb-fold	3/3		3/3			2/2		
Bacterial subfamily	FHbs (flavohaemoglobin)	SDgb (single-domain globins)	GCSs (globin-coupled sensors)	Pgbs (protoglobins)	SSDgbs (sensor single-domain globins)	TrHb1s N	TrHb2s O	TrHb3s P
Archaea	–		HemATs	Pgb	SSDgb	TrHb1	–	–
Eukaryote	FHbs	All animal globins	–	–	SSDgb	TrHb1	TrHb2	–

(Ngbs) and cytoglobins (Cygb), and others with more limited distribution, such as globin X (GbX) and globin Y (GbY), the avian eye-specific globin E (GbE), and the recently discovered chimeric androglobin, present in most deuterostomes, several protostomes, more basal animal clades and in choanoflagellates (Hoogewijs, Ebner, et al., 2012). These are all 3/3 globins as are almost all the metazoan globins. Because of the sequence similarity between vertebrate globins such as Mb, Ngb and Cygb on one hand, and the bacterial SDgbs and FHbs on the other (Vinogradov, 2008), we proposed subsuming the FHb family into the M family (for myoglobin-like), which would therefore encompass all 3/3 eukaryote globins, including all the 3/3 globins of vertebrates, other metazoans, plants and protists. For the bacterial GCS family, we propose S for sensor family, retaining GCS for the chimeric members and using SSDgb for the related single-domain sensor globins. The three classes of the T family are TrHb1s, TrHb2s and TrHb3s. In bacteria these are also sometimes called N, O and P, respectively.

4. MICROBIAL EUKARYOTE GLOBINS

Table 9.3 summarizes the globin subfamilies identified in approximately 120 protist genomes and transcriptomes. A detailed list of protist globins is available as Supplementary Table S1 at http://www.elsevierdirect.com/companions/9780124076938. A diagrammatic view of protist diversity and the globin subfamilies found in protist genomes are shown in Fig. 9.1. In Table 9.3, we also include the globins found in opisthokonts other than metazoans and fungi, namely choanoflagellates and ichtyosporeans.

4.1. Archaeplastida

The Archaeplastida (Viridiplantae in Table 9.3) are represented by 14 chlorophyte genomes, all of which contain globins, TrHb1s, TrHb2s and SDgbs. The chlorophyceans *Coccomyxa subellipsoides* and *Chlorella variabilis* have only TrHb1s and TrHb2s. A chimeric 587-residue TrHb1 in *C. subellipsoides* has an N-terminal domain identified as a member of the bestrophin family (PFAM01062), transmembrane proteins that share an RFP (Arg-Phe-Pro) motif and are thought to function as chloride channels (Hartzell, Qu, Yu, Xiao, & Chien, 2008).

Table 9.3 Summary of globin distribution in protist genomes

Taxon	Number of genomes	Globin subfamilies found	Comments
Alveolata	30		
Apicomplexa	20	SDgb in *Hammonia hammondi*	No globins in 20 genomes
Ciliophora (ciliates)	6	Single and multiple TrHb1 domain (up to $n=6$) in all	*Paramecium tetraurelia* has 92 domains in 42 proteins (18 chimeric)
Dinophyceae (dinoflagellates)	3	SDgb, chimeric SDgbs and TrHb1	Eleven globins in *Lingulodinium polyedrum*
Perkinsea	1	Single and multiple TrHb1 domain (up to $n=6$)	*Perkinsus marinus* has 79 domains in 30 proteins (14 chimeric)
Amoebozoa	14		
Archamoeba	5	No globins in four genomes	*Acanthamoeba castellanii* has a TrHb1, a TrHb2 and an SDgb
Mycetozoa	8	FHbs in all genomes, including *Physarum polycephalum*	
Tubulinea	1	No globins	
Apusozoa	1	FHb	
Cryptophyta	3	No globins	
Euglenozoa	13	Chimeric two-domain SDgb with a C-terminal adenylate cyclase domain in *Leishmania*	No globins in five trypanosomes
Fornicata	1	FHb	Two FHbs in *Giardia*
Glaucocystophyceae	0	?	
Haptophyceae (Primnesiophytes)	1	SDgb, FHb and TrHb1	*Emiliania huxleyi* has a chimeric protein with four TrHb1 domains

Continued

Table 9.3 Summary of globin distribution in protist genomes—cont'd

Taxon	Number of genomes	Globin subfamilies found	Comments
Heterolobosea	1	SSDgb	
Jakobida	0	?	
Katablepharidophyta	0	?	
Malawimonadidae	0	?	
Opisthokonta			
Choanoflagellida	2	SDgbs and FHb	*Monosiga brevicollis* has one FHb
Opisthokonta incertae sedis			
Ichtyosporea	2	SDgbs	*Capsaspora owczarzaki* has chimeras with two SDgb domains
Oxymonadida	0	?	
Parabasalia	1	No globins	
Rhizaria	11		
Acantharea	2	No globins	
Cercozoa	2	TrHb2s in both genomes	
Foraminifera	5	No globins	TrHb1s in *Globobulimina*
Polycystinea	2	No globins	
Rhodophyta (red algae)	4	SDgb and TrHb2	*Chondrus crispus* has only TrHb2s
Stramenopiles	25		
Bacillariophyta (diatoms)	4	SDgb, FHb, TrHb1s and TrHb2s	
Bicosoecida	0	?	
Blastocystis	1	No globins	
Dictyochophyceae (silicoflagellates)	0	?	
Eustigmatophyceae	2	SDgb and FHb	

Continued

Table 9.3 Summary of globin distribution in protist genomes—cont'd

Taxon	Number of genomes	Globin subfamilies found	Comments
Hyphochytriomycete	1	SSDgb	
Oomycetes (water molds)	17	SDgb, FHb, TrHb1, TrHb2 and TrHb3	
Albuginales	2	Incomplete FHbs and TrHb2	
Peronosporales	9	SDgb, FHb, TrHb1 and TrHb2	No globins in *Hyaloperonospora arabidopsidis*
Pythiales	1	SDgb and incomplete FHb	
Saprolegnales	2	SDgb and FHb	
Pelagophyceae	1	SDgb and TrHb1	
PX Clade	1	SDgb and FHb	
Raphidophyceae	1	TrHb2	TrHb2 domain in a nitrite reductase
Viridiplantae (green plants)	14		
Chlorophyta (green algae)			
Chlorophycea	3	TrHb1s and multidomain TrHb1s	
Mamiellophyceae	6	SDgb and TrHb1	
Prasinophyceae	1	SDgb and TrHb1	
Trebouxiophyceae	4	SDgb, TrHb1 and TrHb2	
Total number of genomes	122		
Number of genomes with globins	80		

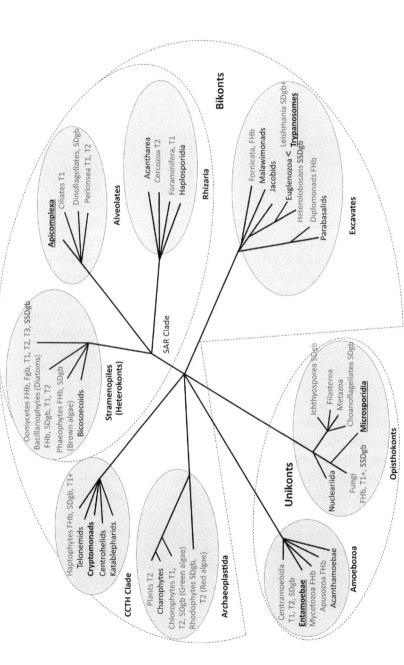

Figure 9.1 Diagrammatic representation of the eukaryote supergroups showing globin presence (red) or absence (bold), based on data in Supplementary Tables S1 and S2 at http://www.elsevierdirect.com/companions/9780124076938. (For interpretation of the references to colour in this figure legend, the reader is referred to the online version of this chapter.)

4.2. SAR clade

The SAR clade is represented by 24 stramenopile, 30 alveolate and 11 rhizarian genomes (Table 9.3). The stramenopile genomes are globin rich, with 2–7 globins, including FHbs, SDgbs, TrHb1s and TrHb2s. Only *Blastocystis hominis* and *Hyaloperonospora arabidopsidis* appear to lack globins, and one, *Phythophtora capsicis*, has a putative TrHb3 (Table 9.3). A curious chimeric nitrite reductase comprising a central TrHb2 domain is found in the raphidophyte *Heterosigma akashiwo* (Stewart & Coyne, 2011). Of the 30 alveolate genomes, 17 are apicomplexans, obligate intracellular animal parasites that share a unique organelle, the apicoplast and an apical complex structure used in host cell penetration. They have complex life cycles that include sexual and asexual reproduction and appear to lack globins, except for the coccidian *Hammonia hammondi* that has a larger than usual genome (67.5 Mbp) and an SDgb (see Supplementary Table S1 at http://www.elsevierdirect.com/companions/9780124076938). Some of these fungi are discussed in more detail in Section 6. The ciliates, the second group of alveolates, are represented by six genomes (Table 9.3), of which only one, *Stylonychia lemnae*, lacks globins. Their salient characteristic is having numerous TrHb1s, including many chimeric multidomain proteins, with up to seven covalently linked TrHb1 domains (see Supplementary Table S1 at http://www.elsevierdirect.com/companions/9780124076938). The common fish parasite *Ichthyophthirius multifiliis* has four TrHb1s, three of which have three to four TrHb1 domains and an N-terminal thioredoxin-like domain, identified specifically as a two-Cys peroxiredoxin (PFAM cd 3015). *Tetrahymena thermophila* has 20 TrHb1 domains in 20 proteins, 3 being chimeric, with an N-terminal peroxiredoxin and 3–7 TrHb1 domains. *Paramecium tetraurelia* comprises 92 globin domains in 42 proteins, of which 18 are chimeric. Of the latter, 2 have an N-terminal serine kinase with a C-terminal TrHb1; the remaining 16 have an unidentified N-terminals linked with 2–4 TrHb1 domains. The genome of *Perkinsus marinus*, a dominant parasite of oysters and clams, has 39 globin domains in 30 proteins, 14 of them chimeric. Of the 16 single-domain proteins, all are TrHb1s, except for 2 TrHb2s, that also score high as TrHb1s in FUGUE searches (see Supplementary Table S1 at http://www.elsevierdirect.com/companions/9780124076938). The largest chimeric protein is a 2145-residue-long protein (XP_002778565.1) with an unknown 140-residue N-terminal followed by a 350-residue serine/threonine metallophosphatase, an unidentified 1050 residues domain, and seven consecutive TrHb1 domains. The Spirotrichean,

Oxytricha trifallax, has three TrHbs, one of them a three-domain chimera (Table 9.3). The dinoflagellates (whirling flagella), represent a major alveolate group and are among the largest microbial eukaryote groups (Guiry, 2012). They are also part of the phytoplankton responsible for algal blooms, such as the 'red tides' that occur off the coastal waters worldwide. The nefarious effects of algal blooms are due to depletion of free oxygen and to the secretion of toxins harmful to fish and shellfish. Furthermore, many dinoflagellates are bioluminescent due to the presence of luciferase in individual cytoplasmic bodies (Haddock, Moline, & Case, 2010). The three dinoflagellates have globins, mostly SDgbs. The genus *Symbiodinium* occurs as an intracellular symbiont of cnidarians, corals, jellyfish and sea anemones, as well as molluscs, flatworms and sponges. Its transcriptome (Bayer et al., 2012) reveals half a dozen SDgbs. The Rhizaria, first introduced by Cavalier-Smith (2002), comprise three main groups, the Cercozoa, Foraminifera and Radiolaria, and are represented by 11 genomes and transcriptomes. The two acanthereans and the two polycystineans lack globins. Of the five foraminifera only one, *Globobulimina turgida*, has only TrHb1s. The two cercozoans each have a TrHb2.

4.3. Excavata

The Excavata, a major supergroup also proposed by Cavalier-Smith (2002) encompasses six phyla, the Euglenozoa, represented by 13 genomes (Table 9.3), the Heterolobosea, Fornicata and Parabasalia, each with one genome, and the Oxymonadida and Jakobida with none. The Euglenozoa, include the Trypanosomatids, that are responsible for three major human diseases, sleeping sickness (African trypanosomiasis), Chagas disease (South American trypanosomiasis) and leishmaniasis. Although all trypanosomatids are exclusively parasitic, and the *Trypanosoma brucei*, *T. congolense*, *T. cruzi* and *T. vivax* genomes lack globins, as does *Crithidia fasciculate* that parasitizes insects, four of the five *Leishmania* genomes have an interesting chimeric, 600–700 residue globin, consisting of two N-terminal SDgb domains in tandem, linked to a C-terminal nucleotidyl cyclase catalytic domain (Table 9.3). The single heterolobosean genome, that of *Naegleria gruberi*, harbours an SSDgb (Fritz-Laylin, Ginger, Walsh, Dawson, & Fulton, 2011). The genome of *Giardia lamblia* representing the Fornicata contains two FHbs while the genome of the human pathogen *Trichomonas vaginalis*, the only Parabasalia genome, lacks globins (Table 9.3). Functions of these and other protozoal globins, insofar as they have been studied, are described in Section 6.

4.4. Amoebozoa

The Amoebozoa are part of the Opisthokonta and are considered to be a sister group to the metazoans and fungi (Eichinger et al., 2005). They are represented by 14 genomes, half of which are the human pathogenic parasites *Entamoeba* and *Acanthamoeba*. Although the former lack globins, the latter contains a TrHb1, a TrHb2 and an SDgb (Table 9.3). The remaining mycetozoan genomes all have one or two FHbs. *Hartmannella vermiformis* lacks globins. In *Dictyostelium*, the FHbs were not found to be essential for growth (Iijima, Shimizu, Tanaka, & Urushihara, 2000).

4.5. Other microbial eukaryotes

In addition to the members of the supergroups considered above, there exists an indeterminate number of microbial eukaryote lineages (see Fig. 9.2). These include the Apusozoa represented by the *Thecamonas trahens* genome that harbours one FHb, and the Cryptophyta with three genomes lacking

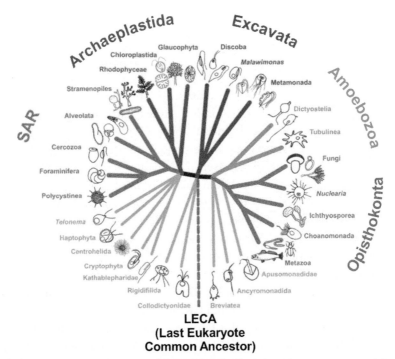

Figure 9.2 A diagrammatic representation of the major protist supergroups. *Adapted from Adl et al. (2012), with permission from John Wiley and Sons.* (For colour version of this figure, the reader is referred to the online version of this chapter.)

globins. No genomes are available for the Centroheliozoa, Glaucocystophyceae, Jakobida, Katablepharidophyta and Malawimonadidae. The Haptophyceae are represented by the genome of *Emiliania huxleyi*, a unicellular phytoplankton covered with extraordinary calcium carbonate discs, one of the most abundant and widely distributed coccolithophores, with a global distribution from the tropics to subarctic waters. It is unique in having three SDgbs, three FHbs and a chimeric protein with an unidentified N-terminus, four TrHb1 domains and a C-terminal von Willebrand factor (vWF) type A domain (Sadler, 1998). The physiological basis of this wealth of globins is unknown.

4.6. Opisthokonta

The Opisthokonta comprise metazoans (animals), fungi and several additional microbial eukaryote lineages, including the Choanoflagellida, Ichthyosporea, Nucleariidae and Capsaspora. It is likely that the closest extant relative of both fungi and metazoans is a member of the Opisthokonta. Furthermore, it appears that the closest relatives of metazoans are the choanoflagellates, followed by the Capsaspora and Ichthyosporea lineages (Ruiz-Trillo, Roger, Burger, Gray, & Lang, 2008). Recently, the genomes of five, close relatives of fungi and metazoa have been sequenced, as part of the Origin of Multicellularity Project at the Broad Institute (Ruiz-Trillo et al., 2007). These include *Capsaspora owczarzaki*, an amoeboid parasite of the pulmonate snail *Biomphalaria glabrata*, which has a relatively small genome about 22–25 Mbp (Ruiz-Trillo, Lane, Archibald, & Roger, 2006), the apusozoan *T. trahens* (formerly *Amastigomonas* sp. ATCC 50062), two choanoflagellates, *Salpingoeca rosetta* (formerly *Proterospongia* sp. ATCC 50818) and *Monosiga brevicollis*, and two basal fungi, *Allomyces macrogynus* and *Spizellomyces punctatus*. Although these genomes contain globins, *T. trahens* has only one FHb, and *A. macrogynus* has two SSDgbs and a chimeric globin with a C-terminal TrHb1 domain. The ichthyosporean *Sphaeroforma arctica* has five SDgbs and an FHb, while *C. owczarzaki* has two chimeric two-SDgb domain proteins (Table 9.3). No nucleariid genomes are available. The two available choanoflagellate genomes, each encode three SDgbs. It is worth pointing out that a hypothetical protein of 1653 amino acids PTSG_01043 (EGD76343.1) can be identified in *S. rosetta* (*Salpingoeca* sp. ATCC 50818) via BLASTP search using *Strongylocentrotus purpuratus* androglobin (isoform 1; XP_001186225.2). Whereas the N-terminal cysteine protease domain is present, the central globin domain is not identified by either BLASTP or FUGUE searches. Although this result is hardly surprising, given the fact that in multicellular metazoans, including all deuterostomes, androglobin appears to be predominantly expressed only in testis tissue (Hoogewijs, Ebner, et al., 2012), it suggests that androglobin was present in the ancestor shared by metazoans and choanoflagellates.

4.7. Fungi

Table 9.4 presents a summary of the globin subfamilies identified in about 240 available fungal genomes, representing an update of an earlier census of fungal globins (Hoogewijs, Dewilde, Vierstraete, Moens, & Vinogradov, 2012). The added fungal genomes have filled the lacunae present in the earlier count, improving the representation of Dothideomycetes and Basidiomycota. A detailed list of fungal globins is provided by Supplementary Table S2 at http://www.elsevierdirect.com/compan ions/9780124076938. The more complete coverage of fungal genomes relative to the other microbial eukaryotes makes it clear that fungi are on the average more globin rich than any other microbial eukaryote group. However, the fungal globins encompass only two of the eight subfamilies, the FHbs and SSDgbs. Furthermore, they are unique in the extensive pairing of the two in most Ascomycota, except the Onygenales, which have only SSDgbs, and in most of the Basidiomycota. This pairing is non-existent in the bacterial genomes (Vinogradov et al., 2013). The Blastocladiomycota and the Chytridiomycota appear to have only SSDgbs (Table 9.4). Many saccharomycete FHbs are incomplete, lacking the C-terminal moiety of the reductase domain, the 1cqx3 domain comprising the NAD-binding site (Ilari & Boffi, 2008). An important issue to resolve is whether the shortened reductase domains are active and whether these incomplete FHbs have any function. Such studies are woefully lacking.

5. PHYLOGENETIC RELATIONSHIPS
5.1. Methods of globin identification and alignment

Two approaches were used in the identification of putative globins and globin domains. In one, we employed the globin gene assignments, based on a library of hidden Markov models (Gough, Karplus, Hughey, & Chothia, 2001), provided by the SUPERFAMILY website (http://supfam.mrc-lmb.cam.ac.uk). In the other approach, we used known globin sequences

Table 9.4 Summary of globin distribution in fungal genomes

Taxon	Number of genomes	Globin subfamilies found	Comments
Ascomycota (ASC)			
Pezizomycotina			
Leotiomyceta			
Dothideomycetes (Dot)	23	FHbs and SSDgbs	Globins in all genomes
Eurotiomycetes (Eur)			
Eurotiales	15	FHbs and SSDgbs	Globins in all genomes
Onygenales (Ony)	15	Only SSDgbs	Globins in all genomes
Leotiomycetes (Leo)	9	FHbs and SSDgbs in seven genomes	
Sordariomycetes (Sor)	51	FHbs and SSDgbs	Globins in all genomes; SDgb in *Colletotrichum higginsianum*
Pezizomycetes (Pez)	2	SSDgb in *Tuber melanosporum*	
Saccharomycotina (Sac)	46	FHbs and SSDgbs	
Taphrinomycotina	8		
Pneumocystidomycetes	4	No globins	
Schizosaccharomycetes (Sch)	3	Only FHbs	
Mitosporic Ascomycota	1	FHb	
Basidiomycota (BAS)	52		
Agaricomycotina (Aga)	37	FHbs and SSDgbs	No globins in eight genomes
Pucciniomycotina (Puc)	10	FHbs and SSDgbs	No globins in seven genomes
Ustilaginomycotina (Ust)	4	FHbs	No globins in two genomes

Continued

Table 9.4 Summary of globin distribution in fungal genomes—cont'd

Taxon	Number of genomes	Globin subfamilies found	Comments
Basidiomycota incertae sedis	1	FHbs	
Blastocladiomycota (BLA)	2	SSDgbs	
Chytridiomycota (CHY)	3	SSDgb	TrHb1 in *Batrachochytrium dendrobatidis*
Glomeromycota	1	No globins	
Fungi incertae sedis (FINS)	7	SDgbs and SSDgbs	No globins in one genome
Neocallimastigomycota	0		
Microsporidia	12	No globins	
Total number of genomes	238		
Genomes with globins	203		

as queries in BLASTP and TBLASTN (Altschul et al., 1997; Schaffer et al., 2001) searches with pairwise alignment of the NCBI non-redundant protein sequence and transcriptome databases (www.ncbi.nlm.nih.gov/BLAST/). Putative sequences shorter than 100 residues were discarded and the remaining sequences were subjected to a FUGUE search (Shi, Blundell, & Mizuguchi, 2001; www.cryst.bioc.cam.ac.uk). FUGUE scans a database of structural profiles, calculates the sequence–structure compatibility scores for each entry, using environment-specific substitution tables and structure-dependent gap penalties, and produces a list of potential homologs and alignments. FUGUE assesses the similarity between the query and a given structure via the Z score, the number of standard deviations above the mean score obtained by chance: the default threshold $Z=6.0$ corresponds to a 99% probability (Shi et al., 2001). We accepted all sequences identified as globins, when the following three criteria were fulfilled: a FUGUE Z score >6, the occurrence of a His residue at position F8 and proper alignment of helices BC through G. Non-globin domains in chimeric proteins were identified using CDD search (Marchler-Bauer et al., 2005) and InterProScan 4.8 (Quevillon et al., 2005).

5.2. Multiple sequence alignments and phylogenetic analysis

Multiple sequence alignments were carried out using MUSCLE (Edgar, 2004), Clustal Omega (Sievers et al., 2011), MAFFT employing the L-INS-i option (Katoh, Kuma, Toh, & Miyata, 2005), ProbCons (Do, Mahabhashyam, Brudno, & Batzoglou, 2005) and T-Coffee (Di Tommaso et al., 2011). To eliminate ambiguous alignments, we used the online version of Gblocks 0.91b (Castresana, 2000) with the 'less stringent selection' parameter set (www.phylogeny.fr). The quality of the alignments was assessed by MUMSA (Lassmann & Sonnhammer, 2005). Maximum likelihood (ML) analyses were carried out using RA × ML (Stamatakiis et al., 2008). Neighbour-joining (NJ) analyses were performed using MEGA version 5.05 (Tamura et al., 2011). Distances were corrected for superimposed events using the Poisson method. All positions containing alignment gaps and missing data were eliminated only in pairwise sequence comparisons (pairwise deletion option). The reliability of the branching pattern was tested by bootstrap analysis with 1000 replications. Bayesian inference trees were obtained employing MrBayes version 3.1.2 (Ronquist & Huelsenbeck, 2003), assuming the WAG model of amino acid substitution and a gamma distribution of evolutionary rates, as determined by the substitution model testing option in MEGA 5.05. Two parallel runs, each consisting of four chains, were run simultaneously for at least 8×10^6 generations and trees were sampled every 1000 generations generating a total of at least 8000 trees. The final average standard deviations of split frequencies were stationary in all analyses and posterior probabilities were estimated on the final 60–80% trees. The CIPRES web portal was used for the Bayesian analyses (Miller, Pfeiffer, & Schwartz, 2010) and MEGA version 5.05 was used to visualize radial trees. With the exception of Fig. 9.3, we employed as outgroups, two non-haem, globin-like stress response regulators RsbR from *Bacillus subtilis* and *B. amyloliquofaciens* (NP_388348.1 and YP_00391940.1). Although they have a globin-like secondary structure, their G and H helices are bent inwards, eliminating the haem-binding cavity (Murray, Delumeau, & Lewis, 2005). Phylogenetic trees were also constructed employing SSU rRNA sequences (Guillou et al., 2013). In compiling the lists of sequences, we identified each sequence by the first three letters of the two portions of the binomial, the number of residues, one or more three-letter abbreviations of the taxon, followed by the identifier. Subcellular localization was identified using PSORT II (Nakai & Horton, 1999).

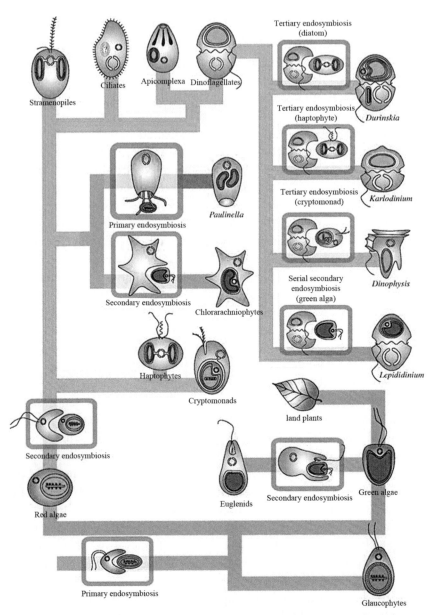

Figure 9.3 Diagrammatic representation of the endosymbiotic origin, diversification and evolution of plastids. *From Keeling (2010), with permission from the Royal Society.* (For colour version of this figure, the reader is referred to the online version of this chapter.)

5.3. Origin of protist globins and their relationship to higher eukaryote globins

Although there is no broad consensus on eukaryote origin, due to substantial disagreement between the interpretations based on the fossil record and the calculations based on molecular clock estimates, a recent estimate based on multigene phylogenies (Parfrey, Lahr, Knoll, & Katz, 2011) suggested that last eukaryote common ancestor (LECA) existed between 1679 and 1866 Ma. Furthermore, the major eukaryote lineages, such as the Amoebozoa, Excavata, SAR and Opisthokonta, are thought to have diverged during the time period 1000–1600 Ma. The lengthy evolution of unicellular eukaryotes is undoubtedly responsible for the diversity in their globin sequences, resulting in a formidable obstacle to the untangling of globin phylogeny.

There is general agreement that the two major events in microbial eukaryote evolution were the endosymbioses of an alpha-proteobacterium and a cyanobacterium (Andersson, Karlberg, Canback, & Kurland, 2003; Gray, 1999; Gray, Lang, & Burger, 2004), estimated to have occurred at 2.3–1.8 billion years ago and 1.5–1.6 billion years ago, respectively (Hedges, Blair, Venturi, & Shoe, 2004), resulting in the emergence of mitochondria and chloroplasts, respectively. In contrast to the evolution of the mitochondria, that of plastids turned out to be extremely complex, leading to extensive diversification involving secondary and tertiary endosymbiotic events, and including plastid loss and replacement within the emerging microbial eukaryote lineages. It is now known that the primary plastids survived in green plants, red and green algae and glaucophytes (Rodriguez-Ezpeleta et al., 2005). Figure 9.3 shows a diagrammatic representation of the many steps in plastid diversification that have been elucidated (Keeling, 2010). We have proposed that the two foregoing endosymbiotic events are also responsible for the transfer of bacterial globin genes to the early eukaryote ancestor (Vinogradov et al., 2007). The extant cyanobacteria have only SDgbs and TrHb1s, while the alpha-proteobacteria contain SDgbs and FHbs, members of the three T subfamilies, and several GCSs (Vinogradov et al., 2013). It is likely that all the TrHb1s in microbial eukaryotes and some of the SDgbs are derived from the ancient endosymbiosis with a cyanobacterium. The complexity of plastid evolution hints at the likely complexity of microbial eukaryote globin phylogeny whose unravelling will probably require the elucidation of the origins of the numerous microbial eukaryote lineages. However, whilst the elucidation of the relations between bacterial and protist globins remains an uncertain goal, the clarification of the relations between the protist globins and the globins of higher eukaryotes represents a more attainable objective. In examining the phylogeny of protist globins, it must be kept in mind that one can only examine the evolution of globin lineages that belong to the same globin subfamily. A major roadblock in globin phylogeny is the unacceptably poor statistics found in support of phylogenetic trees, irrespective of the alignments employed, the type of trees (Bayesian inference, ML or NJ) and the evolutionary models used. On the one hand, the globin sequences are relatively short, and on the other, the low identities found in pairwise alignments of protist and higher eukaryote globins result in the masking of phylogenetic signals coded within the sequences. Figure 9.4 shows a Bayesian tree of a MAFFT alignment of the protist SDgbs and FHb globin domains. It is evident that the node support probabilities are unacceptably low, even though the sequences being compared are from the same family. Nevertheless, this result provides useful information. First of all, it is apparent that the FHb globin domains cluster apart from the SDgbs and related domains in chimeric proteins, in agreement with their having completely separate phylogenies. It is likely that the diversification of ancient bacterial globins into the eight subfamilies observed in extant bacteria would have occurred in the bacterial precursors to LECA, predating the two endosymbiotic events, that resulted in the emergence of mitochondria and chloroplastids and permitted the transfer of bacterial globin genes representing some of the eight bacterial subfamilies to LECA (Vinogradov et al., 2007). Furthermore, the tree assists in the assignment of protist globins. For example, it shows the *Albugo laibachi* FHb globin domain (red arrow) to cluster with the SDgbs, as do the chimeric globins of Leishmanias. Although the former was assigned based on FUGUE search results, it appears to be missing the C–terminal, NADP-binding module (PDB:1cqx3). Such truncated FHbs also occur widely among the fungi and bacteria (Hoogewijs, Dewilde, et al., 2012) and were mentioned earlier, and may thus represent a third subfamily in the FHb family. The same can be said of the *Ectocarpus siliculosus* chimeric FHb (blue arrow); in both cases, the globin domains belong to the SDgb lineage and should not be considered as degraded FHbs.

Very low probabilities of node support are also evident in a Bayesian tree of all the TrHb1s and TrHb1 domains of *P. tetraurelia*, *T. thermophila* and *P. marinus*, shown in Fig. 9.5, underlining the sequence divergence within one globin subfamily, the likely result of more than a billion years of ciliate evolution, accompanied by genome duplications and reductions. It is

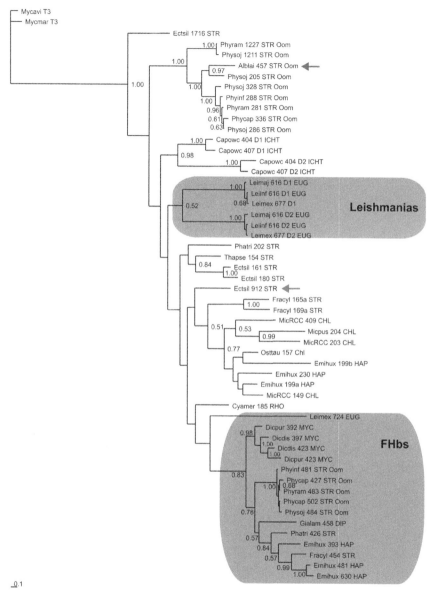

Figure 9.4 Bayesian tree of a MAFFT alignment of protist SDgbs and FHb globin domains. The TrHb3s of *Mycobacterium avium* and *M. marinum* were used as outgroup sequences. Support values at branches represent Bayesian posterior probabilities (>0.5). The sequences are identified by the first three letters of the binary species name, the number of residues and the abbreviated phylum and family names. (For colour version of this figure, the reader is referred to the online version of this chapter.)

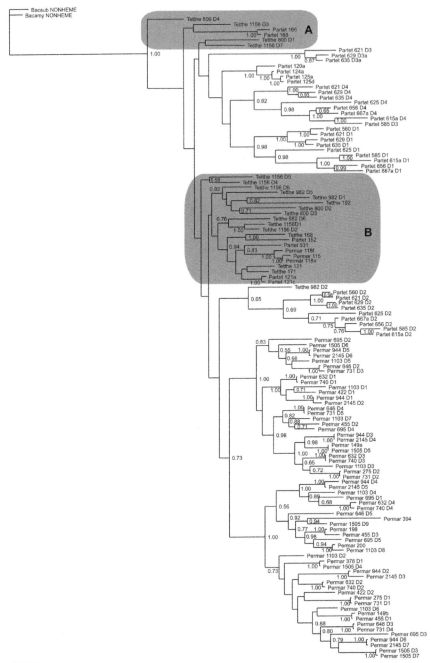

Figure 9.5 Bayesian tree of a ProbCons alignment of the TrHb1 domains of *Paramecium tetraurelia*, *Perkinsus marinus* and *Tetrahymena thermophila*. Support values at branches represent Bayesian posterior probabilities (>0.5). Each sequence is identified by the first three letters of the binomial, number of residues and the identification number.

known that *P. tetraurelia* has undergone three successive whole genome duplications (Aury et al., 2006). Interestingly, some of the *P. tetraurelia* globins cluster with *T. thermophila* globins (boxes A and B in Fig. 9.5), similarly to what has been observed with cytochrome P450 genes (Fu, Xiong, & Miao, 2009). The tree in Fig. 9.5 is consonant with the shared ancestry of the three ciliate genomes.

Despite the inherent weaknesses of molecular reconstruction of globin phylogenies, we believe that useful results can be obtained by repeated analyses with different globin sequences representing the same bacterial subfamilies. The Bayesian tree of protist, plant and bacterial TrHb2s, shown in Fig. 9.6, provides results that agree with the proposed close phylogenetic relationship of plant and chlorophyte TrHb2s with each other and with the bacterial Chloroflexi TrHb2s, inferred earlier, suggesting the possibility of an ancient horizontal gene transfer (HGT) of TrHb2 genes form an ancestor of Chloroflexi to the last common ancestor of the TrHb2 containing protists (Vinogradov, Fernandez, Hoogewijs, & Arredondo-Peter, 2011; Vinogradov, Hoogewijs, & Arredondo-Peter, 2011).

Figure 9.7 shows a Bayesian tree of a MAFFT alignment of 37 protist, and 30 vertebrate SDgbs and 7 plant globins, with the two bacterial non-haem globin-like sequences as outgroups. Each sequence is identified by the first three letters of the binomial, the number of residues, the first three or four letters of the taxon and the identification number. Although the node probabilities are low, the tree reproduces the phylogeny of the major vertebrate globin lineages established by Storz and Hoffmann (Hoffmann, Opazo, Hoogewijs, et al., 2012; Hoffmann, Opazo, & Storz, 2012; Storz, Opazo, & Hoffmann, 2013). Although as expected, there is a clear separation of the microbial eukaryote groups from the vertebrate sequences, it also suggests a close phylogenetic relationship between vertebrate Ngbs and plant symbiotic and non-symbiotic Hbs, on one hand, and with choanoflagellate as well as chlorophyte and haptophyte SDgbs on the other. Note the relatively short phylogenetic distances (marked in red) between the three groups. It is appropriate to point out that robust molecular phylogenies support the monophyly of choanoflagellates (Carr, Leadbeater, Hassan, Nelson, & Baldauf, 2008; Shalchian-Tabrizi et al., 2008). Furthermore, they appear to be among the closest sister groups to the metazoans, and likely share a common marine protistan ancestor (Carr et al., 2008). Interestingly, cadherin and integrin genes, considered to be the hallmark of multicellular animals, are present in the choanoflagellate *M. brevicollis*, *C. owczarzaki* (Nichols, Roberts, Richter, Fairclough, & King, 2012) and other protists

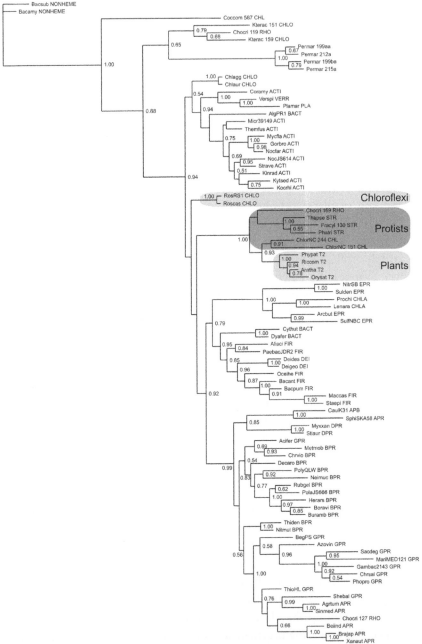

Figure 9.6 Bayesian tree based on a ProbCons alignment of 68 bacterial, 12 protist and 4 plant TrHb2s. Support values at branches represent Bayesian posterior probabilities (>0.5). Each sequence is identified by the first three letters of the binomial, number of residues and first three letters of the taxon.

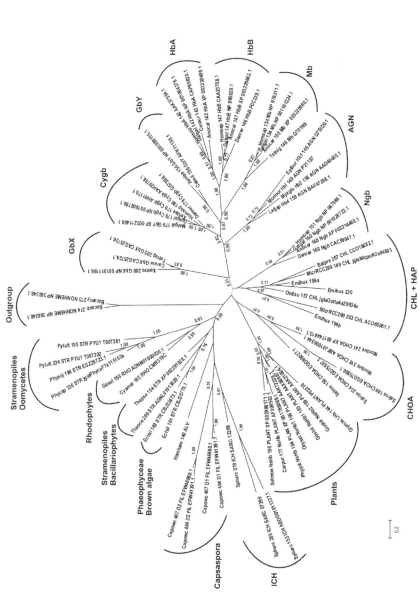

Figure 9.7 Bayesian tree based on MAFFT alignment of 30 microbial eukaryotes, 30 vertebrate and 7 plant globins. Support values at branches represent Bayesian posterior probabilities (>0.5). Each sequence is identified by the first three letters of the binomial, number of residues, the first three or four letters of the taxon and the identification number. Abbreviations: AGN, Agnathan, jawless vertebrates; FIL, Filasterea; ICH, Ichthyosporea; RHO, Rhodophyte; STR, Stramenopile.

(Torruella et al., 2012). Figure 9.8 shows a molecular phylogenetic tree of opisthokonts (Ruiz-Trillo et al., 2007). The overall conclusion, subject to the caveat of low statistical support, indicates a close relationship between protist 3/3 SDgbs and the vertebrate Ngb subfamily. A molecular phylogenetic analysis of deuterostome SDgbs by Hoffmann, Opazo, Hoogewijs, et al. (2012) has suggested at least four separate globin lineages, one of which is related to Ngbs. Another recent study by one of us (X. B.) has provided strong evidence for the presence of Ngb-related globins throughout the metazoans, including placozoans, choanoflagellates, sponges and cnidarians (Lechauve et al., 2013). The results shown in Fig. 9.7 are in agreement with the Bayesian tree, shown in Fig. 9.9, based on the Clustal Omega alignment of 37 protist, 30 vertebrate, 7 plant and 41 lower eukaryotes, representing urochordates, cephalochordates hemichordates, echinoderms, placozoans,

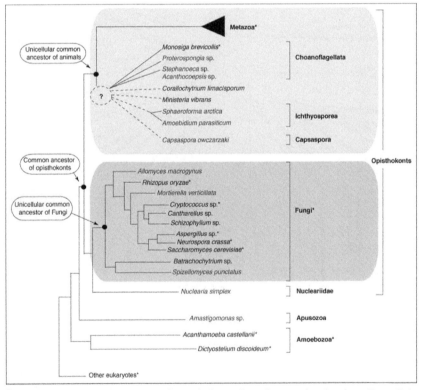

Figure 9.8 A phylogenetic tree of opisthokonts based on molecular data. *From Ruiz-Trillo et al. (2007), with permission from Elsevier.* (For colour version of this figure, the reader is referred to the online version of this chapter.)

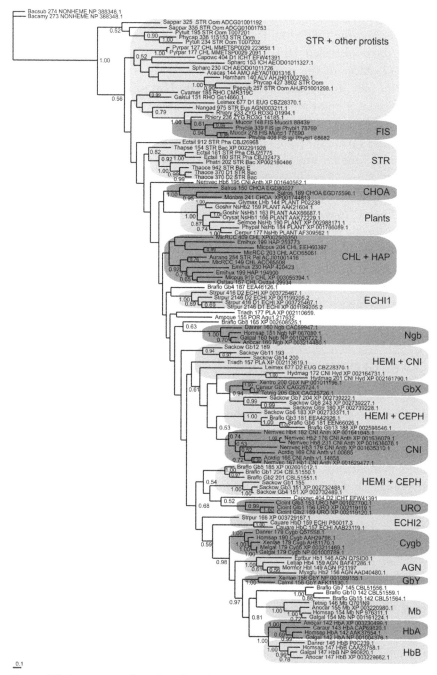

Figure 9.9 Bayesian tree based on Clustal Omega alignment of 50 protist, 28 vertebrate, 43 lower eukaryote and 7 plant Hbs. Support values at branches represent Bayesian

(Continued)

cnidarians and poriferans. Although the node probabilities remain low, it is worth noting the following: (1) The vertebrate globin phylogeny is preserved (boxes labelled HbB, HbA, Mb, GbY, AGN, Cygb, GbX and Ngb). (2) Most of the protist globins are clearly separated from higher eukaryote globins. (3) The proximity of plant Hbs to vertebrate Ngbs with the insertion of chlorophyte, haptophyte and sea-urchin (ECHI1) globins between the latter. (4) The proximity of choanoflagellate (CHOA) and stramenopile (STR) to the plant Hbs. (5) The grouping of echinoderm Hbs into two groups, one next to Ngbs (box ECHI1), and another, vicinal to Cygbs (ECHI2). A number of lower eukaryote globins cluster either with the Ngbs and GbXs, or with the Cygbs, implying the presence of at least two distinct globin lineages in the lower eukaryotes, one related to Ngbs and the other to Cygbs. Notable is the clustering of cnidarian globins with the GbX lineage. The latter result agrees with the recent observations by Blank and Burmester (2012). Furthermore, there is the clustering (though only with 0.85 support value) of the multidomain *Leishmania* chimeric globin with the GbX clade. Our results suggest that opisthokont SDgbs are probably related to the Ngb clade. Obviously, much work remains to be done.

6. GLOBIN FUNCTION
6.1. Overview of globin functions

It has become abundantly clear over the last two decades that Hbs have functions other than oxygen storage and transport, which are probably limited to multicellular organisms in which oxygen must be transported over greater distances (Vinogradov & Moens, 2008). However, we should not forget that physiologically significant gradients of oxygen exist even in prokaryotic cells (see, however, Unden et al., 1995) and that oxygen-binding proteins may play key roles in oxygen delivery and storage. For example, there is a vast literature detailing the beneficial effects of heterologous expression of the *Vitreoscilla* globin (Vgb, VHb) on growth, metabolism and product yields (reviewed by Frey, Shepherd, Jokipii-Lukkari, Haggman, & Kallio, 2011). These effects

---

Figure 9.9—Cont'd posterior probabilities (>0.5). Each sequence is identified by the first three letters of the binomial, number of residues, the first three or four letters of the taxon and the identification number. Abbreviations: AGN, Agnathan, jawless vertebrates; ALV, Alveolate; AMO, Amoebozoa; CEPH, Cephalochordate; CHL, Chlorophyte; CHOA, Choanoflagellate; CNI, Cnidaria; ECHI, Echinoderm; EUG, Euglenozoa; FIL, Filasterea; HEMI, Hemichordate; ICH, Ichtyosporea; POR, Porifera; RHO, Rhodophyte; STR, Stramenopile; URO, Urochordate.

have been reported not only in eukaryotes and multicellular organisms but in many bacteria too. Since there is little evidence that the *Vitreoscilla* globin is important in NO biochemistry, the tacit assumption is that VHb functions in these numerous examples as an oxygen transport protein (Khosla & Bailey, 1989). This conclusion should be borne in mind when considering the roles of globins of all classes in eukaryotic microbes.

Nevertheless, within unicellular organisms, there are plausible alternatives to oxygen storage and transport for the functions of globins. One is the protective role of globins, demonstrated most convincingly for FHbs, against nitrosative stress (Angelo, Hausladen, Singel, & Stamler, 2008; Forrester & Foster, 2012a; Gardner, 2005, 2012). The other, suggested by E. van Holde (Oregon State University), is the sensing of changes in oxygen concentration, leading to a role in regulating chemotaxis, providing the means for the host microorganism to migrate towards optimal oxygen concentrations (Vinogradov, Hoogewijs, Vanfleteren, et al., 2011).

Numerous studies over the last decade have demonstrated that FHbs play a role in the response of pathogenic microorganisms to the nitric oxide generated by the innate immune system (Angelo et al., 2008; Brown, Haynes, & Quinn, 2009; Forrester & Foster, 2012a). The FHbs protect against the toxic effects of nitric oxide via their catalysis of the conversion of NO and oxygen to nitrate anion. The overall reaction discovered by Gardner and his collaborators (Gardner, 2005, 2012; Gardner, Gardner, Martin, & Salzman, 1998) has been the subject of heated debate regarding the mechanism, with advocates for both the dioxygenase route first proposed by Gardner and the denitrosylase mechanism, championed by Stamler and colleagues (see Forrester & Foster, 2012a, 2012b; Hausladen & Stamler, 2012). Whatever the mechanism, the robust abilities of FHbs to remove and detoxify NO is one of the most important defences against reactive nitrogen intermediates in the repertoire of microbial pathogens. Without this protein, the virulence of many pathogens is attenuated, and FHbs have been demonstrated to be important for survival in human macrophages (e.g. in *Salmonella*, Gilberthorpe, Lee, Stevanin, Read, & Poole, 2007; Laver et al., 2010; Stevanin, Poole, Demoncheaux, & Read, 2002). There are extensive reviews of this topic (Bowman, McLean, Poole, & Fukuto, 2011; Laver et al., 2013).

6.2. Yeasts

Before the advent of molecular cloning or sequencing approaches, studies of microbial globins were fragmentary. Nevertheless, two landmark studies stand out, describing the properties of FHb from the bacterium *Alcaligenes eutrophus* (the name later changing to *Ralstonia eutrophus* then *Wausteria eutrophus*, and now *Ralstonia metallidurans* or *Cuprivadus necator*) and the budding yeast *Candida mycoderma*. Earlier, however, in the early 1970s, Chance and the Oshinos described the purification of a 'yeast haemoglobin–reductase complex' (Oshino, Asakura, Tamura, Oshino, & Chance, 1972) from *C. mycoderma* in which haem and flavin were associated with a single polypeptide reducible by NAD(P)H. Subsequent characterization revealed its oxygen affinity and other properties (Oshino, Asakura, et al., 1973; Oshino, Oshino, & Chance, 1971, 1973) but no clear physiological function emerged. A similar protein was purified from *Ralstonia* (Probst & Schlegel, 1976; Probst, Wolf, & Schlegel, 1979) and this protein was later to become the first FHb for which a crystal structure would be solved (Ermler, Siddiqui, Cramm, & Friedrich, 1995).

The literature on yeast globins (YHbs) is extensive yet not always clear regarding protein function. Chance and co-workers did not propose a role in NO biochemistry as is generally accepted now. They were able, however, to demonstrate a robust NADH oxidation rate (Oshino, Asakura, et al., 1973; Oshino et al., 1971), which is a hallmark of the FHb-catalysed NO detoxification reaction of bacteria (Mills, Sedelnikova, Soballe, Hughes, & Poole, 2001; Poole, Ioannidis, & Orii, 1994). Most studies have been conducted on non-pathogenic yeasts, particularly *Saccharomyces cerevisiae* but also *Candida* species (Kobayashi et al., 2002). Prior to the discovery of the up-regulation of *hmp* gene expression in *Escherichia coli* by NO (Poole et al., 1996) and the demonstration of NO consumption, Crawford, Sherman, and Goldberg (1995) had studied *S. cerevisiae* haemoglobin expression ('YHG') and found it to be induced during logarithmic growth and under oxygen-replete conditions. At the time, this was considered to mark a clear distinction from the only bacterial globin that had been studied in depth, namely the *Vitreoscilla* case (Dikshit, Spaulding, Braun, & Webster, 1989; Lamba & Webster, 1980). However, we now know that FHbs in bacteria are also expressed during aerobic growth. Although Crawford et al. (1995) also tested the effects of gene disruption and found no phenotype (growth, viability) under a variety of growth conditions and with various carbon sources, no clues to function emerged. Zhao et al. (1996) showed that intracellular levels of the YHb were greatly increased in cells in which mitochondrial electron transport was impaired and demonstrated maximal expression under hypoxic and anoxic conditions. The response of a *YHB1* deletion mutant to NO was not tested but the mutant was sensitive to oxidative stress, suggesting a role for the globin in the oxidative stress response. This proposal was later rejected (Buisson & Labbe-Bois, 1998) on the grounds that globin expression is unchanged on exposure to antimycin A or menadione, but decreased by hydrogen peroxide and other reagents. Indeed, using a *YHB1* deletion strain, it was shown that YHb1 not only protects cells against Cu(II) and dithiothreitol but also sensitizes them to hydrogen peroxide (Buisson & Labbe-Bois, 1998).

Subsequently, however, other groups clearly established up-regulation of the YHb by NO, and a role in protection from nitrosative stress (Liu, Zeng, Hausladen, Heitman, & Stamler, 2000). Lewinska and Bartosz (2006) later published similar results. It now seems likely that the pre-eminent role of the *S. cerevisiae* globin is a protection against nitrosative stress. FHb is located in both the cytosol and the mitochondrial matrix of normoxic cells but exclusively in the mitochondria in the absence of oxygen. Localization of FHb in the mitochondrial matrix suggests a role for this globin as an NO-detoxifying system under hypoxic conditions by controlling levels of NO that suppress respiration by inhibiting cytochrome *c* oxidase (Cassanova, O'Brien, Stahl, McClure, & Poyton, 2005).

The protective function of FHbs from nitrosative stress also occurs in pathogenic yeasts, such as *Cryptococcus neoformans* (de Jesus-Berrios et al., 2003) and *Candida albicans*, the most prevalent human fungal pathogen (Ullmann et al., 2004). Curiously, only one of the three FHb genes in *C. albicans* is responsible for NO consumption and detoxification (Ullmann et al., 2004). The gene most highly up-regulated in the presence of NO is *YHB1*, and strains with a deletion in this gene show hypersensitivity to NO and are highly filamentous (Hromatka, Noble, & Johnson, 2005). In the case of *C. neoformans*, it was shown recently that production of NADPH via the NADP(+)-dependent isocitrate dehydrogenase played an important role in the maintenance of resistance to nitrosative stress (Brown et al., 2009).

6.3. Other fungi

The true fungi are eukaryotes and all are heterotrophs, that is, they rely on external sources of organic carbon compounds for biosynthesis and survival. Fungi recycle the biomass of other organisms, including biomaterials such as lignin, wood and leaves that other organisms may be unable to digest. Many of the true fungi represented in Tables 9.4 and Supplementary Table S2 at http://www.elsevierdirect.com/companions/9780124076938 are important plant pathogens and therefore exert a high impact economically.

The lifestyles of plant pathogenic fungi are generally considered under the following headings: biotrophs, necrotrophs and hemibiotrophs. Biotrophs are obligate parasites that have evolved elaborate host–pathogen interactions characterized by infection structures called haustoria that enable the pathogen to maintain an intricate and intimate interaction with their plant host without causing host cell death. This ensures that the fungus can maintain a stable interaction that may last for months. Amongst many examples are the *Puccinia* sp. that infect wheat. In contrast, the necrotrophic pathogens feed on and promote the destruction of dead organic matter; they cause major pre- and post-harvest diseases in crops, thus inflicting significant economic losses. These fungi profusely produce toxins or cytolytic enzymes during invasion. A classic example is *Botrytis cinerea* (see below), a widely used model for studies of necrotroph biology. The disease symptoms appear as localized lesions as a spreading necrosis. Finally, hemibiotrophs have an initial biotrophic phase but kill the host plant at a later stage of the association. Examples are *Magnaporthe grisea*, *Phytophthora infectans* and *Colletotrichum* spp. After completion of a biotrophic phase, the fungus switches to necrotrophic development in which the host plant cells are killed. These differences underpin the divergent pathogenesis strategies and plant immune responses evident in biotrophic and necrotrophic infections. To what extent the disparate lifestyles can explain the distribution of globins in fungi is much less clear.

As in the case of bacteria (Vinogradov et al., 2013), a profitable place to begin in attempting to correlate patterns of globin distribution with protein function and organisms lifestyle is the FHbs. Of all globin types, we know more about the function of these globins than any other class found in microbes. It is our contention that the only well-documented function of these proteins is NO detoxification. NO is well recognized as a bioactive signalling molecule and as an important player in regulating a multitude of physiological and pathophysiological responses in mammalian and plant cells. NO generation in mammalian cells arises from nitric oxide synthase (NOS) isoforms, all of which have been intensively studied (Stuehr, Santolini, Wang, Wei, & Adak, 2004). In plants, although no homologues of mammalian NOS genes are clear, several possible pathways for NO synthesis have been put forward. These include nitrate and nitrite reductases, xanthine oxidoreductase, various mitochondrial routes, arginine-dependent synthases (that may or may not involve a 'NO synthase'-like enzyme) and a hydroxylamine-mediated reaction (Gupta, Fernie, Kaiser, & van Dongen, 2011). Processes regulated by NO in plants include seed germination, root growth, respiration, stomatal movements and adaptation to biotic and abiotic stresses. Critically, plant NO has been implicated in playing a role in mediating defences against bacterial pathogens: NO collaborates with reactive oxygen species (ROS) to play crucial roles in the execution of the plant hypersensitive response (HR), a localized programmed of cell death (Delledonne, Zeier, Marocco, & Lamb, 2001; Mur, Carver, & Prats, 2006). NO accumulation in host tissue when challenged by a pathogen is well documented using cytochemical methods (Mur, Mandon, Cristescu, Harren, & Prats, 2011; Prats, Carver, & Mur, 2008).

Although HR-induced cell death is critical in establishing disease resistance, the effect of the plant defence response in necrotrophic interactions is not clearly understood. Necrotrophic fungi are not deterred by the hypersensitive reaction of the host (Govrin & Levine, 2000). In fact, the HR has been reported to stimulate and facilitate the colonization of necrotrophs (Malolepsza & Rozalska, 2005). Thus, certain plant pathogens are able to evade the radical burst from the host and/or exploit the activated host defence for pathogenic development and colonization. A study of the bacterium *Erwinia chrysanthemi* (Boccara et al., 2005) has shed some light on the possibility of the participation of plant pathogens in regulating NO levels during infection. Mutation of the FHb gene, *hmpX*, resulted in a strong burst of NO that accumulated in the infected leaf, which led to a HR-like response in a compatible plant–pathogen interaction. Introduction of *hmpX* into an incompatible strain, *Pseudomonas syringae*, suppressed the HR elicited in *Arabidopsis*. Thus the regulation of NO levels is a prerequisite for a successful infection and the mode of regulation was demonstrated to be via the NO-detoxifying protein FHb. This suggests that the roles of NO-detoxifying globins in plant pathogens are likely to be more complex and subtle than simply eliminating all NO from their environments.

Generalized conclusions from Tables 9.4 and Supplementary Table S2 at http://www.elsevierdirect.com/companions/9780124076938 are hard to discern because the fungi represented have such diverse habitats and hosts. However, it appears to be the case that the true biotrophs lack genes for FHbs (common in other fungal groups) or indeed any globin. Examples in Supplementary Table S2 at http://www.elsevierdirect.com/companions/9780124076938 are *Blumeria graminis* and *Erysiphe pisi* (Sordariamyceta). The former is a biotrophic fungus causing barley powdery mildew. Infection responds to internal and external stimuli. Spore germination follows separation from the spore chain in the mother colony, but recognition of host features is necessary for elongation and differentiation of the appressorial germ tube. Maturation of the appressorial lobe results in a functional unit capable of generating turgor and producing those enzymes necessary for degradation of host cuticle/cell wall complex, allowing the fungal penetration peg to enter the cell lumen and produce the haustorium (or feeding structure). Although the fungus lacks a discernable globin, and globins are generally regarded as the main defence against NO, *B. graminis* f.sp. *hordei* does generate NO, which is a pathogenesis determinant on barley (Prats et al., 2008). Transient NO generation occurs in the appressorium during maturation. The mammalian NO synthase inhibitor L-NAME or the NO scavenger c-PTIO affected the number of appressorial lobes produced by the fungus, indicating that NO plays a key role in formation of *B. graminis* appressoria. Might then the absence of a globin circumvent the problem of uncontrolled NO quenching in this organism? Might the absence of globins in most/all true biotrophs represent a necessary condition for communication via NO with the host plant?

Globins are also absent in certain Pucciniomycotina (Table 9.4), a group that contains a number of important obligately biotrophic plant pathogens. Those lacking globins appear to be *Microbotryum violaceum* (formerly *Ustilago violacea*), *Melampsora laricis-populina* (responsible for poplar leaf rust disease) and *Phakopsora pachyrhizi* (a basidiomycete fungus that causes rust disease in soybean plants). There is no evidence in this group for an FHb, the class of globin most clearly correlated with NO removal. In contrast, in the same group, *Puccinia graminis*—an obligate biotroph that infects wheat—there is evidence from the sequence of two SSDgbs, each containing an unidentified C-terminal extension of 140–150 amino acids. In two members of the Ustilagomycyes, globins also seem to be absent; these are *Sporisorium reilanum* (a biotrophic pathogen of maize) and *Ustilago maydis* (a biotrophic pathogen of maize causing smut disease). There are too few examples and too little published information on NO tolerance in these fungi to draw conclusions but the hypothesis that NO-consuming globins will be absent from fungi that produce NO during infection is worthy of investigation.

An apparent exception is *Cladosporium fulvum* (Supplementary Table S2 at http://www.elsevierdirect.com/companions/9780124076938), a biotroph that is a member of the Dothideomycetes fungi—one of the largest and most diverse groups, which includes many plant pathogens. The group makes a major contribution to local ecosystems by degrading biomass and contributing to regulating the carbon cycle (Ohm et al., 2012). However, *C. fulvum* is atypical in that it can be cultivated *in vitro*. It causes grey leaf mold of tomato and does contain an FHb. In the tomato, it has been shown that application of exogenous NO inhibits synthesis of a proteinase inhibitor protein implicated in wounding in leaves (Orozco-Cardenas & Ryan, 2002). However, we are unaware of evidence for a role of the *C. fulvum* FHb in moderating NO availability in the plant.

In contrast to the obligate biotrophs, among the hemibiotrophs in Supplementary Table S2 at http://www.elsevierdirect.com/companions/ 9780124076938, that is, those that have an initial biotrophic phase but kill the host plant at a later stage of the association, are several that do contain complete FHb sequences. A filamentous fungal FHb was first purified from *Fusarium oxysporum* (Takaya, Suzuki, Matsuo, & Shoun, 1997). This is a cosmopolitan species comprising both pathogenic and non-pathogenic isolates. The pathogenic isolates of *F. oxysporum* cause wilt of several agricultural crops; one of the economically more important and destructive forms is the causal agent of fusarium wilt (Panama disease) of banana (*Musa* spp.). Several other examples can be seen in the Dothideomycetes, including *Dothistoma* spp., *Mycosphaerella* spp. and *Septria musiva*. A variety of globins are also found in *Phytophthora* spp., a hemibiotroph (Supplementary Table S1 at http://www.elsevierdirect.com/companions/9780124076938). *P. capsicis* is a pathogenic oomycete (water mold) infecting pepper plants. Oomycetes are heterokonts (i.e. protists having two flagella of unequal length) whose life cycle resembles that of the true fungi, with which they were formerly classified. This relationship shows a high level of genetic diversity in host resistance and pathogen avirulence proteins. Two varieties of pepper generate NO as a defence and one variety is able to overcome the infection, apparently correlated with the higher levels of NO detected (Requena, Egea-Gilabert, & Candela, 2005). It is interesting, therefore, that *P. capsici* has several globins—an SDgb, another missing the A helix, two FHbs and a TrHb3 (Supplementary Table S1 at http://www.elsevierdirect.com/companions/ 9780124076938).

The necrotrophs, in contrast to the strict biotrophs, commonly contain globin genes and FHb-encoding genes appear widespread, for example, in the Glomerellales, such as *Colletotrichum* sp. and *Glomerella graminicola* (Supplementary Table S2 at http://www.elsevierdirect.com/companions/ 9780124076938). *Acremonium alcalophilum* is the only known cellulolytic fungus that thrives in alkaline conditions and can be readily cultivated in the laboratory. FHb genes are widespread in these sequences, as in the sequences of Hypocreales, which are important plant pathogens. *B. cinerea* is a necrotrophic fungus with a broad host range. It colonizes not only senescent tissues but also healthy plants and is presumably exposed to NO and

ROS produced by the host. *B. cinerea* has a single FHb-coding gene, *Bcfhg1* (Turrion-Gomez, Eslava, & Benito, 2010). Its expression is developmentally regulated, with maximum levels during germination of conidia. Expression is enhanced on exposure to NO and expression *in planta* parallels the expression pattern during saprophytic growth with maximal expression occurring during the very early stages of the infection process. *Bcfhg1* complemented the *S. cerevisiae yhb1* mutation, indicating that the encoded FHb has NO-consuming activity. Studies with *Bcfhg1* deletion mutants suggest that, although BCFHG1 confers protection against NO, the ability of the mutant strains to infect different hosts was not affected.

Aspergillus oryzae (Supplementary Table S2 at http://www.elsevier direct.com/companions/9780124076938) is an asexual, ascomycetous fungus used for hundreds of years in the production of soy sauce, miso and sake. There are conflicting opinions about whether *A. oryzae* can be isolated in nature. *A. oryzae* and *A. flavus* are so closely related that all strains of the former are regarded by some as natural variants of *A. flavus* modified through years of selection for fermenting of foods. *A. oryzae* is regarded as not being pathogenic for plants or animals, though there are a handful of reports of isolation of this fungus from patients. The protective function of FHb has been shown in *A. oryzae* (Zhou et al., 2009) and, more recently, two FHbs were found in *A. oryzae* and *A. niger* (Eurotiomycetes) FHb1 (416/417 residues) and FHb2 (436/439 residues); the N-terminus of the latter contains a potential signal sequence. Furthermore, FHb1 is monomeric, occurs in the cytosol and uses either NADH or NADPH as electron donor, while FHb2 is dimeric, is located in mitochondria, and can use only NADH (Zhou et al., 2011). Interestingly, only the expression of the cytosolic FHb1 is up-regulated in the presence of NO. It was also suggested that the reductases of the two FHbs tend to promote oxidative damage (Zhou et al., 2010). Te Biesebeke et al. (2010) have shown that the *A. oryzae* FHb1 transcription levels appear to be positively correlated with hyphal growth. Moreover, not all the N-terminal extensions have predicted mitochondrial transit peptides as observed for *A. nidulans*, *A. terreus*, *A. fumigatus* and *A. oryzae* (Te Biesebeke et al., 2010), but rather have predicted signal peptides suggesting that localization of the FHbs is species dependent. It should be mentioned that many Ascomycetes have two or more FHbs (Hoogewijs, Dewilde, et al., 2012). Deletion of the FHb gene *fhbA* of *A. nidulans* induces sexual development and decreases sterigmatocystin (a mycotoxin related to aflatoxins) production (Baidya, Cary, Grayburn, & Calvo, 2011).

An FHb (GenBank Accession No. FJ874761) has been briefly reported to be expressed in *Cylindrocarpon* (Kim, Fushinobu, Zhou, Wakagi, & Shoun, 2010), a common soil-borne fungus, which causes root rot in many plant species, but this sequence was not found in our analyses.

6.4. Protozoa

The anti-parasitic effects of nitric oxide have been widely documented in Protozoa and Metazoa (reviewed in Ascenzi, Bocedi, & Gradoni, 2003); however, the defence mechanisms against NO and RNS that must be associated with efficient invasion and colonization of the host by pathogens are incompletely understood. In pathogenic bacteria, the presence of haemoglobins (e.g. FHb in *E. coli* and *Salmonella*, a TrHb in *Mycobacterium tuberculosis* and an SDgb in *Campylobacter jejuni*; for a review see Bowman et al., 2011) represents the main mechanisms involved in resistance to nitrosative stress (Bowman et al., 2011; Vinogradov et al., 2013). On the other hand, a number of pathogenic protozoal genomes lack globin-like sequences (Supplementary Table S1 at http://www.elsevierdirect.com/companions/9780124076938). For instance, production by iNOS of NO from L-arginine by human macrophages is involved in *Leishmania* killing (Green, Meltzer, Hibbs, & Nacy, 1990; Liew, Millott, Parkinson, Palmer, & Moncada, 1990). However, evasion and survival strategies of the protozoan include the suppression of the iNOS induction and the entry into iNOS-negative target cells (Bogdan, Gessner, Solbach, & Rollinghoff, 1996). Clinical manifestations of American tegumentary leishmaniasis (ATL) in the New World are mainly associated with *Leishmania* (*Viannia*) *braziliensis* and *Leishmania* (*Leishmania*) *mexicana* and NO-resistant isolates of these parasites have been reported and related to disease severity (Giudice et al., 2007). The lack of globin-like sequences in the *L. braziliensis* genome indicates a different, unidentified mechanism(s) to detoxify NO but *L. mexicana* does have sequences encoding two globin-like sensors; whether these gene products are related to NO detoxification is unknown.

Activated macrophages kill *Entamoeba histolytica* trophozoites (Supplementary Table S1 at http://www.elsevierdirect.com/companions/9780124076938) mainly by NO production (Denis & Chadee, 1989; Lin, Seguin, Keller, & Chadee, 1994). The presence of NO severely compromises the viability of this parasite by triggering an apoptosis-like mechanism (Ramos et al., 2007) producing profound morphological modifications such as generation of vesicle-like structures of the atypical endoplasmic reticulum. Under nitrosative stress conditions, *E. histolytica* expresses a number of genes related to membrane traffic, extreme stress responses, anaerobic energy production and cysteine metabolism (Santi-Rocca et al., 2012). Nevertheless, NO detoxification mechanisms in this parasite have not been identified. The genome of this parasite lacks globin-like sequences; this might be related to the high susceptibility of the protozoan to NO. Interestingly, *E. histolytica* trophozoites produce NO in culture, and the activity of NO synthase together with recognition of proteins by specific antibodies against NOS leads to the suggestion that NO is produced as a pathogenicity factor during invasive amebiasis (Hernandez-Campos et al., 2003). The absence of NO detoxification mechanisms such as globins in an organism that produces NO may appear logical and consistent with the picture emerging for certain fungi, as discussed above. However, how this organism copes with the toxicity of exogenous and endogenous NO during invasion and pathogenesis is a challenging question. Genomic analyses of *E. histolytica* and *Giardia intestinalis* have revealed the existence of orthologues of the flavorubredoxin system (*nor* genes) (Andersson et al., 2003) of *E. coli*, which represents the main NO detoxification mechanism under anaerobic conditions (Gardner & Gardner, 2002). However, the expression and function of these parasitic putative proteins have not been investigated.

T. vaginalis (Supplementary Table S1 at http://www.elsevierdirect.com/companions/9780124076938) is a flagellated microaerophilic parasite that causes a range of clinical manifestations from asymptomatic infection to severe and extensive damage of the vaginal epithelium. An increased risk of developing invasive cervical cancer is associated with Trichomoniasis (Yap et al., 1995). Chronic infection by *T. vaginalis* is common and implies the survival of the parasite despite the host's defence mechanisms. Even though this parasite lacks haemoglobins, consumption of NO in oxygen-limited conditions has been reported and a flavorubredoxin (NorV) homologue has been identified by immunoblotting using *E. coli* anti-NorV antibodies. This suggests that the reduction of NO to N_2O may act as a detoxification mechanism, independently of any globin (Sarti et al., 2004). Interestingly, in *T. vaginalis*, the arginine deaminase produces ammonia from the conversion of arginine to citrulline in anaerobic conditions (Yarlett, Martinez, Goldberg, Kramer, & Porter, 2000), but in aerobic conditions, NO rather than ammonia is produced in an NOS-dependent reaction (Harris, Goldberg, Biagini, & Lloyd, 2006).

Another example of a pathogenic protozoan lacking globins is *T. brucei* (Supplementary Table S1 at http://www.elsevierdirect.com/companions/ 9780124076938), the causal agent of sleeping sickness. Again, the innate response against trypanosomes involves the production of NO by iNOS (Gobert et al., 2000), playing an essential role as a messenger between trypanosomes and the host immune and nervous system (Antoine-Moussiaux, Magez, & Desmecht, 2008). However, in an infected mice model, proliferation of trypanosomes in the proximity of peripheral macro-phages was associated with decreased NO-dependent cytotoxicity. Indeed, NO levels are decreased in the blood of animals infected with *T.* (*brucei*) *brucei* (Buguet, Banzet, Bouteille, Vincendeau, & Tapie, 2002; Buguet et al., 1996). The requirement for a common substrate, L–arginine, for both arginase and iNOS reduces NO production and favours synthesis of polyamines, compounds required for trypanosome development (Amrouni et al., 2010; Duleu et al., 2004). On the other hand, in the brain, NO concentrations are increased, due perhaps to the presence of iNOS in glial and neuron cells (Amrouni et al., 2010). Utilization of L–arginine might be an indirect protection mechanism. However, how this organism deals with the toxicity of NO in the brain is not clear. NOS activity has been discovered also in *Trypanosoma cruzi* (Paveto et al., 1995) and *Leishmania donovani* (Basu, Kole, Ghosh, & Das, 1997). None of them contains globins.

All species of *Dictyostelium* shown in Supplementary Table S1 at http:// www.elsevierdirect.com/companions/9780124076938 possess one or more FHbs. *D. discoideum* possesses two (Iijima et al., 2000). This chapter draws the surprising conclusion that, for FHbs in general, 'their physiolog-ical significance has not yet been determined', despite a large body of evi-dence predating 2000 showing their key role in NO detoxification. The two *D. discoideum* globin genes (*DdFHa* and *DdFHb*) are in close proximity on the chromosome. Their expression was induced by submerged culture con-ditions. Simultaneous disruption of both genes, but not of the genes individ-ually, led to a striking sensitivity to GSNO and sodium nitroprusside. These are not effective NO-releasing molecules but the growth sensitivity to these compounds does suggest a role for the globins in nitrosative stress tolerance. Paraquat and hydrogen peroxide were not effective in reducing viability of the mutants.

A functional FHb is encoded by the genome of *G. lamblia* (Excavata), a mammalian parasite of the small intestine (Rafferty, Luu, March, & Yee,

2010). This is despite the absence in this organism of the enzymes for haem biosynthesis, raising interesting questions about haem acquisition. The recombinant FHb, however, was purified from *E. coli* extracts with haem and FAD, as expected for FHbs from numerous other species. *In vitro*, the FHb was capable of metabolizing NO efficiently in the presence of oxygen, as well as exogenous FAD to compensate for the cofactor lost on purification (Mastronicola et al., 2010). The occupancy of haem in the purified protein was not reported. Although it might be argued that the globin acquired haem from the host *E. coli* and that the protein must be inactive in *Giardia*, given the absence of haem, the authors also showed that, under anaerobic conditions, nitrite induces a NO-consuming activity that was sensitive to cyanide and CO, implicating the participation of a haem protein (Mastronicola et al., 2010) in NO metabolism. Interestingly, NO production and NOS activity by this parasite has been reported and the presence of an NOS-like sequence in the *G. lamblia* genome has been identified (Harris et al., 2006).

One of the earliest reports of a globin in any microbe was that of a haemoglobin- or myoglobin-like protein in the ciliated protozoan *Paramecium caudatum*, initially by Sato and Tamiya. This was confirmed by Keilin and Ryley (1953), who also showed that another ciliate *Tetrahymena pyriformis* possessed a similar protein, judging by whole cell spectroscopy. The *T. pyriformis* and *T. thermophila* proteins have been isolated and characterized (Korenaga, Igarashi, Matsuoka, & Shikama, 2000). Recently, O_2 association and dissociation rates and autoxidation rate constants were determined for the *T. pyriformis* protein (Igarashi, Kobayashi, & Matsuoka, 2011) and shown to be similar to those reported for the *M. tuberculosis* HbN, a bacterial globin implicated in NO tolerance. A Fe (III)–H_2O complex was formed following the reaction of NO with the Fe(II)–O_2 complex in the crystal state. Although this was interpreted by the authors as indicating a function for this globin in NO detoxification, we are unaware of any studies that clearly show this *in vivo* or that measure protein turnover in the presence of NO. In contrast, the O_2 kinetics of the *P. caudatum* protein (Irie & Usuki, 1980; Iwaasa et al., 1989; Smith, George, & Preer, 1962; Steers, Barnett, & Lee, 1981; Usuki, Hino, & Ochiai, 1989), with which the *Tetrahymena* protein shares high amino acid identity (Yamauchi, Mukai, Ochiai, & Usuki, 1992; Yamauchi, Ochiai, & Usuki, 1992; Yamauchi, Tada, Ochiai, & Usuki, 1993; Yamauchi et al., 1995), are similar to those of sperm whale myoglobin

(Das et al., 2000). Its function is unknown. However, the high concentrations of globins seen in ciliates (up to 0.9% of total protein) and the extremely high oxygen affinities (Smith et al., 1962) suggest that their role may be to trap O_2 under O_2-limited conditions.

6.5. Algae

The only eukaryotic alga from which a globin has been characterized in any detail is *Chlamydomonas*, a genus of unicellular green algae (Chlorophyta). These algae are found universally. *Chlamydomonas* is a unicellular flagellate and used as a model organism for molecular biology. The most widely used laboratory species is *Chlamydomonas reinhardtii*. Two genes of *Chlamydomonas eugametos* encode chloroplast haemoglobins related to those of *P. caudatum*, *T. pyriformis* and the cyanobacterium *Nostoc commune* (Couture, Chamberland, St-Pierre, Lafontaine, & Guertin, 1994). The dissociation rate for O_2 is extremely low and results in part from at least two hydrogen bonds stabilizing the bound ligand (Couture et al., 1999). However, within the chloroplast the globin concentration is estimated to be only 130 nM, far too low to store significant O_2 or facilitate O_2 transport. The high oxygen affinity argues against a metabolic function and the function of the globin remains unknown. It is suggested that its role may be to safeguard photosynthesis from oxygen 'leaks' in the chloroplast (Das et al., 1999).

6.6. Perspectives

The protective function against nitrosative stress makes the FHbs of pathogenic microorganisms attractive targets for antibiotic design. In particular, antimicrobial imidazoles bind to *S. cerevisiae* and *C. albicans* FHbs and inhibit their NOD function (Helmick et al., 2005). Recently, the structural aspects of this inhibition have been investigated by El Hammi et al. (2011), El Hammi, Houee-Levin, et al. (2012) and El Hammi, Warkentin, et al. (2012).

A very recent report by Rosic, Leggat, Kaniewska, Dove, and Hoegh-Guldberg (2013) has suggested a role for the SDgbs of *Symbiodinium* sp. *clade 206D*, a coral endosymbiont, as stress biomarkers, based on the increase in transcription levels in the presence of thermal and nutrient stress. The proposed function is reminiscent of some of the functions suggested for the non-symbiotic Hbs of plants (Hill, 2012).

7. CONCLUSION

The most salient finding of the present study, but which should be viewed as preliminary in nature, is the relative paucity of globins in most microbial eukaryotes relative to multicellular eukaryotes. The frequency of their occurrence appears to be similar to that found in prokaryotes. It is evident that many more microbial eukaryote genomes, other than fungi, would have to be sequenced before a definitive description of their globins can be achieved. Furthermore, the elucidation of the deep phylogeny of protists will also be required for understanding the relationship of their globins to those of higher eukaryotes. Provisionally, our molecular phylogenies, while statistically uncertain, agree on a close relationship of opisthokont SDgbs to vertebrate Ngbs. There are also intriguing relationships between the occurrence of NO-detoxifying FHbs and the lifestyles of many pathogenic fungi; a tentative conclusion is that the obligately biotrophic fungi, which rely on the production of NO as a signalling molecule between pathogen and host, lack those globins that might remove the NO.

In the field of eukaryotic globins, as in most other areas of globin science, there is a glaring lack of reliable information on *function*. Sadly, it is the case that increasingly sophisticated methods for analysing ligand binding to these proteins, analysis of intramolecular ligand channels by structural biology and molecular modelling, and globin distributions are not matched by studies of *function*. We suggest that a more focused approach to studying those globins whose function can be studied *in vivo* and in carefully designed studies with mutants will pay rich dividends in this rapidly expanding field. Nevertheless, the field of microbial eukaryotes holds much promise of future discoveries, and not least, their beguiling and exotic attractiveness, illustrated in Fig. 9.10.

ACKNOWLEDGEMENTS

We would like to acknowledge the assistance provided by the Consejo Nacional de Ciencia y Tecnologia (CONACyT) (Mexico) through Grant No. 99171 and Consejo Estatal de Ciencia, Tecnología e Innovación de Michoacán (CECTI) through Grant No. 007 (Mariana Tinajero-Trejo).

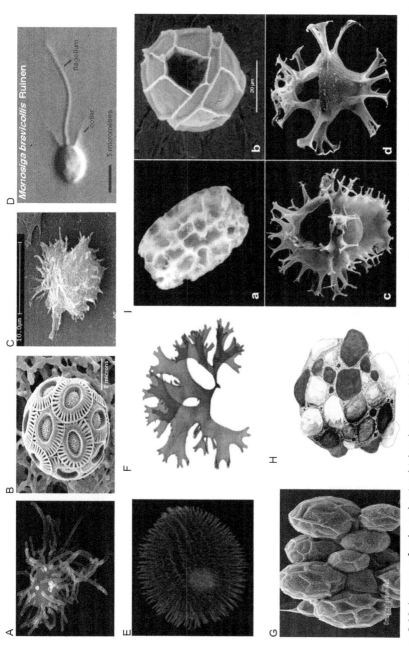

Figure 9.10 Images of selected microbial eukaryotes: (A) *Capsaspora owczazaki*, one of the closest unicellular relatives of animals; (B) haptophyte coccolithiophore *Emiliania huxleyi*; (C) amoebozoan *Acanthamoeba castellani*; (D) choanoflagellate *Monosiga brevicollis*; (E) ciliate *Tetrahymena thermophila*; (F) Rhodophyte *Chondrus crispus* (Irish moss seaweed); (G) the ichtyosporean *Sphaeroforma arctica*; (H) foraminiferan *Astrammina rara*; and (I) dinoflagellates. (For colour version of this figure, the reader is referred to the online version of this chapter.)

REFERENCES

Adl, S. M., Leander, B. S., Simpson, A. G., Archibald, J. M., Anderson, O. R., Bass, D., et al. (2007). Diversity, nomenclature, and taxonomy of protists. *Systematic Biology, 56*, 684–689.

Adl, S. M., Simpson, A. G., Farmer, M. A., Andersen, R. A., Anderson, O. R., Barta, J. R., et al. (2005). The new higher level classification of eukaryotes with emphasis on the taxonomy of protists. *Journal of Eukaryotic Microbiology, 52*, 399–451.

Adl, S. M., Simpson, A. G., Lane, C. E., Lukes, J., Bass, D., Bowser, S. S., et al. (2012). The revised classification of eukaryotes. *Journal of Eukaryotic Microbiology, 59*, 429–493.

Altschul, S. F., Madden, T. L., Schaffer, A. A., Zhang, J., Zhang, Z., Miller, W., et al. (1997). Gapped BLAST and PSI-BLAST: A new generation of protein database search programs. *Nucleic Acids Research, 25*, 3389–3402.

Amrouni, D., Gautier-Sauvigne, S., Meiller, A., Vincendeau, P., Bouteille, B., Buguet, A., et al. (2010). Cerebral and peripheral changes occurring in nitric oxide (NO) synthesis in a rat model of sleeping sickness: Identification of brain iNOS expressing cells. *PLoS One, 5*, e9211.

Andersson, S. G., Karlberg, O., Canback, B., & Kurland, C. G. (2003). On the origin of mitochondria: A genomics perspective. *Philosophical Transactions of the Royal Society of London. Series B, Biological Sciences, 358*, 165–177(discussion 177-169).

Angelo, M., Hausladen, A., Singel, D. J., & Stamler, J. S. (2008). Interactions of NO with hemoglobin: From microbes to man. *Methods in Enzymology, 436*, 131–168.

Antoine-Moussiaux, N., Magez, S., & Desmecht, D. (2008). Contributions of experimental mouse models to the understanding of African trypanosomiasis. *Trends in Parasitology, 24*, 411–418.

Ascenzi, P., Bocedi, A., & Gradoni, L. (2003). The anti-parasitic effects of nitric oxide. *IUBMB Life, 55*, 573–578.

Aury, J. M., Jaillon, O., Duret, L., Noel, B., Jubin, C., Porcel, B. M., et al. (2006). Global trends of whole-genome duplications revealed by the ciliate *Paramecium tetraurelia. Nature, 444*, 171–178.

Baidya, S., Cary, J. W., Grayburn, W. S., & Calvo, A. M. (2011). Role of nitric oxide and flavohemoglobin homolog genes in *Aspergillus nidulans* sexual development and mycotoxin production. *Applied and Environmental Microbiology, 77*, 5524–5528.

Basu, N. K., Kole, L., Ghosh, A., & Das, P. K. (1997). Isolation of a nitric oxide synthase from the protozoan parasite, *Leishmania donovani. FEMS Microbiology Letters, 156*, 43–47.

Bayer, T., Aranda, M., Sunagawa, S., Yum, L. K., DeSalvo, M. K., Lindquist, E., et al. (2012). *Symbiodinium* transcriptomes: Genome insights into the dinoflagellate symbionts of reef-building corals. *PLoS One, 7*, e35269.

Bik, H. M., Porazinska, D. L., Creer, S., Caporaso, J. G., Knight, R., & Thomas, W. K. (2012). Sequencing our way towards understanding global eukaryotic biodiversity. *Trends in Ecology & Evolution, 27*, 233–243.

Blank, M., & Burmester, T. (2012). Widespread occurrence of N-terminal acylation in animal globins and possible origin of respiratory globins from a membrane-bound ancestor. *Molecular Biology and Evolution, 29*, 3553–3561.

Boccara, M., Mills, C. E., Zeier, J., Anzi, C., Lamb, C., Poole, R. K., et al. (2005). Flavohaemoglobin HmpX from *Erwinia chrysanthemi* confers nitrosative stress tolerance and affects the plant hypersensitive reaction by intercepting nitric oxide produced by the host. *The Plant Journal, 43*, 226–237.

Bogdan, C., Gessner, A., Solbach, W., & Rollinghoff, M. (1996). Invasion, control and persistence of Leishmania parasites. *Current Opinion in Immunology, 8*, 517–525.

Bowman, L. A., McLean, S., Poole, R. K., & Fukuto, J. M. (2011). The diversity of microbial responses to nitric oxide and agents of nitrosative stress close cousins but not identical twins. *Advances in Microbial Physiology, 59*, 135–219.

Brown, A. J., Haynes, K., & Quinn, J. (2009). Nitrosative and oxidative stress responses in fungal pathogenicity. *Current Opinion in Microbiology, 12*, 384–391.

Buguet, A., Banzet, S., Bouteille, B., Vincendeau, P., & Tapie, P. (2002). NO a cornerstone in sleeping sickness: Voltammetric assessment in mouse and man. *Adaptation Biology and Medicine, 3*, 222–230.

Buguet, A., Burlet, S., Auzelle, F., Montmayeur, A., Jouvet, M., & Cespuglio, R. (1996). Action duality of nitrogen oxide (NO) in experimental African trypanosomiasis. *Comptes Rendus de l'Académie des Sciences. Série III, Sciences de la vie, 319*, 201–207.

Buisson, N., & Labbe-Bois, R. (1998). Flavohemoglobin expression and function in *Saccharomyces cerevisiae*. No relationship with respiration and complex response to oxidative stress. *The Journal of Biological Chemistry, 273*, 9527–9533.

Burki, F., Shalchian-Tabrizi, K., Minge, M., Skjaeveland, A., Nikolaev, S. I., Jakobsen, K. S., et al. (2007). Phylogenomics reshuffles the eukaryotic supergroups. *PLoS One, 2*, e790.

Burmester, T., Haberkamp, M., Mitz, S., Roesner, A., Schmidt, M., Ebner, B., et al. (2004). Neuroglobin and cytoglobin: Genes, proteins and evolution. *IUBMB Life, 56*, 703–707.

Carr, M., Leadbeater, B. S., Hassan, R., Nelson, M., & Baldauf, S. L. (2008). Molecular phylogeny of choanoflagellates, the sister group to Metazoa. *Proceedings of the National Academy of Sciences of the United States of America, 105*, 16641–16646.

Cassanova, N., O'Brien, K. M., Stahl, B. T., McClure, T., & Poyton, R. O. (2005). Yeast flavohemoglobin, a nitric oxide oxidoreductase, is located in both the cytosol and the mitochondrial matrix: Effects of respiration, anoxia, and the mitochondrial genome on its intracellular level and distribution. *The Journal of Biological Chemistry, 280*, 7645–7653.

Castresana, J. (2000). Selection of conserved blocks from multiple alignments for their use in phylogenetic analysis. *Molecular Biology and Evolution, 17*, 540–552.

Cavalier-Smith, T. (2002). The phagotrophic origin of eukaryotes and phylogenetic classification of Protozoa. *International Journal of Systematic and Evolutionary Microbiology, 52*, 297–354.

Chatton, É. (1925). Pansporella perplexa, amoebiens à spores protégées parasite de daphnies. Réflections sur la biologie et la phylogénie des protozoaires. *Annales Des Sciences Naturelles Zoologie, 8*, 5–84.

Couture, M., Chamberland, H., St-Pierre, B., Lafontaine, J., & Guertin, M. (1994). Nuclear genes encoding chloroplast hemoglobins in the unicellular green alga *Chlamydomonas eugametos*. *Molecular and General Genetics, 243*, 185–197.

Couture, M., Das, T. K., Lee, H. C., Peisach, J., Rousseau, D. L., Wittenberg, B. A., et al. (1999). *Chlamydomonas* chloroplast ferrous hemoglobin. Heme pocket structure and reactions with ligands. *The Journal of Biological Chemistry, 274*, 6898–6910.

Crawford, M. J., Sherman, D. R., & Goldberg, D. E. (1995). Regulation of *Saccharomyces cerevisiae* flavohemoglobin gene expression. *The Journal of Biological Chemistry, 270*, 6991–6996.

Das, T. K., Couture, M., Lee, H. C., Peisach, J., Rousseau, D. L., Wittenberg, B. A., et al. (1999). Identification of the ligands to the ferric heme of *Chlamydomonas* chloroplast hemoglobin: Evidence for ligation of tyrosine-63 (B10) to the heme. *Biochemistry, 38*, 15360–15368.

Das, T. K., Weber, R. E., Dewilde, S., Wittenberg, J. B., Wittenberg, B. A., Yamauchi, K., et al. (2000). Ligand binding in the ferric and ferrous states of *Paramecium* hemoglobin. *Biochemistry, 39*, 14330–14340.

de Jesus-Berrios, M., Liu, L., Nussbaum, J. C., Cox, G. M., Stamler, J. S., & Heitman, J. (2003). Enzymes that counteract nitrosative stress promote fungal virulence. *Current Biology, 13*, 1963–1968.

Del Campo, J., & Ruiz-Trillo, I. (2013). Environmental survey meta-analysis reveals hidden diversity among unicellular opisthokonts. *Molecular Biology and Evolution, 30*, 802–805.

Delledonne, M., Zeier, J., Marocco, A., & Lamb, C. (2001). Signal interactions between nitric oxide and reactive oxygen intermediates in the plant hypersensitive disease resistance response. *Proceedings of the National Academy of Sciences of the United States of America, 98*, 13454–13459.

Denis, M., & Chadee, K. (1989). Human neutrophils activated by interferon-gamma and tumour necrosis factor-alpha kill *Entamoeba histolytica* trophozoites in vitro. *Journal of Leukocyte Biology, 46*, 270–274.

Dikshit, K. L., Spaulding, D., Braun, A., & Webster, D. A. (1989). Oxygen inhibition of globin gene transcription and bacterial haemoglobin synthesis in Vitreoscilla. *Journal of General Microbiology, 135*, 2601–2609.

Di Tommaso, P., Moretti, S., Xenarios, I., Orobitg, M., Montanyola, A., Chang, J. M., et al. (2011). T-Coffee: A web server for the multiple sequence alignment of protein and RNA sequences using structural information and homology extension. *Nucleic Acids Research, 39*, W13–W17.

Do, C. B., Mahabhashyam, M. S., Brudno, M., & Batzoglou, S. (2005). ProbCons: Probabilistic consistency-based multiple sequence alignment. *Genome Research, 15*, 330–340.

Duleu, S., Vincendeau, P., Courtois, P., Semballa, S., Lagroye, I., Daulouede, S., et al. (2004). Mouse strain susceptibility to trypanosome infection: An arginase-dependent effect. *Journal of Immunology, 172*, 6298–6303.

Edgar, R. C. (2004). MUSCLE: Multiple sequence alignment with high accuracy and high throughput. *Nucleic Acids Research, 32*, 1792–1797.

Eichinger, L., Pachebat, J. A., Glockner, G., Rajandream, M. A., Sucgang, R., Berriman, M., et al. (2005). The genome of the social amoeba *Dictyostelium discoideum*. *Nature, 435*, 43–57.

El Hammi, E., Houee-Levin, C., Rezac, J., Levy, B., Demachy, I., Baciou, L., et al. (2012). New insights into the mechanism of electron transfer within flavohemoglobins: Tunnelling pathways, packing density, thermodynamic and kinetic analyses. *Physical Chemistry Chemical Physics, 14*, 13872–13880.

El Hammi, E., Warkentin, E., Demmer, U., Limam, F., Marzouki, N. M., Ermler, U., et al. (2011). Structure of *Ralstonia eutropha* flavohemoglobin in complex with three antibiotic azole compounds. *Biochemistry, 50*, 1255–1264.

El Hammi, E., Warkentin, E., Demmer, U., Marzouki, N. M., Ermler, U., & Baciou, L. (2012). Active site analysis of yeast flavohemoglobin based on its structure with a small ligand or econazole. *The FEBS Journal, 279*, 4565–4575.

Ermler, U., Siddiqui, R. A., Cramm, R., & Friedrich, B. (1995). Crystal structure of the flavohemoglobin from *Alcaligenes eutrophus* at 1.75 Å resolution. *The EMBO Journal, 14*, 6067–6077.

Ford, B. J. (1995). First steps in experimental microscopy, Leeuwenhoek as practical scientist. *Microscope, 43*, 47–57.

Forrester, M. T., & Foster, M. W. (2012a). Response to: "Is flavohemoglobin a nitric oxide dioxygenase?" *Free Radical Biology & Medicine, 53*, 1211–1212.

Forrester, M. T., & Foster, M. W. (2012b). Protection from nitrosative stress: A central role for microbial flavohemoglobin. *Free Radical Biology & Medicine, 52*, 1620–1633.

Freitas, T. A., Saito, J. A., Hou, S., & Alam, M. (2005). Globin-coupled sensors, protoglobins, and the last universal common ancestor. *Journal of Inorganic Biochemistry, 99*, 23–33.

Frey, A. D., Shepherd, M., Jokipii-Lukkari, S., Haggman, H., & Kallio, P. T. (2011). The single-domain globin of *Vitreoscilla*: Augmentation of aerobic metabolism for biotechnological applications. *Advances in Microbial Physiology, 58*, 81–139.

Fritz-Laylin, L. K., Ginger, M. L., Walsh, C., Dawson, S. C., & Fulton, C. (2011). The *Naegleria* genome: A free-living microbial eukaryote lends unique insights into core eukaryotic cell biology. *Research in Microbiology, 162,* 607–618.

Fu, C., Xiong, J., & Miao, W. (2009). Genome-wide identification and characterization of cytochrome P450 monooxygenase genes in the ciliate *Tetrahymena thermophila*. *BMC Genomics, 10,* 208.

Gardner, P. R. (2005). Nitric oxide dioxygenase function and mechanism of flavohemoglobin, hemoglobin, myoglobin and their associated reductases. *Journal of Inorganic Biochemistry, 99,* 247–266.

Gardner, P. R. (2012). Hemoglobin, a nitric oxide dioxygenase. *Scientifica, 2012,* 1–34.

Gardner, A. M., & Gardner, P. R. (2002). Flavohemoglobin detoxifies nitric oxide in aerobic, but not anaerobic, *Escherichia coli*. Evidence for a novel inducible anaerobic nitric oxide-scavenging activity. *The Journal of Biological Chemistry, 277,* 8166–8171.

Gardner, P. R., Gardner, A. M., Martin, L. A., & Salzman, A. L. (1998). Nitric oxide dioxygenase: An enzymic function for flavohemoglobin. *Proceedings of the National Academy of Sciences of the United States of America, 95,* 10378–10383.

Gilberthorpe, N. J., Lee, M. E., Stevanin, T. M., Read, R. C., & Poole, R. K. (2007). NsrR: A key regulator circumventing *Salmonella enterica* serovar *Typhimurium* oxidative and nitrosative stress in vitro and in IFN-gamma-stimulated J774.2 macrophages. *Microbiology, 153,* 1756–1771.

Giudice, A., Camada, I., Leopoldo, P. T., Pereira, J. M., Riley, L. W., Wilson, M. E., et al. (2007). Resistance of *Leishmania* (*Leishmania*) *amazonensis* and *Leishmania* (*Viannia*) *braziliensis* to nitric oxide correlates with disease severity in Tegumentary Leishmaniasis. *BMC Infectious Diseases, 7,* 7.

Gobert, A. P., Daulouede, S., Lepoivre, M., Boucher, J. L., Bouteille, B., Buguet, A., et al. (2000). L-Arginine availability modulates local nitric oxide production and parasite killing in experimental trypanosomiasis. *Infection and Immunity, 68,* 4653–4657.

Gough, J., Karplus, K., Hughey, R., & Chothia, C. (2001). Assignment of homology to genome sequences using a library of hidden Markov models that represent all proteins of known structure. *Journal of Molecular Biology, 313,* 903–919.

Govrin, E. M., & Levine, A. (2000). The hypersensitive response facilitates plant infection by the necrotrophic pathogen *Botrytis cinerea*. *Current Biology, 10,* 751–757.

Gray, M. W. (1999). Evolution of organellar genomes. *Current Opinion in Genetics and Development, 9,* 678–687.

Gray, M. W., Lang, B. F., & Burger, G. (2004). Mitochondria of protists. *Annual Review of Genetics, 38,* 477–524.

Green, S. J., Meltzer, M. S., Hibbs, J. B., Jr., & Nacy, C. A. (1990). Activated macrophages destroy intracellular *Leishmania major* amastigotes by an L-arginine-dependent killing mechanism. *Journal of Immunology, 144,* 278–283.

Guillou, L., Bachar, D., Audic, S., Bass, D., Berney, C., Bittner, L., et al. (2013). The protist ribosomal reference database (PR2): A catalog of unicellular eukaryote small sub-unit rRNA sequences with curated taxonomy. *Nucleic Acids Research, 41,* D597–D604.

Guiry, M. D. (2012). How many species of algae are there? *Journal of Phycology, 48,* 1057–1063.

Gupta, K. J., Fernie, A. R., Kaiser, W. M., & van Dongen, J. T. (2011). On the origins of nitric oxide. *Trends in Plant Science, 16,* 160–168.

Haddock, S. H., Moline, M. A., & Case, J. F. (2010). Bioluminescence in the sea. *Annual Review of Marine Science, 2,* 443–493.

Hardison, R. C. (1996). A brief history of hemoglobins: Plant, animal, protist, and bacteria. *Proceedings of the National Academy of Sciences of the United States of America, 93,* 5675–5679.

Harris, K. M., Goldberg, B., Biagini, G. A., & Lloyd, D. (2006). *Trichomonas vaginalis* and *Giardia intestinalis* produce nitric oxide and display NO-synthase activity. *Journal of Eukaryotic Microbiology, 53*(Suppl. 1), S182–S183.

Hartzell, H. C., Qu, Z., Yu, K., Xiao, Q., & Chien, L. T. (2008). Molecular physiology of bestrophins: Multifunctional membrane proteins linked to best disease and other retinopathies. *Physiological Reviews, 88*, 639–672.

Hausladen, A., & Stamler, J. S. (2012). Is flavohemoglobin a nitric oxide dioxygenase? *Free Radical Biology & Medicine, 53*, 1209–1210.

Hedges, S. B., Blair, J. E., Venturi, M. L., & Shoe, J. L. (2004). A molecular timescale of eukaryote evolution and the rise of complex multicellular life. *BMC Evolutionary Biology, 4*, 2.

Helmick, R. A., Fletcher, A. E., Gardner, A. M., Gessner, C. R., Hvitved, A. N., Gustin, M. C., et al. (2005). Imidazole antibiotics inhibit the nitric oxide dioxygenase function of microbial flavohemoglobin. *Antimicrobial Agents and Chemotherapy, 49*, 1837–1843.

Hernandez-Campos, M. E., Campos-Rodriguez, R., Tsutsumi, V., Shibayama, M., Garcia-Latorre, E., Castillo-Henkel, C., et al. (2003). Nitric oxide synthase in *Entamoeba histolytica*: Its effect on rat aortic rings. *Experimental Parasitology, 104*, 87–95.

Hill, R. D. (2012). Non-symbiotic haemoglobins—What's happening beyond nitric oxide scavenging? *AoB Plants, 2012*, pls004.

Hoffmann, F. G., Opazo, J. C., Hoogewijs, D., Hankeln, T., Ebner, B., Vinogradov, S. N., et al. (2012). Evolution of the globin gene family in deuterostomes: Lineage-specific patterns of diversification and attrition. *Molecular Biology and Evolution, 29*, 1735–1745.

Hoffmann, F. G., Opazo, J. C., & Storz, J. F. (2012). Whole-genome duplications spurred the functional diversification of the globin gene superfamily in vertebrates. *Molecular Biology and Evolution, 29*, 303–312.

Hoogewijs, D., Dewilde, S., Vierstraete, A., Moens, L., & Vinogradov, S. N. (2012). A phylogenetic analysis of the globins in fungi. *PLoS One, 7*, e31856.

Hoogewijs, D., Ebner, B., Germani, F., Hoffmann, F. G., Fabrizius, A., Moens, L., et al. (2012). Androglobin: A chimeric globin in metazoans that is preferentially expressed in mammalian testes. *Molecular Biology and Evolution, 29*, 1105–1114.

Hou, S., Freitas, T., Larsen, R. W., Piatibratov, M., Sivozhelezov, V., Yamamoto, A., et al. (2001). Globin-coupled sensors: A class of heme-containing sensors in Archaea and Bacteria. *Proceedings of the National Academy of Sciences of the United States of America, 98*, 9353–9358.

Hromatka, B. S., Noble, S. M., & Johnson, A. D. (2005). Transcriptional response of *Candida albicans* to nitric oxide and the role of the YHB1 gene in nitrosative stress and virulence. *Molecular Biology of the Cell, 16*, 4814–4826.

Hunt, P. W., Watts, R. A., Trevaskis, B., Llewelyn, D. J., Burnell, J., Dennis, E. S., et al. (2001). Expression and evolution of functionally distinct haemoglobin genes in plants. *Plant Molecular Biology, 47*, 677–692.

Igarashi, J., Kobayashi, K., & Matsuoka, A. (2011). A hydrogen-bonding network formed by the B10-E7-E11 residues of a truncated hemoglobin from *Tetrahymena pyriformis* is critical for stability of bound oxygen and nitric oxide detoxification. *Journal of Biological Inorganic Chemistry, 16*, 599–609.

Iijima, M., Shimizu, H., Tanaka, Y., & Urushihara, H. (2000). Identification and characterization of two flavohemoglobin genes in *Dictyostelium discoideum*. *Cell Structure and Function, 25*, 47–55.

Ilari, A., & Boffi, A. (2008). Structural studies on flavohemoglobins. *Methods in Enzymology, 436*, 187–202.

Irie, T., & Usuki, I. (1980). Disparity of native oxyhemoglobin components isolated from *Paramecium caudatum* and *Paramecium primaurelia*. *Comparative Biochemistry and Physiology. Part B, Biochemistry & Molecular Biology, 67*, 549–554.

Iwaasa, H., Takagi, T., & Shikama, K. (1989). Protozoan myoglobin from *Paramecium caudatum*. Its unusual amino acid sequence. *Journal of Molecular Biology, 208*, 355–358.

Katoh, K., Kuma, K., Toh, H., & Miyata, T. (2005). MAFFT version 5: Improvement in accuracy of multiple sequence alignment. *Nucleic Acids Research, 33,* 511–518.

Keeling, P. J. (2010). The endosymbiotic origin, diversification and fate of plastids. *Philosophical Transactions of the Royal Society of London. Series B, Biological Sciences, 365,* 729–748.

Keilin, D., & Ryley, J. F. (1953). Haemoglobin in Protozoa. *Nature, 172,* 451.

Khosla, C., & Bailey, J. E. (1989). Evidence for partial export of *Vitreoscilla* hemoglobin into the periplasmic space in *Escherichia coli.* Implications for protein function. *Journal of Molecular Biology, 210,* 79–89.

Kim, S. W., Fushinobu, S., Zhou, S., Wakagi, T., & Shoun, H. (2010). The possible involvement of copper-containing nitrite reductase (NirK) and flavohemoglobin in denitrification by the fungus *Cylindrocarpon tonkinense. Bioscience, Biotechnology, and Biochemistry, 74,* 1403–1407.

Kobayashi, G., Nakamura, T., Ohmachi, H., Matsuoka, A., Ochiai, T., & Shikama, K. (2002). Yeast flavohemoglobin from *Candida norvegensis.* Its structural, spectral, and stability properties. *The Journal of Biological Chemistry, 277,* 42540–42548.

Korenaga, S., Igarashi, J., Matsuoka, A., & Shikama, K. (2000). A primitive myoglobin from *Tetrahymena pyriformis:* Its heme environment, autoxidizability, and genomic DNA structure. *Biochimica et Biophysica Acta, 1543,* 131–145.

Lamba, P., & Webster, D. A. (1980). Effect of growth conditions on yield and heme content of *Vitreoscilla. Journal of Bacteriology, 142,* 169–173.

Lassmann, T., & Sonnhammer, E. L. (2005). Automatic assessment of alignment quality. *Nucleic Acids Research, 33,* 7120–7128.

Laver, J. R., McLean, S., Bowman, L. A., Harrison, L. J., Read, R. C., & Poole, R. K. (2013). Nitrosothiols in bacterial pathogens and pathogenesis. *Antioxidants & Redox Signaling, 18,* 309–322.

Laver, J. R., Stevanin, T. M., Messenger, S. L., Lunn, A. D., Lee, M. E., Moir, J. W., et al. (2010). Bacterial nitric oxide detoxification prevents host cell S-nitrosothiol formation: A novel mechanism of bacterial pathogenesis. *The FASEB Journal, 24,* 286–295.

Lechauve, C., Jager, M., Laguerre, L., Kiger, L., Correc, G., Leroux, C., et al. (2013). Neuroglobins, pivotal proteins associated with emerging neural systems and precursors of metazoan globin diversity. *The Journal of Biological Chemistry, 288,* 6957–6967.

Lewinska, A., & Bartosz, G. (2006). Yeast flavohemoglobin protects against nitrosative stress and controls ferric reductase activity. *Redox Report, 11,* 231–239.

Liew, F. Y., Millott, S., Parkinson, C., Palmer, R. M., & Moncada, S. (1990). Macrophage killing of *Leishmania* parasite *in vivo* is mediated by nitric oxide from L-arginine. *Journal of Immunology, 144,* 4794–4797.

Lin, J. Y., Seguin, R., Keller, K., & Chadee, K. (1994). Tumor necrosis factor alpha augments nitric oxide-dependent macrophage cytotoxicity against *Entamoeba histolytica* by enhanced expression of the nitric oxide synthase gene. *Infection and Immunity, 62,* 1534–1541.

Liu, L., Zeng, M., Hausladen, A., Heitman, J., & Stamler, J. S. (2000). Protection from nitrosative stress by yeast flavohemoglobin. *Proceedings of the National Academy of Sciences of the United States of America, 97,* 4672–4676.

Lopez-Garcia, P., Rodriguez-Valera, F., Pedros-Alio, C., & Moreira, D. (2001). Unexpected diversity of small eukaryotes in deep-sea Antarctic plankton. *Nature, 409,* 603–607.

Malolepsza, U., & Rozalska, S. (2005). Nitric oxide and hydrogen peroxide in tomato resistance. Nitric oxide modulates hydrogen peroxide level in o-hydroxyethylorutin-induced resistance to *Botrytis cinerea* in tomato. *Plant Physiology and Biochemistry, 43,* 623–635.

Marchler-Bauer, A., Anderson, J. B., Cherukuri, P. F., DeWeese-Scott, C., Geer, L. Y., Gwadz, M., et al. (2005). CDD: A conserved domain database for protein classification. *Nucleic Acids Research*, *33*, D192–D196.

Mastronicola, D., Testa, F., Forte, E., Bordi, E., Pucillo, L. P., Sarti, P., et al. (2010). Flavohemoglobin and nitric oxide detoxification in the human protozoan parasite *Giardia intestinalis*. *Biochemical and Biophysical Research Communications*, *399*, 654–658.

Miller, M. A., Pfeiffer, W., & Schwartz, T. (2010). Creating the CIPRES Science Gateway for inference of large phylogenetic trees. In *Proceedings of the gateway computing environments workshop (GCE), 14 November, New Orleans, LA* (pp. 1–8).

Mills, C. E., Sedelnikova, S., Soballe, B., Hughes, M. N., & Poole, R. K. (2001). *Escherichia coli* flavohaemoglobin (Hmp) with equistoichiometric FAD and haem contents has a low affinity for dioxygen in the absence or presence of nitric oxide. *The Biochemical Journal*, *353*, 207–213.

Moon-van der Staay, S. Y., De Wachter, R., & Vaulot, D. (2001). Oceanic 18S rDNA sequences from picoplankton reveal unsuspected eukaryotic diversity. *Nature*, *409*, 607–610.

Mur, L. A. J., Carver, T. L. W., & Prats, E. (2006). NO way to live; the various roles of nitric oxide in plant–pathogen interactions. *Journal of Experimental Botany*, *57*, 489–505.

Mur, L. A., Mandon, J., Cristescu, S. M., Harren, F. J., & Prats, E. (2011). Methods of nitric oxide detection in plants: A commentary. *Plant Science*, *181*, 509–519.

Murray, J. W., Delumeau, O., & Lewis, R. J. (2005). Structure of a nonheme globin in environmental stress signaling. *Proceedings of the National Academy of Sciences of the United States of America*, *102*, 17320–17325.

Nakai, K., & Horton, P. (1999). PSORT: A program for detecting sorting signals in proteins and predicting their subcellular localization. *Trends in Biochemical Sciences*, *24*, 34–36.

Nichols, S. A., Roberts, B. W., Richter, D. J., Fairclough, S. R., & King, N. (2012). Origin of metazoan cadherin diversity and the antiquity of the classical cadherin/beta-catenin complex. *Proceedings of the National Academy of Sciences of the United States of America*, *109*, 13046–13051.

Ohm, R. A., Feau, N., Henrissat, B., Schoch, C. L., Horwitz, B. A., Barry, K. W., et al. (2012). Diverse lifestyles and strategies of plant pathogenesis encoded in the genomes of eighteen *Dothideomycetes* fungi. *PLoS Pathogens*, *8*, e1003037.

Orozco-Cardenas, M. L., & Ryan, C. A. (2002). Nitric oxide negatively modulates wound signaling in tomato plants. *Plant Physiology and Biochemistry*, *130*, 487–493.

Oshino, R., Asakura, T., Takio, K., Oshino, N., Chance, B., & Hagihara, B. (1973). Purification and molecular properties of yeast hemoglobin. *European Journal of Biochemistry*, *39*, 581–590.

Oshino, R., Asakura, T., Tamura, M., Oshino, N., & Chance, B. (1972). Yeast hemoglobin–reductase complex. *Biochemical and Biophysical Research Communications*, *46*, 1055–1060.

Oshino, R., Oshino, N., & Chance, B. (1971). The oxygen equilibrium of yeast hemoglobin. *FEBS Letters*, *19*, 96–100.

Oshino, R., Oshino, N., & Chance, B. (1973). Studies on yeast hemoglobin. The properties of yeast hemoglobin and its physiological function in the cell. *European Journal of Biochemistry*, *35*, 23–33.

Parfrey, L. W., Lahr, D. J., Knoll, A. H., & Katz, L. A. (2011). Estimating the timing of early eukaryotic diversification with multigene molecular clocks. *Proceedings of the National Academy of Sciences of the United States of America*, *108*, 13624–13629.

Paveto, C., Pereira, C., Espinosa, J., Montagna, A. E., Farber, M., Esteva, M., et al. (1995). The nitric oxide transduction pathway in *Trypanosoma cruzi*. *The Journal of Biological Chemistry*, *270*, 16576–16579.

Pawlowski, J., Christen, R., Lecroq, B., Bachar, D., Shahbazkia, H. R., Amaral-Zettler, L., et al. (2011). Eukaryotic richness in the abyss: Insights from pyrotag sequencing. *PLoS One*, *6*, e18169.

Pesce, A., Couture, M., Dewilde, S., Guertin, M., Yamauchi, K., Ascenzi, P., et al. (2000). A novel two-over-two alpha-helical sandwich fold is characteristic of the truncated hemoglobin family. *The EMBO Journal, 19*, 2424–2434.

Poole, R. K., Anjum, M. F., Membrillo-Hernandez, J., Kim, S. O., Hughes, M. N., & Stewart, V. (1996). Nitric oxide, nitrite, and Fnr regulation of hmp (flavohemoglobin) gene expression in *Escherichia coli* K-12. *Journal of Bacteriology, 178*, 5487–5492.

Poole, R. K., Ioannidis, N., & Orii, Y. (1994). Reactions of the *Escherichia coli* flavohaemoglobin (Hmp) with oxygen and reduced nicotinamide adenine dinucleotide: Evidence for oxygen switching of flavin oxidoreduction and a mechanism for oxygen sensing. *Proceedings of the Royal Society B: Biological Sciences, 255*, 251–258.

Prats, E., Carver, T. L., & Mur, L. A. (2008). Pathogen-derived nitric oxide influences formation of the appressorium infection structure in the phytopathogenic fungus *Blumeria graminis*. *Research in Microbiology, 159*, 476–480.

Probst, I., & Schlegel, H. G. (1976). Respiratory components and oxidase activities in *Alcaligenes eutrophus*. *Biochimica et Biophysica Acta, 440*, 412–428.

Probst, I., Wolf, G., & Schlegel, H. G. (1979). An oxygen-binding flavohemoprotein from *Alcaligenes eutrophus*. *Biochimica et Biophysica Acta, 576*, 471–478.

Quevillon, E., Silventoinen, V., Pillai, S., Harte, N., Mulder, N., Apweiler, R., et al. (2005). InterProScan: Protein domains identifier. *Nucleic Acids Research, 33*, W116–W120.

Rafferty, S., Luu, B., March, R. E., & Yee, J. (2010). *Giardia lamblia* encodes a functional flavohemoglobin. *Biochemical and Biophysical Research Communications, 399*, 347–351.

Ramos, E., Olivos-Garcia, A., Nequiz, M., Saavedra, E., Tello, E., Saralegui, A., et al. (2007). *Entamoeba histolytica*: Apoptosis induced *in vitro* by nitric oxide species. *Experimental Parasitology, 116*, 257–265.

Requena, M.-E., Egea-Gilabert, C., & Candela, M.-E. (2005). Nitric oxide generation during the interaction with *Phytophthora capsici* of two *Capsicum annuum* varieties showing different degrees of sensitivity. *Physiologia Plantarum, 124*, 50–60.

Rodriguez-Ezpeleta, N., Brinkmann, H., Burey, S. C., Roure, B., Burger, G., Loffelhardt, W., et al. (2005). Monophyly of primary photosynthetic eukaryotes: Green plants, red algae, and glaucophytes. *Current Biology, 15*, 1325–1330.

Roger, A. J., & Simpson, A. G. (2009). Evolution: Revisiting the root of the eukaryote tree. *Current Biology, 19*, R165–R167.

Ronquist, F., & Huelsenbeck, J. P. (2003). MrBayes 3: Bayesian phylogenetic inference under mixed models. *Bioinformatics, 19*, 1572–1574.

Rosic, N. N., Leggat, W., Kaniewska, P., Dove, S., & Hoegh-Guldberg, O. (2013). New-old hemoglobin-like proteins of symbiotic dinoflagellates. *Ecology and Evolution, 3*, 822–834.

Rothschild, L. J. (1989). Protozoa, Protista, Protoctista: What's in a name? *Journal of the History of Biology, 22*, 277–305.

Ruiz-Trillo, I., Burger, G., Holland, P. W., King, N., Lang, B. F., Roger, A. J., et al. (2007). The origins of multicellularity: A multi-taxon genome initiative. *Trends in Genetics, 23*, 113–118.

Ruiz-Trillo, I., Lane, C. E., Archibald, J. M., & Roger, A. J. (2006). Insights into the evolutionary origin and genome architecture of the unicellular opisthokonts *Capsaspora owczarzaki* and *Sphaeroforma arctica*. *Journal of Eukaryotic Microbiology, 53*, 379–384.

Ruiz-Trillo, I., Roger, A. J., Burger, G., Gray, M. W., & Lang, B. F. (2008). A phylogenomic investigation into the origin of metazoa. *Molecular Biology and Evolution, 25*, 664–672.

Sadler, J. E. (1998). Biochemistry and genetics of von Willebrand factor. *Annual Review of Biochemistry, 67*, 395–424.

Santi-Rocca, J., Smith, S., Weber, C., Pineda, E., Hon, C. C., Saavedra, E., et al. (2012). Endoplasmic reticulum stress-sensing mechanism is activated in *Entamoeba histolytica* upon treatment with nitric oxide. *PLoS One, 7*, e31777.

Sarti, P., Fiori, P. L., Forte, E., Rappelli, P., Teixeira, M., Mastronicola, D., et al. (2004). *Trichomonas vaginalis* degrades nitric oxide and expresses a flavorubredoxin-like protein: A new pathogenic mechanism? *Cellular and Molecular Life Sciences, 61*, 618–623.

Scamardella, J. M. (1999). Not plants or animals: A brief history of the origin of Kingdoms Protozoa, Protista and Protoctista. *International Microbiology, 2*, 207–216.

Schaffer, A. A., Aravind, L., Madden, T. L., Shavirin, S., Spouge, J. L., Wolf, Y. I., et al. (2001). Improving the accuracy of PSI-BLAST protein database searches with composition-based statistics and other refinements. *Nucleic Acids Research, 29*, 2994–3005.

Shalchian-Tabrizi, K., Minge, M. A., Espelund, M., Orr, R., Ruden, T., Jakobsen, K. S., et al. (2008). Multigene phylogeny of choanozoa and the origin of animals. *PLoS One, 3*, e2098.

Shi, J., Blundell, T. L., & Mizuguchi, K. (2001). FUGUE: Sequence–structure homology recognition using environment-specific substitution tables and structure-dependent gap penalties. *Journal of Molecular Biology, 310*, 243–257.

Sievers, F., Wilm, A., Dineen, D., Gibson, T. J., Karplus, K., Li, W., et al. (2011). Fast, scalable generation of high-quality protein multiple sequence alignments using Clustal Omega. *Molecular Systems Biology, 7*, 539.

Simpson, A. G., & Roger, A. J. (2002). Eukaryotic evolution: Getting to the root of the problem. *Current Biology, 12*, R691–R693.

Smith, M. H., George, P., & Preer, J. R., Jr. (1962). Preliminary observations on isolated *Paramecium* hemoglobin. *Archives of Biochemistry and Biophysics, 99*, 313–318.

Sogin, M. L., Morrison, H. G., Hinkle, G., & Silberman, J. D. (1996). Ancestral relationships of the major eukaryotic lineages. *Microbiología, 12*, 17–28.

Sogin, M. L., & Silberman, J. D. (1998). Evolution of the protists and protistan parasites from the perspective of molecular systematics. *International Journal for Parasitology, 28*, 11–20.

Stamatakis, A., Hoover, P., & Rougemont, J. (2008). A rapid bootstrap algorithm for the RAxML Web servers. *Systematic Biology, 57*, 758–771.

Steers, E., Jr., Barnett, A., & Lee, C. E. (1981). Isolation and characterization of the hemoglobin from *Paramecium caudatum*. *Comparative Biochemistry and Physiology. Part B, Biochemistry & Molecular Biology, 70*, 185–191.

Stevanin, T. M., Poole, R. K., Demoncheaux, E. A., & Read, R. C. (2002). Flavohemoglobin Hmp protects *Salmonella enterica* serovar *typhimurium* from nitric oxide-related killing by human macrophages. *Infection and Immunity, 70*, 4399–4405.

Stewart, J. J., & Coyne, K. J. (2011). Analysis of raphidophyte assimilatory nitrate reductase reveals unique domain architecture incorporating a 2/2 hemoglobin. *Plant Molecular Biology, 77*, 565–575.

Storz, J. F., Opazo, J. C., & Hoffmann, F. G. (2013). Gene duplication, genome duplication, and the functional diversification of vertebrate globins. *Molecular Phylogenetics and Evolution, 66*, 469–478.

Stuehr, D. J., Santolini, J., Wang, Z.-Q., Wei, C.-C., & Adak, S. (2004). Update on mechanism and catalytic regulation in the NO synthases. *The Journal of Biological Chemistry, 279*, 36167–36170.

Takaya, N., Suzuki, S., Matsuo, M., & Shoun, H. (1997). Purification and characterization of a flavohemoglobin from the denitrifying fungus *Fusarium oxysporum*. *FEBS Letters, 414*, 545–548.

Tamura, K., Peterson, D., Peterson, N., Stecher, G., Nei, M., & Kumar, S. (2011). MEGA5: Molecular evolutionary genetics analysis using maximum likelihood, evolutionary distance, and maximum parsimony methods. *Molecular Biology and Evolution, 28*, 2731–2739.

Te Biesebeke, R., Levasseur, A., Boussier, A., Record, E., van den Hondel, C. A., & Punt, P. J. (2010). Phylogeny of fungal hemoglobins and expression analysis of the *Aspergillus oryzae* flavohemoglobin gene fhbA during hyphal growth. *Fungal Biology, 114*, 135–143.

Torruella, G., Derelle, R., Paps, J., Lang, B. F., Roger, A. J., Shalchian-Tabrizi, K., et al. (2012). Phylogenetic relationships within the Opisthokonta based on phylogenomic analyses of conserved single-copy protein domains. *Molecular Biology and Evolution, 29*, 531–544.

Turrion-Gomez, J. L., Eslava, A. P., & Benito, E. P. (2010). The flavohemoglobin BCFHG1 is the main NO detoxification system and confers protection against nitrosative conditions but is not a virulence factor in the fungal necrotroph *Botrytis cinerea*. *Fungal Genetics and Biology, 47*, 484–496.

Ullmann, B. D., Myers, H., Chiranand, W., Lazzell, A. L., Zhao, Q., Vega, L. A., et al. (2004). Inducible defense mechanism against nitric oxide in *Candida albicans*. *Eukaryotic Cell, 3*, 715–723.

Unden, G., Becker, S., Bongaerts, J., Holighaus, G., Schirawski, J., & Six, S. (1995). O_2-sensing and O_2-dependent gene regulation in facultatively anaerobic bacteria. *Archives of Microbiology, 164*, 81–90.

Usuki, I., Hino, A., & Ochiai, T. (1989). Reinvestigation of the hemoglobins from *Paramedium jenningsi, P. multimicronucleatum* and *P. caudatum*. *Comparative Biochemistry and Physiology. Part B, Biochemistry & Molecular Biology, 93*, 555–559.

Vasudevan, S. G., Armarego, W. L., Shaw, D. C., Lilley, P. E., Dixon, N. E., & Poole, R. K. (1991). Isolation and nucleotide sequence of the hmp gene that encodes a haemoglobin-like protein in *Escherichia coli* K-12. *Molecular and General Genetics, 226*, 49–58.

Vinogradov, S. N. (2008). Tracing globin phylogeny using PSIBLAST searches based on groups of sequences. *Methods in Enzymology, 436*, 571–583.

Vinogradov, S. N., Fernandez, I., Hoogewijs, D., & Arredondo-Peter, R. (2011). Phylogenetic relationships of 3/3 and 2/2 hemoglobins in Archaeplastida genomes to bacterial and other eukaryote hemoglobins. *Molecular Plant, 4*, 42–58.

Vinogradov, S. N., Hoogewijs, D., & Arredondo-Peter, R. (2011). What are the origins and phylogeny of plant hemoglobins? *Communicative and Integrative Biology, 4*, 443–445.

Vinogradov, S. N., Hoogewijs, D., Bailly, X., Arredondo-Peter, R., Gough, J., Dewilde, S., et al. (2006). A phylogenomic profile of globins. *BMC Evolutionary Biology, 6*, 31.

Vinogradov, S. N., Hoogewijs, D., Bailly, X., Arredondo-Peter, R., Guertin, M., Gough, J., et al. (2005). Three globin lineages belonging to two structural classes in genomes from the three kingdoms of life. *Proceedings of the National Academy of Sciences of the United States of America, 102*, 11385–11389.

Vinogradov, S. N., Hoogewijs, D., Bailly, X., Mizuguchi, K., Dewilde, S., Moens, L., et al. (2007). A model of globin evolution. *Gene, 398*, 132–142.

Vinogradov, S. N., Hoogewijs, D., Vanfleteren, J. R., Dewilde, S., Moens, L., & Hankeln, T. (2011). Evolution of the globin superfamily and its function. In M. Nagai (Ed.), *Hemoglobin: Recent developments and topics* (pp. 232–254). Kerala (India): Research Signpost.

Vinogradov, S. N., & Moens, L. (2008). Diversity of globin function: Enzymatic, transport, storage, and sensing. *The Journal of Biological Chemistry, 283*, 8773–8777.

Vinogradov, S. N., Tinajero-Trejo, M., Poole, R. K., & Hoogewijs, D. (2013). Bacterial and archaeal globins—A revised perspective. *Biochimica et Biophysica Acta, 1834*, 1789–1800.

Vuletich, D. A., & Lecomte, J. T. (2006). A phylogenetic and structural analysis of truncated hemoglobins. *Journal of Molecular Evolution, 62*, 196–210.

Wittenberg, J. B., Bolognesi, M., Wittenberg, B. A., & Guertin, M. (2002). Truncated hemoglobins: A new family of hemoglobins widely distributed in bacteria, unicellular eukaryotes, and plants. *The Journal of Biological Chemistry, 277*, 871–874.

Woese, C. R., Kandler, O., & Wheelis, M. L. (1990). Towards a natural system of organisms: Proposal for the domains Archaea, Bacteria, and Eucarya. *Proceedings of the National Academy of Sciences of the United States of America, 87*, 4576–4579.

Yamauchi, K., Mukai, M., Ochiai, T., & Usuki, I. (1992). Molecular cloning of the cDNA for the major hemoglobin component from *Paramecium caudatum*. *Biochemical and Biophysical Research Communications, 182,* 195–200.

Yamauchi, K., Ochiai, T., & Usuki, I. (1992). The unique structure of the *Paramecium caudatum* hemoglobin gene: The presence of one intron in the middle of the coding region. *Biochimica et Biophysica Acta, 1171,* 81–87.

Yamauchi, K., Tada, H., Ochiai, T., & Usuki, I. (1993). Structure of the *Paramecium caudatum* gene encoding the B-type of the major hemoglobin component. *Gene, 126,* 243–246.

Yamauchi, K., Tada, H., & Usuki, I. (1995). Structure and evolution of *Paramecium* hemoglobin genes. *Biochimica et Biophysica Acta, 1264,* 53–62.

Yap, E. H., Ho, T. H., Chan, Y. C., Thong, T. W., Ng, G. C., Ho, L. C., et al. (1995). Serum antibodies to *Trichomonas vaginalis* in invasive cervical cancer patients. *Genitourinary Medicine, 71,* 402–404.

Yarlett, N., Martinez, M. P., Goldberg, B., Kramer, D. L., & Porter, C. W. (2000). Dependence of *Trichomonas vaginalis* upon polyamine backconversion. *Microbiology, 146*(Pt 10), 2715–2722.

Zhao, X. J., Raitt, D., Burke, P. V., Clewell, A. S., Kwast, K. E., & Poyton, R. O. (1996). Function and expression of flavohemoglobin in *Saccharomyces cerevisiae*. Evidence for a role in the oxidative stress response. *The Journal of Biological Chemistry, 271,* 25131–25138.

Zhou, S., Fushinobu, S., Kim, S. W., Nakanishi, Y., Maruyama, J., Kitamoto, K., et al. (2011). Functional analysis and subcellular location of two flavohemoglobins from *Aspergillus oryzae*. *Fungal Genetics and Biology, 48,* 200–207.

Zhou, S., Fushinobu, S., Kim, S. W., Nakanishi, Y., Wakagi, T., & Shoun, H. (2010). *Aspergillus oryzae* flavohemoglobins promote oxidative damage by hydrogen peroxide. *Biochemical and Biophysical Research Communications, 394,* 558–561.

Zhou, S., Fushinobu, S., Nakanishi, Y., Kim, S. W., Wakagi, T., & Shoun, H. (2009). Cloning and characterization of two flavohemoglobins from *Aspergillus oryzae*. *Biochemical and Biophysical Research Communications, 381,* 7–11.

Zhu, H., & Riggs, A. F. (1992). Yeast flavohemoglobin is an ancient protein related to globins and a reductase family. *Proceedings of the National Academy of Sciences of the United States of America, 89,* 5015–5019.

AUTHOR INDEX

Note: Page numbers followed by "*f*" indicate figures, and "*t*" indicate tables, and "*np*" indicate footnote.

A

Abbruzzetti, S., 85, 90, 91, 92, 93, 94, 362–363
Abdelghany, T.M., 254
Abel, S., 19
Abell, G.C., 334–335
Abergel, C., 353
Abugo, O., 363
Ackley, L., 15–16
Adachi, S., 15–16, 17, 282–283, 284–285
Adak, S., 424–425
Adams, L.B., 66–67
Adl, S.M., 394–395, 404*f*
Adney, E.M., 57, 251–252
Aerts, T., 364
Agogué, H., 335
Agron, P.G., 282–283
Ahlin, A., 108
Airola, M.V., 310*f*, 311
Akaishi, T., 11
Akimoto, S., 14–15, 283–284, 285
Akiyama, S., 9
Aklujkar, M., 25
Al Dahouk, S., 294–295
Alam, M., 7–8, 18, 19, 25, 80–81, 82, 85, 87–88, 90–92, 93, 201–202, 299–300, 304, 393
Albeck, S., 2
Albrecht, D., 348–350, 354, 355, 357
Aldenbratt, A., 110
Alderton, W.K., 106
Ali, S.K., 3, 23
Alimenti, C., 336
Allemand, F., 351–352
Allen, M.A., 350
Alley, R.B., 332
Allos, B.M., 99
Aloni, Y., 19
Aloy, P., 255–256
Alpman, A., 12–13
Al-Salloom, F.S., 99

Altman, B., 109
Altschul, S.F., 406–408
Alvarez, B., 258
Álvarez-Salgado, E., 66
Alving, K., 110
Amann, R., 334
Amaral-Zettler, L.A., 336, 394–395
Amikam, D., 7–8, 19, 298–299
Amiot, N., 19
Amrouni, D., 431
Anand, A., 63, 101*t*, 156–157, 160–162, 183–184, 362, 372
Anantharaman, V., 12–13, 23–24, 295–296, 300–301
Ancoli-Israel, S., 12–13
Andersen, R.A., 394–395
Anderson, J.B., 406–408
Anderson, O.R., 394–395
Anderson, R.C., 109–110
Andersson, S.G., 411–412, 429–430
Andrade, M., 19
Andreev, V.P., 91
Andrews, D., 339*t*, 340
Andrews, J.T., 332
Angelo, M., 421
Angeloni, S.V., 199–200, 217, 222, 228, 229*t*, 254–255
Angers, M., 11
Angove, H.C., 129
Angulo, F.J., 99
Anjum, M.F., 101*t*, 127, 130, 342, 343, 370, 422–423
Anthamatten, D., 15, 16
Anthony, C., 120–121
Antoch, M.P., 280
Antoine-Moussiaux, N., 431
Antoniani, D., 20
Antonini, E., 226–228, 227*np*, 248*t*, 364–365
Antonini, G., 55–56, 101*t*, 241*t*, 346–347
Anxolabéhère-Mallart, E., 16, 285–286, 289
Anzi, C., 425

Aono, S., 6, 7–8, 9, 11, 275, 277, 278–279, 295–296, 307, 308, 309–311, 314–315
Appelberg, R., 66–67
Apweiler, R., 406–408
Araki, Y., 20, 304
Aranda, M., 402–403
Aravind, L., 23–24, 295–296, 300–301, 406–408
Archibald, J.M., 394–395, 405–406
Arcovito, A., 115*f*
Arenghi, F., 3, 24
Argueta, C., 219–220
Arisio, R., 12–13
Armarego, W.L., 393
Armitage, J.P., 19
Arndt, J., 16, 17–18, 285–286, 287
Aromaa, A., 12–13
Arredondo-Peter, R., 28–29, 66, 80–81, 120, 197–198, 200, 201–203, 205–210, 212–216, 223–224, 247*t*, 340–341, 344–345, 393, 415
Arroyo Mañez, P., 58, 59–60, 123, 149–150
Arya, S., 63, 101*t*, 156–157, 160–162, 183–184, 362, 372
Asakura, T., 421–423
Asamizu, E., 200
Ascenzi, P., 51, 52–56, 57–60, 61–62, 63–64, 65, 66–67, 68–71, 80–81, 82, 85–86, 88–89, 90, 92–94, 101*t*, 114, 115*f*, 120, 121–122, 122*f*, 149–150, 153, 155–156, 159, 160–161, 163–167, 168–169, 173, 185–186, 188–189, 201, 201*f*, 213*f*, 214*f*, 223, 233–235, 234*f*, 241*t*, 242, 248*np*, 248*t*, 259, 261–262, 341, 345–347, 360–361, 362–363, 364, 367, 368, 369, 370–372, 371*f*, 393, 429
Assumpção, M.C., 294–295
Aste, E., 86, 88–89, 90, 93
Atteia, A., 222–224, 259
Attili, A., 3, 24
Audic, S., 409
Aurilia, V., 357–358
Aury, J.M., 412–415
Ausmees, N., 19
Aussel, L., 108
Autore, F., 351, 357–358
Auzelle, F., 431

Avila-Ramirez, C., 101*t*, 106, 111–112, 113, 121, 126, 128–130, 131–132
Axmann, I.M., 204–205
Ayala-del-Rio, H.L., 350, 352–353
Ayers, R.A., 16, 17–18, 276
Ayub, N.D., 337–340, 339*t*, 356
Azam, F., 334–335

B

Ba, Y., 12–13
Babior, B.M., 108
Babst, M., 16
Bachar, D., 394–395, 409
Baciou, L., 433
Bacon, F., 331–332
Baharuddin, A., 210, 211*f*
Baidya, S., 428
Baik, H.S., 296
Bailey, J.E., 101*t*, 111, 420–421
Bailly, X., 23, 28–29, 66, 80–81, 120, 200, 201–202, 206–207, 212, 340–341, 344–345, 393, 411–412
Baker, S.C., 126–127, 302
Bakermans, C., 337–340, 339*t*, 350, 351–352, 353, 355, 356
Bakhiet, M., 99
Balasubramanian, V., 148–149
Baldauf, S.L., 415–420
Baldwin, E.P., 27
Baldwin, R.L., 354
Balland, V., 16, 285–286, 288*t*, 289–290
Ballif, B.A., 224, 260
Ballou, D.P., 7–8
Bamford, V.A., 129
Banas, P., 235*f*
Banerjee, P., 12–13
Bankston, L.A., 3, 28
Banzet, S., 431
Baraquet, C., 308
Barbe, V., 337–340, 339*t*, 348–350, 351–352, 355, 356, 357–358
Barbieri, P., 358
Barbiero, G., 69–70, 149–150, 355, 358, 360, 361
Barnard, M., 11
Barnes, D.K.A., 332, 334
Barnett, A., 432–433

Barrett, J., 101*t*, 106, 125
Barron, J.A., 14–15
Barry, J.K., 247*t*
Barry, K.W., 426–427
Barta, J.R., 394–395
Bartek, I., 173
Bartlett, D.H., 336
Bartosz, G., 423
Baruh, D., 362–363, 365
Barynin, V., 101*t*, 113, 114–116, 114*f*, 115*f*,
    117, 117*f*, 118
Basham, D., 108–109, 111, 123–125
Baskakov, I.V., 204–205
Basle, A., 3, 18
Bass, D., 394–395, 404*f*, 409
Bassam, D., 66, 199–200, 204, 206,
    217–219, 222, 228, 229*t*, 254–255
Basu, M.K., 339*t*, 340
Basu, N.K., 431
Bates, N.R., 334
Battistoni, A., 246
Batut, J., 7–8, 14, 15, 282–283
Batzoglou, S., 409
Baudin-Creuza, V., 364
Bauer, C.E., 282
Baumann, K., 351
Bayer, T., 402–403
Beales, P.L., 224
Beard, D.A., 245–246
Becher, D., 354
Beck, C., 204–205
Becker, K.G., 12–13, 20
Becker, S., 420–421
Bedmar, E.J., 16
Bedzyk, L.A., 106, 123–125
Behr, S., 281–282
Belbin, T.J., 66, 204, 206, 217–219
Belisle, C., 5, 201–202
Bellelli, A., 51, 57, 58–59, 63–64, 360–361
Belohradsky, B.H., 108
Beltramini, M., 23
Benabbas, A., 11
Benelli, E.M., 294–295
Benfatto, M., 115*f*, 344
Benitez, M., 109
Benito, E.P., 427–428
Benziman, M., 19
Bergholz, P.W., 350, 351–353, 355

Berka, V., 244, 249, 300
Berks, B.C., 110
Berlier, Y., 129
Bernardet, J.F., 337–340, 339*t*, 350, 356
Bernatowska, E., 108
Berney, C., 409
Berney, M., 148–149
Bernhardt, J., 354
Bernhardt, P.V., 70–71, 101*t*, 113, 114–116,
    114*f*, 115*f*, 117, 117*f*, 118, 119
Bernsel, A., 337–340, 339*t*, 348–350,
    351–352, 355, 356, 357–358
Berriman, M., 404
Berrisford, J.M., 3, 18
Berry, M.B., 247*t*
Bertero, M.G., 357
Bertin, P.N., 337–340, 339*t*, 348–350,
    351–352, 355, 356, 357–358
Bertini, I., 231–232
Bertolacci, L., 84–85
Bertolucci, C.M., 19–20
Bertoni, G., 358
Bevan, D.R., 149–150, 228–231, 229*t*,
    240–242, 247*t*, 248*t*, 250
Bhattacharya, D., 204–205
Bhattacharyya, A.J., 364
Biagini, G.A., 430, 431–432
Bidon-Chanal, A., 61–62, 63, 64–65, 101*t*,
    155–157, 159–162, 183–184, 360, 362,
    372
Bie, X., 27
Bik, H.M., 394–395
Bikiel, D.E., 85, 86–88, 89, 91–92, 94
Bindayna, K., 99
Bindschadler, R., 332–333
Binhtran, N.T., 307
Birck, C., 14
Bird, T.H., 282
Bittner, L., 409
Bitz, C.M., 331
Bizebard, T., 351–352
Blackwell, T., 12–13
Blair, D.F., 307
Blair, J.E., 411–412
Blaise, V., 339*t*, 340
Blank, M., 28, 415–420
Blankenship, L.E., 336
Blaser, M.J., 99

Blattner, F.R., 19–20
Blizman, A., 15–16
Bloch, C.A., 19–20
Blom, N., 336
Blonski, C., 282–283
Blumberg, W.E., 232–233
Blundell, T.L., 406–408
Bobkov, A., 3, 28
Bobrov, A.G., 19
Boccara, M., 425
Bocci, P., 20
Bocedi, A., 69, 168–169, 248np, 248t, 429
Bodenmiller, D.M., 101t
Boechi, L., 57–58, 59–60, 61–62, 63–64, 65, 66, 70, 85, 86–88, 89, 91–92, 94, 123, 149–150, 159–160, 166–167, 360, 361, 361f, 362, 364–365, 373–374
Boehringer, D., 354
Boeve, M., 126–127
Boffi, A., 51, 55–56, 57, 58–59, 63–64, 65, 101t, 114, 174, 241t, 341, 342, 344, 346–347, 346f, 360–361, 406
Bogdan, C., 429
Boistard, P., 15, 282–283
Bolli, A., 21, 51, 52–55, 59–60, 65, 70–71, 101t, 120, 121–122, 122f, 169, 299, 345–347, 370–372
Bollinger, C.J.T., 101t, 111–112, 343, 344–345
Bolognese, F., 358
Bolognesi, M., 50–51, 52–54, 55–56, 57–59, 60, 61–62, 63–65, 66–67, 68–69, 70–71, 80–81, 82, 85–87, 89, 91–94, 101t, 114, 115f, 120, 148–150, 152–153, 155–158, 159, 160–167, 168–172, 182–183, 200, 201–202, 206–207, 212–213, 214f, 233, 235, 248np, 248t, 261, 301, 341, 345–347, 348, 360–361, 362–363, 367, 393
Bonamore, A., 3, 24, 51, 57, 58–59, 63–64, 66, 70, 114, 149–150, 174, 341, 342, 344, 360–361
Bongaerts, J., 420–421
Bonnard, N., 16
Bonomi, G., 362–363
Bonvalet, A., 289–290
Boon, E.M., 300–301
Borbat, P.P., 311

Borden, N.J., 109–110
Bordi, E., 431–432
Bordo, D., 53–54, 55, 80–81, 82, 93–94, 341
Borjigin, M., 11
Boron, I., 362–363
Borrelli, K., 57–58, 64–65, 163–166, 168
Botta, G.A., 99
Boubeta, F.M., 58, 66, 70
Boucher, J.L., 431
Boudko, D., 5, 308
Bourassa, J.L., 9–11
Boussac, A., 16, 285–286, 289
Boussaha, M., 337–340, 339t, 350, 356
Boussier, A., 428
Bouteille, B., 431
Bouzhir-Sima, L., 16, 19–20, 285–286, 288t, 289–290, 304
Bowman, J.P., 334–335, 336
Bowman, J.S., 336
Bowman, L.A.H., 100–106, 107, 125, 252, 342–343, 344–345, 421, 429
Bowser, S.S., 394–395, 404f
Boxer, S.G., 255–256
Boyle, J., 108
Boyle, P., 12–13
Bradfield, C.A., 7–8, 11, 12–13
Brandish, P.E., 7–8
Brantley, R.E. Jr, 249–250, 368
Brashear, W.T., 112, 343, 368
Brass, S., 66, 204, 206, 217–219
Braun, A., 101t, 422–423
Brautigam, C., 285, 286, 288t
Brautigan, D.L., 232–233
Brem, K.L., 315
Brencic, A., 19
Brettin, T.S., 350, 351–352, 359
Brigham-Grette, J., 332
Brinkhoff, T., 334–335
Brinkmann, H., 205–206, 411–412
Brinkmeyer, R., 334
Britigan, B.E., 108
Brogioni, S., 58
Brokx, S.J., 357
Brom, S., 15
Brondijk, T.H., 110
Brosch, R., 148–150
Brown, A.J., 423

Brown, H.N., 12–13
Brown, J.M., 224–225, 260
Brown, M.V., 335, 336
Brown, S.L., 336
Brown, Z., 332–333
Brucker, E.A., 247t
Brudno, M., 409
Brugiere, S., 222–224, 259
Brumell, J.H., 108
Brumlik, M.J., 314–315
Brunger, A.T., 352
Bruno, S., 85, 90, 91, 94
Brunori, M., 60, 85–86, 91–92,
   128–129, 226–228, 227np, 248t,
   362–363, 364–365
Brunzelle, J.S., 123–124
Bryant, D.A., 51, 56–57, 67–68, 200, 203,
   204–205, 220–221, 229t, 239, 246,
   362–363
Bubacco, L., 23
Bubenzer, C., 255–256
Buchner, J., 351
Buguet, A., 431
Buisson, N., 422–423
Bukau, B., 354
Burden, L.M., 19
Burey, S.C., 205–206, 411–412
Burg, D.W., 350
Burger, G., 205–206, 405–406, 411–412,
   415–420, 418f
Burke, P.V., 422–423
Burki, F., 205–206, 394–395
Burland, V., 19–20
Burlat, B., 129
Burlet, S., 431
Burmester, T., 28, 201–202, 362–363,
   364, 365–366, 366t, 395–397,
   415–420
Burnell, J., 393
Burnside, K., 307
Burstyn, J.N., 9, 11, 13, 278–280
Bush, M., 295–296
Butcher, J., 123–124
Butler, C.S., 110
Butt, J.N., 129
Buttner, K., 354
Butzler, J.P., 99
Bydalek, P., 101t

C

Caceres, A.I., 285, 286, 288t
Cai, T., 111–112
Calhoon, R., 7–8
Calvo, A.M., 428
Camada, I., 429
Camilli, A., 19, 298–299,
   300–301
Campanaro, S., 350
Campbell, D.H., 249
Campbell, E.A., 18
Campbell, E.L., 219–220
Campbell, L., 336
Campbell, L.P., 252, 253, 343
Campos-Rodriguez, R., 429–430
Canback, B., 411–412, 429–430
Candela, M.-E., 427
Cannillo, E., 85–86
Cannistraro, V.J., 5
Cao, L., 356
Capece, L., 63
Caporaso, J.G., 394–395
Carballido-Lopez, R., 5
Carlioz, A., 108
Carlson, C.A., 334
Carlson, H.K., 300–301
Carpenter, J.F., 352
Carr, M., 415–420
Carrica Mdel, C., 294–295
Cartier, G., 351–352
Carver, T.L.W., 424–426
Cary, C., 339t, 350, 355–356
Cary, J.W., 428
Casamayor, E.O., 335
Casanueva, A., 339t, 350, 355–356
Casassa, G., 332
Case, J.F., 402–403
Case, M.A., 9–11
Casero, D., 225–226, 257–258
Cassanova, N., 423
Castello, P., 343–344
Castillo-Henkel, C., 429–430
Castresana, J., 409
Caswell, C.C., 294–295
Catacchio, B., 3, 24, 51, 57, 58–59, 63–64,
   360–361
Catarino, T., 313–314
Cavalier-Smith, T., 403

Cavicchioli, R., 337–340, 339t, 350, 351–352, 354, 355, 357–358, 362
Cawthraw, S., 124
Ceci, P., 101t
Cermakian, N., 11–12
Cespuglio, R., 431
Ceulemans, H., 255–256
Chadee, K., 429–430
Chain, P.S., 350, 352–353
Chamberland, H., 200, 206, 222–223, 433
Champine, J.E., 25
Champion, P.M., 11
Chan, M.K., 16, 17–18, 285–286, 287
Chan, Y.C., 430
Chance, B., 421–423
Chanfon, A., 16, 282
Chang, A.L., 7–8, 23, 197, 288t, 289, 301, 305
Chang, J.M., 409
Chang, Z., 68–69, 101t, 186–187
Channa, A., 101t, 342
Chapman, S.K., 252, 253, 343
Chastain, R.A., 336
Chateline, R., 99
Chatton, 394–395
Chauhan, S., 339t
Chavan, M.A., 25
Cheesman, M.R., 19–20, 110, 129
Chen, C.C., 3, 26–27
Chen, L., 356
Chen, Z., 356
Cheng, C.H.C., 356
Cheng, Y., 12–13
Chenivesse, S., 353
Cherney, M.M., 9
Cherukuri, P.F., 406–408
Chiancone, E., 3, 24, 51, 57, 58–59, 63–64, 65, 101t, 346f, 360–361
Chien, E.Y., 290
Chien, L.T., 397
Chien, P., 19
Chintalapati, S., 337–340
Chiranand, W., 423
Cho, H.J., 296, 297
Cho, H.Y., 296, 297
Chothia, C., 406–408
Chou, J.H., 108
Chou, K.J., 125–126

Chown, S.L., 334
Christen, R., 394–395
Christie Vu, B., 362–363
Christman, H.D., 219–220
Chu, D., 66–67, 173
Chung, S.Y., 9
Churcher, C., 108–109, 111, 123–125, 148–150
Chythanya, R., 19
Ciaccio, C., 57, 69, 70–71, 84–85, 86, 88–89, 90, 92–93, 94, 120, 149–150, 188–189, 345–346, 360, 361, 361f, 362–363, 364–365, 366, 366t, 370–372, 373–374
Cianci, M., 115f
Ciniglia, C., 205
Clark, R.W., 9, 11
Clarke, A., 332
Claverie, P., 339t, 340
Clewell, A.S., 422–423
Cocco, M.J., 352
Cochran, D., 224, 260
Coda, A., 58, 101t, 114, 115f
Cogne, G., 222–224, 259
Cokus, S., 225–226
Cole, J.A., 70–71, 101t, 110, 111–112, 121, 123–125, 126–127, 129, 131, 344
Cole, S.T., 148–150, 153
Coletta, M., 57, 68–69, 70–71, 88–89, 92–93, 101t, 120, 149–150, 188–189, 345–346, 360, 362–363, 370–372
Collins, T., 337, 339t, 340
Comandini, A., 58, 66, 70
Comiso, J.C., 332
Compagno, C., 351
Conley, M., 124–125
Connelly, J.E., 307
Connelly, T.L., 336
Conrad, M., 8, 9, 275, 277
Constantinou, J., 336
Contreras, M.L., 120–121
Contreras-Zentella, M.L., 101t, 113
Convey, P., 332–333, 334
Cook, G.M., 101t, 148–149
Coopamah, M.D., 101t, 342, 343, 370
Cooper, A.M., 66–67
Cooper, C.E., 106
Coppi, M.V., 25

Coppola, D., 69–70, 360, 362, 368, 370–372, 371*f*
Corbeel, L., 108
Corker, H.A., 101*t*, 107, 130, 342, 369
Correc, G., 415–420
Costa, C., 129
Costantino, G., 101*t*, 106, 107, 113
Cotter, P.A., 298–299
Cottrell, M.T., 335
Courtois, P., 431
Couture, J.F., 123–124
Couture, M., 51, 52–53, 54, 55–58, 61–62, 63–64, 65, 66, 101*t*, 118, 120, 149–150, 153, 155, 157–158, 159, 161–162, 166, 169–172, 182–183, 200, 201, 201*f*, 203, 206, 213*f*, 222–223, 229*t*, 232–235, 234*f*, 236, 241*t*, 242, 247*t*, 248*t*, 250, 259, 261, 301, 345–346, 365, 393, 433
Cowan, D.A., 334, 339*t*, 350, 355–356
Cowley, M.J., 350, 351–352, 355
Cox, G.M., 423
Coyne, K.J., 253, 402–403
Crack, J., 101*t*, 106
Craft, J.M., 224–225, 260
Cramm, R., 113, 421–422
Crawford, M.J., 422–423
Creer, S., 394–395
Crespo, A., 61–62, 63, 64–65, 101*t*, 155–159, 160–161, 163, 360
Crick, D.C., 296
Cristescu, S.M., 424–425
Cronin, M.T., 357
Cronin, T.M., 332
Crook, Z., 343–344
Crosson, S., 15, 19, 276
Crowley, D., 124–125
Cruz-Ramos, H., 101*t*, 106
Culbertson, D.S., 226–228
Cullen, J.J., 334–335
Curmi, P.M., 337–340, 350, 352–353, 355
Curnutte, J., 108
Curtis, A.M., 12–13
Curtis, B.A., 205–206
Curtis, J.E., 101*t*
Cusano, A.M., 351, 357–358
Cutruzzolà, F., 60, 128–129
Czeluzniak, J., 28–29

D
Da Re, S., 282–283
Daigle, R., 55–56, 61, 63, 101*t*, 155–159, 160–162, 261
Dakappagari, N., 288*t*
Dalton, D.K., 66–67
Damborsky, J., 235*f*
D'Amico, S., 337–340, 339*t*, 348–350, 349*f*, 353, 354, 355–356, 357, 373
Danchin, A., 348, 349*f*, 350, 351–352, 353, 355, 356, 357, 359
D'Angelo, P., 344
Daniel, R.M., 352
Dankster, D., 55
Dantsker, D., 55–56, 57–58, 61, 66–67, 68–69, 155–156, 157–158, 161–162, 163–166, 167–168, 187, 348, 360–361, 365–366, 366*t*
Darby, D.A., 332
Dardente, H., 11–12
D'Ari, R., 107–108
Darst, S.A., 18
Das, P.K., 431
Das, T.K., 56–57, 59–60, 61, 200, 223, 229*t*, 232–233, 234–235, 236, 248*t*, 261, 296, 301, 365, 432–433
Daskalakis, V., 6
Daulouede, S., 431
D'Auria, S., 357–358
D'Ausilio, C.A., 311–312
D'Autréaux, B., 295–296
Davalos, A., 15
Daveran, M.L., 7–8, 14, 15, 282
David, M., 7–8, 14, 15, 282–283
David, P., 343–344
Davidge, K.S., 101*t*, 106, 111–112, 113, 121, 126, 128–130, 131–132
Davila, J.R., 120–121
Davis, F.C., 7–8
Davis, J.H., 300–301
Davis, R.C., 255–256
Dawson, J.H., 227*np*
Dawson, S.C., 403
Dayton, P.K., 331–332
de Bruijn, F.J., 294–295
De Candia, A.G., 313–314
De Francisci, D., 350

de Jesus-Berrios, M., 423
de Klerk, M.A., 99
de la Longrais, I.A., 12–13
de Los Reyes, M., 307
De Marinis, E., 68–69, 188–189, 360, 364
de Pascale, D., 351, 357–358
de Philip, P., 15
de Ropp, J.S., 198–199, 231, 233–234
de Sanctis, D., 248np, 248t, 362–363
De Sanctis, G., 57, 92, 93, 94, 149–150, 362–363
De Wachter, R., 394–395
Debruine, L., 99
Deckers, H.M., 314–315
Decking, U.K., 368
Dedieu, A., 7–8, 14, 15, 282
DeFlaun, M., 334–335
Del Campo, J., 394–395
DeLano, W.L., 235f
DelArenal, I.P., 120–121
Delgado, M.J., 16
Delgado-Nixon, V.M., 7–8, 19–20, 301, 304
Della Longa, S., 344
Delledonne, M., 258, 424–425
DeLong, E.F., 334
Delumeau, O., 3, 18, 26–27, 197, 409
Demachy, I., 433
DeMaere, M.Z., 335–336
Dementieva, I.S., 311–312
Deming, J.W., 336, 337–340, 339t, 350, 351–353, 356, 357
Deming, W., 336
Demmer, U., 433
DeModena, J., 101t
Demoncheaux, E.A., 421
Demoncheaux, E.A.G., 185, 187–188, 343
Demple, B., 108
Denery, J.R., 21–22, 299
Deng, P., 364
Deng, Y., 19, 352–353
Denis, M., 429–430
Dennis, E.S., 362–363, 393
Dent, R., 223
Depner, M., 12–13
Derelle, R., 415–420
DeSalvo, M.K., 402–403
Desbois, A., 246, 315

Desmecht, D., 431
Desmet, F., 25–26, 82–84, 85, 86–87, 90, 91, 94
Deterding, L.J., 368
Dettaï, A., 69–70, 149–150, 355, 358, 360, 361
DeWeese-Scott, C., 406–408
Dewez, D., 223
Dewilde, S., 25–26, 28–29, 51, 52–53, 54–56, 59–60, 61–62, 66–67, 80–81, 82, 84–86, 87–89, 90, 91–92, 93, 101t, 120, 149–150, 155, 200, 201–202, 201f, 203, 206–207, 212, 213f, 223, 233–235, 234f, 241t, 242, 254, 259, 340–341, 344–346, 347t, 360–361, 362–363, 364, 365–366, 366t, 393, 406, 411–412, 421, 428, 432–433
Dey, D., 296
Dhiman, R.K., 296
D'Hooghe, I., 15
Di Donato, A., 358
Di Giuseppe, G., 336
di Masi, A., 370–372
Di Matteo, A., 128–129
di Prisco, G., 69–70, 332–333, 351, 357–358, 360, 362–363, 364–367, 366t, 368, 370–372, 371f, 373–374
Di Tommaso, P., 409
Dickinson, J.H., 124
Dikshit, K.L., 61–62, 65, 66–67, 68–69, 101t, 112–113, 148–149, 150, 153, 160–166, 167–168, 169–172, 173–174, 180, 183–188, 345, 369, 422–423
Dikshit, R.P., 101t, 112–113
Dineen, D., 409
Dines, M., 352
Ding, Y.H., 25
Dini, F., 336
Dioum, E.M., 7–8, 12, 16, 17, 19–20, 25, 28, 80–81, 90–91, 280, 283–284, 285
DiRita, V.J., 123
DiSpirito, A.A., 253–254
Ditta, G.S., 7–8, 14–15, 282–284
Divoky, G.J., 332–333
Dixon, N.E., 342, 393
Dixon, R., 295–296
Djordjevic, S.A., 296
D'mello, R.A.M., 120–121, 342

Do, C.B., 409
Dolin, P., 182
Dolla, A., 314–315
Domergue, O., 7–8, 14, 15, 282
Donné, J., 86, 88–89, 90, 93
Donohue, T.J., 8
Dorward, D.W., 256
Dou, Y., 112, 149–150, 228, 229t, 247t, 253, 343
Douglas, D.J., 334–335
Dove, S., 433
Dow, J.M., 19
Doyle, M.P., 254
Dragosits, M., 351
Dreyfus, M., 351–352
Droghetti, E., 58, 149–150
Duchaud, E., 337–340, 339t, 350, 356
Ducklow, H.W., 334–336
Dudley, C.A., 7–8, 11–13, 280
Duerig, A., 19
Duff, S.M., 203
Dufresne, A., 336
Duilio, A., 351, 357–358
Duke, N.E., 312, 313–314
Duleu, S., 431
Dunbar, J.M., 351–352
Duncan, K., 173
Duner, M., 225, 257–258
Dunham, C.M., 16, 17, 285
Dunman, P.M., 106, 124–125
Durack, D.T., 343
Duret, L., 412–415
Durnford, D.G., 204–205
Dussaq, A.M., 335–336
Dutcher, S.K., 260
Dye, C., 182
Dym, O., 2

E

Eastman, J.T., 331, 333
Ebel, C., 282–283
Ebel, R.E., 199–200, 217, 222, 228–231, 229t, 240–242, 248t, 250, 254–255
Ebner, B., 201–202, 362–363, 364, 395–397, 396t, 405–406, 415–420
Edgar, R.C., 409
Eejtar, T., 91

Egawa, T., 57–58, 59–60, 64–65, 70–71, 117, 118, 120, 123, 163–166, 168, 346–347
Egea-Gilabert, C., 427
Ehlers, S., 66–67
Ehrenreich, A., 354
Ehrt, S., 66–67, 148–149
Eich, R.F., 149–150, 228, 229t, 247t
Eichinger, L., 404
Eiglmeier, K., 153
Eisen, J.A., 25
Ekholm, J., 12–13
Ekman, A., 110
El Hammi, E., 433
El Mkami, H., 25–26
Elhai, J., 219
El-Mahdy, M.A., 254
El-Mashtoly, S.F., 6, 20, 303–304
Elmerich, C., 15
Elsen, S., 282
Elvers, K.T., 70–71, 101t, 108–110, 111–112, 120–121, 122–125, 126–127, 129–130, 131, 132, 149–150, 344
Engel, P.C., 107–108
Engelman, D.M., 352
Erbel-Sieler, C., 11–13
Erbil, W.K., 300
Erickson, M., 335–336
Eriksson, S., 340
Ermler, U., 113, 421–422, 433
Ernst, A., 66, 204, 206, 217–219
Erra, F., 336
Ertan, H., 350, 355
Esaki, N., 350, 353, 355
Escamilla, E., 101t, 113
Escamilla, J.E., 120–121
Eslava, A.P., 427–428
Espelund, M., 415–420
Espinosa, J., 431
Esteva, M., 431
Esteve-Nunez, A., 25
Estill, S.J., 11–13
Estrin, D.A., 57–58, 61–62, 63–64, 65, 149–150, 155–156, 159, 161, 166–167, 261
Evans, D.S., 12–13
Evans, F., 335–336
Evans, J.E., 224–225, 260
Eymann, C., 354

F

Fabian, M., 203, 248*t*
Fabozzi, G., 57, 66–67, 68–69, 149–150, 153, 168–169, 173, 185–186, 362–363, 369
Fabrizius, A., 396*t*, 405–406
Fabry, B., 14
Fago, A., 254
Fahrbach, E., 332–333
Fairclough, S.R., 415–420
Falzone, C.J., 51, 56–57, 61, 66, 67–68, 200, 203–204, 206, 220–222, 229*t*, 236, 237–238, 239, 241*t*, 250–251, 252, 253, 254, 255, 257, 258, 261, 362–363
Fanali, G., 370–372
Fang, F.C., 106, 108, 124–125
Fang, H., 12–13
Farah, C.S., 19
Farber, M., 431
Farid, I., 99
Farina, A., 114
Farmer, M.A., 394–395
Farr, S.B., 107–108
Farres, J., 101*t*, 111–112, 343, 344–345
Fartmann, B., 337–340, 339*t*, 350, 352–353, 356
Fasano, M., 370–372
Fear, A.L., 7–8
Feau, N., 426–427
Feesche, J., 354
Feinberg, B.A., 232–233
Feis, A., 58, 66, 70, 344
Feller, G., 337, 339*t*, 340, 348, 349*f*, 350, 353, 355, 356, 357–358, 362
Feng, J.X., 19
Feng, L., 363
Ferguson, S.J., 110, 126–127, 302
Fernandes, A.P., 313–314
Fernandez, I., 197–198, 200, 201–202, 205–206, 207–210, 212–216, 223–224, 294–295, 415
Fernie, A.R., 424–425
Ferrari, D., 101*t*
Ferry, J.G., 91
Fetherston, J.D., 19
Filenko, N., 110
Finel, M., 108–109
Finkel, T., 355–356

Fiocchetti, M., 364
Fiori, P.L., 430
Firbank, S.J., 18, 27
Fischer, B.B., 225–226
Fischer, H.M., 16, 126, 282
Fisher, M., 99
Fitzgerald, G.A., 12–13
Fitzgerald, L.M., 12–13
Flatley, J., 101*t*, 106
Fletcher, A.E., 433
Flögel, U., 368
Flores, E., 204–205, 218
Foggi, P., 58
Folcher, M., 19
Ford, B.J., 394–395
Forrester, M.T., 421
Forsell, Y., 12–13
Forte, E., 430, 431–432
Forti, F., 61–62, 85, 86–88, 89, 91–92, 94, 159–160, 362–363
Fortier, E.E., 11–12
Foster, M.W., 421
Fouhy, Y., 19
Fourment, J., 282–283
Fourrat, L., 67
Foussard, M., 282
Fouts, D.E., 109
Francke, U., 11
Frank, J., 12–13
Franken, P., 11–13
Franzese, M., 362–363
Fraser, G.M., 300–301
Frawley, E.R., 251–252
Frazer, T.K., 332–333
Freeman, K.M., 13, 278–280
Freitas, T., 6, 80–81, 91
Freitas, T.A.K., 7–8, 18, 21–22, 25, 28, 80–81, 90–91, 201–202, 299, 393
Frey, A.D., 101*t*, 111–112, 118, 120–121, 174, 343, 344–345, 420–421
Frickey, T., 337–340, 339*t*, 350, 352–353, 356
Fridovich, I., 107–108
Fried, L., 281–282
Friedman, C.R., 99
Friedman, J., 60, 61–62, 66–67, 101*t*, 235, 242
Friedman, J.M., 167–168, 184–185, 231–232, 290

Friedrich, B., 113, 421–422
Frigaard, N.U., 204–205, 246, 334
Fritz-Laylin, L.K., 403
Frizzell, S., 254
Fu, C., 412–415
Fu, R., 314–315
Fuchs, C., 201–202, 203
Fujii, H., 15–16, 285–286, 288*t*
Fujisawa, T., 9
Fukuto, J., 100–106, 107, 125
Fukuto, J.M., 252, 342–343, 344–345, 421, 429
Fülöp, V., 302
Fulton, C., 403
Fulton, D.B., 254
Furukawa, K., 99
Fushinobu, S., 428, 429

G

Gaal, T., 8
Gagné, G., 222–223, 224–225
Gahagan, L.M., 333
Gaidenko, T.A., 27
Galand, P.E., 335
Galizzi, A., 101*t*, 114, 115*f*
Galli, E., 358
Gallin, J.I., 108
Galperin, M., 298–299
Gambhir, V., 66–67, 101*t*, 172, 185
Gao, Q.J., 220
Garcia, J.A., 7–8, 11–13, 280
García-Cánovas, F., 55
Garcia-Latorre, E., 429–430
Garcia-Pichel, F., 220
Garczarek, L., 336
Gardner, A.M., 61–62, 66–67, 68–69, 101*t*, 107, 112, 153, 161–162, 183–186, 253, 343, 345, 368, 369, 421, 429–430, 433
Gardner, P.R., 61–62, 66–67, 68–69, 101*t*, 106, 107, 112, 113, 153, 161–162, 183–186, 252, 253, 343, 345, 368, 369, 421, 429–430
Garman, E.F., 302
Garnier, T., 148–150
Garrocho-Villegas, V., 207–210
Gasser, B., 351
Gautier-Sauvigne, S., 431
Gazur, B., 252, 253, 343
Geer, L.Y., 406–408

Gekakis, N., 7–8
Gelfand, D., 7–8
Gell, D., 363
Gelpi, J.L., 101*t*, 156–159, 160–162, 163, 183–184
Gelpí, J.L., 63, 362, 372
Genkov, T., 222–223
George, P., 249–250, 432–433
Georlette, D., 339*t*, 340
Gerashchenko, D., 12–13
Gerday, C., 337, 340
Germani, F., 396*t*, 405–406
Gersten, R., 332
Gessner, A., 429
Gessner, C.R., 433
Geuens, E., 244, 368
Gevers, D., 99
Ghai, J., 7–8, 14, 15, 282
Ghiglione, J.F., 335–336
Ghosh, A., 431
Ghosh, T., 295–296
Giacometti, G.M., 85–86
Giangiacomo, A., 360–361
Giangiacomo, L., 51, 57, 58–59, 63–64, 65, 346*f*, 360–361
Giardina, G., 128–129
Gibson, Q.H., 60, 86, 91–92, 114–116, 244
Gibson, T.J., 409
Gidley, M.D., 106, 109, 125
Giebel, H.A., 334–335
Gilany, K., 362–363
Gilbert, M., 99
Gilberthorpe, N.J., 70–71, 101*t*, 111–112, 120, 123–124, 131, 132, 342, 343, 344, 421
Gilevicius, L., 242–244, 251–252
Gilles-González, M.A., 7–8, 12, 14–15, 16–17, 19–20, 24, 197, 275, 280, 282–284, 285, 286, 288*t*, 289, 296–297, 301, 304, 305
Ginger, M.L., 403
Giometti, C.S., 25, 337–340, 339*t*, 350, 353, 356
Giordano, D., 66, 69–70, 149–150, 350, 355, 358, 360, 361, 361*f*, 362–363, 364–367, 366*t*, 368, 369, 370–372, 371*f*, 373–374
Giovannoni, S.J., 25
Girard, L., 15

Giudice, A., 429
Giuliani, M., 66, 69–70, 350, 351, 360, 369
Gladwin, M.T., 254
Glagla, S., 348–350, 354, 355, 357
Glekas, G.D., 5
Glockner, G., 404
Gobert, A.P., 431
Gödecke, A., 368
Goldbaum, F.A., 294–295
Goldberg, A.L., 354
Goldberg, B., 430, 431–432
Goldberg, D.E., 422–423
Goldschmidt-Clermont, M., 220–221
Goldstein, S., 69, 370–372
Gollan, B., 342
Gomelsky, M., 298–299
Gomes, C.M., 312
Gondim, A.C., 7–8, 14–15, 285
Gong, W., 16, 17, 285, 287
Gonzalez, C., 285, 286, 288t
Gonzalez, G., 7–8, 12, 14–15, 16–17,
    19–20, 23, 24, 25, 28, 80–81, 90–91, 197,
    275, 280, 282–284, 285, 288t, 289,
    296–297, 299–300, 301, 304, 305
González, G., 282–283
Gonzalez-Ballester, D., 225–226
Goodchild, A., 350, 354, 355
Goodson, C.M., 109–110
Gopalasubramaniam, S.K., 207–210
Gorby, Y.A., 25
Gottfried, D.S., 290
Gottschalk, G., 354
Gouet, P., 14
Gough, J., 28–29, 66, 80–81, 120, 201–202,
    206–207, 340–341, 344–345, 393,
    406–408
Gourlay, L., 25–26, 82–84, 85
Gourse, R.L., 8
Gouveia, Z., 313–314
Govrin, E.M., 425
Gow, A.J., 101t, 106, 112, 113, 119, 343,
    360
Gradoni, L., 429
Grange, J.M., 148–149
Granger, D.L., 343
Grant, T., 18, 27
Gratia, E., 339t, 340
Gray, H.B., 315

Gray, M.W., 405–406, 411–412
Grayburn, W.S., 428
Green, J., 70–71, 101t, 106, 125–127,
    130–132, 344
Green, S.J., 343, 429
Greenberg, J.T., 108
Grimes, J., 109–110
Grogan, S., 124–125
Groot, M.L., 289–290
Gross, S.S., 343
Grossman, A.R., 223, 225–226
Gruber, A., 205–206
Grubina, R., 254
Gruner, S.M., 346f
Grzymski, J.J., 334, 335–336, 350, 352–353
Gu, L., 363
Gu, Y., 6
Gu, Y.Z., 7–8, 11, 12
Guallar, V., 57–58, 63, 64–65, 163–166, 168
Guccione, E., 109–110
Guerinot, M.L., 16
Guertin, M., 28–29, 50–51, 52–53, 54,
    55–56, 57–59, 60, 61–62, 63–64, 66–67,
    68–69, 101t, 120, 148–150, 152–153,
    155–159, 160–161, 162–172, 182–183,
    184–185, 186–187, 200, 201–202, 201f,
    203, 206–207, 212–213, 213f, 214f,
    222–223, 224–225, 229t, 231–232,
    233–235, 234f, 241t, 242, 250, 259, 301,
    345–347, 348, 360–361, 362–363, 367,
    393, 433
Guest, J.R., 126–127
Guilhaus, M., 350, 351–352, 354, 355
Guilhon, A.A., 108
Guillet, V., 14
Guillou, L., 409
Guiry, M.D., 402–403
Gullotta, F., 370–372
Gupta, K.J., 424–425
Gupta, S., 148–149, 150, 173–180,
    187–188, 189
Gusarov, I., 107
Guss, C., 12–13
Gustin, M.C., 433
Gutt, J., 332–333
Gutteridge, J.M., 100–106
Guy, P.A., 364
Gwadz, M., 406–408

H

Haberkamp, M., 201–202, 203, 395–397
Hackett, D.P., 101t
Hackett, J.D., 205
Haddock, S.H., 402–403
Hade, M.D., 148–149, 150, 174–180, 187–188, 189
Haggman, H., 120–121, 344–345, 420–421
Hagihara, B., 421–423
Hajdu, J., 302
Halder, P., 238, 241t, 250–251, 252
Hall, S.J., 108–110, 129
Hallam, S.J., 334
Halliwell, B., 100–106
Hamdane, D., 362, 365–366, 366t
Hamel, B., 223
Han, Z., 107–108
Hankeln, T., 201–202, 203, 362–363, 364, 365–366, 366t, 393, 395–397, 415–420, 421
Hansen, J.N., 125
Hao, B., 16, 17–18, 285–286, 287
Happe, T., 225, 257–258
Harari, E., 226–228
Harder, A., 294–295
Hardison, R.C., 393
Hargrove, M.S., 51, 56–57, 61, 67–68, 149–150, 203, 204, 207–210, 228, 229t, 237–238, 238f, 241t, 242, 245–246, 247t, 248np, 248t, 250–251, 252, 253–254, 259, 261, 362–363, 364
Harren, F.J., 424–425
Harris, D., 148–150
Harris, E.E., 223, 225
Harris, E.H., 223, 257, 260
Harris, K.M., 430, 431–432
Harris, R., 3, 26–27
Harrison, L.J., 125, 421
Harrison, W.G., 334–335
Harte, N., 406–408
Hartl, F.U., 354
Hartzell, H.C., 397
Haruki, M., 355
Hasnain, S.S., 115f
Hassan, R., 415–420
Hauser, C., 223, 225
Hauser, L.J., 350, 351–352, 359

Hausladen, A., 101t, 106, 112, 113, 119, 343, 360, 421, 423
Hawari, J., 352–353
Hayakawa, Y., 19, 22–23, 299
Hayasaka, K., 13
Hayer-Hartl, M., 354
Haynes, K., 423
Hayward, A., 99
Hazemann, J.L., 344
He, Y.Q., 7–8, 19, 68–69, 101t, 186–187
He, Z., 7–8, 350, 351–352, 359
Hebrard, M., 108
Hedges, S.B., 411–412
Heermann, R., 281–282
Heib, V., 203
Heidelberg, J.F., 25
Heidrich, J., 101t
Heikaus, C.C., 295–296
Heim, J.M., 363
Heinnickel, M., 223
Heitman, J., 343, 423
Helinski, D.R., 7–8, 14–15, 282–284
Helmann, J.D., 106, 123–125
Helmick, R.A., 433
Helmke, E., 334
Hemann, C., 254
Hemminki, K., 12–13
Hemschemeier, A., 225, 257–258
Hendrickson, W.A., 293–294, 354
Hendrixson, D.R., 123
Hengge, R., 20, 298–299, 300–301
Henle, E.S., 107–108
Hennecke, H., 15, 16, 282
Henri, S., 108
Henrissat, B., 426–427
Henry, J.T., 19, 276
Herbrink, P., 99
Hernandez-Campos, M.E., 429–430
Hernandez-Urzua, E., 101t, 113, 124–125
Herndl, G.J., 335
Herold, S., 368, 370–372
Herrero, A., 204–205, 218
Herzik, M.A. Jr, 244, 300
Hescheler, J., 355–356
Hess, D.T., 106, 125
Hess, W.R., 336
Hibbs, J.B. Jr, 343, 429
Higuchi, Y., 11, 277

Hikage, N., 13
Hilbert, J.L., 247t
Hill, D.R., 66, 204, 206, 217–219
Hill, K., 91
Hill, R.D., 203, 364, 433
Hill, S., 120–121
Hille, R., 288t, 357
Hillesland, K.L., 7–8
Hills, T.L., 357
Hillson, N.J., 7–8, 15
Hinds, J., 70–71, 101t, 111–112, 121, 123–125, 126–127, 131, 344
Hiner, A.N., 55
Hinkle, G., 394–395
Hino, A., 432–433
Hirata, S., 7–8, 19–20
Hiruma, Y., 15–16, 290, 291
Hitchcock, A., 109–110
Hixson, K.K., 25
Ho, L.C., 430
Ho, T.H., 430
Ho, Y.S., 19
Hoashi, Y., 9
Hodgson, D.A., 332–333
Hoegh-Guldberg, O., 433
Hoekstra, R.M., 99
Hoffman, A.E., 12–13
Hoffman, B.M., 232–233
Hoffman, F.G., 203, 248t
Hoffman, J.M., 335
Hoffmann, A., 354
Hoffmann, F.G., 396t, 405–406, 415–420
Hofreuter, D., 109
Hogenesch, J.B., 7–8, 11, 12
Hogg, N., 125–126
Hol, W.G.J., 173
Hola, K., 16–17, 289–290
Holden, H.M., 60
Holford, T., 12–13
Holighaus, G., 420–421
Holland, P.W., 405–406, 415–420, 418f
Hollocher, T.C., 369
Holmes, K., 124–125
Holmfeldt, P., 308
Holmqvist, S., 308
Hon, C.C., 429–430
Honaker, R.W., 296

Hong, C.S., 308
Honma, Y., 128–129, 277, 278
Hoogewijs, D., 3, 21, 28–29, 50–51, 66, 80–81, 82, 99–100, 111, 120, 197–198, 200, 201–203, 205–210, 212–216, 217, 223–224, 244, 299, 340–341, 344–345, 368, 393, 395–397, 396t, 405–406, 411–412, 415–420, 421, 424–425, 428, 429
Hormaeche, C.E., 108
Horton, P., 409
Horwitz, B.A., 426–427
Hou, S., 5, 6, 7–8, 18, 25, 28, 80–81, 90–91, 93, 201–202, 308, 393
Houee-Levin, C., 433
House, C.H., 91
Howard, A.J., 112–113
Howell, L.G., 107–108
Howes, B.D., 360, 361, 361f, 362, 364–365, 366, 366t, 373–374
Hoy, J.A., 51, 56–57, 61, 67–68, 207–210, 237–238, 238f, 241t, 242, 247t, 248np, 250, 252, 259
Hoyoux, A., 339t, 340
Hromatka, B.S., 423
Hu, S.Z., 290, 315
Hu, Z., 12–13
Huang, J., 108
Huang, S.G., 225–226
Huang, S.H., 300–301
Huebner, M., 350, 351–352, 359
Huelsenbeck, J.P., 409
Hughes, K., 334
Hughes, M.N., 28, 100–106, 101t, 127, 130, 148–149, 182–183, 187–188, 342–343, 370, 422–423
Hughey, R., 406–408
Huh, D., 311
Hui Bon Hoa, G., 360, 362, 365–367, 366t, 368, 373–374
Hunt, D.E., 348–350
Hunt, P.W., 362–363, 393
Hunte, C., 174–180
Hurley, J.H., 19
Hurme, R., 340
Hutchings, M.I., 107
Hutchinson, F., 107–108

Hvitved, A.N., 9–11, 112, 247t, 261, 343, 362–363, 368, 433
Hwa, T., 354
Hyduke, D.R., 125–126
Hyodo, M., 19, 22–23, 299

I

Iddar, A., 67
Igarashi, J., 12, 13, 18, 19–20, 51, 55–56, 61, 63, 67, 280–281, 304, 432–433
Iijima, M., 404, 431
Iizuka, T., 15–16, 282–283, 284–285, 292–293
Ikeda, T., 308
Ikeda-Saito, M., 19–20, 197, 288t, 289, 301, 305
Ilari, A., 51, 57, 58–59, 63–64, 65, 101t, 114, 346f, 360–361, 406
Ilari, L., 360–361
Iliuk, A., 307
Imai, K., 282–283, 292–293
Imhoff, F., 336
Imlay, J.A., 101t, 107–108
Inagaki, S., 11, 277
Iniesta, A.A., 7–8, 15
Ioanitescu, A.I., 203
Ioannidis, N., 101t, 106, 127, 182–183, 187–188, 342, 370, 422–423
Ioanoviciu, A., 296, 297–298
Iobbi-Nivol, C., 308
Iomini, C., 260
Irie, T., 432–433
Irimia, M., 205–206
Irvine, A.S., 126–127
Irwin, B., 334–335
Isaza, C., 16, 17–18, 285–286, 287
Ishida, M.L., 12, 281, 294–295
Ishigami, I., 20, 299–300
Ishikawa, H., 6, 8, 9, 277, 278, 309–311
Ishikawa, Y., 9
Ishimori, K., 8, 9, 12, 13, 18, 19, 22–23, 277, 278, 280–281, 299
Ishitsuka, Y., 19–20, 304
Islam, A., 99
Ito, O., 20, 304
Ito, S., 20, 304
Itoh, M., 295–296
Ivanova, N., 350, 351–353, 359

Iwaasa, H., 199–200, 217, 222, 393, 432–433
Iyengar, R., 343
Iyer, L.M., 23–24, 300–301

J

Jackson, R.J., 109
Jacob, A., 288t
Jacobs, B.C., 99
Jacobs, W.R. Jr, 66–67, 173
Jacoby, C.A., 332–333
Jaenicke, V., 203
Jagannadham, M.V., 339t, 340
Jager, M., 415–420
Jagtap, P., 339t, 340
Jaillon, O., 412–415
Jain, S., 7–8, 11
Jakobsen, K.S., 394–395, 415–420
James, K.D., 153
Jang, S., 107–108
Janiyani, K., 339t, 340
Jarboe, L.R., 125–126
Jasaitis, A., 16–17, 167–168, 184–185, 289–290
Jeffrey, W.H., 334–335
Jenal, U., 298–299
Jentzen, W., 236
Jetten, A.M., 11
Ji, Q., 305–307
Ji, X.B., 369
Jia, S.L., 236
Jiang, B.L., 19
Joachimiak, M.P., 7–8
Johansson, C., 12–13
Johnson, A.D., 423
Johnson, E.C., 224, 260
Johnson, K.A., 114, 128–129
Johnson, M.S., 308, 309
Johnson, R.J., 334
Johnston, R.B. Jr, 108
Jokipii-Lukkari, S., 120–121, 344–345, 420–421
Jones, A.D., 203–204, 230np, 236–237, 237f, 241t, 248t
Jones, M.A., 101t, 109–110, 111–112, 121, 129–130
Jones, R.A., 106, 125
Joseph, J., 125–126

Joseph, S.V., 101*t*, 173
Josephy, P.D., 108
Jouhet, J., 222–223
Jourlin-Castelli, C., 308
Jouvet, M., 431
Jubin, C., 412–415
Jung, K., 281–282
Junge, K., 336
Jungersten, L., 110
Jürgens, J., 334
Justino, M.C., 101*t*
Juszczak, L., 57–58, 66–67, 68–69,
    163–166, 167–168, 187, 348, 360–361,
    365–366, 366*t*

K

Kaasik, K., 13
Kader, A., 19
Kadkhodayan, S., 227*np*
Kadono, E., 226–228
Kaever, V., 19
Kahn, D., 14, 282–283
Kaiser, W.M., 424–425
Kakar, S., 203, 207–210, 248*t*
Kalinga, S., 368, 370–372
Kalko, S.G., 156–159, 160–161, 163
Kallberg, M., 128–129
Kallio, P.T., 101*t*, 111–112, 118, 120–121,
    174, 343, 344–345, 420–421
Kalyanaraman, B., 125–126
Kaminski, P.A., 15
Kamiya, T., 7–8, 9
Kanai, A., 126
Kanai, M., 282–283, 284
Kanaya, S., 355
Kanbe, M., 210, 211*f*
Kandler, O., 394–395
Kandror, O., 354
Kaneko, T., 200
Kang, B.S., 297
Kang, C., 20, 302, 303–304
Kang, H.S., 11
Kaniewska, P., 433
Kannan, K., 339*t*, 340
Kanthasamy, A., 12–13
Kapoor, S., 12–13
Kapp, O.H., 28–29, 57
Karaolis, D.K., 19

Karatan, E., 5, 298–299, 308
Karlberg, O., 411–412, 429–430
Karls, R., 8
Karnovsky, N.J., 332–333
Karoly, D., 332
Karow, D.S., 300–301
Karplus, K., 406–408, 409
Karpowicz, S.J., 223
Karunakaran, V., 11
Kathariou, S., 350, 355
Kato, I., 350, 353, 355
Kato, J., 308
Kato, T., 6, 128–129, 277, 278
Katoh, K., 409
Katsanis, N., 224
Katz, L.A., 411
Kawamoto, J., 350, 353, 355
Kawamura, Y., 20, 299–300
Keeling, P.J., 205, 410*f*, 411–412
Keilin, D., 199–200, 432–433
Keller, K., 429–430
Kelly, A., 101*t*, 342, 343, 370
Kelly, D.J., 101*t*, 106, 108–110, 111–112,
    113, 121, 126, 128–130, 131–132
Kelly, R.M., 352
Kendall, S.L., 173, 296
Kennett, J.P., 333
Kenney, C., 254
Kerby, R.L., 7–8, 9, 11, 13, 275, 277,
    278–280
Kerouault, B., 337–340, 339*t*, 350, 356
Kersten, M., 294–295
Ketley, J.M., 108–109, 111, 123–125
Key, J.M., 16, 17–18, 285–286, 289–290
Khan, S., 108
Khare, T., 25
Khosla, C., 101*t*, 420–421
Kiefhaber, T., 351
Kiehl, J.T., 332
Kieseppa, T., 12–13
Kiger, L., 15, 16, 203, 244, 246, 282–283,
    288*t*, 362, 364, 368, 415–420
Kikuchi, A., 15–16, 290, 291
Kikuchi, T., 66
Kim, B.C., 25
Kim, C.H., 14
Kim, K.J., 101*t*, 112–113
Kim, M.J., 296–297

Kim, S.-O., 101*t*, 106, 125, 127, 130, 182–183, 187–188, 342, 343, 370, 422–423
Kim, S.W., 428, 429
Kim, W.K., 109–110
Kim, Y.M., 296–297
King, D.P., 7–8, 280
King, G.M., 278
King, J.C., 332
King, N., 405–406, 415–420, 418*f*
King, S.M., 260
Kiran, M.D., 337–340
Kirby, J.R., 281–282
Kirchman, D.L., 334, 335
Kirillina, O., 19
Kirov, G., 12–13
Kitagawa, M., 350, 353, 355
Kitagawa, T., 6, 11, 12, 20, 280–281, 303–304, 314–315
Kitamoto, K., 428
Kitanishi, K., 13, 18, 19, 20, 299–300
Kjelgaard, P., 51, 57, 58–59, 63–64, 343–344, 360–361
Klappenbach, J.A., 351–352
Klevit, R.E., 295–296
Klose, K.E., 3, 23
Klosse, U., 294–295
Klotz, L.O., 368
Klumpp, S., 354
Knap, A.H., 334
Knight, R., 394–395
Knittel, K., 334
Knoll, A.H., 411
Knoop, H., 204–205
Knowles, R.G., 106
Ko, I.J., 296–297
Kobayashi, G., 422–423
Kobayashi, K., 6, 13, 18, 19–20, 51, 55–56, 61, 63, 67, 299–300, 304, 314–315, 432–433
Kobayashi, M., 291–294, 293*f*, 294*f*
Kobayashi, N., 20
Koca, J., 235*f*
Kock, H., 348–350, 354, 355, 357
Koga, M., 99
Koga, N., 249
Kogoma, T., 107–108
Kohler, H., 351

Kohno, S., 66–67, 173
Koike, S., 99
Kole, L., 431
Komori, H., 11, 277
Koonin, E.V., 295–296
Koot, M.G., 126–127
Korenaga, S., 432–433
Korner, H., 123–124, 126–127, 128
Korolik, V., 108–109
Kosinova, P., 235*f*
Koskenkorva, T., 101*t*
Kotani, H., 200
Koudo, R., 12
Kovanen, L., 12–13
Kramer, D.L., 430
Kramer, G., 354
Kranz, R.G., 251–252
Kraulis, P., 198*f*
Kraut, J., 115*f*, 117*f*
Krieger-Liszkay, A., 259
Kriegl, J.M., 364
Krin, E., 337–340, 339*t*, 348–350, 351–352, 355, 356, 357–358
Krogfelt, K.A., 99
Krogh, A., 126–127
Kroncke, K.D., 110
Kruglik, S.G., 16–17, 289–290
Kubo, M., 6
Kuczek, E., 350, 352–353
Kugelstadt, D., 201–202
Kulasakara, H., 19
Kullik, I., 16
Kuma, K., 409
Kumar, A., 297–298
Kumar, G.S., 339*t*, 340
Kumar, R.A., 101*t*, 173
Kumar, S., 409
Kumita, H., 282–283, 292–293
Kundu, S., 51, 56–57, 61, 67–68, 237–238, 238*f*, 241*t*, 242, 247*t*, 248*np*, 250, 259
Kuriakose, S.A., 67–68, 203–204
Kurihara, T., 350, 353, 355
Kuriyan, J., 300
Kurland, C.G., 411–412, 429–430
Kuroda, A., 308
Kurokawa, H., 12, 302, 303
Kurtz, D.M. Jr, 3, 23
Kurz, T.A., 354

Kustu, S., 14
Kwast, K.E., 422–423
Kwon, E.M., 12–13

L

La Clair, C., 3, 28
La Mar, G.N., 198–199, 228–231, 233–234
Labarre, M., 51, 52–55, 57–60, 61–62,
    63–64, 65, 66–67, 70–71, 101t, 120,
    121–122, 122f, 157–158, 161–163, 166,
    169–172, 182–183, 184, 188–189,
    345–347, 360, 369
Labbe-Bois, R., 422–423
LaCourse, R., 66–67
Lafontaine, J., 200, 206, 222–223, 433
Lagroye, I., 431
Lagüe, P., 55–56, 61, 63, 101t, 155–159,
    160–162, 261
Laguerre, L., 415–420
Lahr, D.J., 411
Laine, T., 207–210
Lam, S., 5, 201–202
Lama, A., 63, 66–67, 101t, 156–157,
    160–163, 167–168, 169–172, 183–184,
    185, 362, 372
Lamb, C., 424–425
Lamba, P., 422–423
Lambry, J.C., 167–168, 184–185, 289–290
Lamerdin, J., 219
Lamy, D., 335
Lancaster, J.R. Jr, 297–298
Landfried, D.A., 255–256
Landini, P., 20
Landry, M.R., 336
Lane, C.E., 394–395, 404f, 405–406
Lang, B.F., 405–406, 411–412, 415–420,
    418f
Lanz, N.D., 11
Lanzilotta, W.N., 7–8, 9, 277
Lapini, A., 58
Lara, F., 362–363, 365
Lardinois, O.M., 368
Larimer, F., 219
LaRossa, R.A., 106, 108, 123–125
Larsen, R.W., 5, 6, 80–81, 91, 308, 393
Larsen, T.S., 126–127
Larsson, J.T., 70, 343–344
Lassmann, T., 409

Laub, M.T., 19
Lauro, F.M., 335–336, 350, 351–352, 355
Lavebratt, C., 12–13
Laver, J.R., 125, 421
Lawver, L.A., 333
Lay, P.A., 118
Lazdunski, C., 353
Lazzell, A.L., 423
Le Gall, J., 129
Leach, E.R., 129
Leadbeater, B.S., 415–420
Leaderer, D., 12–13
Leaf, C.D., 343
Leander, B.S., 394–395
Leang, C., 25
Leboffe, L., 261–262
Lechauve, C., 19–20, 244, 246, 304, 368,
    415–420
Lechtreck, K.F., 224–225, 260
Lecomte, J.T.J., 50–51, 53–54, 55–57,
    59–60, 61–62, 64–65, 66, 67–68, 70–71,
    80–81, 149–150, 200, 203–204, 212–215,
    219, 220–221, 229t, 230np, 233,
    236–238, 237f, 239–240, 240np, 241t,
    242–244, 248t, 250–252, 261, 347,
    360–361, 362–363, 393
Lecroq, B., 394–395
Ledford, H.K., 225–226
Lee, A.J., 11
Lee, C.C., 13
Lee, C.E., 432–433
Lee, D.S., 302, 303
Lee, H.C., 57, 61, 223, 229t, 232–233,
    234–235, 248t, 365, 433
Lee, J.M., 296
Lee, K.F., 289–290
Lee, K.S., 93
Lee, L.J., 101t, 109–110, 111–112, 121,
    129–130
Lee, M.E., 342, 343, 421
Lee, V., 19
Lefebvre, P.A., 223, 225
Leggat, W., 433
Lehnert, N., 100–106
Lehrmann, E., 12–13
Lembo, A., 307
Leopoldo, P.T., 429
Lepoivre, M., 431

Leprince, P., 348, 349f, 350, 353, 355, 356, 357
Leroux, C., 415–420
Lesk, A.M., 212, 255–256
Lessner, D.J., 91
Levasseur, A., 428
Levi, R., 343
Levine, A., 425
Levy, A., 363
Levy, B., 433
Lewinska, A., 423
Lewis, M.E.S., 342
Lewis, P.J., 18, 27
Lewis, R.J., 3, 18, 26–27, 197, 409
Li, D., 118
Li, H., 11, 254
Li, J., 99
Li, L., 91, 260
Li, T.S., 112, 343
Li, W., 409
Li, W.K.W., 334–335
Li, X., 11
Li, Z., 51, 56–57, 61, 66, 67–68, 204, 206, 220–222, 241t, 252, 253, 254, 255, 257, 258
Liang, Z.X., 23, 305–307
Liao, J.C., 125–126
Liao, R.P., 173
Liberati, N., 19
Liebl, U., 16–17, 19–20, 285–286, 288t, 289–290, 304
Liew, F.Y., 343, 429
Lightfoot, J., 109
Lilley, P.E., 393
Lim, J., 351–352
Limam, F., 433
Lin, J.Y., 429–430
Lin, W.C., 25
Lin, Y., 112, 114–116
Lind, J., 69, 370–372
Lindquist, E., 402–403
Linn, S., 107–108
Liu, C., 68–69, 101t, 186–187
Liu, H., 336
Liu, L., 343, 423
Liu, P.T., 182
Liu, Y., 12–13
Liu, Y.X., 101t, 112–113
Ljungkvist, G., 110

Llewelyn, D.J., 393
Lloyd, D., 343, 430, 431–432
Loewen, P.C., 124–125
Loffelhardt, W., 205–206, 411–412
Lohrman, J., 203, 247t
Lois, A.F., 14–15
Loisel-Meyer, S., 294–295
Londer, Y.Y., 311–312, 313–314
Long, E., 335–336
Lonnqvist, J., 12–13
Lopez, N.I., 337–340, 339t, 356
Lopez, O., 15
Lopez-Garcia, P., 394–395
Lorieux, F., 351–352
Loschi, L., 357
Loux, V., 337–340, 339t, 350, 356
Love, N., 57
Lovejoy, C., 335
Lovering, A.L., 357
Lovley, D.R., 25
Lowenstein, C.J., 106
Lu, C., 57–58, 59–60, 64–65, 70–71, 101t, 112, 113, 114–116, 114f, 115f, 117, 117f, 118, 120, 123, 148–149, 150, 163–166, 168, 173–180, 187–188, 189, 346–347
Lu, H., 128–129
Lubman, D.M., 350, 351–352, 355
Lucarelli, D., 344
Lucas, T.G., 312
Lucey, J.F., 19
Luchinat, C., 231–232
Lukat, G.S., 307
Lukat-Rodgers, G.S., 14–15, 289
Lukes, J., 394–395, 404f
Lund, L., 60
Lunn, A.D., 421
Luo, W., 19
Luo, Y., 107–108
Luo, Z., 23
Luque, F.J., 57–58, 61–62, 63–64, 65, 87–88, 155–156, 159, 161, 166–167
Luscombe, N.M., 300–301
Lutz, Z., 15–16
Luu, B., 431–432

M

Ma, J.G., 236
Macedo, A., 129

Machado, H.B., 294–295
Maciag, A., 20
Mackay, J., 363
Mackichan, C., 5
MacMicking, J.D., 66–67
Madden, T.L., 406–408
Mäder, U., 354
Madhavilatha, G.K., 101t, 173
Magez, S., 431
Magliozzo, R.S., 188–189, 360
Mahabhashyam, M.S., 409
Maier, T., 354
Majumdar, A., 57, 67–68, 239–240, 240np, 250, 251–252
Makino, R., 118
Malfatti, F., 334–335
Malhotra, V., 296
Mallick, N., 253
Malolepsza, U., 425
Malone, J., 298–299
Mandhana, N., 107
Mandon, J., 424–425
Mandrell, R.E., 109
Mañez, P.A., 58, 65
Manno, D., 352–353
Manoff, D., 15–16
Manson, S.R., 352
Mansy, S.S., 17, 285
March, R.E., 431–432
Marchadier, E., 5
Marchler-Bauer, A., 406–408
Marden, M.C., 15, 16, 19–20, 203, 246, 282–283, 288t, 304, 360, 362, 365–367, 366t, 368, 373–374
Margesin, R., 334, 357–358
Margoliash, E., 232–233
Marina, A., 293–294
Marino, G., 66, 69–70, 149–150, 350, 351, 355, 357–358, 360, 361, 369
Marino, M.C., 108, 364
Markert, S., 348–350, 354, 355, 357
Markovic, D., 255–256
Marles-Wright, J., 3, 18, 27
Marletta, M.A., 7–8, 244, 300–301, 343
Marocco, A., 424–425
Marsden, G., 70–71, 101t, 111–112, 121, 123–125, 126–127, 131, 344
Marshall, H.E., 106, 125

Martì, M.A., 57–58, 59–60, 61–62, 63–65, 85, 86–88, 89, 91–92, 94, 101t, 123, 155–161, 163, 166–167, 261, 294–295, 360
Martin, E.E., 244, 249, 333
Martin, J.L., 16–17, 167–168, 184–185, 289–290
Martin, L.A., 101t, 112, 253, 343, 421
Martin, M.E., 219–220
Martineau, V., 11–12
Martinez, D., 223, 225
Martinez, M.P., 430
Martinez-Hackert, E., 354
Maruyama, J., 428
Maruyama, S., 205–206
Marvin, K.A., 13, 278–280
Marx, J.C., 337
Marzella, L., 148–149
Marzouki, N.M., 433
Mason, R.P., 368
Massey, V., 107–108
Mastroeni, P., 108
Mastronicola, D., 430, 431–432
Masuda, C., 11
Matejka, M., 110
Mathies, R.A., 300
Matilla, M.A., 19
Matsubara, H., 148–149, 199–200
Matsui, M., 126
Matsui, T., 20, 368, 370–372
Matsuki, M., 6
Matsumoto, A., 106, 125
Matsumoto, D., 250
Matsuo, M., 427
Matsuo, T., 277, 278–279
Matsuoka, A., 51, 55–56, 61, 63, 67, 206–207, 250, 422–423, 432–433
Matthews, R.G., 107–108
Mattick, J.S., 350, 352–353
Maurelli, S., 86–87
Mayer, R., 7–8, 19, 23, 288t, 305
Mayhew, S.G., 107–108
Mazard, S., 336
Mazzarella, L., 362–363
Mazzucchelli, G., 337–340, 339t, 348–350, 349f, 353, 354, 355, 356, 357
McAdams, H.H., 2, 15
McCarthy, Y., 19

McClure, T., 343–344, 423
McGrath, P.T., 15
McKinnie, R.E., 114–116, 244
McKnight, S.L., 7–8, 11–12, 280
McLean, S., 100–106, 107, 125, 252, 342–343, 344–345, 421, 429
McNicholl-Kennedy, J., 124–125
Mead, G.C., 109–110
Medforth, C.J., 236
Médigue, C., 337–340, 339t, 348–350, 351–352, 355, 356, 357–358
Meeks, J.C., 219–220
Meged, R., 2
Megson, I.L., 111–112
Meharenna, Y.T., 297
Meiller, A., 431
Meindre, F., 101t, 157, 158–159, 160–161, 261
Méjean, V., 308
Melamud, E., 337–340, 339t, 350, 351–353, 356, 357
Meltzer, M.S., 343, 429
Membrillo-Hernández, J., 101t, 127, 130, 342, 343, 344, 370, 422–423
Mendz, G.L., 108–109
Meng, J., 99
Merchant, S.S., 225–226, 257–258
Meredith, M.P., 332
Merényi, G., 69, 370–372
Merico, A., 351
Merli, A., 85–86
Merlino, A., 362–363
Merx, M.W., 368
Merz, F., 354
Mesa, S., 16
Messenger, S.L., 421
Metcalf, W.W., 91
Methé, B.A., 25, 337–340, 339t, 350, 351–353, 356, 357
Meuwly, M., 63, 158–159, 163
Meyer, M.T., 222–223
Miao, W., 412–415
Michaeli, D., 19
Michaels, A.F., 334
Michel, C., 337–340, 339t, 350, 356
Michiels, J., 15
Michnoff, C., 12–13
Mikami, B., 302, 303

Milani, M., 50–51, 52–53, 54–56, 57–59, 60, 61–62, 63–65, 66–67, 68–69, 80–81, 85–86, 101t, 149–150, 153, 155–158, 159, 160–162, 163–167, 168–169, 182–183, 188–189, 201–202, 214f, 233, 235, 241t, 242, 301, 345, 346–347, 347t, 360–361, 362, 365, 367
Miles, M., 15–16
Miller, G.H., 332
Miller, M.A., 409
Miller, M.R., 111–112
Miller, R.A., 108
Miller, W.G., 109, 406–408
Millott, S., 343, 429
Mills, C.E., 101t, 106, 113, 116, 118, 174, 182–183, 187–188, 253, 370, 422–423, 425
Mincer, T., 334
Minch, K.J., 148–149
Minge, M.A., 394–395, 415–420
Minkow, O., 364
Minton, N., 300
Mishra, S., 63, 158–159, 163
Mitchell, M.C., 222–223
Mitchell, M.J., 249
Miteva, V., 334
Mitz, S., 395–397
Miyata, S., 19
Miyata, T., 409
Miyatake, H., 15–16, 17, 128–129, 277, 278, 282–283, 284–285
Mizuguchi, K., 28–29, 80–81, 200, 201–202, 206–207, 212, 393, 406–408, 411–412
Mizutani, Y., 6, 15–16, 19–20, 290, 291, 303–304, 309–311
Modlin, R.L., 182
Moënne-Loccoz, P., 296, 297–298
Moens, L., 2–3, 25–26, 28–29, 57, 66, 80–81, 82, 85–86, 196–197, 200, 201–202, 203, 206–207, 212, 244, 340–341, 343–344, 348, 362–363, 365–366, 366t, 368, 393, 396t, 405–406, 411–412, 420–421, 428
Moffat, K., 16, 17–18, 276, 285–286
Möglich, A., 276
Mohan, V.P., 66–67, 173
Mohn, F.H., 253
Moir, J.W., 129, 302, 421
Moline, M.A., 332–333, 402–403

Momen, B., 337–340, 339*t*, 350, 351–353, 356, 357
Monach, P., 108
Moncada, S., 343, 429
Mongkolsuk, S., 123–124
Mongodin, E.F., 109
Monk, C.E., 70–71, 101*t*, 106, 111–112, 113, 121, 123–126, 127, 128–132, 344
Monson, E.K., 14–15, 282–283
Montagna, A.E., 431
Montanyola, A., 409
Monteferrante, C.G., 5
Montmayeur, A., 431
Moon-van der Staay, S.Y., 394–395
Moore, C.M., 106, 124–125
Moran, J.F., 203, 247*t*
Morán, X.A.G., 334, 335
Mordida, B.J., 331–332
Morea, V., 3, 24, 51, 57, 58–59, 63–64, 65, 360–361
Moreira, D., 394–395
Morett, E., 124–125
Moretti, S., 409
Morikawa, M., 355
Morin, S., 101*t*, 157, 158–159, 160–161, 261
Morishima, I., 8, 9, 277, 278
Moritz, R.E., 331
Morr, M., 19
Morreale, A., 156–159, 160–161, 163
Morris, H., 357
Morris, I.K., 315
Morris, S.L., 125
Morrison, H.G., 394–395
Motohashi, K., 337–340, 339*t*
Moura, I., 129
Moura, J.J., 129
Mourey, L., 14
Movahedzadeh, F., 173, 296
Mo,W., 260
Mowat, C.G., 252, 253, 343
Mozzarelli, A., 101*t*
Mu, L., 12–13
Mucciacciaro, S., 360, 361, 361*f*, 362, 364–365, 373–374
Mudgett, J.S., 66–67
Muhs, D.R., 332

Muir, L., 335
Mukai, M., 15–16, 17, 68–69, 112, 114–116, 118, 157–158, 163–166, 167–168, 174, 186–187, 253, 282–283, 284–285, 301, 360–361, 432–433
Mukaiyama, Y., 12, 280–281
Mukhopadhyay, P., 106, 123–125
Mukhopadhyay, S., 101*t*
Mulder, N., 406–408
Mulholland, F., 101*t*, 106, 109–110, 111–112, 123–126, 127, 128–129, 344
Mundayoor, S., 101*t*, 173
Mungall, K., 108–109, 111, 123–125
Mur, L.A.J., 424–426
Murad, F., 106
Muro-Pastor, A.M., 204–205
Murphy, E.J., 332
Murray, A.E., 334, 335–336
Murray, J.W., 27, 197, 409
Myers, H., 423
Myers, J.D., 109–110
Myers, R.A.M., 148–149

N

Nacy, C.A., 343, 429
Nadra, A.D., 261
Nagao, S., 250
Nakai, K., 409
Nakajima, H., 6, 7–8, 9, 11, 128–129, 277, 278–279, 295–296
Nakajima, K., 20, 299–300
Nakajima, S., 66
Nakamura, H., 14–16, 17, 282–286, 288*t*, 291–294, 293*f*, 294*f*
Nakamura, K., 14–16, 283–285
Nakamura, S., 253
Nakamura, T., 422–423
Nakamura, Y., 200
Nakanishi, Y., 428
Nakano, M.M., 106, 124–125, 296, 297–298
Nakashima, S., 6, 20, 304
Narayanasamy, P., 296
Nardini, M., 25–26, 50–51, 52–56, 57, 58–60, 63–64, 65, 66–67, 70–71, 80–81, 82–84, 85–88, 89, 90, 91–93, 94, 101*t*, 120, 121–122, 122*f*, 169, 201–202, 233,

235, 248np, 248t, 345–347, 347t, 360–361, 362–363, 365
Nash, S., 12–13
Nathan, C., 66–67
Nathan, C.F., 66–67, 343
Navani, N.K., 61–62, 65, 66–67, 68–69, 101t, 153, 161–162, 163–166, 167–168, 183–187, 345, 369
Neal, P.R., 335
Negrerie, M., 289–290
Negrisolo, E., 23
Nei, M., 409
Neilson, J.A.D., 204–205
Neimann, J., 99
Nellen-Anthamatten, D., 16
Nelson, K.E., 25, 337–340, 339t, 350, 351–353, 356, 357
Nelson, M., 415–420
Nelson, W., 25
Neofotis, P., 332
Nequiz, M., 429–430
Newell, D.G., 124
Newhouse, J.S., 21–22, 25, 28, 80–81, 90–91, 210, 211f, 299
Newman, J.A., 3, 18
Ng, G.C., 430
Nguyen, H.B., 7–8
Nicholas, B., 12–13
Nichols, S.A., 415–420
Nicol, S., 331–332
Nicoletti, F.P., 58, 66, 70, 149–150, 362–363, 370–372
Nicollier, M., 19
Nielsen, S.B., 9–11
Nienhaus, G.U., 85–86, 115f, 362–363, 364
Nienhaus, K., 85–86, 115f, 362–363, 364
Nievergelt, C.M., 12–13
Nikolaev, S.I., 394–395
Nilavongse, A., 110
Nimtz, M., 19
Nioche, P., 300
Nisbet, D.J., 109–110
Nishimura, N., 226–228
Nishimura, R., 250
Noble, S.M., 423
Noel, B., 412–415

Nolla, H.A., 336
North, R.J., 66–67
Nothnagel, H.J., 51, 53–54, 56–57, 59–60, 70–71, 251–252
Novik, V., 109
Nudler, E., 107
Nuernberger, P., 289–290
Nussbaum, J.C., 423

O
O'Brien, K.M., 423
O'Brien, S.J., 99
Ochiai, T., 422–423, 432–433
Oelgeschläger, E., 91
Ogura, T., 6, 20, 299–300
Oh, J.I., 296–297
Ohana, P., 7–8
Ohkubo, K., 8, 277, 278–279
Ohm, R.A., 426–427
Ohmachi, H., 422–423
Ohno, A., 291–294, 293f, 294f
Ohno, H., 66–67, 173
Ohta, T., 6, 11
Ohtake, H., 308
Okada, M., 7–8, 277
Olea, C. Jr, 244, 300
Olin, A.C., 110
Oliveira, A., 61–62, 159–160
Oliver, T.A., 336
Olivos-Garcia, A., 429–430
Olson, J.S., 112, 114–116, 203, 226–228, 244, 247t, 248t, 249–250, 253, 343, 365–366, 368
Olson, J.W., 109–110
On, S.L., 126–127
Opazo, J.C., 415–420
Ordal, G.W., 5, 307
Orii, Y., 342, 343, 422–423
Orobitg, M., 409
Orozco, M., 61–62, 63, 64–65, 101t, 155–159, 160–161, 163, 360
Orozco-Cardenas, M.L., 426–427
Orr, R., 415–420
Ortiz de Montellano, P.R., 296, 297–298
Oshino, N., 421–423
Oshino, R., 421–423
Osterberg, S., 308
Ostrander, E.A., 12–13

Ostrowski, M., 336
Otto-Bliesner, B.L., 332
Otyepka, M., 235f
Oudega, B., 249–250
Ouellet, H., 51, 52–53, 54–56, 57–59,
    60, 61–62, 63–64, 66–67, 68–69,
    86, 101t, 157–158, 161–168, 169–172,
    182–183, 184–185, 186–187, 188–189,
    214f, 235, 241t, 301, 345, 346–347,
    347t, 348, 360–361, 365–366,
    366t, 367
Ouellet, Y.H., 51, 54–58, 60, 61–62, 65,
    66–67, 68–69, 86, 101t, 118, 149–150,
    153, 155–156, 157–158, 159, 160–166,
    167–168, 169–172, 182–183, 184, 187,
    200, 229t, 231–232, 235, 236, 241t, 242,
    247t, 261, 301, 345, 346–347, 347t, 348,
    360–361, 362, 365–366, 366t, 367, 369
Overland, J.E., 332
Overpeck, J.T., 332
Owen, M.J., 12–13

P

Pachebat, J.A., 404
Padalko, E., 106
Page, K.M., 16
Pahl, R., 16
Palmer, R.M.J., 343, 429
Palomares, A., 14
Palsson, B.O., 2
Pan, D., 300–301
Pandit, J., 295–296
Pant, N., 296
Papa, R., 358
Papin, J.A., 2
Paps, J., 415–420
Parales, R.E., 308
Parfrey, L.W., 411
Parigi, G., 231–232
Parimi, N., 12–13
Paris, G., 294–295
Parish, T., 173
Park, H., 20, 302, 303–304
Park, K.J., 296–297
Park, K.W., 112–113
Park, M.K., 148–149
Park, S.F., 70–71, 101t, 108–110, 111–112,
    120–121, 122–124, 129–130, 131–132,
    149–150, 344

Park, S.W., 296
Park, S.Y., 15–16, 17, 128–129, 277, 278,
    284–285
Parkhill, J., 108–109, 111, 123–125,
    148–150, 153
Parkinson, C.L., 332, 343, 429
Parks, R.B., 9, 11
Parracino, A., 357–358
Parrilli, E., 66, 69–70, 149–150, 350, 351,
    355, 357–358, 360, 361, 369
Partonen, T., 12–13
Pascal, G., 337–340, 339t, 348–350,
    351–352, 355, 356, 357–358
Pastore, A., 255–256
Pasumarthi, R.K., 12–13
Pasunooti, S., 306, 307
Patel, R.P., 297–298
Pathak, S., 12–13
Pathania, R., 61–62, 65, 66–67, 68–69,
    101t, 153, 161–162, 163–166, 167–168,
    183–187, 345, 369
Pathania, V., 182
Patzelt, H., 354
Paul, J., 334–335
Paulsen, I.T., 25
Paunio, T., 12–13
Paveto, C., 431
Pawaria, S., 63, 66–67, 101t, 148–149, 150,
    156–157, 160–163, 167–168, 169–172,
    173–180, 183–184, 185, 187–188, 189,
    362, 372
Pawlowski, J., 394–395
Pawlowski, K., 294–295
Peacock, W.J., 362–363
Pearlstein, R.M., 255–256
Pearson, B.M., 101t, 106, 111–112, 113,
    123–126, 127, 128–129, 344
Peck, L.S., 331, 332, 334, 372–373
Pedrini, B., 336
Pedrosa, F.O., 294–295
Pedros-Alio, C., 394–395
Peeters, K., 28–29
Pegg, D.E., 337–340, 339t
Peisach, J., 57, 61, 223, 229t, 232–233,
    234–235, 248t, 365, 433
Pellegrini, M., 225–226
Pelletier, H., 115f, 117f
Pellicena, P., 300
Peng, J., 128–129

Pereira, C., 431
Pereira, J.M., 429
Perfect, J.R., 343
Perna, N.T., 19–20
Perpetua, L.A., 25
Perutz, M.F., 14–15, 52–53, 80–81, 82, 282–283
Pesavento, C., 20, 298–299, 300–301
Pesce, A., 25–26, 50–51, 52–56, 57, 58–60, 61–62, 63–64, 65, 66–67, 70–71, 80–81, 82–85, 87–89, 90, 91–93, 94, 101t, 120, 121–122, 122f, 149–150, 155–156, 159, 160–161, 169, 201–202, 201f, 213f, 223, 233–235, 234f, 241t, 242, 248np, 248t, 259, 261, 345–347, 347t, 360–361, 362–363, 365, 367, 393
Pessanha, M., 312, 313–314
Pessi, G., 19
Petach, H.H., 352
Petersen, I., 99
Petersen, L., 126–127
Petersen, M.G., 254
Peterson, D., 409
Peterson, E.S., 290
Peterson, N., 409
Petrek, M., 235f
Petrucca, A., 57, 149–150, 362–363
Pezzella, C., 351
Pfeiffer, W., 409
Pham, T.H., 306, 307
Phillips, E.J., 25
Phillips, G.N. Jr, 6–7, 25, 82–84, 85, 244, 247t, 248t, 249–250, 346f, 365–366
Phillips, S.E., 198f
Piatibratov, M., 5, 6, 80–81, 91, 201–202, 393
Piette, F., 337–340, 339t, 348–350, 349f, 353, 354, 355, 356, 357
Pilak, O., 350
Pillai, S., 406–408
Pin, C., 124–125
Pinakoulaki, E., 6
Pineda, E., 429–430
Pinto, G., 205
Piperno, G., 260
Pirt, S.J., 108–109
Pisciotta, J.M., 204–205
Pittman, M.S., 101t, 109–110, 111–112, 121, 129–130

Pitts, S., 11–13
Plunkett, G. III., 19–20
Podust, L.M., 297
Pointing, S., 334
Pokkuluri, P.R., 311–312, 313–314
Poljak, A., 350
Poljakovic, M., 101t
Polticelli, F., 261–262
Polyak, L., 332
Polz, M.F., 348–350
Ponce-Coria, J., 124–125
Pond, M.P., 51, 53–54, 57, 59–60, 67–68, 70–71, 239–240, 240np, 242–244, 250, 251–252
Ponder, M.A., 350, 351–353, 355
Ponting, C.P., 295–296
Poock, S.R., 129
Poole, R.K., 3, 28, 50–51, 59–60, 69–71, 80–81, 99–106, 101t, 107, 108–110, 111–112, 113, 114–116, 117, 118, 119, 120–121, 122–126, 127, 128–132, 148–150, 174, 182–183, 185, 187–188, 200, 201–203, 206–207, 210, 212–213, 217, 224, 252, 253, 340–341, 342–343, 344–345, 346–347, 360, 362, 368, 369, 370–372, 371f, 393, 395–397, 396t, 406, 411–412, 421, 422–423, 424–425, 429
Porazinska, D.L., 394–395
Porcel, B.M., 412–415
Porter, C.W., 430
Pörtner, H.O., 373
Possling, A., 20
Potts, M., 149–150, 199–200, 217, 219, 222, 228–231, 229t, 240–242, 247t, 248t, 250, 254–255
Poulos, T.L., 7–8, 9, 11, 116, 277, 297
Poyart, C., 15, 282–283
Poyton, R.O., 343–344, 422–423
Prats, E., 424–426
Pratt, J.T., 19, 298–299, 300–301
Preer, J.R. Jr, 432–433
Preimesberger, M.R., 57, 242–244, 250, 251–252
Preisig, O., 16
Preissler, S., 354
Prejean, M.V., 5
Premer, S.A., 247t, 248np, 261

Presley, A., 300
Preston, C.M., 334
Price, C.W., 27
Price, M.S., 300
Price, P.A., 19
Price, T.S., 12–13
Probst, I., 421–422
Proll, S., 255–256
Ptitsyn, O.B., 57
Pucillo, L.P., 431–432
Pullan, S.T., 101*t*, 106, 125
Punt, P.J., 428
Puranik, M., 9–11
Purdy, D., 124

Q

Qareiballa, A., 99
Qi, Y., 23
Qian, H., 245–246
Qiu, Y., 350, 355
Qu, Z., 397
Quevillon, E., 406–408
Quin, M.B., 3, 18
Quinn, J., 423

R

Rabus, R., 337–340, 339*t*, 350, 352–353, 356
Rachmilewitz, E.A., 226–228
Radi, R., 258
Raffaelli, N., 20
Rafferty, S., 431–432
Raftery, M.J., 335–336, 350, 351–352, 354, 355
Ragsdale, S.W., 278
Rahat, O., 2
Rai, L.C., 253
Rai, P., 107–108
Raitt, D., 422–423
Rajamohan, G., 65, 66–67, 68–69, 101*t*, 153, 163–166, 167–168, 172, 185, 186–187
Rajandream, M.A., 404
Raje, M., 101*t*, 169–172
Rajeev, L., 7–8
Rajeevan, M.S., 12–13
Raman, C.S., 300, 302, 303
Ramandeep, Hwang, K.W., 101*t*

Ramaswamy, S., 51, 56–57, 61, 237–238, 238*f*, 241*t*, 259
Ramos, E., 429–430
Ramos, J.L., 19
Ramos-Gonzalez, M.I., 19
Ranguelova, K., 188–189, 360
Rao, C.V., 5
Rao, F., 23, 305–307
Rao, V.J., 339*t*, 340
Rappelli, P., 430
Rashid, M.H., 19
Rasko, D.A., 109
Rasmussen, S., 336
Rattei, T., 337–340, 339*t*, 350, 352–353, 356
Rauch, T., 354
Rausch-Fan, X., 110
Raushel, F.M., 60
Ravel, J., 109
Raven, E.L., 55
Raviglione, R.C., 182
Ray, A., 55
Ray, M.K., 339*t*, 340
Rayner, B.S., 118
Razzera, G., 362–363, 365
Read, R.C., 106, 125, 185, 187–188, 342, 343, 421
Rebbapragada, A., 308
Record, E., 428
Redline, S., 12–13
Reed, J.L., 2
Reichlen, M., 91
Reichmann, D., 2
Reick, M., 7–8, 11–13, 280
Reilly, D.F., 12–13
Reinhardt, S., 201–202, 364
Renaut, J., 337–340, 339*t*, 348–350, 353, 354, 355, 356
Renzi, S.D., 66–67, 68–69, 153, 168–169, 173, 185–186, 369
Requena, M.-E., 427
Rexine, J.L., 289
Reyes-Prieto, A., 204–205
Reynolds, M.F., 9, 11, 15–16
Reyrat, J.M., 282–283
Rezac, J., 433
Rhen, M., 340
Riccio, A., 362–363

Rich, V., 334

Richard, C., 51, 52–55, 57, 58–60, 61–62, 63–64, 65, 66–67, 70–71, 101t, 120, 121–122, 122f, 157–158, 161–163, 169–172, 182–183, 184, 345–347, 369

Richard-Fogal, C., 251–252

Richards, T.A., 205–206

Richardson, A.R., 106, 124–125

Richardson, D.J., 110, 129

Richter, D.J., 415–420

Ricke, S.C., 109–110

Riesenfeld, C.S., 335–336

Rifkind, J.M., 363

Riggs, A.F., 393

Riley, C.W., 5, 308

Riley, L.W., 429

Riley, M., 19–20, 350, 351–352, 359

Rinaldo, S., 128–129

Rinas, U., 101t, 351

Ring, H.Z., 11

Rison, S.C.G., 173, 296

Rizzi, M., 53–54, 55, 58, 80–81, 82, 93–94, 341

Robert, B., 255–256

Roberts, B.W., 415–420

Roberts, D.M., 173

Roberts, G.P., 7–8, 9, 11, 13, 275, 277, 278–280

Robinson, H., 207–210

Robles, E.F., 16

Rock, C.O., 356

Rodgers, K.R., 14–15, 289

Rodrigues, D.F., 337–340, 339t, 350, 351–352, 359

Rodrigues, L.C., 99

Rodríguez-Carmona, E., 351

Rodriguez-Ezpeleta, N., 205–206, 411–412

Rodríguez-López, J.N., 55

Rodriguez-Valera, F., 394–395

Roesner, A., 203, 395–397

Roger, A.J., 394–395, 405–406, 415–420, 418f

Rogers, N.J., 342

Rogstam, A., 70, 343–344

Rohlfs, R.J., 114–116, 244

Rokhsar, D., 223, 225

Rolland, N., 222–224, 259

Rollinghoff, M., 429

Romero, D., 15

Römling, U., 19, 298–299

Ron, E.Z., 354

Ronquist, F., 409

Roop, R.M., 294–295

Rosen, R., 354

Rosenbaum, J.L., 224, 260

Rosenzweig, C., 332

Rosic, N.N., 433

Ross, E., 247t

Ross, P., 19

Rossi, G.L., 101t

Rossi, M., 357–358

Rossi, P., 16

Rothbarth, P.H., 99

Rother, M., 91

Rothschild, L.J., 394–395

Roure, B., 205–206, 411–412

Rousseau, D.L., 51, 55–56, 57, 61–62, 65, 101t, 118, 149–150, 153, 155, 157–158, 161–162, 169–172, 223, 229t, 232–233, 234–235, 247t, 248t, 301, 365, 433

Rousseau, J.A., 63

Rousseau, P., 282–283

Rowe, J., 352

Roy, S.L., 99

Rozalska, S., 425

Ruden, T., 415–420

Rudolph, R., 351

Rudrasingham, V., 12–13

Ruepp, A., 337–340, 339t, 350, 352–353, 356

Ruiz-Trillo, I., 394–395, 405–406, 415–420, 418f

Rush, J., 224, 260

Russ, W.P., 352

Russell, N.J., 337–340, 339t

Russell, R.B., 255–256

Russo, R., 66, 69–70, 149–150, 350, 355, 358, 360, 361, 361f, 362, 364–367, 366t, 368, 369, 373–374

Rustad, T.R., 148–149

Rutter, J., 7–8, 11–13, 280

Ryan, C.A., 426–427

Ryan, R.P., 19

Ryjenkov, D.A., 19

Ryley, J.F., 199–200, 432–433

Rysgaard, S., 336

S

Saarikoski, S.T., 12–13
Saavedra, E., 429–430
Sadkowska- Todys, M., 99
Sadler, J.E., 404–405
Saengkerdsub, S., 109–110
Sagami, I., 11, 12, 19–20, 280–281, 302, 303–304
Saiful, I., 13
Saini, D.K., 296
Saito, J.A., 7–8, 18, 19, 21–22, 25, 28, 80–81, 82, 85, 87–88, 90–92, 93, 201–202, 210, 211f, 299–300, 304, 393
Saito, K., 7–8, 277
Sakai, T., 224, 260
Sakihama, Y., 253
Sakuma, I., 343
Salgueiro, C.A., 312
Saloheimo, M., 351
Salter, M.D., 85–86
Salvato, B., 23
Salzman, A.L., 101t, 106, 107, 112, 113, 343, 421
Samadani, A., 348–350
Sampaio, J.L., 224–225, 260
Samuni, U., 55, 57–58, 66–67, 68–69, 155–156, 163–166, 167–168, 187, 231–232, 348, 360–361, 365–366, 366t
Sancar, A., 12–13
Sanchez, C., 16
Sangoram, A.M., 280
Sannia, G., 351, 358
Santelli, E., 3, 28
Santero, E., 14
Santi-Rocca, J., 429–430
Santolini, J., 424–425
Santucci, R., 370–372
Sapir, S., 7–8
Saraiva, L.M., 101t
Saralegui, A., 429–430
Sarand, I., 308
Sarath, G., 203, 247t
Sardiwal, S., 296
Sarti, P., 430, 431–432
Sarvan, S., 123–124
Sasaki, A., 12, 280–281
Sasakura, Y., 7–8, 19–20, 303–304

Sato, A., 11, 19–20, 280–281, 303–304
Sato, E., 11, 12, 280–281
Sato, S., 200
Sato, T., 199–200
Sato, Y., 12, 280–281
Satterlee, J.D., 20, 198–199, 231, 233–234, 302, 303–304
Sauer, H., 355–356
Saunders, N.F., 126–127, 302, 350, 352–353, 354, 355
Savard, P.-Y., 51, 52–53, 56–59, 60, 63–64, 66–67, 68–69, 101t, 157, 158–159, 160–161, 163–168, 186–187, 200, 214f, 229t, 236, 261, 301, 348, 360–361, 365–366, 366t, 367
Save, S., 101t
Saviano, M., 357–358
Savino, C., 60
Sawai, H., 6, 22–23, 299, 307, 309–311
Scallan, E., 99
Scamardella, J.M., 394–395
Scanlan, D.J., 336
Schaffer, A.A., 406–408
Schaffitzel, C., 354
Schaller, R.A., 3, 23
Schalling, M., 12–13
Scharf, C., 354
Scheele, S., 182
Scheer, H., 255–256
Scheidt, W.R., 100–106
Schellhorn, H.E., 101t
Scher, H.D., 333
Scherb, B., 15, 16
Schewe, T., 110
Schiffer, M., 311–312
Schinner, F., 357–358
Schirawski, J., 420–421
Schirmer, T., 19
Schlegel, H.G., 421–422
Schmid, N., 19
Schmid, R., 354
Schmidt, M., 203, 395–397
Schmidt, T.M., 351–352
Schmitz, P.I., 99
Schnappinger, D., 66–67, 148–149
Schoch, C.L., 426–427
Scholz, H.C., 294–295

Schon, T., 101t
Schoolnik, G.K., 66–67, 148–149, 173
Schrader, J., 368
Schreiber, G., 2
Schuller, D.J., 7–8, 9, 277
Schumacher, J., 282–283
Schwartz, T., 409
Schwarzenbacher, R., 3, 28
Schwede, T., 19
Schweder, T., 354
Sciamanna, N., 58
Sciotti, M.A., 16, 282
Scott, C., 101t, 106, 127
Scott, E.E., 244
Scott, N.L., 51, 56–57, 61, 66, 67–68, 200,
    203, 204, 206, 220–222, 229t, 236–238,
    239, 241t, 250–251, 252, 253, 254, 255,
    257, 258, 362–363
Scott, W.G., 16, 17, 285
Sebilo, A., 101t, 157, 158–159, 160–161, 261
Sedelnikova, S., 370, 422–423
See, R.Y., 305–306
Seguin, R., 429–430
Selje, N., 334–335
Sellars, M.J., 108–110, 129
Semballa, S., 431
Serate, J., 9
Serpa, V.I., 362–363, 365
Serrano, A., 67
Seshasayee, A.S., 300–301
Setchell, K.D., 112, 343, 368
Seward, H.E., 129
Seymour, J.R., 348–350
Shah, S.K., 66–67
Shahbazkia, H.R., 394–395
Shalchian-Tabrizi, K., 394–395, 415–420
Shapiro, L., 2, 7–8, 15
Shatalin, K., 107
Shavirin, S., 406–408
Shaw, D.C., 393
Shearer, N., 109–110
Sheikh, A.G., 290
Shelnutt, J.A., 236
Shelver, D., 7–8, 9, 11
Shen, G., 51, 56–57, 61, 66, 67–68, 204,
    206, 220–222, 241t, 252, 253, 254, 255,
    257, 258

Shepherd, M., 59–60, 70–71, 101t, 113,
    114–116, 114f, 115f, 117, 117f, 118, 119,
    120–121, 123, 125–126, 127, 130–132,
    344–345, 420–421
Sherman, D.R., 148–149, 173, 422–423
Sherrid, A.M., 148–149
Shevchenko, A., 224–225, 260
Shi, J., 406–408
Shibasaki, Y., 226–228
Shibata, T., 250
Shibayama, M., 429–430
Shih, J.Y., 350, 351–352, 355
Shikama, K., 199–200, 206–207, 217, 222,
    249–250, 393, 422–423, 432–433
Shilo, Y., 7–8
Shiloh, M.U., 66–67
Shimizu, H., 404, 431
Shimizu, T., 11, 12, 13, 18, 19–20,
    280–281, 303–304
Shingler, V., 308
Shiro, Y., 14–16, 282–286, 288t, 290,
    291–294, 293f, 294f, 307, 309–311
Shitashiro, M., 308
Shivaji, S., 337–340, 339t
Shively, J.E., 352
Shoe, J.L., 411–412
Shoun, H., 427, 428, 429
Siani, L., 358
Sicheritz-Ponten, T., 336
Siddiqui, K.S., 339t, 340, 350, 357–358, 362
Siddiqui, R.A., 113, 421–422
Sies, H., 107–108, 110
Sievers, F., 409
Silberman, J.D., 394–395
Silva, M.A., 312
Silventoinen, V., 406–408
Simm, R., 19
Simon, M., 334–335
Simonsen, J., 99
Simpson, A.G., 394–395, 404f
Singel, D.J., 421
Singh, C., 148–149, 150, 174–180,
    187–188, 189
Singh, J.P., 315
Singh, R.J., 99, 125–126
Singh, S., 61–62, 159–160
Singleton, E.W., 249–250, 368

Sinibaldi, F., 370–372
Sircar, R., 311
Sisinni, L., 25–26, 82–84, 85
Sivaramakrishnan, S., 297–298
Sivozhelezov, V., 5, 6, 80–81, 91, 201–202, 393
Six, S., 420–421
Sjoholm, L.K., 12–13
Skärfstad, E., 308
Skepper, J.N., 222–223
Skirrow, M.B., 99
Skjaeveland, A., 394–395
Slade, R., 350, 352–353
Sligar, S.G., 249–250, 290, 365–366
Smagghe, B.J., 204, 207–210, 238, 241t, 247t, 252, 253, 362, 364
Smedh, C., 12–13
Smerdon, S.J., 249–250, 368
Smetacek, V., 331–332
Smith, A., 120–121
Smith, A.K., 12–13
Smith, A.T., 13, 278–280
Smith, D.D., 352
Smith, D.W., 148–149
Smith, G., 25–26
Smith, H.K., 70–71, 101t, 106, 111–112, 123–126, 127, 128–129, 130–132, 344
Smith, J.C., 334–335
Smith, K.M., 290, 315
Smith, M.A., 108–109
Smith, M.H., 432–433
Smith, S., 429–430
Smulevich, G., 360, 362–363, 364–365, 366, 366t, 373–374
Smulski, D.R., 108
Søballe, B., 370, 422–423
Soberon, M., 15
Sochirca, O., 12–13
Soeder, C.J., 253
Soehano, I., 305–307
Sofia, H.J., 123–124, 126–127, 128
Sogin, M.L., 335, 336, 394–395
Solbach, W., 429
Soldatova, A.V., 249
Soltis, M., 16, 17–18, 285–286, 287
Soman, J., 244
Somero, G.N., 331, 373

Sommerfeldt, N., 20
Song, W., 99
Song, X.Z., 236
Sonnhammer, E.L., 409
Sono, M., 227np
Soukri, A., 67
Soule, T., 220
Soupène, E., 282
Sousa, E.H., 7–8, 14–15, 16–17, 19, 282–283, 285, 286, 288t, 296–297, 299–300, 304
Souza, E.M., 294–295
Sowa, A.W., 364
Sowa, S., 364
Sowers, K.R., 339t, 340
Spaulding, D., 101t, 422–423
Specht, M., 222–224, 259
Spiro, S., 101t, 107, 126–127, 295–296, 342, 360
Spiro, T.G., 9, 118, 249, 290, 315
Spouge, J.L., 406–408
Spreitzer, R.J., 222–223
Springer, B.A., 249–250, 365–366, 366t
Srajer, V., 16
Stahl, B.T., 423
Staknis, D., 7–8
Staley, J.T., 336, 350, 351–352, 359
Stamatakiis, 409
Stamler, J.S., 101t, 106, 112, 113, 119, 125, 343, 360, 421, 423
Stanfield, S.W., 7–8, 14
Stanley, A., 25
Staples, C.R., 9
Stark, A., 255–256
Stark, B.C., 101t, 112–113
Stark, M., 337–340, 339t, 350, 352–353, 356
Starodubtseva, M., 107
Stecher, G., 409
Steers, E. Jr, 432–433
Steig, E.J., 331
Steinberg, D.K., 334
Stephens, C., 15
Steuer, R., 204–205
Stevanin, T.A., 106, 125, 343
Stevanin, T.M., 101t, 106, 182–183, 185, 187–188, 342, 343, 370, 421

Stevens, R.G., 12–13
Stewart, J.J., 253, 402–403
Stewart, V., 101*t*, 127, 130, 342, 422–423
Steyn, A.J., 297–298
Stibitz, S., 298–299
Stintzi, A., 123–124
Stock, A.M., 307
Stock, J.B., 307
Stock, L., 332
Stocker, R., 118, 348–350
Stoker, N.G., 296
Storz, G., 106, 108, 123–125
Storz, J.F., 203, 248*t*, 415–420
Stout, V., 220
St-Pierre, B., 200, 206, 222–223, 433
Stranzl, G.R., 3, 28
Stratmann, C.J., 249–250
Stratton, L.P., 352
Strickland, S., 107–108
Strid, S., 101*t*
Struvay, C., 337–340, 339*t*, 348–350, 353, 354, 355, 356
Stuehr, D.J., 106, 343, 424–425
Sturms, R., 253–254
Stynes, D.V., 249
Subba Rao, D.V., 334–335
Sucgang, R., 404
Sugawara, Y., 226–228
Sugimoto, H., 291–294, 293*f*, 294*f*, 307, 309–311
Sugiyama, S., 7–8, 19–20, 303–304
Sukomon, N., 311
Sul, W.J., 336
Summers, M.L., 219–220
Sunagawa, S., 402–403
Suquet, C., 20, 302, 303–304
Suschek, C.V., 110
Suzuki, A., 226–228
Suzuki, S., 7–8, 19–20, 427
Suzuki, T., 57, 112, 343, 368
Suzuki, Y., 355
Svensson, L., 101*t*
Swartz, J.R., 101*t*
Switala, J., 124–125
Szabo, C., 106
Szurmant, H., 307

T

Tada, H., 393, 432–433
Tagliabue, L., 20
Taguchi, S., 7–8, 19–20
Tai, H., 250
Takagi, T., 199–200, 217, 222, 393, 432–433
Takahashi, H., 19–20, 303–304
Takahashi, M., 99
Takahashi, S., 8, 9, 277, 278
Takahashi, Y., 295–296
Takami, H., 201–202
Takano, K., 355
Takasaki, H., 7–8, 9
Takatani, N., 295–296
Takaya, N., 427
Takeda, Y., 11
Takiguchi, N., 308
Takio, K., 421–423
Tal, R., 7–8
Tam, C.C., 99
Tamayo, R., 19, 298–299, 300–301
Tamiya, H., 199–200
Tamura, K., 282–283, 284–285, 409
Tamura, M., 421–422
Tan, E., 306, 307
Tanaka, A., 14–16, 19–20, 200, 283–284, 285–286, 288*t*, 290, 291, 303–304
Tanaka, M., 280
Tanaka, Y., 404, 431
Tang, J., 12–13
Tang, L., 15
Tang, N., 107–108
Tang, P., 342
Tang, X., 111–112
Tanida, H., 282–283, 284
Tanifuji, G., 205–206
Tapie, P., 431
Tardif, M., 222–224, 259
Tarricone, C., 53–54, 55, 80–81, 82, 93–94, 101*t*, 114, 115*f*, 341
Tauxe, R.V., 99
Tavares, C., 362–363, 365
Taveirne, M.E., 109–110
Tawara, T., 128–129, 277, 278
Taylor, B.L., 308, 309
Taylor, B.T., 148–149
Taylor, J.-S., 251–252

Te Biesebeke, R., 428
Teh, A.-H., 210, 211f
Teixeira, M., 101t, 430
Tejero, J., 254
Tello, E., 429–430
Templeton, L.J., 108
Testa, F., 431–432
Teunis, P., 99
Théraulaz, L., 308
Thiel, T., 219
Thijs, L., 21, 25–26, 82–84, 85, 87–88, 90, 91–92, 93, 299
Thoden, J.B., 60
Thomas, M.T., 113
Thomas, T., 335, 337–340, 350, 351–353
Thomas, W.K., 394–395
Thomashow, M.F., 337–340, 339t, 350, 351–352, 353, 355, 356
Thompson, C.L., 12–13
Thomson, A.J., 101t, 106, 110, 129
Thomson, M.R.A., 333
Thomson, N.R., 153
Thong, T.W., 430
Thorneley, R.N., 55
Thorsteinsson, M.V., 7–8, 9, 11, 66, 149–150, 204, 206, 217–219, 228–231, 229t, 240–242, 247t, 248t, 250, 277
Tian, H., 11
Tiedje, J.M., 337–340, 339t, 350, 351–352, 353, 355, 356, 359
Tilburg, C.M., 336
Tilleman, L., 84–85, 86–87, 88–89, 90, 91, 92, 93, 94, 244, 368
Tinajero-Trejo, M., 3, 50–51, 69–70, 80–81, 99–100, 101t, 106, 111–112, 113, 120, 121, 126, 128–130, 131–132, 200, 201–203, 206–207, 210, 212, 217, 224, 340–341, 344–345, 360, 362, 368, 370–372, 371f, 393, 395–397, 396t, 406, 411–412, 424–425, 429
Ting, K.L., 57
Ting, L., 350, 351–352, 355
Tischler, A.D., 19
Tiso, M., 254
Tobin, J.L., 224
Toh, D.C., 305–306
Toh, H., 409
Toledo, J.C., 297–298

Tollaksen, S.L., 337–340, 339t, 350, 353, 356
Tomaso, H., 294–295
Tomchick, D.R., 285, 286, 288t
Tomer, K.B., 368
Tomisugi, Y., 9
Tomita, M., 126
Tomita, T., 197, 288t, 289, 301, 305
Torruella, G., 415–420
Touati, D., 107–108
Tran, L.M., 125–126
Tran, R., 244, 300–301
Trandafir, F., 21, 86–87, 299
Travaglini-Allocatelli, C., 60
Traylor, T.G., 249
Trent, J.T. III., 51, 56–57, 61, 67–68, 203, 204, 207–210, 237–238, 238f, 241t, 242, 247t, 248np, 250–251, 253, 259, 261, 362, 364
Treu, L., 350
Treutlein, J., 12–13
Trevaskis, B., 393
Tribelli, P.M., 337–340, 339t, 356
Trotman, C.N., 57
Truchet, G., 282
Tsai, A.L., 244, 249, 300
Tsai, J., 109
Tsai, P.S., 101t
Tsaneva, I.R., 108
Tschowri, N., 20
Tsuchiya, S., 249
Tsutsumi, V., 429–430
Tucker, N., 295–296
Tucker, N.P., 295–296
Tuckerman, J.R., 7–8, 12, 14–15, 16–17, 19, 21–22, 23, 210, 211f, 280, 282–284, 285, 288t, 296–297, 299–300, 304, 305
Tuffin, M., 334, 339t, 350, 355–356
Tundo, G.R., 84–85, 370–372
Turner, J., 332–333
Turner, M.S., 306, 307
Turner, S.M., 70–71, 101t, 111–112, 121, 123–125, 126–127, 131, 344
Turrion-Gomez, J.L., 427–428
Tutino, M.L., 337–340, 339t, 348–350, 351, 353, 354, 355, 356, 357–358
Tyagi, J.S., 296

U

Uchida, T., 6, 8, 9, 11, 12, 22–23, 280–281, 299, 314–315
Ueha, T., 281
Ullmann, B.D., 423
Unden, G., 420–421
Unger, E.R., 12–13
Ungerechts, G., 203
Unno, H., 7–8, 9
Uno, T., 9
Urayama, P., 346*f*
Urushihara, H., 404, 431
Ussery, D.W., 126–127
Usuki, I., 393, 432–433
Uzan, J., 362, 365–366, 366*t*

V

Valko, M., 357
Vallesi, A., 336
Vallone, B., 60, 362–363
Valverde, F., 67
van den Berg, J.M., 108
van den Hondel, C.A., 428
van der Giezen, M., 205–206
van der Meche, F.G., 99
van der Zeijst, B.A., 126–127
van Dijl, J.M., 5
van Dongen, J.T., 424–425
Van Doorslaer, S., 85, 86–87, 90, 91, 94, 203
van Duynhoven, Y., 99
Van Gelder, B.F., 249–250
van Grondelle, R., 289–290
van Koppen, E., 108
Van Leuven, W., 362–363
van Nuenen, A.C., 126–127
van Pelt, W., 99
Van Stokkum, I.H., 289–290
van Vliet, A.H., 109–110, 113
van Wilderen, L.J., 289–290
van Wonderen, J.H., 129
Vance, R.E., 300–301
Vandamme, P., 99
Vandelle, E., 258
Vanderleyden, J., 15
Vanfleteren, J., 57
Vanfleteren, J.R., 393, 421
Vanin, S., 23

Varotsis, C., 6
Varshney, G.C., 66–67, 101*t*, 172, 185
Vasudevan, S.G., 101*t*, 113, 342, 393
Vaulot, D., 394–395
Vazquez-Limon, C., 202–203, 206–210
Vazquez-Torres, A., 108
Vega, L.A., 423
Venturi, M.L., 411–412
Verde, C., 66, 69–70, 350, 360, 362–363, 365–367, 366*t*, 368, 369, 373–374
Vergara, A., 362–363
Vernal, J., 362–363, 365
Véron, M., 99
Verrept, B., 86–87
Verreth, C., 15
Viala, J.P., 108
Viappiani, C., 85, 90, 91, 94
Vicarelli, M., 332
Vicente, J.B., 101*t*
Vierstraete, A., 82, 200, 201–202, 406, 411–412, 428
Vigentini, I., 351
Vincendeau, P., 431
Vinck, E., 21, 299, 362–363
Vinogradov, S.N., 2–3, 28–29, 50–51, 57, 66, 80–81, 82, 99–100, 111, 120, 148–149, 196–198, 200, 201–203, 205–210, 212–216, 217, 223–224, 340–341, 343–345, 348, 362–363, 393, 395–397, 396*t*, 406, 411–412, 415–421, 424–425, 428, 429
Vipond, J., 300
Virtanen, A.I., 207–210
Virts, E., 14
Virts, E.L., 7–8, 14
Visca, P., 57, 66–67, 68–69, 149–150, 153, 168–169, 173, 185–186, 188–189, 360, 362–363, 369
Visconti, K.C., 148–149, 173
Vitagliano, L., 362–363
Vlassak, K., 15
Vogel, K.M., 9
Voigt, B., 348–350, 354, 355, 357
Volkert, M.R., 124–125
Volman, G., 7–8, 19, 23, 288*t*, 305
von Wachenfeldt, C., 51, 57, 58–59, 63–64, 70, 343–344, 360–361
Voordouw, G., 314–316

Vorderwulbecke, S., 354
Vos, M.H., 16, 246, 285–286, 288t,
    289–290, 304
Voskuil, M.I., 148–149, 173, 296
Vothknecht, U.C., 204–205
Vu, B.C., 51, 56–57, 61, 67–68, 203–204,
    230np, 236–238, 237f, 241t, 248t,
    250–251, 261
Vuletich, D.A., 50–51, 53–54, 55–57,
    59–60, 61–62, 64–65, 66, 67–68, 70–71,
    80–81, 149–150, 200, 203–204, 206,
    212–215, 219, 220–222, 229t, 233, 239,
    241t, 252, 253, 254, 255, 257, 258, 347,
    360–361, 362–363, 393

W

Wada, T., 304
Waelkens, F., 15
Wainwright, L.M., 59–60, 70–71, 101t,
    108–109, 111–112, 117, 120–121,
    122–125, 126–127, 131–132, 149–150,
    212–213, 344, 346–347
Wakabayashi, K., 260
Wakabayashi, S., 148–149, 199–200
Wakagi, T., 428, 429
Wakao, S., 225–226
Walburger, A., 353
Waldburger, C.D., 293–294
Walisser, J.A., 12–13
Walker, F.A., 231–232
Wall, J.D., 314–315
Walsh, C., 403
Wan, X., 19, 21–22, 93, 299–300, 304
Wang, C., 19
Wang, F., 12–13
Wang, H., 128–129
Wang, J., 19, 290
Wang, M., 332
Wang, P.G., 111–112
Wang, S., 128–129
Wang, T.Z., 106, 124–125, 350, 351–352,
    359
Wang, X., 108
Wang, Y., 120, 131–132
Wang, Z.-Q., 128–129, 424–425
Wardman, P., 221
Warkentin, E., 433
Wartenberg, M., 355–356

Wasbotten, I.H., 118
Wassenaar, T.M., 99
Wassmann, P., 19
Watanabe, M., 7–8, 19–20, 302, 303
Watanabe, Y., 337–340, 339t, 368, 370–372
Watanabe, Y.H., 339t
Waters, C.M., 19
Watnick, P., 298–299
Watson, R.O., 109
Watts, K.J., 308, 309
Watts, R.A., 362–363, 393
Weber, A.P.M., 204–205
Weber, C.F., 278, 429–430
Weber, R.E., 59–60, 61, 148–149, 228–231,
    229t, 240–242, 248t, 250, 432–433
Webster, D.A., 101t, 112–113, 148–149,
    199–200, 422–423
Weeks, A., 60
Weerakoon, D.R., 109–110
Wegener, H.C., 99
Wei, C.-C., 424–425
Weich, B., 201–202, 364
Weiland, T.R., 250
Weinberger-Ohana, P., 19
Weingarten, R.A., 109–110
Weinhouse, H., 7–8, 19, 23, 288t, 305
Weinstein, M., 14–15
Weintraub, S.J., 352
Weiss, B., 108, 130
Weiss, M., 363
Wemmer, D.E., 300
Wen, Z., 111–112
Wenke, B.B., 242–244, 251–252
Wernisch, L., 173
Westblade, L.F., 18
Westgate, E.J., 12–13
Westhoff, P., 204–205
Wever, R., 249–250
Weyland, H., 334
Wheeler, P.R., 153
Wheelis, M.L., 394–395
Whistler, T., 12–13
White, G.P., 101t, 113
White, J.M., 226–228
Widdowson, M.A., 99
Widom, J., 311
Wiegeshaus, E.H., 148–149
Wilhelm, B., 255–256

Wilkerson, C., 337–340, 339*t*, 350, 353, 356
Wilkins, D., 335–336
Wilkinson, A.J., 249–250, 368
Williams, P.A., 302
Williams, T.J., 335–336, 350, 351–352, 355
Willis, A.C., 302
Wilm, A., 409
Wilmes, B., 348–350, 354, 355, 357
Wilsbacher, L.D., 280
Wilson, J.L., 59–60, 123
Wilson, M.E., 429
Wilton, R., 312, 313–314
Wimpory, D.C., 12–13
Winer, B.Y., 51, 53–54, 57, 59–60, 70–71, 251–252
Winkelstein, J.A., 108
Winkler, M., 225, 257–258
Winkler, W.C., 288*t*
Winter, M.B., 244
Wisedchaisri, G., 173
Wishnok, J.S., 343
Wisor, J.P., 12–13
Witman, G.B., 224, 260
Wittenberg, B.A., 50–51, 55–57, 58–60, 61–62, 63–64, 65, 66–67, 101*t*, 118, 120, 148–150, 152–153, 155–156, 157–158, 161–163, 169–172, 182–183, 184, 188–189, 200, 206–207, 212–213, 223, 229*t*, 232–233, 234–235, 236, 247*t*, 248*t*, 261, 345, 347, 348, 360–361, 362–363, 365, 369, 393, 432–433
Wittenberg, J.B., 50–51, 55–57, 58–60, 61–62, 63–64, 65, 66–67, 101*t*, 118, 120, 148–150, 152–153, 155–156, 157–158, 161–163, 169–172, 182–183, 184, 188–189, 200, 203, 206–207, 212–213, 229*t*, 236, 247*t*, 261, 288*t*, 345, 347, 348, 360–361, 362–363, 369, 393, 432–433
Witting, P.K., 118
Woese, C.R., 394–395
Wolf, G., 421–422
Wolf, Y.I., 406–408
Wong, H.C., 7–8
Wood, S.J., 312, 313–314
Wosten, M.M., 126–127
Wren, B.W., 108–109, 111, 123–125
Wright, K.A., 255–256
Wu, D., 19

Wu, G.H., 70–71, 101*t*, 106, 111–112, 113, 114–116, 114*f*, 115*f*, 117, 117*f*, 118, 120, 123–124, 131, 132, 212–213, 342, 344
Wu, L.C., 11–12
Wu, Q., 332
Wu, X., 111–112

X

Xenarios, I., 409
Xian, M., 111–112
Xiao, Q., 397
Xiong, J., 412–415
Xu, J., 128–129
Xu, Y., 51, 56–57, 61, 66, 67–68, 108, 204, 206, 220–222, 241*t*, 252, 253, 254, 255, 257, 258

Y

Yager, P.L., 336
Yamada, H., 118
Yamada, S., 291–294, 293*f*, 294*f*
Yamamoto, A., 80–81, 91, 393
Yamamoto, K., 9, 277, 278
Yamanaka, M., 307
Yamasaki, H., 253
Yamashita, T., 9, 289–290, 304
Yamauchi, K., 51, 52–53, 54, 55–56, 59–60, 61, 120, 149–150, 155, 201, 201*f*, 213*f*, 223, 233–235, 234*f*, 241*t*, 242, 259, 345–346, 393, 432–433
Yamazaki, I., 118
Yang, J., 101*t*
Yang, M., 290
Yang, R., 12–13
Yang, Y., 12–13
Yap, E.H., 430
Yarlett, N., 430
Yayanos, A.A., 336
Ye, R.W., 106, 124–125
Yee, J., 431–432
Yeh, D.C., 228–231
Yeh, S.-R., 51, 55–56, 57–58, 59–60, 61–62, 64–65, 68–69, 70–71, 101*t*, 112, 113, 114–116, 114*f*, 115*f*, 117, 117*f*, 118, 120, 123, 131–132, 148–150, 153, 155, 157–158, 161–162, 163–166, 167–168, 169–172, 173–180, 186–188, 189, 229*t*,

231–232, 247*t*, 253, 301, 346–347, 360–361
Yi, C.H., 12–13
Yohda, M., 337–340, 339*t*
Yokota, N., 304
Yoon, H.S., 205
Yoon, J., 244
Yoon, P.S., 343
Yoshida, M., 337–340, 339*t*
Yoshida, Y., 6
Yoshimitsu, K., 295–296
Yoshimura, H., 6, 314–315
Yoshimura-Suzuki, T., 303
Yoshioka, S., 6, 22–23, 277, 299, 314–315
You, X., 101*t*, 108
Youn, H., 8, 9–11, 275, 277, 278–279
Yu, H., 12–13
Yu, K., 397
Yudkin, M.D., 3, 26–27
Yukl, E.T., 296, 297–298
Yukuta, Y., 226–228
Yum, L.K., 402–403

Z

Zagari, A., 362–363
Zaghloul, N.A., 224
Zamorano-Sanchez, D.S., 124–125
Zane, G.M., 7–8
Zeier, J., 424–425
Zeng, M., 343, 423
Zhang, D., 305–306
Zhang, G., 357
Zhang, J., 356, 406–408

Zhang, W., 6–7, 25, 82–84, 85
Zhang, X., 295–296, 337–340, 339*t*, 350, 351–353, 356, 357
Zhang, Y., 12–13
Zhang, Y.M., 356
Zhang, Z., 354, 406–408
Zhao, J., 51, 56–57, 67–68, 200, 203, 220–221, 229*t*, 239
Zhao, J.S., 352–353, 362–363
Zhao, Q., 423
Zhao, S., 99
Zhao, W., 108
Zhao, X.J., 422–423
Zhao, Y., 7–8, 280
Zheng, J., 99
Zheng, M., 106, 108, 123–125
Zheng, S., 350, 351–352, 355
Zheng, T., 12–13
Zhou, A., 7–8
Zhou, F.X., 352
Zhou, J., 350, 351–352, 359
Zhou, L., 356
Zhou, S., 363, 428, 429
Zhou, Y.D., 11
Zhu, G., 66–67, 173
Zhu, H., 393
Zhu, Y., 12–13
Zhuang, C., 227*np*
Zimmer, M., 308
Zou, Y., 204–205
Zumft, W.G., 123–124, 126–127, 128
Zweier, J.L., 254
Zwisler, W., 334–335

SUBJECT INDEX

Note: Page numbers followed by "*f*" indicate figures, and "*t*" indicate tables.

A

Acetobacter xylinum phosphodiesterase A1
(*Ax*PDEA1), 305
Aerotactic sensors
description, 4–5
HemAT, 5–7
Algae, 433
Amoebozoa, 398*t*, 404
Archaeplastida, 397–401, 398*t*
AxPDEA1. *See Acetobacter xylinum*
phosphodiesterase A1 (*Ax*PDEA1)

B

Bacterial chemotaxis system, 307–308
Biochemical and functional characterisation,
Pgb
*Ap*Pgb globin domains, 93
Archaea, 90–91
CO complexes, 91–92
ferric and ferrous MaPgb nitrosylation,
92–93
ligation/deligation, 91–92
methane production, 91
Trp(60)B9 and Tyr(61)B10, 92
Bis-(30,50)-cyclic dimeric guanosine
monophosphate (c-di-GMP),
298–299
Burkholderia xenovorans (BxRcoM), 13
BxRcoM. *See Burkholderia xenovorans*
(BxRcoM)

C

Campylobacter jejuni
bacterial gastroenteritis, 99
bacterial globins, 100, 101*t*
description, 99
globin expression control
nitrite and nitrate, 130–131
NO resistance, oxygen-limited
conditions, 131
NrfA, 129–130
NssR, 123–126

globins function, 100
integrated response, Cgb and Ctb,
131–132
Mb-fold family, 99–100
NO and RNS, biology, 100–107
oxygen and reactive oxygen species,
biology, 107–108
poultry products contamination, 99
respiratory metabolism, 108–110
single-domain globin, 111–119
truncated globin, 120–123
Campylobacter nitrosative stress-responsive
regulator (NssR)
nitrite, 129
nitrosative stress, 111–112
protein, 114
structural modelling, 128–129
CAPs. *See* Cold-adapted protein (CAPs)
Carbon monoxide oxidation activator
(CooA)
crystal structure, 9
evolutionary studies, 8
F-helix, 9
kinetics measurements, 9–11
Rhodospirillum rubrum, 8
RNA polymerase, 8
RrCooA homologue, 11
spectroscopic and mutagenesis
techniques, 9
Chemical reactivity and possible function,
cyanobacteria and algae
endogenous hexacoordination, 251
haem post-translational modification,
251–252
hydroxylamine, 254
nitrite consumption, 254
NO dioxygenation, 252–253
peroxidase activity, 254
reduction, hydroxylamine, 254
Chlamydomonas eugametos LI637 Hb
CtrHb, 223
CtrHb cyanomet structure, 234

Chlamydomonas eugametos LI637 Hb
    (*Continued*)
  CtrHb three-dimensional structure, 233
  cyanomet CtrHb cavities, 235, 235*f*
  description, 222
  distal hydrogen bond network, CtrHb,
      234, 234*f*
  ferrous state, 232–233
  haem orientational isomerism, 233–234
  haem proteins, 233
  LI637, 222–223
  myoglobin, 232
Chlamydomonas reinhardtii
  advantages, 225
  nuclear genome, 223–224
  sulphur deprivation, 225–226
  THB, 223–224
Cold-adapted marine microorganisms
  Antarctic polar regions, 332–333
  Arctic characters, 332
  bacterial globins
    classification, 341
    description, 340–341
    evolutionary relationship, 341
    FHbs, 341–344
    non-vertebrate, 341
    SDgb, 344–345
    TrHbs, 345–348
  evolutionary adaptations
    adequate reaction rates, 340
    challenges, cellular function and
        structural integrity, 337–340
    comparative genome analysis, 337
    enzyme molecular flexibility, 340
    organism to sense temperature changes,
        340
    protein, 337
    survive and grow, strategies, 337–340,
        339*t*
  low temperatures, 331
  molecular adaptation, 331
  PhTAC125 (*see Pseudoalteromonas
      haloplanktis* TAC125 (*Ph*TAC125))
  phylogeny and biogeography
    bacterial and archaeal phylogenetic
        tree, 337, 338*f*
    bacterial, archaeal and viral diversity,
        334

  bipolar biogeographical patterns, 336
  culture-independent studies, 337
  data acquisition, 334–335
  denaturing gradient gel electrophoresis,
      334–335
  environmental genomics, 335
  geographic separation, 335
  global warming, 334
  heterotrophic microbes, 335
  sea-ice microbiota, 336
  seasonal changes, environmental
      parameters, 335–336
  polar oceans, 331
  structure and function, proteins, 373
  temperature, 373
Cold-adapted protein (CAPs), 352–353
CooA. *See* Carbon monoxide oxidation
    activator (CooA)
Cyanobacteria and algae
  biophysical studies
    electron transfer, 260–261
    structures and dynamic properties,
        261–262
  canonical vertebrate myoglobin, 203
  description, 197–198
  globins, 196–197
  haem axial ligand, 197, 198*f*
  in vitro characterization
    auto-oxidation, 249–250
    C. eugametos LI637 Hb, 232–236
    chemical reactivity and possible
        function, 251–254
    electron transfer and redox potential,
        250–251
    endogenous and exogenous ligands,
        246–249
    ligand-binding properties,
        244–249
    nostoc commune GlbN, 226–232
    structural information, 240–244
    Synechococcus sp. strain PCC 7002
        GlbN, 239–240
    Synechocystis sp. strain PCC 6803 GlbN,
        236–238
  ligand binding, 199
  light-induced (LI) globins, 200
  lineages, 202–203
  myoglobin, 201–202

nitrogen metabolism, 199–200
NMR, 198–199
P. caudatum, 201, 201*f*
phylogeny (*see* Phylogeny)
physiological characterization
    bacterial globin research, 217
    Chlamydomonas eugametos, 222–223
    Chlamydomonas reinhardtii, 223–226
    Nostoc commune, 217–219
    Nostoc punctiforme, 219–220
    Synechococcus sp. strain PCC 7002,
        220–222
physiological studies
    genetic manipulation and phenotypic
        analysis, 257–258
    globin function, photosynthetic
        organisms, 255
    globins purification, source organisms,
        255–257
    nitrogen metabolism, 258
    photosynthesis, 259
    signal transduction, 260
PTM, 203–204
Synechocystis sp, 203
TrHb1s, 202–203
truncated globins, 200
unicellular photosynthetic organisms,
    201–202
vertebrate haemoglobins and myoglobins,
    197

D

DcrA
    amino acid sequence alignments, 315,
        316*f*
    CO-binding affinity, 315–316
    Desulfovibrio vulgaris, 314–315
    ferric haem, 315
    PAS domains, 315
Diguanylate cyclases (DGC) activity
    *Bpe*Greg, 299
    c-di-GMP promote, 298–299
    GGDEF domain, 299
    globin-coupled sensor proteins, 299
    H-NOX domains, 300–301
    human colon cancer cells, 298–299
    mutagenesis and resonance Raman
        studies, 299

YddV (dosC) and *EcDos*, 299–300
Direct oxygen sensor from *Escherichia coli*
    (*Ec*Dos)
    dynamic changes, haem, 302
    PDEs activity, 304
    spectroscopic properties, 301
    structures, haem-containing PAS domain,
        302–304
    yddV (dosC) gene, 299–300
DosS and DosT
    crystal structures, GAF-A domains,
        295–297
    ligand binding to haem, 297–298
    Mycobacterium tuberculosis, 295–297

E

E. coli direct O_2 sensor (EcDos)
    dosCP operon, 19
    gas-binding affinity, 20
    PNAG, 20
    signal transduction, 20
    Tyr43 and Leu65, 20
Endogenous and exogenous ligands
    CO and NO, 249
    gaseous ligand binding, 246–248, 247*t*
    oxygen affinity and partition coefficient,
        248*t*, 249
Enzymatic function
    chimeric proteins, 24
    tGCSs, 24
Excavata, 398*t*, 403

F

FixL/FixJ two-component systems
    ADP production, 283
    amino acid residues, haem pocket
        Arg220, 287
        *Bj*FixLH, 286, 287
        electrostatic interactions network,
            285–286
        ligand-binding kinetics parameters,
            287, 288*t*
        "the FG loop mechanism," 287–289
    ligand-binding properties, 289
    nodule bacteria, 282
    O_2 regulation, 283–284
    PAS domain, 284–285
    phosphoryl group transfer, 282–283

FixL/FixJ two-component systems
(*Continued*)
sensor kinase, 281–282
structural model
CA domain, 293–294
domain organisation, 291–292
His547, 292–293, 294*f*
β3, ThkA PAS domain, 291–292, 293*f*
ThkA/TrrA, 291, 292*f*
time-resolved spectroscopy, 289–291
Flavohaemoglobins (FlavoHbs)
description, 182
MtbFHb homologues, 189
mycobacteria
genomes and protein, 173–174
species, 174
structure-based sequence alignment,
174–180, 175*f*
nitric oxide scavenging, 182–186
oxygen metabolism, 186–188
peroxidase activity, 188–189
redox signalling and stress management,
188
FlavoHbs. *See* Flavohaemoglobins
(FlavoHbs)
Fungi genomes
Aspergillus oryzae, 428
Cladosporium fulvum, 426–427
Dothideomycetes, 427
Erwinia chrysanthemi, 425
FHb, 428, 429
globin distribution, 406, 407*t*
Glomerellales, 427–428
hemibiotrophs, 424
necrotrophic interactions, 425
NOS, 424–425
Oomycetes, 427
organic carbon compounds, 423
plant pathogenic, 424
Puccinia graminis, 426
pucciniomycotina, 426
spore germination, 425–426

G

β-Galactosidase assay, 173
GCSs. *See* Globin-coupled sensors (GCSs)
Gene-regulating function
description, 7–8

messenger biosynthesis
*Av*GReg, 21
AxPDEA1, 23
biosynthesis and degradation, 19
*Bpe*GReg, 21–22
cyclic dimeric, 19
EcDos, 19–20
haem NO-binding sensors, 23–24
HemDGC, 22
*Vc*Bhr-DGC, 23
protein–DNA interaction
BxRcoM, 13
CooA, 8–11
description, 8
mPER2, 13
NPAS2, 11
protein–protein interaction
*Af*GcHK, 18–19
nitrogen fixation gene expression
regulator, 14–18
*Vv*GReg and *Cv*GReg, 18
2-on-2 globin (2/2Hb)
amino acid sequence analysis, 50–51
evolutionary/engineering process, 50
fold and fold variation
CD region, 53–54
haem crevice, 54
helices, 52–53, 52*f*
structural differences and group specific
residues, 54
functions
Campylobacter jejuni, 70–71
cyanide binding, 71
cyanobacterium, 67–68
homologous system, 66–67
infection, 66–67
metabolic pathways, 70
M. tuberculosis, 68–69
mycobacterial species, 66
nitric oxide detoxification, 66
peroxidase assay, 67–68
peroxynitrite, 69
reactive oxygen and nitrogen species,
69–70
haem environment, 54–60
Mb and Hbs, 50
protein matrix, 60–65
structures, 51

Globin-coupled sensors (GCSs), 393, 395–397
Globin nomenclature, 395–397, 396t
GSU0935 and GSU0582
  conformational stability and signalling properties, proteins, 312
  coordination structures, haem, 312–313, 314f
  crystal structures, 312, 313f
  cytoplasmic N-terminal tail, 312
  distal haem pocket accessibility, 314
  EPR spectra, 313–314
  ferrous haem, 314
  Geobacter sulfurreducens, 311–312
  His143, 312–313
  signal transduction/chemotaxis proteins, 311–312

H

Haem-based aerotactic transducer (HemAT)
  cellular stoichiometry experiments, 5
  crystal structures, 6–7, 7f
  deletion and overexpression, 5
  gaseous ligands, 6
  Halobacterium salinarum and *Bacillus halodurans*, 5
  hydrogen bonds, 6
  N-terminal globin domain, 5
  participation, PhoD secretion, 5
  spectroscopic evidence, 6
  Tyr133, 6
Haem-based sensors
  chimeric multi-domain proteins, 2
  evolution, 28–29
  function
    aerotactic sensors, 4–7
    classification, 4, 4f
    enzymatic function, 24
    gene-regulating function, 7–24
    Geobacter sulfurreducens and *metallireducens*, 25–26
  globin-coupled proteins, 3
  haem-iron atom, 2–3
  hydrophobic pocket, 3
  Streptomyces avermitilis and *Frankia* sp, 3
  transduction pathways, 2
Haem-containing diguanylate cyclase (HemDGC), 22

Haem-containing PAS
  bacterial chemotaxis system, 307–308
  cognate physiological signals, 316–317
  DGC activity (*see* Diguanylate cyclases (DGC) activity)
  MCP
    b-type haem, 308–311
    c-type haem, 311–316
  "on-state" and "off-state," protein, 316–317
  PDEs (*see* Phosphodiesterases (PDEs))
  YybT, 305–307
Haem environment
  A. tumefaciens, 58–59
  C. eugametus 2/2-HbN, 57
  characterisation, 87
  C. jejuni 2/2HbP, 59–60
  distal site residues, 89–90
  E-helix, 55
  E7 route entry system, 59
  ferric MaPgb*, 88–89
  ferrous derivative, 57
  genome database, 59
  2/2 helical sandwich, 54–55
  hexacoordinate *Synechocystis* sp., 57
  hexacoordination, 56–57
  HisF8, 55
  ligand, 89
  molecular dynamics simulation, 58
  M. tuberculosis, 57–58
  O_2 dissociation rate, 87–88
  oxygen-binding affinity, 59–60
  Phe(93)E11, 89
  porphyrin ring system, 87–88
  T. fusca 2/2HbO, 58
  Trp(60)B9 insertion, 90
  TyrB10 side chain, 55–56
Haemoglobins (Hbs)
  Chlamydomonas eugametos, 433
  ciliated protozoan, 432–433
  horse heart, 393
  oxygen storage and transport, 420–421
  pathogenic bacteria, 429
  Saccharomyces cerevisiae, 422–423
  'yeast haemoglobin-reductase complex,' 421–422
Hbs. *See* Haemoglobins (Hbs)
Heat-shock proteins (HSPs), 354

HemDGC. *See* Haem-containing
    diguanylate cyclase (HemDGC)
Hexacoordinated globins
    absorption spectra, 364–365, 364*f*
    electronic absorption bands, 364–365
    exogenous ligands, 362
    functions and physiological role *in vivo*,
        374
    haemichromes, 363
    iron coordination, 362–363, 363*f*
    O_2 stabilising interaction, 362
    Ph-2/2HbO, 362
    spectroscopic data and molecular-
        dynamics simulation, 365
HSPs. *See* Heat-shock proteins (HSPs)

L
Last eukaryote common ancestor (LECA),
    404–405, 404*f*, 411
LECA. *See* Last eukaryote common ancestor
    (LECA)
Ligand-binding properties, cyanobacteria
    and algae
    exogenous ligand binding, 244–245, 245*f*
    haemoglobins bind dioxygen, 244
    rapid-mixing approach, 244
    rapid-mixing stopped-flow experiments,
        244

M
Mammalian Period protein 2 (mPER2), 13
Methylaccepting chemotaxis domain
    (MCP)
    PAS domain with b-type haem
        Aer2, 308
        amino acid sequence, 309
        crystal structure, Aer2 N-terminal
            domain, 309, 310*f*
        α1 helix, 309
        HemAT and Aer, 308
        hydrogen bond, 311
        intramolecular signal transductions,
            311
        non-polar side chains, 309–311
        PH-Aer2, 309, 310*f*
    PAS domain with c-type haem
        DcrA, 314–316
        GSU0935 and GSU0582, 311–314

Microbial eukaryote globins
    algae, 433
    Amoebozoa, 398*t*, 404
    antimicrobial imidazoles, 433
    Archaeplastida, 397–401, 398*t*
    bioinformatic survey, 393
    diagrammatic representation, eukaryote
        supergroups, 397, 401*f*
    estimation, species, 394–395, 395*t*
    Excavata, 398*t*, 403
    FHbs, 421
    fungi (*see* Fungi genomes)
    GCS, 393
    haemoglobins (Hbs), 393
    heterologous expression, *Vitreoscilla*,
        420–421
    images, 434, 435*f*
    LECA, 404–405, 404*f*
    molecular phylogenies, 434
    molecular phylogeny, 394–395
    nomenclature, 395–397, 396*t*
    Opisthokonta, 398*t*, 405–406
    oxygen storage and transport, 420–421
    phylogenetic relationships
        identification and alignment, 406–408
        protist globins (*see* Protist globins)
        sequence alignments and phylogenetic
            analysis, 409–410
    protozoa (*see* Protozoa)
    SAR clade, 398*t*, 402–403
    SDgbs, 433
    subfamilies identification, 397, 398*t*
    yeasts, 421–423
M lineage, phylogeny
    cyanobacteria and algae, 210
    18 cyanobacterial M globins, *Synechococcus*
        sp, 210, 211*f*
    green algae, 210
    land plants, 207–210
mPER2. *See* Mammalian Period protein 2
    (mPER2)
Mycobacteria
    description, 148–149
    FlavoHbs, 173–182
    phylogenetic analysis, 149–150
    species, 150, 151*t*
    trHbs (*see* Truncated haemoglobins
        (trHbs))

N

Neuronal PAS domain protein 2 (NPAS2)
  distal histidine residue, 12
  DNA-binding region, 12
  evidence, 11–12
  functional mutation, 12–13
  transcription function, 11
Nitric oxide (NO)
  and RNS
    generation, 106
    immune system, 106
    microorganisms, 107
    resistance, bacteria, 107
  scavenging
    detoxification activity, 182–183
    electrode, 183–184
    glbN mutant, 184
    M. leprae genome, 185–186
    M. smegmatis cells, 183–184
    Salmonella enterica serovar
      Typhimurium, 185
    trHbN expression, 184–185
    TyrB10, 182–183
Nitric oxide dioxygenation (NOD) reaction
  ferric and ferrous states, 180
  protein environment, 163
  trHbN, 163
Nitric oxide synthase (NOS), 424–425, 431
Nitrogen fixation gene expression regulator
  bacterial strains, 15
  *Bj*PAS domain, 17–18, 17*f*
  evidence, 14
  FixL and FixJ, 14, 14*f*
  signal transduction, 16
  SmFixL, 15
NO detoxification, *Ph*TAC125, 369–372
Non-haem globin sensors
  B. subtilis, 26–27
  moorella thermoacetica and sporulation
    inhibitory proteins, 28
NOS. *See* Nitric oxide synthase (NOS)
Nostoc commune
  cross-reacting protein, 218–219
  GlbN
    b haem and nomenclature, 231, 231*f*
    cyanobacterial and green algal
      haemoglobins, 228, 229*t*
    deoxymyoglobin, 226–228

    haem methyl substituents,
      231–232
    lower intensity bands, 226
    sperm whale myoglobin, 226–228,
      227*t*
    UV-visible, 228
  glbN gene, 218
  ORF, 217
  polyclonal antibodies, 218
  resulting protein, 217–218
Nostoc punctiforme
  NpR0416, 219
  RNA isolation, 219–220
NPAS2. *See* Neuronal PAS domain protein
  2 (NPAS2)
NssR. *See Campylobacter* nitrosative stress-
  responsive regulator (NssR)
NtrY/NtrX two component system,
  294–295

O

Open reading frame (ORF), 217
Opisthokonta, 398*t*, 405–406
ORF. *See* Open reading frame (ORF)
Oxygen and reactive oxygen species
  anaerobes and microaerophiles, 107
  bacterial infection, 108
  oxidative stress, 108
  oxygen molecule, 107
  transition metals, 107–108
Oxygen metabolism
  aerobic metabolism, 187
  flavoHbs, 187–188
  trHbO, 186–187

P

PAS domain. *See* Per-Arnt-Sim (PAS)
    domain
PDEs. *See* Phosphodiesterases (PDEs)
Per-Arnt-Sim (PAS) domain. *See
    also* Haem-containing PAS
  amino acid sequence homologies, 276
  α/β structure, 276
Phosphodiesterases (PDEs)
  c-di-GMP, 298–299
  haem-containing PAS and EAL domains
    *Ax*PDEA1, 305
    dynamic change, *Ed*Dos, 302

Phosphodiesterases (PDEs) (*Continued*)
  regulation, 304
  spectroscopic properties, *Ec*Dos, 301
  structures, 302–304
  sensor domain, 299
  YybT, 305–307
Phosphotransferase system (PTS),
  348–350
Photosynthesis, cyanobacterial and algal
  globins
  activities, 204
  endosymbiotic event, 205–206
  green algae, 205
  molecular oxygen, 206
  prokaryotes and eukaryotes, 204–205
*Ph*TAC125. *See Pseudoalteromonas*
  haloplanktis TAC125 (*Ph*TAC125)
Phylogeny
  and evolution
    globin genes, 207
    green algae, 207
    M lineage, 207–212
    phylogenetic tree, proteins,
      207, 208*f*
    SDgbs, 206–207
    T lineage, 212–217
    photosynthesis, 204–206
Poly-*N*-acetylglucosamine (PNAG), 20
Post-translational modification (PTM),
  203–204
Protein matrix, 2/2Hb
  B. subtilis, 65
  cavities, 61
  E7 route, 64–65
  haem–distal-site gating, 65
  ligand migration pathways, 63
  nitric-oxide dioxygenase activity, 63
  PheE15, 61–62, 62*f*
  protein matrix tunnel, 61
  residues, core, 60
  T. pyriformis 2/2HbN, 63
  tunnel, 61, 62*f*
  tunnel/cavity system, 63–64
Protist globins
  alpha-proteobacterium and
    cyanobacterium, 411–412
  Bayesian tree
    Clustal Omega alignment, 415–420, 419*f*

MAFFT alignment, 415–420, 417*f*
  ProbCons alignment, 415, 416*f*
  SDgbs and FHb globin domains,
    MAFFT alignment, 411–412, 413*f*
  TrHb1 domains, ProbCons alignment,
    412–415, 414*f*
  fossil record and calculations, 411
  node probabilities, 415–420
  phylogenetic tree, opisthokonts,
    415–420, 418*f*
  plastid diversification, 410*f*, 411–412
Protoglobin (Pgb)
  biochemical and functional
    characterisation, 90–93
  description, 80–81
  diatomic ligands, 81
  domain variants, 81
  GCS globin domain, 81
  haem environment (*see* Haem
    environment)
  structure
    B-and E-helices, 82
    MaPgb*, 82, 83*f*
    molecular dynamics simulations, 85
    N-terminal region, 84
    O_2 sensor, 82–84
    quaternary structure level, 85
    Z-helix, 82
  two-tunnel system (*see* Two-tunnel
    system, Pgb)
Protozoa
  anti-parasitic effects, nitric oxide, 429
  ciliate *Tetrahymena pyriformis*, 432–433
  Dictyostelium, 431
  Entamoeba histolytica, 429–430
  FHb, 431–432
  haemoglobins, 429
  inducible nitric oxide synthase (iNOS),
    429
  Paramecium caudatum, 432–433
  Trichomonas vaginalis, 430
  Trypanosoma brucei, 431
Pseudoalteromonas haloplanktis TAC125
  (*Ph*TAC125)
  biotechnological applications, 357–358
  doubling time, 348–350
  Escherichia coli, 348–350
  genomic and post-genomic insights

CAPs, 352–353
chaperonins, 355
cold adaptation, proteomic and
    transcriptomic approaches, 350
cytoplasmic proteome, 354
down-regulated proteins, 355
HSPs, 354
MaGe annotation platform, 350
NO metabolism, 355
proteins, 352
protein-secretion mechanisms, 351
ribosomal and translation-specific
    proteins, 354
RNA helicases, 351–352
rRNA and d tRNA genes, 351–352
sequence analyses, 352
survive at low temperature, 351
trigger factor (TF), 354, 355
yeast α-glucosidase, 351
globins
    C. psychrerythraea and P. cryohalolentis
        K5 genomes, 359
    distribution, FHb and TrHbs, 358, 359t
    hexacoordination, Ph-2/2HbO,
        362–365
    mesophilic organisms, 359
    Mycobacterium tuberculosis, 360
    NO detoxification, Ph-2/2HbO,
        369–372
    reactivity, Ph-2/2HbO, 365–368
    sequence identity, 358
    structure, Ph-2/2HbO, 360–362
    transcriptional analysis, 360
micro-aerobiosis, 350
O₂ and metabolic constraints
    antioxidant capacity, 357
    cell protection, 357
    genome annotations, 355
    iron-related proteins, 357
    molybdate biosynthetic and transport
        genes, 357
    nitrosative and oxidative stress, 357
    pentose phosphate pathway, 357
    permanent oxidative stress, 356
    proteomic analyses, 356
    redox buffering capacity, 356
    RNS and ROS, 355–356
    solubility, 357

γ-proteobacteria, 348
transport-related proteins, 350
PTM. See Post-translational modification
    (PTM)
PTS. See Phosphotransferase system (PTS)

R

Reactive nitrogen species (RNS)
    cold-adapted bacteria, 356
    detoxification, 370
    dioxygenases and deletion, 356
    O₂ supply, 355–356
    PhTAC125, 356
    toxic activity, macrophage, 369
Reactive oxygen species (ROS)
    cold-adapted bacteria, 356
    concentration, 368
    O₂ supply, 355–356
    oxidation, 368
    PhTAC125, 356
    protection against, 339t
    stability, 357
Redox signalling and stress management
    haem proteins, 188
    peroxidase activity, 188
Respiratory metabolism
    electron transport system, 108–109
    microaerobic respiration, 109
    nitrosative stress impact, 110
    respiration, oxygen-limited conditions,
        109–110
RNS. See Reactive nitrogen species (RNS)
ROS. See Reactive oxygen species (ROS)

S

SAR clade, 398t, 402–403
SDgbs. See Single-domain globins (SDgbs)
Single-domain globins (SDgbs)
    biophysical and mechanistic
        characterisation
        F8His, 116
        peroxidase-like character, 118
        positive polar environment, 117–118
        'push–pull' model, 116, 117f
    functional characterisation
        catalyses, NO, 112
        description, 111
        resistance, nitrosative stress, 111–112

Single-domain globins (SDgbs) (*Continued*)
  redox partner conundrum
    FHbs, 113
    haem-binding cleft, 114–116, 115*f*
    lactate dehydrogenase enzyme, 113
    structural overlays, 114, 115*f*
    topology, 114, 114*f*
Structural information, cyanobacteria and
    algae
  covalent haem attachment, 242–244
  distal hydrogen bond network, 242
  endogenous hexacoordination, 240–242
  globins, 240, 241*t*
  2/2 topology intact, 240
  tunnels and cavities, 242
Synechococcus sp. strain PCC 7002
  DglbN and wild-type strains, 221
  nitrogen metabolism, 221–222
  TrHb1-1protein, 220–221
Synechococcus sp. strain PCC 7002 GlbN
  bis-histidine coordination, 239
  ^{15}N NMR relaxation, 239–240, 240*t*
Synechocystis sp. strain PCC 6803 GlbN
  bis-histidine state, 238, 238*f*
  cyanomet and azidomet complexes, 238
  gene product, 236
  haemichrome, 236
  haemin chloride, 236–237, 237*f*
  TrHb1s, 238, 239*t*

T

T lineage
  2/2 and 3/3 proteins, 212
  bacteria, 212–213
  endosymbiotic event, 215–216
  eukaryotic hypothetical proteins, 216
  18 green algae and cyanobacterial,
      212–213, 213*f*
  representative TrHb2s, 212–213, 214*f*
  SDgb type globins, 216–217
  TrHbs, 212
  TrHb1-1s and TrHb1-2s, 213–215, 215*f*
  TrHb1s cluster, 216
Transcriptional regulation
  single-component systems
    CO oxidation operon (CooA),
        276–278
    NPAS2, 280–281

  regulator, CO metabolism, 278–280
  two-component systems
    DosS and DosT, 295–298
    FixL/FixJ (*see* FixL/FixJ two-
        component systems)
    haem-based redox sensor, NtrY/NtrX,
        294–295
trHbN
  crystal structure, 160–161
  GlnE11, 158–159
  haem, 161
  helices, 157
  ligand binding, 157–158
  molecular dynamics, 156–157
  M. smegmatis, 161
  NO and O_2, 155–156
  NO detoxification reaction, 161–162
  NOD reaction, 163
  PheE15, 159
  PheE15Tyr mutant, 159–160
  proteins, 162–163
  structural features and dynamics, 153
  TyrB10, 155
trHbO
  CO photodissociation, 168
  functional characteristics, 168–169
  M. tuberculosis, 167–168
  TrpG8, 166
  tunnel system, 166–167
  TyrCD1 and TyrB10, 163–166
trHbP
  mycobacterial genome, 169
  X-ray structure, 169
trHbs. *See* Truncated haemoglobins (trHbs)
Truncated globin, *Campylobacter jejuni*
  biophysical and mechanistic
      characterisation, 123
  functional characterisation
    microaerobic growth, 121
    Vgb ability, 120–121
  structural characterisation
    dimeric cyanide-bound, 121–122
    distal- and proximal-binding pockets,
        122, 122*f*
    peroxide decomposition, 122–123
Truncated globin-coupled sensors (tGCSs)
  ligand-binding properties, 24
  S. avermitilis and *Frankia* sp., 24

Truncated haemoglobins (trHbs)
  FlavoHbs, 182–189
  genetic regulation
    alcohol dehydrogenase, 169–172
    β-galactosidase assay, 173
    DosS and DosT, 173
    genomic organisation, 169–172,
      171*f*
    glbN and *glbO*, 172
  helix pairs, 152–153
  mycobacterial species, 153
  trHbN, 153–163
  trHbO, 163–169
  trHbP, 169

Two-tunnel system, Pgb
  apolar protein matrix, 86
  diatomic ligands, 85–86
  haem distal site, 86–87, 87*f*
  molecular dynamics simulations, 86–87

V

Vgb. *See Vitreoscilla* single-domain
        haemoglobin (Vgb) ability
Vitreoscilla single-domain haemoglobin
        (Vgb) ability, 120–121

Y

Yeast globins (YHbs), 422–423
YHbs. *See* Yeast globins (YHbs)

Printed and bound by CPI Group (UK) Ltd, Croydon, CR0 4YY

08/05/2025

01864958-0002